Neural Theory and Modeling

Neural Theory and Modeling

Proceedings of the 1962 Ojai Symposium

***Edited by* Richard F. Reiss**

Associate Editors: H. J. Hamilton, Leon D. Harmon, Graham Hoyle, Donald Kennedy, O. Schmitt, C. A. G. Wiersma

Stanford University Press, Stanford, California 1964

STANFORD UNIVERSITY PRESS
STANFORD, CALIFORNIA

Library of Congress Catalog Card Number: 64-13359
Printed in the United States of America

Preface

Various developments in science and engineering have in recent years combined to produce a surge of activity in neural theory and modeling. As in other interdisciplinary fields, research has been carried forward by a very mixed company of biologists, chemists, engineers, physicists, mathematicians, and computer specialists; these men have brought with them the research philosophies and methodologies peculiar to their own disciplines. The result has been a potpourri of partial theories and models, ideas and data, that is difficult to comprehend. The numerous studies have not in general been coordinated, and communication between individuals has often been confounded by jargon and conflicting attitudes. The confusion has been further amplified by the immense complexities of the research subject: the nervous systems of man and animal.

In the Spring of 1962, Dr. Harvey Savely, of the Air Force Office of Scientific Research, suggested to me that a small symposium be organized to facilitate communication between diverse types of theorists and modelers scattered over the United States and Europe. It was hoped that such a symposium would mitigate the present confusion by producing some agreements on aims, problems, and methods.

Under the joint sponsorship of the Air Force Office of Scientific Research and General Precision, Incorporated, a symposium was duly held at Ojai, California, on December 4, 5, and 6, 1962. The main objective in the organization of the Ojai symposium was to bring together the greatest variety of theorists and modelers possible under existing restrictions of size, time, place, and so on. As deliberately planned, the participants were almost evenly divided with respect to their backgrounds; slightly more than half of them were professional biologists, the remainder coming primarily from various fields of engineering and the physical sciences. There were twenty-four formal papers presented at the symposium, of which twenty-one appear in this volume. The subject matter of these papers ranges from highly speculative theories to rather detailed reports of physiological experiments thought to be particularly suggestive in the further development of theory.

All the papers in this volume were put into their final form after the symposium was over. Some of them have been modified in varying degrees as a result of discussions at the symposium. However, none of the discussions themselves have been preserved here.

The papers have been divided into two distinct groups. Part I is composed of papers which are predominantly theoretical or general in nature; with the exceptions of Pringle and Rall, all of their authors are non-biologists. The papers in Part II are concerned with theories and data derived from very specific nervous systems; with the exception of Fender, the authors of these papers are all biologists. The emphasis on invertebrate nervous systems in Part II is largely the result of an invitational policy followed in organizing the symposium. Invertebrate systems are simpler and generally more accessible than the much-publicized vertebrate central nervous system, and, in the opinion of the organizers (not to be confused with the sponsors), they are more promising subjects for detailed theoretical analysis in the near future.

Because of the rather sharp differences in subject and approach, the two parts of this volume have received separate and independent introductions. In the introduction to Part I, I have concerned myself primarily with the methodologies and objectives of papers which present formal theories and physical models. Since a few of these papers are in Part II, my introduction is not strictly limited to Part I.

The introduction to Part II has been written by a biologist, Donald Kennedy, who is an associate editor of as well as a contributor to this volume. The majority of papers in Part II are so closely connected with the experimental techniques and data of biologists that it was desirable to have this material reviewed by a professional biologist. Although Kennedy's specific comments are confined to the papers in Part II, his general remarks are relevant to the volume as a whole.

Thus to some extent both introductions are concerned with both parts of this volume; on a small scale they reflect the differing interests and criteria of the biologist and the non-biologist that are commonly met in the field of neural theory and modeling.

I wish first to acknowledge the invaluable assistance rendered by the associate editors, each of whom carefully edited a subset of papers. I also wish to thank the authors for their efforts and patience in revising their papers subsequent to editorial review.

Although no transcripts of discussions are included in this volume, all formal presentations and discussions at Ojai were recorded by R. Vinetz; these recordings supplied valuable documentary support for editing.

It is fitting that I acknowledge here the services of T. H. Bullock; he provided aid in organizing the symposium and acted as a constant, effective, and tireless catalyst throughout the long and exhausting days at Ojai.

Finally, without the financial and moral support of the sponsors, the Air Force Office of Scientific Research and General Precision, Incorporated, the production of this volume would have been impossible, and their support would have been fruitless without the guidance and extensive aid provided by the staff of the Stanford University Press.

RICHARD F. REISS

Contents

Participants in the Ojai Symposium, December 4–6, 1962

1. E. R. Lewis, General Precision, Incorporated
2. R. Vinetz, General Precision, Incorporated
3. Lawrence Stark, Massachusetts Institute of Technology
4. Wilfrid Rall, National Institutes of Health
5. Charles Molnar, Cambridge Research Laboratories, USAF
6. Harvey Savely, Air Force Office of Scientific Research, USAF
7. Robert K. Josephson, University of Minnesota
8. James C. Bliss, Stanford Research Institute
9. Belmont G. Farley, Massachusetts Institute of Technology
10. H. D. Crane, Stanford Research Institute
11. M. R. DeLucchi, Air Force Office of Scientific Research, USAF
12. J. Thorson, University of California, Los Angeles
13. H. Mittelstaedt, Max-Planck-Institut für Verhaltensphysiologie
14. Ervin J. Nalos, Boeing Aircraft Corporation
15. H. J. Hamilton, General Precision, Incorporated
16. Leon D. Harmon, Bell Telephone Laboratories
17. Melvin J. Cohen, University of Oregon
18. Peter Kunze, Universität Freiburg
19. G. E. MacGinitie, California Institute of Technology
20. Derek H. Fender, California Institute of Technology
21. W. Van Bergeijk, Bell Telephone Laboratories
22. B. S. D. Raj, University of Oregon
23. Donald M. Wilson, University of California, Berkeley
24. G. D. McCann, California Institute of Technology
25. F. Jenik, Technische Hochschule, Darmstadt
26. Richard F. Reiss, General Precision, Incorporated
27. D. M. Vowles, University of Oxford
28. T. H. Bullock, University of California, Los Angeles
29. Otto Schmitt, University of Minnesota
30. J. W. S. Pringle, University of Oxford
31. L. Butsch, Aerospace Medical Laboratory, USAF
32. Graham Hoyle, University of Oregon
33. E. R. Caianiello, Università di Napoli
34. Donald Kennedy, Stanford University
35. C. A. G. Wiersma, California Institute of Technology
36. Ann Biederman, University of California, Los Angeles

Part I

General Theory and Modeling

Introduction to Part I

RICHARD F. REISS, *General Precision, Inc.*

Although man's knowledge of his own and other animal nervous systems is still quite limited, the diversity of neural structures and functions already uncovered has led to the creation of similarly diverse theories and models. Like the fabled blind men who examined an elephant, each observer tends to generate a theory which reflects the particular type, level, and function of the nervous system which, for one reason or another, has attracted his attention. Thus it is not surprising that current neural theories and models, taken together, present a confusing and often exasperating jumble of logical forms and empirical correlations; many generations may pass before a remotely satisfactory unified picture of the whole beast or beasts can be constructed.

The present disorder is not entirely due to the chameleon-like qualities of nervous systems, the fragmentary experimental data, or the necessarily limited perceptual grasp of the individual human observer. In recent years there has been an accelerating increase in the number of neural theorists who initially trained and worked in nonbiological fields. These poachers in the domain of biology bring with them a wide variety of research attitudes, habits, prejudices, and analytic tools developed in their other occupations. Some of this baggage may prove to be a liability rather than an aid—it is too early to say—but it is clear that the influx of nonbiologists is contributing to the ferment and confusion. Speaking as a poacher myself, I believe that the confusion is a transient disturbance which will soon subside into well-ordered research; professional biologists are beginning to absorb the poacher's methods and ideology, and the poachers themselves are gradually being initiated into the often painful mysteries of biology. The chief short-term danger is that narrow enthusiasms in one faction will inadvertently mislead members of the other faction into wasteful or inappropriate projects.

As noted in the Preface, these factions are represented here by almost equal numbers of papers which have accordingly been organized into two groups. I have implied sharp factional differences which, in many individual cases, do not in fact exist. However, in general attitude and methodology, the papers in Part I are noticeably different from those in Part II. The nonbiologist modelers usually prefer to risk irrelevance by postulating universal principles which will satisfy their needs for economy and logic in model building. The biologists in Part II, on the other hand, tend to be conservative empiricists who are haunted by the diversity and variability of neural processes; their discussions are tied closely to specific experimental data obtained from particular neural preparations.

As a poacher, a nonbiologist, I shall confine most of my remaining remarks to the papers in Part I. Donald Kennedy's introduction to Part II provides a balancing commentary from the standpoint of a biologist.

There is a semantic difficulty inherent in both the title and many papers of this volume: What is the difference between a "theory" and a "model"? A physical device intended to explain or predict some aspect of neural behavior is, by more or less general agreement, called a "model." Such physical models are based, explicitly or implicitly, on some "theory" of neural behavior. The difficulty arises when the two terms are used interchangeably. On the one hand, a theory is often called a model, meaning a *conceptual* or *mathematical* model; for brevity the adjectives "conceptual" or "mathematical" are dropped. On the other hand, a physical model is sometimes considered to *be* a theory, i.e., it is the *definitive* description of a "theory." This viewpoint is most often taken when a physical model, although it may be constructed from components whose behavior is well understood, is so complex in structure or behavior that the human mind is incapable of "understanding" it in any detail.

Thus there are no hard and sharp criteria for the use of "model" and "theory"; these terms remain ambiguous in their usage, and the reader must determine from the context in each particular case what the author intends.

The initial paper of Part I outlines a number of problems which are of great importance in the construction of neural models (under any definition of "model"). With a background in communications engineering coupled with extensive experience and knowledge of neural modeling, L. D. Harmon is a good example of the nonbiologist who has become deeply involved in a study of biological systems. Both his detailed remarks and general attitude are of special interest because they are the result of a sustained clash between the ideology of engineering, with its emphasis on formal precision and free-wheeling synthesis, and the harsh realities of biological research: fragmented, incomplete, uncertain data acquired by painfully slow experiments on systems of bewildering complexity.

In the latter half of his paper Harmon rapidly surveys a wide variety of current and recent neural models. Although he has not attempted an exhaustive survey, with its accompanying bibliography the paper provides an excellent introduction for newcomers to the field of neural modeling whether they are biologists or otherwise. A careful reading of Harmon's paper will also provide a general framework for evaluating much of the remainder of this volume.

The following paper by the eminent zoologist J. W. S. Pringle concentrates on one aspect of the problem of achieving realism in neural models. In the context of specific types of sensors and effectors he examines certain general properties that a model's input and output should have; he argues that without these properties, the inputs and outputs constitute arbitrary simplifications which may destroy the value, or at least the biological relevance, of the model as a whole. This possibility has often been ignored in the past, presumably because many modelers have concerned themselves with very tentative models of fragments of central nervous systems. Even though a central nervous system, from one viewpoint, exists only to serve the organism's sensors and effectors, it is tempting to make a few hasty, simplifying assumptions about these peripheral elements in order to get on with the ostensibly more exciting business of analyzing central networks. Perhaps this research strategy is, and for some time will remain, legitimate owing to our extensive ignorance of *all* parts of the nervous system, but it must eventually lead to folly.

Both Harmon and Pringle have directed their attention to problems of establishing correlations between current neurophysiological data and models; by contrast B. G. Farley has approached the problems of model validity from an entirely different direction. Starting with a more or less positivistic conception of the nature of scientific theory, he argues that it will ultimately be possible, if not desirable or necessary, to develop a formal schema for defining and evaluating biological "theories," this schema to be based on the "theoretical properties of ideal computers." Since our ignorance of the future capabilities and properties of computers, ideal or real, is perhaps as great or greater than our ignorance of neural mechanisms, Farley has been forced to engage in hazardous speculations; but whatever his errors may be, he has here opened a subject that has received little consideration among modelers.

In order to illustrate some of his points, Farley describes recent digital simulation experiments he has carried out on a large, special-purpose machine. A similar application of digital computers to network analysis is outlined by R. K. Josephson in Part II of this volume. There is this marked difference, however: Farley's experiments were conceived as explorations of the possible behavior, under various assumptions, of cortical networks—at the highest level of the vertebrate central nervous system—whereas Josephson's simulation experiments were aimed at testing very specific hypotheses regarding the behavior of particular coelenterate networks—at one of the lowest levels of the invertebrate peripheral nervous system. It is interesting to note that systems at two extremes of organizational level have suggested similar conceptual models, hence similar simulation programs.

This curious methodological and formal similarity of models at opposite extremes of the organizational spectrum is also apparent in the following two papers. In both cases rather elaborate mathematical techniques, involving equations taken from the physical sciences, are applied to neural problems. Wilfrid Rall analyzes one of the smallest elements of cortical systems, the dendritic structure of an individual neuron, whereas E. R. Caianiello is concerned with the cortical mass as a whole, a system estimated to contain from 10^9 to 10^{10} neurons. At both extremes the behavior of the subject appears to be sufficiently continuous and linear in form to permit the application of well-known mathematical techniques; at many intermediate levels of organization the behavior is so nonlinear as to generally defy mathematical description.

Rall's paper is but one of the latest in his exceptionally consistent and methodical series of studies of the individual neuron. He presents a persuasive argument to the effect that the behavior of a cortical neuron can be dominated by dendritic synapses in certain spatial distributions of a type that has been observed; if this is true, then any cortical network model which ignores the peculiar properties of dendritic structures (and most such models have to date been guilty of this simplification) is likely to be in serious error.

Rall is one of those still rather rare physiologists with a good grounding in mathematics. In order to get a grip on dendrites with his differential equations he makes use of a "compartmental model" which he has "borrowed" from the fields of chemical and radioactive tracer kinetics. Caianiello, a theoretical physicist, has similarly borrowed equations from the field of thermodynamics and modified them in a manner which, he believes, will make them applicable to masses of cortical neurons. However he does not go into details here; he mentions them only as examples of a methodological approach that he champions.

The next two papers also have much in common. Both authors are connected with the computer field, and both use combinations of simple logical, numerical, and physical arguments to explore possible signal-processing mechanisms that could be theoretically realized by small networks of neurons. R. F. Reiss argues that by means of "resonant networks," a nervous system could take advantage of valuable pulse interval coding techniques which would be forbidden to it if only the commonly assumed statistical decoding mechanisms are available. He examines several types of hypothetical network that could perform various switching, filtering, and decoding tasks; deduces the ranges of neural parameters required; and suggests how the presence of such networks may be detected by physiologists.

In his introductory section Hugh Crane provides the reader with a glimpse of the line of reasoning that has drawn a computer engineer into neural modeling. He shows why he has been led to examine the possible roles of axon–axon interactions in neural signal processing, a subject that has received very little attention from other theorists. Such interactions may occur quite naturally in ganglia, and Crane proposes several kinds of switching and frequency-modulation functions that could result. As in Reiss's paper, Crane's material is highly speculative but predicts patterns of neural behavior which, if detected, would indicate the existence of the hypothesized networks.

The papers by Lewis, Jenik, and Hamilton describe existing physical models of all or parts of the individual neuron. Working within the framework of the ionic hypothesis and guided specifically by the experimental data of Hodgkin, Huxley, and others, E. R. Lewis has developed an electronic circuit which is capable of reproducing most of the subliminal responses that have been observed in neurons. His results suggest that the ionic hypothesis may be sufficient to account for somatic behavior as well as spike initiation and propagation in axons; one need only vary a few parameters to transform the model's behavior from that of an axon membrane, through a variety of semi-active membranes (including the pacemaker type), to that of electrically inexcitable synaptic membranes.

Franz Jenik's model is based on the Hodgkin-Huxley mathematical theory, and it has several features in common with the Lewis model. However, Jenik has focussed his attention on the spike initiation process as it affects the statistical relations between trains of input and output pulses; he displays a very tidy method for characterizing the behavior of a neuron in terms of ratios of mean input and output frequencies.

The circuits of Lewis and Jenik are probably the most flexible and realistic neuron models yet devised. Although they were both intended primarily for use in the study of individual cell behavior, groups of them can be connected together to simulate networks of neurons. They are too costly for use in large quantities, but they could well lead to a new era in the study of small networks. Even if they were cheap, there is reason to doubt that a network containing a dozen or more such complex and protean models could be rationally managed at the present time.

Looking forward to the time when we can manage large systems of realistic model neurons in meaningful experiments, H. J. Hamilton has been developing a special type of analog device which he believes will serve as a realistic model that is both small and cheap. There appear to be some striking similarities between the behavior of gas ions, under certain conditions, and the evident behavior of sodium and potassium ions in a neural membrane. Hamilton describes a class of simple gas-ion devices which, though still in an early stage of development, have several of the most important properties required in a model neuron. If these devices can be further refined and mass produced in convenient and flexible organizations, then they may provide serious competition for digital computers in the simulation of large neural networks.

But even if gas-ion or other special types of device turn out to be superior vehicles for neural modeling, the digital computer is likely to be the best tool for controlling and analyzing the behavior of large simulation systems. The problem of managing experiments on complex systems appears to be exceedingly difficult, and digital computers offer the only promising means of solution in the foreseeable future. From the standpoint of the computer there may ultimately be little difference between the management of experiments on real nervous systems and on realistic models of nervous systems. Thus if the problem can be solved for either type of experimental subject, it may be largely solved for the other.

For this reason efforts such as that described by G. D. McCann are of interest to the modeler as well as the experimental biologist. In his paper McCann outlines the organization of a rather elaborate system composed of digital computers, measuring instruments, and special machines which can control certain types of physiological experiment, make and record observations, and perform data-reduction tasks which will facilitate the analysis of results. As an example of one of the first applications of this system he reports some automated experiments on opto-motor responses of the fly.

This type of control and recording system has to date proven most useful in the study of semi-independent servomotor mechanisms in various nervous systems. The "cybernetics" approach to biology has been influential in drawing attention to such mechanisms and in developing appropriate conceptual models. These are often called "black-box" models because the neural mechanism is represented by a set of interconnected units—the "boxes"—whose functions are typically defined in mathematical terms. The specific neural sub-systems responsible for the generation of these functions may not even be postulated, much less known, and therefore the boxes are "black."

The black-box model is an important and increasingly popular type of theory which is not discussed in any detail in Part I of this volume. However, two examples may be found in Part II. H. Mittelstaedt outlines a family of equations which can be considered black-box models intended to explain orientation responses to various kinds of stimuli observed to occur in certain insects. D. H. Fender not only presents a black-box model of the eye movement control system in humans; he also gives a step-by-step account of how this model was developed. Here we see explicitly what kinds of questions and answers are entertained by a man who entered into neural modeling from the field of engineering. It might be noted in passing that black-box models are a stock in trade of electrical engineers; as more such persons enter the field of neural theory we can thus expect to see a great proliferation of black boxes.

The note by James Bliss is connected with a black-box model, but is essentially a brief summary and introduction to the literature of a special mathematical function which has been proposed to explain certain opto-motor responses in insects.

1. Problems in Neural Modeling

LEON D. HARMON, *Bell Telephone Laboratories, Inc.*

The making of models is universal in the search for a consistent and instructive picture of nature. Models of neural action have been frequently employed in the past; their relative importance, however, has been minor. Early neurophysiological research was oriented principally to experimental investigation of gross structure, metabolism, and relatively simple electrochemical stimulus–response properties. Now, as greater emphasis is placed on the view of the nervous system as a processor of information, theoretical models assume greater importance and have become increasingly numerous.

In this paper I shall attempt to sketch out several fundamental considerations and problems which seem to me particularly relevant to an understanding of signal processing in the nervous system. This will be followed by a survey, illustrative rather than exhaustive, of the principal types of neural models. I hope to show how experimental and theoretical neurophysiology now stand in relation to each other and where concentrated effort can be significantly applied in both areas to produce a more satisfactory elucidation of nervous action.

Problems in Experimental and Theoretical Neurophysiology

Single Units

In recent years the discovery of increasing numbers of single-unit properties, many of which are apparently nonlinear besides being continuously variable, has greatly intensified a number of traditional problems. One especially difficult question is the following: To what extent does a particular response depend on a *given* stimulus and to what extent does this response reflect the influence of *preceding* stimuli? Moreover, in what ways does response depend not only on instantaneous values of the stimulus parameters but also on their directions and rates of change?

A well-known example of sensitivity to past events is to be found in postsynaptic inhibition. If a neuron's postsynaptic membrane potential at a given time is at or near the resting level because the unit has been inactive for a long while, then an applied excitatory stimulus will have one particular effect. If, however, this present membrane potential results from a conductance change due to a recent inhibitory stimulus (clamping the membrane to or near the resting level), and if the time course of that conductance change has not yet run out, then the response to the same applied excitatory stimulus can be quite different.

Long-lasting effects such as facilitation and defacilitation (which are not reflected in membrane potential changes) provide clear evidence of dependence on temporally remote events (Terzuolo and Bullock, 1958). Similar dependencies are found in postinhibitory rebound, where spikes may subsequently arise from a *greater* level of membrane polarization than was found permissible before inhibition (Otani and Bullock, 1959); a given level of membrane potential does not always result in spike generation.

Further, an action may depend not only on prior stimuli but on the order in which stimuli are applied, being different for different sequences. An example of sensitivity to sequence is seen in the response of *Aplysia* to a triad of spikes (Bullock, 1962). A single unit fires only for one particular temporal pattern of three presynaptic spike stimuli, although each of the patterns has the same average frequency. Simple temporal summation apparently is not involved.

At present we lack adequate knowledge of what or how many state variables are involved in the moment-to-moment behavior of a neuron; as a consequence we often attempt to "understand" neuronal action in historical terms. In general, however, unless the history of past events (and their interactions) back to some limit is known in relatively great detail, single-unit response to a given stimulus can be quite unpredictable. Even if we had the theoretical structure to make fully reliable predictions on the basis of stimulus trajectory and sequence dependencies, the extent and significance of these dependencies in normal neural action is essentially unknown; it constitutes a prime subject for both theoretical and experimental study.

A very large and important class of problems concerns form–function relationships. Considerable significance has been attached to synaptic geometry. The intricate elaboration of synaptic surfaces revealed by electron microscopy strongly suggests highly differentiated function following from specific form* (Palay, 1958; Szentagothai, 1962). In studies of presynaptic inhibition (Eccles, Eccles, and Magni, 1961; Eccles, Kostyuk, and Schmidt, 1962), evidence is developed for bouton–bouton interaction. In other work, differences in the invasion of somatic membrane by elicited spikes as a function of particular synaptic location have been held accountable for distinctly different ways of modulating over-all unit activity (Purpura, 1959; Frank and Fuortes, 1961; Rall, 1962b).

Besides the many ways in which single-unit response depends on spatial arrangement and temporal sequence among synapses, additional effects arise from ephaptic interaction between adjacent neurons† (Arvanitaki, 1942; Grundfest, 1959) and from gross electric-field influences (Bullock, 1957).

* Gross structural differences alone, however, are apparently not necessary for functional variety, since synapses which appear to be quite similar morphologically may yield strikingly different electrophysiological responses (Robertson, 1953).

† Recent theoretical studies (Crane, 1962, and this volume) suggest that proximity relationships in axon-like structures may be profoundly important. Discrete-pulse interactions have been shown to be sufficient for developing a complete set of logical operations.

These implications for form-dependent differentiated function emphasize the importance of the interconnections between "functional elements" for information processing in nervous tissue. With the exception of some recent developments, electronic digital-computer function depends essentially only on the properties of the discrete, *lumped* elemental parts and the simple directed pathways connecting them. In contrast, nerve networks seem to be *distributed* computers; i.e., there are no passive elements comparable to the wires and joints of an electronic computer. Neural logical processing appears to depend heavily on the nature of the interconnections. Viewed in this way, the operations specified by the microstructure of the "wiring" may be no fewer and no less important than those specified by the "components."

Problems of *intracellular* structure and function have received the most sustained experimental and theoretical attention, although increasing emphasis is being placed on synaptic mechanisms. It appears now that an extensive and systematic theoretical investigation of *intercellular* functions, particularly as they are influenced by spatial parameters, is greatly needed. Several steps already taken in this direction are of interest. Rall (1959, 1962a, b, and this volume) has explored the physiological consequences of geometrical variation in dendritic structures, and van Bergeijk (1961) has studied the effect of dendritic tree size on the dynamic range of input stimuli that can be handled by a single neuron. Lettvin* proposes a theoretical formulation for synaptic interaction as a function of geometrical configuration. He shows how somatic-membrane response resulting from sequences of stimuli to two or more terminals varies radically but regularly with different spatial orderings of these terminals.

There has been considerable study of spike-interval coding (Wiersma and Adams, 1949; Hagiwara, 1954; FitzHugh, 1958; Gerstein and Kiang, 1960; Amassian, Macy, and Waller, 1961; Viernstein and Grossman, 1961; Gerstein, 1962). A bit of controversy among physiologists is apparent here, some believing strongly that discrete-interval information is important, others feeling that in most cases average frequency information alone is utilized.

It is of interest to inquire to what extent spike-interval coding *can* be preserved transneuronally. In some very simple, linear, and noise-free situations it is clear that timing information can be transmitted precisely. For example, no matter how severely monosynaptic temporal integration smears signals (up to the limit of perfect integration, which apparently is never observed), the original message is (in principle) recoverable to any specified degree of accuracy.† Even with the simplest of nonlinear operations, however, incoming spike-interval information can be destroyed beyond recovery at the output. Consider, for example, multispike temporal integration (linear) which causes membrane potential to cross the spike-producing threshold (nonlinear). In the case where the output is a single spike, no unique information has been preserved about the incoming spike intervals, or even (probably) about their average frequency; for each output signal there are many possible antecedents. (Such timing

* Personal communication, 1962.

† In multisynaptic situations, however, or even in multiple dendritic branching, such deduction of input from output may be completely impossible. See, for example, Rall (1962a, b).

information can be preserved, of course, in an ensemble of neurons where different units respond in volleys.)

Undoubtedly some classes of information coding are readily passed through single units, but there has been no systematic investigation of this important subject. The formulation of even a few theoretical constraints would be valuable. If it could be shown, for instance, that a limited or negligible amount of input spike-interval information is representable (by any transformation) in the cell's output signals, then a large amount of effort could be either focused more precisely or else saved entirely.

The role of random processes in organism behavior has long been a subject of controversy.* One seldom hears the term "free will" applied to a single neuron, but most assuredly any assumption of deterministic behavior is challenged by the evidence for random threshold fluctuations, internal noise, and irregular spontaneous activity (Fatt and Katz, 1952; Frishkopf and Rosenblith, 1958; Verveen, 1961). Besides such obvious considerations as reliable network performance in the presence of unpredictable processes, it is of fundamental importance to know the extent of noise effects on single-unit operation.

One class of experiments which would be helpful is the systematic examination of single-unit input-output functions in the presence of deliberately introduced noise. Although several of the modeling studies have made a start in this direction (Derksen, 1962; ten Hoopen and Verveen, 1963), there appear to be no reports of similar biological experiments. One might proceed in the following way: Consider a monosynaptic two-unit system. Ignoring spike amplitude variation for the moment, we may view information traversing the system as being temporally coded in the first axon, amplitude coded in the postsynaptic membrane, and temporally coded again in the second axon. It would be of interest to catalog classes of input-output (i.e., first axon–second axon) signal relationships in the absence of superimposed noise, then to find to what extent these functions are preserved with the introduction of temporal uncertainty (e.g., jitter in an otherwise regular train of spikes) into the first axon, or of level uncertainty (amplitude-modulated noise current) into the postsynaptic region.

In this way it may be possible to learn the extent to which small temporal changes (i.e., variations in pulse-to-pulse intervals) are reliably preserved transneuronally, and perhaps also to establish a "safety factor" for the threshold fluctuations already known to exist (i.e., the relative insensitivity of the input-output functions to threshold variation). It may be possible also to gain some insight into the significance of the postsynaptic-potential overshoots and undershoots.

Networks

Our insufficient knowledge prevents us from deducing function from form in networks just as it does in single units. The enormous number of degrees

* No distinction is made here between truly random processes and what are known as "quasi-random" processes, i.e., those that are so complex that they must be treated by statistical methods. Hence the term "deterministic" refers to processes whose behavior can *in fact* (not just in principle) be predicted with specified precision.

of freedom possible even in a small assembly of neurons makes it possible to find almost any set of stimulus–response characteristics one can imagine. For this reason alone there is a philosophy often adopted by theorists that I should like to urge on experimental neurophysiologists. It is this: Rather than acquiring data where and how it can be found, cataloging complexity after complexity, why not go into a system with a hypothesis, a prejudice? Why not base a hypothesis of function on, say, observed gross behavior, and then look for neural operations which can lead to elucidation of the bases for this behavior? This point of view has been employed with profit in the last few years (e.g., Hubel and Wiesel, 1959, 1962; Maturana, Lettvin, McCulloch, and Pitts, 1960; Maturana, 1962). I suspect that further extensive work which jointly considers physiological and psychological (or even teleological) mechanisms will be tremendously rewarding.

The point that network operations cannot necessarily be deduced from elemental function and network form is illustrated by a homely example from electronic circuits. Given a moderately complicated schematic diagram and no idea of what the circuit is intended to do, even an engineer skilled in circuit analysis is often totally unable to give its function. This will happen even though he knows, *in great detail*, all the properties of the elemental parts. In short, even if there is sufficient information for complete analysis, the situation may be too complicated for contemporary logical or mathematical treatment.

Let me make an even stronger assertion. The circuit of Fig. 1 contains three "neuromimes," neuron models whose properties have been described in detail (Harmon, 1961). Note that there are but four "interneuronal" connections, one of which is fed-back inhibition. The input signal is specified, as are all the time constants. The diodes serve only for "synaptic" isolation. The assertion is this: Given a complete description of the structure of this simple network and the properties of its elements, it is extremely unlikely that anyone will be able to specify its output function.

In general, then, we can no more easily deduce ensemble function from single-unit functions than a physicist can deduce the physical properties of water from his knowledge of hydrogen and oxygen molecules. Both difficulties

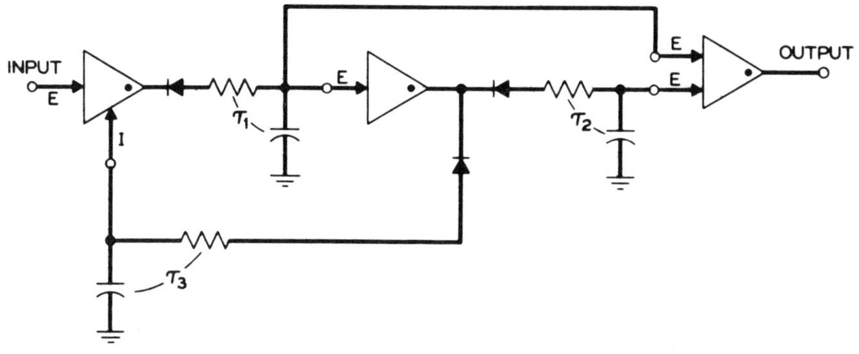

FIG. 1. Three analog neurons connected in a network. Diodes and integrators (τ_1, τ_2, τ_3) simulate synaptic action.

are practical ones at our present levels of understanding; they may not be in-principle impossibilities. The point is that for the time being we may find it most profitable to seek an understanding of nervous system operation on the basis of highly specific a priori conjectures.

Closely related to the preceding discussion is the class of problems dealing with successive transformations of sensory afferent information. Let us consider some findings concerning the early stages of visual systems. Enhanced image contrast results from simple mutually inhibitory networks (Hartline, Wagner, and Ratliff, 1956); even precise image reconstruction has been shown theoretically to be obtainable from such networks despite receptor-field overlap (Reichardt, 1961a). Concentrically regular arrays of inhibitory and excitatory elements give rise to spatial and temporal discontinuity detection (Kuffler, 1953; Barlow, FitzHugh, and Kuffler, 1957; Brown and Wiesel, 1959; Hubel and Wiesel, 1959); such arrays have been shown to be able to detect particular shapes (van Bergeijk and Harmon, 1960). Reasonably regular network connections have been held accountable for detection of visual stimuli such as lines and edges in preferred directions (Hubel and Wiesel, 1962), even to the extent of position invariance, thus affording a physiological toehold on Gestalt psychology.

Each of these findings has an intriguing common element. Spatial arrangements are still important, but extensive information processing can result from fairly *simple* and *regular* network connection patterns. Particular features of the sensory inputs are extracted by relatively homogeneous arrays without highly differentiated, explicitly computer-like logical organization. One is tempted to ask the following: How far into the central nervous system may such regular structures account for increasingly more elaborated information processing? It may well be possible that hierarchies of very simply structured arrays can extract extremely complicated features of input information with no more anatomical specification than a few general laws of neighborhood orderliness. A modest theoretical start has been made in exploring such possibilities (Pitts and McCulloch, 1947; Uttley, 1954; von Foerster, 1958; Babcock, 1960), but considerable additional effort is surely in order.

It is often speculated that in going from the peripheral to the central nervous system specific connections and precise single-unit properties may diminish in importance. One is tempted to view central processes in terms of mass action, slow waves, and statistical averaging. Certainly there appears to be considerable redundancy, or at least immunity to the partial loss of structure; this, however, by no means implies that normal operation of the system is probabilistic. Evidence for highly specific single-unit response in cortical and subcortical areas (Morrell, 1961; John, 1962) clearly warns against speculating only on mass action or average response. Incidentally, it may be noted that there is no more (or less) evidence for critical dependence on individual cells in central structures than in peripheral areas.

Of course, responses are averaged at many places in the system. Undoubtedly two fibers rarely have precisely the same response, and a single fiber rarely responds to the same stimulus in exactly the same way at different times.

Furthermore, average responses appear to be used frequently in sensory processing. Even in a statistically determined system, however, the importance of the *properties* of the elements is by no means diminished. Although an individual contribution to the end result may be small enough to be dispensable, it is precisely the average but *particular* properties of the individual units which account for the aggregate behavior.

It is in suggesting functional relationships between single-unit action and group behavior that theoretical neurophysiology may be most potent. The study of information processing in nervous systems gives rise to a number of problems which are not readily solved by investigation *in vivo*; some questions are more appropriately explored theoretically. For example, there are at present few techniques for controlled stimulation of a single unit via multiple synaptic pathways simultaneously, and for observing the response as well; i.e., the integrated input-output functions of just one cell cannot, in general, be determined experimentally. At the other extreme the situation is equally difficult; simultaneous measurement of activity among an ensemble of units is in most cases either formidable or else impossible.

There are many problems of rationale and technique peculiar to model-making (as well as a large number shared in common with experimental neurophysiology); some of the particularly troublesome ones are discussed in the following sections.

Models

Any model has utility only insofar as it raises questions, suggests new relationships, and serves to focus one's thoughts. This is accomplished by using it to test hypotheses, to compute, to extrapolate and predict, and to suggest useful experiments. In this way models accelerate the process of learning about the real world.

A model must be reasonably accurate, and it must be extensible, i.e., new, relevant properties must be deducible from it. Only under such conditions is model-making nontrivial. Unfortunately the incomplete state of neurophysiological knowledge and the controversies over interpretation preclude precision in model-making.

How does one proceed in a search for acceptable models? The "black box" technique is useful and is widely employed. Systems or processes are modeled by specifying a number of boxes which yield functional equivalence. Thus, for example, if temporal integration is required, a given set of parameters (time, amplitude, and the like) can be equivalently obtained to any desired degree of accuracy by ion-membrane interactions, mathematical expressions, electronic circuits, or wheels and levers. The contents of the black box are immaterial as long as its input-output functions are equivalent to those of the system being modeled.

It is sometimes argued that *all* black boxes must be opened or fragmented into yet smaller black boxes if incompleteness or error is to be avoided. It has been asserted, for instance, that a precise replication of membrane kinetics is required for a satisfactorily accurate single-unit analog. This condition is met, of course, in the most widely known model of all, the Hodgkin-Huxley description of membrane action. Note, however, that even this model (a mathematical abstraction of physical events) contains unopened black boxes, such as ion pumps.

For any finite system there exist an infinite number of black-box equivalents. Every one is valid for all input-output functions that are a priori defined, but unless the mechanisms internal to a given black box are identical to the mechanisms of the system being modeled, there is no assurance that other functions (not considered in the original specifications) will be similar. That is, down to *any* arbitrary level, no "equivalent" system can be guaranteed completely to replicate the functions of a given system under *all* conditions. This contention carried to its logical extreme means that only the original can simulate the original. To go so far is to deny the value of any model.

Clearly what is implied here is the necessity for wise choice. Model-makers must seek (perhaps divine) the *essential* ingredients of a given system. This need for selection presents particularly difficult problems in theoretical neurophysiology where one must cope with an immense array of experimental results. One must begin by inquiring which are the pertinent parameters and which are of secondary importance.

There are essentially two distinct philosophies in the selection of parameters in neural modeling, particularly for single-unit analogs. In one (e.g., Jenik, 1962, and this volume) a large number of neural properties are reproduced with high accuracy (e.g., precise replication of spike waveform). It is presumed that the model has been initially overspecified to some extent, and there is a tacit intention to simplify if and when it seems *reasonable* to do so. In the other approach (e.g., Babcock, 1960) a more restricted set of properties is used (e.g., rectangular output pulses), but the restrictions have been made on the basis of a calculated a priori "best guess". It is assumed that the important, first-order functions have been retained, and there is a tacit intention to introduce more complication if and when it becomes *necessary* to do so.

There are modest advantages and disadvantages to each approach. Models of the first type are more difficult to realize, and more costly. Models of the second type, having less realistic-looking waveforms, are less appealing to experimental neurophysiologists. The behavior of the simpler model is more amenable to analysis; the more complicated analog exhibits some aspects of single-unit activity not possible with the other. The really important question is not yet answerable: namely, whether either or both are adequate. So long as there is continued testing for appropriateness and so long as neither seriously violates physiological evidence, the choice is largely esthetic.

One question, perhaps the most fundamental of all, is easily stated: What are the *relevant* parameters of neural action for information processing? Which of the large number of observed single-unit properties are essential and which are second-order? The catalog of properties is impressive (Bishop, 1956; Grundfest, 1959; Bullock, 1960; Eccles, 1961).

There are two ways in which answers can be developed. The first arises from prejudiced experimentation, experiments based on a priori conjectures. Most of the few results thus far obtained by this process have been extremely illuminating, and much more work of this sort needs to be done.

The second arises from extensive model-making. Given a moderately accurate model of a single unit which replicates a large number of neural input-output functions, one can perform two classes of experiments. First, one finds to what extent the model's specification can be simplified before its performance deteriorates. Second, one determines to what extent the properties of the model predict yet other neural action, and to what extent its behavior can be extrapolated and still accord with the actions of groups of neurons.

The procedures required for both classes of experiments are routine in all worthwhile modeling; they constitute the process of seeking convergent qualitative and quantitative agreement between analog and prototype. In such a process of progressive evolution of a model, hypotheses are freely adopted but are retained only so long as function remains realistic. In this way errors and omissions in parameter choice become apparent; thus seen, limitations are profitable.

It is of interest to inquire whether among various classes of neurons many *different* sets of parameters must be considered; that is, are there a large number of fundamentally different neurons? Undoubtedly there are several kinds of synaptic transmitter substances, several varieties of inhibitory mechanisms, synapses, ephapses, and so on. Undoubtedly the vertebrate motoneuron and the crustacean stretch receptor differ, but in what ways and by how much? Some modeling studies have shown that with only *quantitative* parametric adjustments a rich array of single-unit functions can be demonstrated (e.g., Harmon, 1961; Lewis, this volume). It is entirely conceivable that a neuron is a neuron wherever it is found, basically the same but differing from place to place or system to system only by virtue of different time constants, threshold levels, and spatial configuration.

The naive concept of a generic neuron was one of the earliest held in neurophysiology. In time the gradual acquisition of knowledge appeared to require its abandonment. Perhaps an enlightened return is now possible, and the idea of a generic neuron can mark a significant advance by helping to unify our knowledge.

Contemporary Neural Modeling

Techniques

Models of neural function can be categorized into four broad classes: mathematical, ionic, electronic, and computer-simulated. Considerable advantages and serious shortcomings are found in each, although for a given modeling problem there is generally no difficulty in selecting the most appropriate technique. Since these advantages and disadvantages are seldom stated explicitly, it may be of some use to examine them briefly.

Mathematical models are of great utility in highly restricted areas. Perhaps they have been employed to best advantage for the description and analysis of membrane kinetics (Hodgkin and Huxley, 1952; FitzHugh, 1955, 1960, 1961; Rall, 1955a, b, 1957, 1960, 1962a, b). In such cases a fairly complete model is possible so long as the number of variables is reasonably small and the nonlinearities do not present severe analytical problems. However, when the domain of the model is extended and more complex operations are considered, the limitations of our mathematical tools quickly become apparent. This is especially true for adequate representations of the entire neuron.

There have been numerous examples of single-unit models (e.g., McCulloch and Pitts, 1943; Minsky, 1954; Kleene, 1956; von Neumann, 1956); each of these "formal neurons" is (of mathematical necessity) a highly oversimplified, constrained, essentially unrecognizable vestige of the biological prototype. Although it is true that these elements are sufficient to obtain a complete set of logical functions, they are not necessarily more appropriate than digital-computer components as representations of biological neurons.

The physiological accuracy of such models was in serious doubt even for the old "switchboard "neuron; they are hardly more appropriate for the neuron as we see it now. The fault lies, I emphasize, not in the modelers, but in the nature of the vehicle. Formal mathematical description is simply unable in its present state to deal with the multivariable nonlinear complexities of entire neurons.

If mathematical models are unsuitable for single-unit representation, what of their competence to deal with aggregates? Here, curiously enough, their position can be made considerably more defensible. *If* it is assumed that some aspects of network behavior are amenable to statistical analysis (e.g., if noise and averaging play important roles), then mathematical modeling can be appropriate (or, in the case of extremely large networks, essential).

Among the earliest of the neural analogs were ionic systems (Lillie, 1936).* Similar electrochemical devices were subsequently investigated in more detail (Bonhoeffer, 1948; Yamagiwa, 1950; Tasaki and Bak, 1959) and are currently being reinvestigated (Aldrich, McCulloch, Lettvin, and Pitts, 1962). Recent studies with gas-discharge systems (Hamilton, *this volume;* Hamilton and Vinetz, 1962) appear to hold promise. These techniques have the advantage of readily providing necessary nonlinear parameters and of working in real time. The making of detailed measurements in an acid vat or a vacuum chamber presents some difficulties, but undoubtedly they can be overcome. Although the utility of ionic systems has not yet been adequately demonstrated, these systems seemingly have considerable potential.

Electronic circuits (e.g., Derksen, 1961; Harmon, 1961; Jenik, 1962) and computer simulation programs (e.g., Reiss, 1960; Farley and Clark, 1961; FitzHugh, 1962; Smith and Davidson, 1962) constitute the two most active classes of neural modeling. Since the two approaches often have mutually exclusive advantages and disadvantages, I will discuss them together.

For either single-unit or network studies, digital computers simulate continuous nonlinear interactions inelegantly and uneconomically. Continuous variables are realized only by discrete approximation; the practical constraints of present digital techniques for serial computation tend (as do the mathematical

* References to an interesting assortment of very early analogs may be found in Biedermann (1898) and Ebbecke (1928).

models) to compel oversimplification. Analog computers, though cumbersome, are free of these evils. Electronic models (which are in fact special-purpose analog computers) readily provide the required operations.

Digital computers are slow (well-specified membrane or single-unit representations operate thousands of times more slowly than real time). Analog computers are faster, and hardware models operate comfortably in real time.* Digital computers readily provide delay, a parameter generally obtained only with great difficulty by the other two techniques.

In a modeling situation which may require frequent change of parameter values, punching a few new cards is easy, turning knobs entails little more hardship, and unsoldering components is least desirable. Large-network modeling is generally considered to be a natural for digital-computer simulation, but by employing some sophisticated connection and control techniques, one can easily manipulate 100-unit networks of electronic models (Derksen, 1962). Digital computers, employing discrete symbols, readily permit complete control and observation of assigned variables. They provide useful "snapshots" of the state of a network at any arbitrarily selected time, whereas multiple-state representations are less easily derived from either analog computers or electronic units.

There is one extremely strong objection to the digital-computer simulation of neural models: it prevents effective use of our two most valuable analytic tools, our eyes and ears. There is considerable advantage to observing real-time data *while* stimuli and model parameters are changed. None of our present computer print-out techniques can compete with this. Direct observations of waveforms, phase relationships, modulations, and time-dependent interactions are revealing and suggestive.

Examples and Results

Overt neural modeling is currently being applied to a wide range of problem areas. (By overt modeling I mean studies explicitly designed to complement experimental neurophysiology, not the ubiquitous tacit modeling which always accompanies physiological measurements.) Representative examples of recent work will be cited in this section. More complete inventories may be found in a number of recent surveys and reviews (Sweitzer, 1960, 1961; Minsky, 1961; Simmons and Simmons, 1961, 1962; Harmon, 1962a, b). Figure 2 catalogs many of the models which have appeared.

Membrane Models. Most of the present models in this category are based on the nonlinear differential equation representation of membrane kinetics (Hodgkin and Huxley, 1952) and on the related electrical circuit representations of excitatory and inhibitory mechanisms (Fatt and Katz, 1953; Coombs, Eccles, and Fatt, 1955).

* Some typical examples: For a Hodgkin-Huxley membrane model in an IBM 704 computer, 5 msec of action takes 2 min of computer time to simulate, thus slowing down physiological time by a factor of 24,000 (Jenik, personal communication, 1962); the same model in a Pace analog computer requires 25 sec ($\times$ 5000); a net of 1296 highly simplified single units requires 1 sec of TX-2 digital-computer operation to represent 0.5 msec; i.e., the time dilation is $\times$2000 (Farley and Clark, 1961).

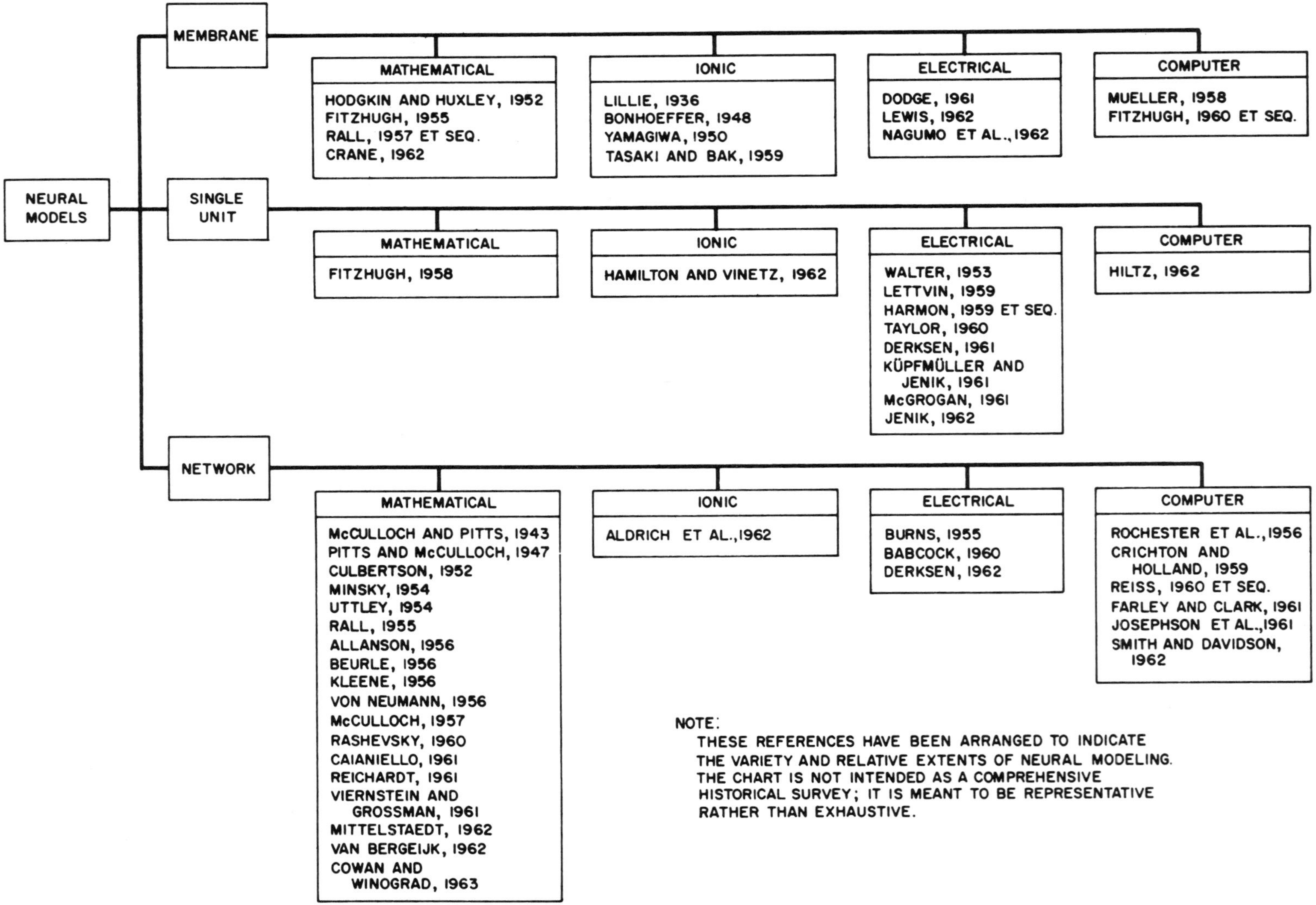

NOTE:
THESE REFERENCES HAVE BEEN ARRANGED TO INDICATE THE VARIETY AND RELATIVE EXTENTS OF NEURAL MODELING. THE CHART IS NOT INTENDED AS A COMPREHENSIVE HISTORICAL SURVEY; IT IS MEANT TO BE REPRESENTATIVE RATHER THAN EXHAUSTIVE.

In developing mathematical models of transient electrical response characteristics of motoneuron membrane, Rall (1957, 1960) showed that *active* conduction was not required to account for the observed time courses of dendritic and somatic response (contrary to previous belief). Earlier physiological measurements had led to the contention that the membrane time constant was too short to be responsible for the rather long decay of postsynaptic potential; active (though nonregenerative) conduction was thought to be responsible. Rall's model satisfactorily explained the apparent discrepancy by showing that the dendritic membrane time constants had to be treated as *distributed* rather than as *lumped*; in that case *passive* postsynaptic decay alone is sufficient to account for the data. Such purely mathematical modeling as this has great utility in clarifying physiological problems and in leading to more concise and accurate views.

In an extended model of dendritic membrane geometry, Rall (1959, 1962a, b) developed a theory of the integrative functions of dendritic trees as they influence somatic response. Studies of this kind are essential for bridging the rather considerable gap between local-membrane kinetics and entire-neuron behavior. The nature of these relationships is such that direct physiological experimentation alone is either impossible or insufficient; formal model-making appears to be essential.

Another important class of membrane models deals with extended analyses of the basic nonlinear systems describing membrane excitation and inhibition. Analog-computer simulation is used to examine the consequences of stimulus and membrane parameter variation that would be difficult or impossible *in vivo*. An example of such modeling is that of FitzHugh (1960, 1961), in which graphical analysis derived from analog-computer modeling lucidly demonstrates the quantitative relationships between physiological states such as resting, active, refractory, enhanced, and depressed. Stability, instability and spike-train characteristics are presented in an organized manner which clarifies the picture of complex interactions in axon physiology. An additional contribution of this work is that the functional relationships of membrane models and various similar physical systems are readily determined and compared; thus in this case a model is used to test and evaluate other models.

Analog computers are also effectively employed to explore some of the subtle implications of subthreshold phenomena. For example, Jenik* has simulated the basic Hodgkin-Huxley model *plus* input channels (unspecified by that model) to examine the implications of membrane potential overshoot for information processing. A considerable extension of this type of investigation is essential to a better understanding of the relevant aspects of membrane kinetics.

A recent important contribution is that of Lewis (1962, and *this volume*). His electronic model accurately replicates a large class of postsynaptic membrane phenomena. Based on parameters of ion-membrane mechanisms, the model demonstrates how such complex phenomena as facilitation, threshold

* Personal communication, 1962.

shift, and pacemaker potentials can arise from simple variations in parameter values. It provides quantitative insight into many of the hitherto obscure determinants of subthreshold action.

Single-Unit Models. In studying the properties of the entire neuron, considerably more is involved than the action of a homogeneous patch of excitable membrane. Parameters of synaptic action and integrated postsynaptic response must be introduced. With the inclusion of these multiple-input-channel operations and their various temporal and geometrical concomitants, entirely new properties (and problems) emerge.

The single-unit models can be classified into two distinct categories, those that are extended representations of the Hodgkin-Huxley model and those that start from a simplified, reduced set of functional input-output relationships.

In the first group are the models of Derksen (1961, 1962) and of Küpfmüller and Jenik (1961) and Jenik (1962, and *this volume*). Each of these transistor-circuit real-time analogs reproduces membrane action with high accuracy, and the input-output functions of single units are readily produced. They illustrate in part the wide range of study possible with such models. Jenik has explored the logical processing of single units stimulated by two repetitive spike-train stimuli applied to two synaptic inputs (either excitatory-excitatory or excitatory-inhibitory). This procedure provides, as does FitzHugh's (1961), a novel technique for comparing and evaluating different models; graphical representation of the responses to a variety of well-defined classes of inputs (with amplitude and frequency as variables) establishes a number of extremely interesting parametric relationships. This is one of the very few quantitative tools now available which provides a ready means for categorizing classes of single-unit action. Jenik's analysis of temporal interactions of spike trains in a realistic model constitutes a significant start in a domain of study which has been conspicuously unexplored.

Derksen's electronic analogs are used to investigate the effects of noise, threshold variations, and inhibition on statistical firing properties (both in single units and in networks of up to 100 units). Since interactions between excitation, inhibition, and threshold in the biological neuron are directly influenced by fluctuations of stimulus and of excitability, it is highly desirable to know their extent and significance. Physiological data on this subject are scarce and are difficult to obtain; theoretical studies and extensive exploration with accurate models can be very useful.

Another investigation of the influence of noise on firing illustrates the physiological accuracy obtainable with simpler models. In a series of experiments on both artificial and real neurons, ten Hoopen and Verveen (1962) showed excellent quantitative agreement between the responses of a "minimum-parameter" model (Harmon, 1961) and biological single units.

This minimum-parameter model is an example of the second kind of single-unit analog, that which is based on a preselected set of functional input-output relationships rather than on a replication of the parameters of membrane kinetics. We have employed an electronic model principally to investigate single-unit input-output functions and to study information processing in peripheral sensory systems. It has been used to show how variation of a small number of parameters (e.g., excitatory and inhibitory time constants) can account for a wide variety of the observed physiological behavior of single units in different species. Such phenomena as accommodation, burst formation, spike-train input-output relationships, and noise effects have been explored (Harmon, 1961). The modeling of sensory systems has been used to study retinal pattern processing (van Bergeijk and Harmon, 1960), dynamic-range properties of the cochlea (van Bergeijk, 1961), flicker-fusion mechanisms (Levinson and Harmon, 1961), and auditory temporal resolution (Harmon, Levinson, and van Bergeijk, 1963).

Network Models. The interactions between units in a network have been studied by two rather different techniques. The first is more abstract and theoretical, typically being a mathematical analysis (often statistical) of group properties. The elements analyzed are generally geometrical (connection patterns) or greatly reduced sets of parameters which implicitly constitute "formal" neurons. In the second approach discrete single units (either hardware- or computer-simulated) are connected in networks, and their individual and group behavior is studied.

Most network models are necessarily somewhat unreal, since we have considerably less physiological knowledge of group behavior than of single-unit behavior. Hence the implications of network modeling can seldom be immediately tested and evaluated. This is no reason to discourage such modeling; on the contrary, the most important theoretical questions for the behavior of nervous systems arise here.

Two examples of mathematical models designed to examine interactions between units in a pool will serve to illustrate the range of the formal-analysis approach. These are the models of Rall and of Beurle.* In a statistical study of input-output relations in a monosynaptic reflex pathway, Rall (1955a) was able to show that the input-output activity of a pool of single units can be adequately understood (and predicted) in terms of a few relatively simple variables, namely, synaptic-knob density and activity, and distribution of motoneuron thresholds. The advantage of this kind of problem and model is that experimental validation is immediately possible. Physiological investigation (Rall, 1955b) established at least the sufficiency of the model.

An early theoretical model which underlies a number of subsequent studies of network activity is that of Beurle (1956). A mathematical analysis was developed for a three-dimensional array of randomly distributed and connected single units. Each element was given properties (principally axonal) such as spatial and temporal summation, absolute refractoriness (time variable), delay, and all-or-none output. Distributions of cell dimensions and packing densities,

---

* A rather different approach should also be mentioned. This is the investigation of reliable computation in the presence of noise or with unreliable components (see, for example, McCulloch, 1957; Cowan and Winograd, 1963). The elements of the networks employed are formal, logical modules rather than overt representations of neurons, and as such are designed for automata studies rather than for neurophysiological modeling. It is important to note, however, that quite similar questions concerning redundancy, diffuse structure, and diverse function are germane to an understanding of nervous-system operation; consequently these investigations have considerable potential significance.

fiber counts, and dendrite and axon densities were specified. In the analysis which followed, quantitative measures were obtained for threshold distributions, proportions of active and inactive units, wave propagation (pattern and velocity), wave interaction, and reverberatory activity.

Models of this kind may perhaps be useful for elucidating neurophysiological behavior; unfortunately, however, many of these purely theoretical essays are not extended or brought to trial in the real world. Others in this category are those of McCulloch and Pitts (1943), Culbertson (1952), Minsky (1954), Kleene (1956), von Neumann (1956), Rashevsky (1960).

A typical example of the second type of network modeling (studies which explore the behavior of explicitly constructed single units connected into large, regular assemblies) is that of Farley and Clark (1961). Simplified-parameter (formal) neurons are simulated in a digital computer. A network of 1296 units occupies a planar array whose connections are well defined. (Other, similar studies, e.g., Smith and Davidson (1962), use random connections.) Experiments are run to examine oscillation and stability characteristics, the net exhibiting such physiologically noted activity as irregular firing, rhythmic patterns, and "driven" firing. These aspects of the network's behavior are functions of controlled variables such as stimulus, connection pattern, and threshold.

The question is open, of course, whether the mass action studied in these models reflects a significant aspect of nervous-system behavior. In the absence of such knowledge the physiological pertinence of this kind of modeling is not easily established; it is interesting, however, that some of the results bear a strong qualitative resemblance to cortical activity. Although the "neurons" used in these models are exceedingly simplified representations of biological neurons, it is entirely conceivable that the gross behavior in both model and prototype is essentially independent of finer unit specification.

A somewhat different approach was taken in the digital-computer network simulation study of Josephson, Reiss, and Worthy (1961). Patterns of activity were analyzed in a diffuse, randomly connected planar network similar to that of a coelenterate. Quantitative analysis of the spread of excitation as a function of particular types and densities of junctions, average length of conducting elements, and successive stimulation disclosed a number of unsuspected relationships. A notable result was that the observed modes of excitation spread were generalizable in such a way that similar action could be expected of other types of nets containing more complex units.

Network models have also been used to study highly *structured* configurations. Reiss (1960) introduced several novel and potent techniques for simulating continuous-variable, asynchronously firing networks in a digital computer. Simplified and idealized single units are employed, but they are more nearly complete representations than are generally found in digital simulation. The model represents portions of a neuromuscular control system (including proprioception). Dynamic behavior, tracking, and oscillation are studied. Such work can be instructive in obtaining some insight into systemic behavior and could be of especial value in suggesting the mechanisms of anomalous behavior (e.g., Parkinsonian tremor), where so much remains to be known.

In a subsequent study, Reiss (1962) used both digital- and analog-computer simulation to explore the consequences of reciprocal inhibition in two units. He demonstrated mechanisms by which continuous streams of impulses can be transformed into rhythmic bursts, and showed some extremely interesting effects of dominance and stability despite changes in parameter values.*

There is a category of network modeling which lies between the two classes discussed above, consisting neither of formal, abstract analyses nor of explicitly constructed nets of idealized single units. This is the domain of *Gedanken* experiments, in which purely conceptual or verbal models are constructed. The binaural interaction model of van Bergeijk (1962) exemplifies the approach. In this case an attempt is made to develop a structural rationale for a known function—the interdependence of stimulus time and intensity found in human binaural perception. The model postulates a particular functional organization of auditory neurons to account for the perceptual data, concurrently taking advantage of known physiological properties of single units and their gross anatomical organization. In this way details of possible structural arrangements are inferred, and plausible relationships between form and function are suggested.

Such models are sufficiently well specified to permit ultimate validation or rejection by direct physiological experiment. Irrespective of the validity of a particular model, however, this approach is valuable because it proceeds inductively from psychophysical data to neural action, seeking at least the *sufficient* connections between the two.

Non-neural Models. By non-neural models I mean models of nervous-system operation which are expressly not based on neurons *per se*. Although this category is outside the intended scope of this paper, it should at least be briefly mentioned because of its relevance.

There are two classes of non-neural models which occupy a level between neurophysiology and over-all system behavior. One is exemplified by the work of Reichardt (1961a, b), of Kunze (*this volume*), and of Mittelstaedt (1962) on models of systemic *mechanisms* which underlie behavior. Analyses of precisely quantified organism responses lead to the elaboration of models whose parts are organizations of neurons, but the individual neural actions are not of immediate interest. Hence the black boxes in these analogs are functional mechanisms. This type of modeling helps to bridge the gap between neural phenomena and over-all system operation, thereby extending our knowledge of the behavioral consequences of nervous action.

The other class of non-neural models attempts to deal conceptually or mathematically with the *entire* nervous system. These models are highly speculative and arbitrary; their use is to suggest a framework for information derived from studies at lower levels of organization. Such a model is found in the work of Caianiello (1961). Models of this sort are essential to the full development

* Such effects in only two units are of more than academic interest. For example, in the Mauthner cells of teleosts the states of two mutually inhibitory cells uniquely determine the direction of tail flexion (Retzlaff and Fontaine, 1960).

of an understanding of brain function; thus they may ultimately be no less important than studies of single-unit action.

Conclusion

With growing emphasis placed on the information-processing aspects of nervous systems, theoretical studies assume greater importance. In some instances questions pertaining to neurophysiological action appear to be unanswerable by any experimental techniques now known or easily envisioned. In such cases theoretical models can significantly complement direct physiological experimentation.

The use of models to help elucidate neuronal behavior has expanded rapidly during the last decade. Theoretical neurophysiologists have applied a wide assortment of tools and techniques to problems at many different levels of complexity. Realistic analogs of membrane, single-unit, and network action, which have already made significant contributions, may be expected to exert even more influence on the course of neurophysiological research.

References

ALDRICH, J. A., W. S. MCCULLOCH, J. Y. LETTVIN, and W. H. PITTS, 1962. "The Iron Wire Neural Nets," in Quarterly Progress Report No. 64, Jan. 15, 1962, Research Laboratory of Electronics, Mass. Inst. Technol., Cambridge, Mass., pp. 303–5.

ALLANSON, J. T., 1956. Some properties of randomly connected neural networks, in *Information Theory*, Colin Cherry, ed., New York: Academic Press, pp. 303–13.

AMASSIAN, V. E., J. MACY, Jr., and H. J. WALLER, 1961. Patterns of activity in simultaneously recorded neurons in midbrain reticular formation, *Ann. N.Y. Acad. Sci.*, **89**(5), 883–95.

ARVANITAKI, A., 1942. Effects evoked in an axon by the activity of a contiguous one, *J. Neurophysiol.*, **5**, 89–108.

BABCOCK, M. L., 1960. "Reorganization by Adaptive Automation," Report TR-1, Nonr 1834(21), Electrical Engineering Research Laboratory, Univ. Illinois, Urbana, Ill.

BARLOW, H. B., R. FITZHUGH, and S. W. KUFFLER, 1957. Change of organization in the receptive fields of the cat's retina during dark adaption, *J. Physiol.*, **137**(3), 338–54.

BERGEIJK, W. A. VAN, 1961. Studies with artificial neurons: II: Analog of the external spiral innervation of the cochlea, *Kybernetik*, **1**(3), 102–7.

BERGEIJK, W. A. VAN, 1962. Variation on a theme of Békésy: a model of binaural interaction, *J. Acoust. Soc. Am.*, **34**(8), 1431–37.

BERGEIJK, W. A. VAN, and L. D. HARMON, 1960. "What Good are Artificial Neurons?," WADD TR 60-600, 395–406.

BEURLE, R. L., 1956. Properties of a mass of cells capable of regenerating pulses, *Phil. Trans. Roy. Soc. (London)* **B-240**, 55–94.

BIEDERMANN, W., 1898. *Electro-Physiology*, London: Macmillan, vol. 2, pp. 227–356.

BISHOP, G. H., 1956. Natural history of the nerve impulse, *Physiol. Rev.*, **36**(3), 376–99.

BONHOEFFER, K. F., 1948. Activation of passive iron as a model for the excitation of nerve, *J. Gen. Physiol.*, **32**, 69–91.

BROWN, K. T., and T. N. WIESEL, 1959. Intraretinal recording with micropipette electrodes in the intact cat eye, *J. Physiol.*, **149**, 537–62.

BULLOCK, T. H., 1957. Neuronal integrative mechanisms, in *Recent Advances in Invertebrate Physiology*, B. T. Scheer, ed., Eugene, Oregon: Univ. Oregon Pubs., pp. 1–20.

BULLOCK, T. H., 1960. Comparative physiology of synaptic junctions, in "Information Processing in Neurones and Small Nets," A. Rapoport and W. J. Horvath, eds., WADD TR 60-652, pp. 153–205.

BULLOCK, T. H., 1962. Paper presented at the Twenty-Second International Congress of Physiology, Leiden, Sept. 1962 (to be published in *Proceedings of the Symposium on Information Processing in the Nervous System*, Excerpta Medica Foundation, in press).

BURNS, B. D., 1955. The mechanism of after-bursts in cerebral cortex, *J. Physiol.*, **127,** 168–88.

CAIANIELLO, E. R., 1961. Outline of a theory of thought processes and thinking machines, *J. Theoret. Biol.*, **2,** 204–35.

COOMBS, J. S., J. C. ECCLES, and P. FATT, 1955. The inhibitory suppression of reflex discharges from motoneurones, *J. Physiol.*, **130,** 396–413.

COWAN, J. D., and S. WINOGRAD 1963. Reliable computation in the presence of noise, *Phil. Trans. Roy. Soc.* (*London*) (in press).

CRANE, H. D., 1962. Neuristor—A novel device and system concept, *Proc. IRE*, **50**(10), 2048–60.

CRICHTON, J. W., and J. H. HOLLAND, 1959. "A New Method of Simulating the Central Nervous System Using an Automatic Digital Computer," Report No. 2144–1195-M University of Michigan Willow Run Laboratories, Willow Run, Mich.

CULBERTSON, J. C., 1952. "Hypothetical Robots and the Problem of Neuroeconomy," RAND Corp. Report P-296. Santa Monica, Calif.

DERKSEN, H. E., 1961. A neurone analog simulating the electrical properties of the cat's motoneurone, *Acta Physiol. et Pharmacol.*, **10,** 164–80.

DERKSEN, H. E., 1962. Paper presented at the Twenty-Second International Congress of Physiology, Leiden, Sept. 1962 (to be published in *Proceedings of the Symposium on Information Processing in the Nervous System*, Excerpta Medica Foundation, in press).

DODGE, F. A., 1961. "A Quantitative Description of Membrane Excitation at a Node of Ranvier," paper presented at the fifth annual meeting of the Biophysical Society, St. Louis, Mo., Feb. 16–18, 1961.

EBBECKE, VON U., 1928. Modellversuche zur Erläuterung der Nervenreizung, in *Handbuch der biologischen Arbeitsmethoden*, E. Abderhalden; ed. Berlin: Urban & Schwarzkopf, Abt V, t. 6, pp. 679–718.

ECCLES, J. C., 1961. The mechanism of synaptic transmission, *Ergebnisse der Physiologie*, **51,** 299–431.

ECCLES, J. C., R. M. ECCLES, and F. MAGNI, 1961. Central inhibitory action attributable to presynaptic depolarization produced by muscle afferent volleys, *J. Physiol.*, **159,** 147–66.

ECCLES, J. C., P. G. KOSTYUK, and R. F. SCHMIDT, 1962. The effect of electric polarization of the spinal cord on central afferent fibres and on their excitatory synaptic action, *J. Physiol.*, **162,** 138–50.

FARLEY, B. G., and W. A. CLARK, 1961. Activity in networks of neuron-like elements, in *Information Theory*, Colin Cherry, ed., New York: Academic Press, 242–251.

FATT, P., and B. KATZ, 1952. Spontaneous subthreshold activity at motor nerve endings, *J. Physiol.*, **117,** 109–28.

FATT, P., and B. KATZ, 1953. The effect of inhibitory nerve impulses on a crustacean muscle fibre, *J. Physiol.*, **121,** 374–89.

FITZHUGH, R., 1955. Mathematical models of threshold phenomena in the nerve membrane, *Bull. Math. Biophys.*, **17,** 257–78.

FITZHUGH, R., 1958. A statistical analyzer for optic nerve messages, *J. Gen. Physiol.*, **41**(4), 675–92.

FITZHUGH, R., 1960. Thresholds and plateaus in the Hodgkin-Huxley nerve equations, *J. Gen. Physiol.*, **43**(5), 867–96.

FITZHUGH, R., 1961. Impulses and physiological states in theoretical models of nerve membrane, *Biophys. J.*, **1**(6), 445–66.

FITZHUGH, R., 1962. Computation of impulse initiation and saltatory conduction in a myelinated nerve fiber, *Boiphys. J.*, **2**(1), 11–21.

FOERSTER, H. VON, 1958. Some aspects in the design of biological computers, in *Proceedings of the Second International Congress on Cybernetics*, Namur, Belgium: International Cybernetic Association, 240–55.

Frank, K., and M. G. F. Fuortes, 1961. Excitation and conduction, in *Annual Review of Physiology*, vol. 23, Victor E. Hall, ed., Palo Alto, Calif.: Annual Reviews, pp. 357–86.

Frishkopf, L. S., and W. A. Rosenblith, 1958. Fluctuations in neural thresholds, in *Symposium on Information Theory in Biology*, H. Quastler, ed., London: Pergamon, pp. 153–68.

Gerstein, G. L., 1962. Mathematical models of the all-or-none activity of some neurons, *IRE Trans. Inf. Theory*, **IT-8**(5), 137–43.

Gerstein, G. L., and N. Y.-S. Kiang, 1960. An approach to the quantitative analysis of electrophysiological data from single neurons, *Biophys. J.*, **1**(1), 15–28.

Grundfest, H., 1959. Synaptic and ephaptic transmission, in *Handbook of Physiology, Section I: Neurophysiology*, J. Field, ed., Washington, D.C.: American Physiological Society, vol. 1, pp. 147–97.

Hagiwara, S., 1954. Analysis of interval fluctuations of the sensory nerve impulse, *Jap. J. Physiol.* **4,** 234–40.

Hamilton, H., and R. Vinetz, 1962. "Note on the analogical simulation of neural behavior," in Semi-Annual Report No. 4, Jan. 1962, Laboratory for Automata Research, General Precision Inc., Librascope Div., Glendale, Calif., pp. 81–93.

Harmon, L. D., 1959. Artificial neuron, *Science*, **129**(3354), 962–63.

Harmon, L. D., 1961. Studies with artificial neurons, I: properties and functions of an artificial neuron, *Kybernetik*, **1**(3), 89–101.

Harmon, L. D., 1962a. Neural analogs, in *Proceedings of the 1962 A.F.I.P.S. Spring Joint Computer Conference*, Palo Alto, Calif.: National Press, pp. 153–58.

Harmon, L. D., 1962b. Natural and artificial synapses, in *Self-Organizing Systems* 1962, M. C. Yovits, G. T. Jacobi, and G. D. Goldstein, eds., Washington, D.C.: Spartan, pp. 177–202.

Harmon, L. D., J. Levinson, and W. A. van Bergeijk, 1963. Studies with artificial neurons: IV. Binaural temporal resolution of clicks, *J. Acoust. Soc. Am.* (in press).

Harmon, L. D., and R. M. Wolfe, 1959. An electronic model of a nerve cell, *Semiconductor Products*, **2,** 36–40.

Hartline, H. K., H. G. Wagner, and F. Ratliff, 1956. Inhibition in the eye of *Limulus*, *J. Gen. Physiol.*, **39**(5), 651–73.

Hiltz, F. F., 1962. Analog computer simulation of a neural element, *IRE Trans. Bio-Med. Elect.*, **BME-9**(1), 12–20; **BME-9**(3), 210–11.

Hodgkin, A. L., and A. F. Huxley, 1952. A quantitative description of membrane current and its application to conduction and excitation in nerve, *J. Physiol.*, **117,** 500–44.

Hoopen, M. ten, and A. A. Verveen, 1963. "Nerve-model experiments on fluctuation in excitability," in *Nerve, Brain, and Memory Models*, N. Wiener and J. P. Schadé, eds.; New York: Elsevier, pp. 8–21.

Hubel, D. H., and T. N. Wiesel, 1959. Receptive fields of single neurones in the cat's striate cortex, *J. Physiol.*, **148,** 574–91.

Hubel, D. H., and T. N. Wiesel, 1962. Receptive fields, binocular interaction and functional architecture in the cat's visual cortex, *J. Physiol.*, **160,** 106–54.

Jenik, F., 1962. Electronic neuron models as an aid to neurophysiological research, *Ergebnisse der Biologie*, **25,** 206–45.

John, E. R., 1962. Paper presented at the Twenty-Second International Congress of Physiology, Leiden, Sept. 1962 (to be published in *Proceedings of the Symposium on Information Processing in the Nervous System*, Excerpta Medica Foundation, in press).

Josephson, R. K., R. F. Reiss, and R. M. Worthy, 1961. A simulation study of a diffuse conducting system based on coelenterate nerve nets, *J. Theoret. Biol.*, **1,** 460–87.

Kleene, S. C., 1956. Representation of events in nerve nets and finite automata, in *Automata Studies*, C. E. Shannon and J. McCarthy, eds., Princeton, N.J.: Princeton Univ. Press, pp. 3–41.

Kuffler, S. W., 1953. Discharge patterns and functional organization of mammalian retina, *J. Neurophysiol.*, **16,** 37–68.

Küpfmüller, K., and F. Jenik, 1961. Über die Nachrichtenverarbeitung in der Nervenzelle, *Kybernetik*, **1**(1), 1–6.

LETTVIN, J. Y., 1959. Nerve models, in Quarterly Progress Report No. 52, Jan. 15, 1959, Research Laboratory of Electronics, Mass. Inst. Technol., Cambridge, Mass., pp. 178–79.

LEVINSON, J., and L. D. HARMON, 1961. Studies with artificial neurons, III: Mechanisms of flicker-fusion, *Kybernetik*, **1**(3), 107–17.

LEWIS, E. R., 1962. Neural analogue studies, in Semi-Annual Report No. 4, Jan. 1962, and Semi-Annual Report No. 5, July 1962, Laboratory for Automata Research, Librascope Div., General Precision, Inc., Glendale, Calif., pp. 9–26 and 87–113.

LILLIE, R. S., 1936. The passive iron wire model of protoplasmic and nervous transmission and its physiological analogues, *Biol. Rev.*, **11**, 181–209.

MATURANA, H. R., 1962. Functional organization of the pigeon retina, paper presented at the Twenty-Second International Congress of Physiology, Leiden, Sept. 1962 (to be published in *Proceedings of the Symposium on Information Processing in the Nervous System*, Excerpta Medica Foundation, in press).

MATURANA, H. R., J. Y. LETTVIN, W. S. MCCULLOCH, and W. H. PITTS, 1960. Anatomy and physiology of vision in the frog (*Rana pipiens*), *J. Gen. Physiol.*, **43** (6, part 2), 129–75.

MCCULLOCH, W. S., 1957. The stability of biological systems, in *Homeostatic Mechanisms* (Brookhaven Symposium on Biology No. 10), Upton, N.Y.: Brookhaven National Laboratories, pp. 207–15.

MCCULLOCH, W. S., and W. PITTS, 1943. A logical calculus of the ideas immanent in nervous activity, *Bull. Math. Biophys.*, **5**, 115–33.

MCGROGAN, E. P., 1961. Improved transistor neuron models, in *Proceedings of the National Electronics Conference*, vol. 17, Chicago, Ill.: National Electronics Conference, Inc., pp. 302–10.

MINSKY, M. L., 1954. Theory of Neural-Analog Reinforcement Systems and its Application to the Brain-Model Problem," Ph.D. dissertation, Princeton University, Ann Arbor, Mich.: University Microfilms.

MINSKY, M. L., 1961. A selected descriptor-indexed bibliography to the literature on artificial intelligence, *IRE Trans. Human Factors in Elect.*, **HFE-2**(1), 39–55.

MITTELSTAEDT, H., 1962. Control systems of orientation in insects, *Annual Review of Entomology*, vol. 7, Palo Alto, Calif.: Annual Reviews, pp. 177–98.

MORRELL, F., 1961. "Lasting Changes in Synaptic Organization Produced by Continuous Neuronal Bombardment," in *Brain Mechanisms and Learning*, A. Fessard, R. W. Gerard, and J. Konorski, eds., Springfield, Ill.: C. C Thomas, pp. 375–92.

MUELLER, P., 1958. On the kinetics of potential, electromotance, and chemical change in the excitable system of nerve, *J. Gen. Physiol.*, **42**(1), 193–229.

NAGUMO, J., S. ARIMOTO, and S. YOSHIZAWA, 1962. An active pulse transmission line simulating nerve action, *Proc. IRE* **50**(10), 2061–70.

NEUMANN, J. VON, 1956. Probabilistic logics and the synthesis of reliable organisms from unreliable components, in *Automata Studies*, C. E. Shannon and J. McCarthy, eds., Princeton, N.J.: Princeton Univ. Press, pp. 43–98.

OTANI, T., and T. H. BULLOCK, 1959. Effects of presetting the membrane potential of the soma of spontaneous and integrating ganglion cells, *Physiol. Zool.*, **32**(2), 104–14.

PALAY, S. L., 1958. The morphology of synapses in the central nervous system, in *The Submicroscopic Organization and Function of Nerve Cells* H. Fernández-Morán and R. Brown, eds. (*Exp. Cell Res.*, suppl. 5), New York: Academic Press, pp. 275–93.

PITTS, W., and W. S. MCCULLOCH, 1947. How we know universals—the perception of auditory and visual forms, *Bull. Math. Biophys.*, **9**, 127–47.

PURPURA, D. P., 1959. Nature of electrocortical potentials and synaptic organizations in cerebral and cerebellar cortex, in *International Review of Neurobiology*, Vol. 1, C. C. Pfeiffer and J. R. Smythies, eds., New York: Academic Press, 47–163.

RALL, W., 1955a. A statistical theory of monosynaptic input-output relations, *J. Cell. Comp. Physiol.*, **46**(3), 373–411.

RALL, W., 1955b. Experimental monosynaptic input-output relations in the mammalian spinal cord, *J. Cell. Comp. Physiol.*, **46**(3), 413–37.

RALL, W., 1957. Membrane time constant of motoneurons, *Science*, **126**(3271), 454.

Rall, W., 1959. Branching dendritic trees and motoneuron membrane resistivity, *Exptl. Neurol.*, **1**, 491–527.

Rall, W., 1960. Membrane potential transients and membrane time constant of motoneurons, *Exptl. Neurol.*, **2**(5), 503–32.

Rall, W., 1962a. Theory of physiological properties of dendrites, in *Mathematical Theories of Biological Phenomena*, N. Rashevsky, ed., *Ann. N.Y. Acad. Sci.*, **96**(4), 1071–92.

Rall, W., 1962b. Electrophysiology of a neuron model, *Biophys. J.*, **2**(2), 145–67.

Rashevsky, N., 1960. *Mathematical Biophysics*, 3rd ed., New York: Dover.

Reichardt, W., 1961a. Über das optische Auflösungsvermögen der Facettenaugen von *Limulus*, *Kybernetik*, **1**(2), 57–69.

Reichardt, W., 1961b. Autocorrelation, a principle for the evaluation of sensory information by the central nervous system, in *Sensory Communication*, W. A. Rosenblith, ed., Cambridge, Mass.: M.I.T. Press, and New York: Wiley, pp. 303–17.

Reiss, R. F., 1960. The digital simulation of neuro-muscular organisms, *Behavioral Science*, **5**(4), 343–58.

Reiss, R. F., 1962. A theory and simulation of rhythmic behavior due to reciprocal inhibition in small nerve nets, in *Proceedings of the 1962 A.F.I.P.S. Spring Joint Computer Conference*, vol. 21, Palo Alto, Calif.: National Press, pp. 171–94.

Retzlaff, E., and J. Fontaine, 1960. Reciprocal inhibition as indicated by a differential staining reaction, *Science*, **131**(3393), 104–5.

Robertson, J. D., 1953. Ultrastructure of two invertebrate synapses, *Proc. Soc. Exp. Biol.*, **82**, 219–23.

Rochester, N., J. H. Holland, L. H. Haibt, and W. L. Duda, 1956. Tests on a cell assembly theory of the action of the brain, using a large digital computer, *IRE Trans. Inf. Theory*, **IT-2**(3), 80–93.

Simmons, P. L., and R. F. Simmons, 1961. The simulation of cognitive processes: An annotated bibliography, *IRE Trans. Elect Comp.*, **EC-10**(3), 462–83.

Simmons, P. L. and R. F. Simmons, 1962. The simulation of cognitive processes. II: An annotated bibliography, *IRE Trans. Elect. Comp.*, **EC-11**(4), 535–52.

Smith, D. R., and C. H. Davidson, 1962. Maintained activity in neural nets, *J. Assn. Comp. Mach.*, **9**(2), 268–79.

Sweitzer, D. I., 1960. Biological and artificial intelligence, Jet Propulsion Laboratory, information search No. 254, Pasadena, Calif.: California Institute of Technology.

Sweitzer, D. I., 1961. Biological and artificial intelligence, Jet Propulsion Laboratory, supplement to literature search No. 254, Pasadena, Calif.: California Institute of Technology.

Szentagothai, J., 1962. "Anatomical Aspects of Junctional Transformation," paper presented at the Twenty-Second International Congress of Physiology, Leiden, Sept. 1962 (to be published in *Proceedings of the Symposium on Information Processing in the Nervous System*, Excerpta Medica Foundation, in press).

Tasaki, I., and A. F. Bak., 1959. Voltage clamp behavior of iron-nitric acid system as compared with that of nerve membrane, *J. Gen. Physiol.*, **42**(5), 899–915.

Taylor, W. K., 1960. Computers and the nervous system, in *Models and Analogues in Biology*, J. W. L. Beament, ed., New York: Academic Press, pp. 152–68.

Terzuolo, C. A., and T. H. Bullock, 1958. Acceleration and inhibition in crustacean ganglion cells, *Arch. Ital. Biol.*, **96**, 117–34.

Uttley, A. M., 1954. The classification of signals in the nervous system, *EEG Clin. Neurophysiol.*, **6**, 479–94.

Verveen, A. A., 1961. *Fluctuation in Excitability*, Amsterdam: Drukkerij Holland N.V.

Viernstein, L. J., and R. G. Grossman, 1961. "Neural Discharge Patterns in the Transmission of Sensory Information," in *Information Theory*, C. Cherry, ed., New York: Academic Press, 252–69.

Walter, W. G., 1953. *The Living Brain*, New York: W. W. Norton, p. 280.

Wiersma, C. A. G., and R. T. Adams, 1949. The influence of nerve impulse sequence on the contractions of different crustacean muscles, *Physiol. Comp. Oecol.*, **II** (1), 20–33.

Yamagiwa, K., 1950. Iterative excitability in Lillie's nerve model, *Jap. J. Physiol.*, **1**, 269–76.

2. Input and Output Elements of Behavioral Models

J. W. S. PRINGLE, *Oxford University*

The interaction between engineering technology and neurophysiology is a reciprocal one; each group of workers can learn much from the other. The neurophysiologist gains by a study of the methodology and refined analytical techniques of the engineer, and the latter, applying his way of thinking to the extremely complicated dynamic systems found in living organisms, can not only sharpen his wits but may discover, in the properties of his biological models, types of behavior which have their uses in other fields. I shall attempt here to outline some of the ways in which physiologists have attempted to analyze the biological machinery of animal behavior, and to show how the methods we use are related to those of the engineer.

The Analysis of Animal Behavior

When a biologist studies the behavioral reactions of an animal or the physiology of the component parts of its neuromuscular machinery, his results are usually expressed in the form of a dynamic relationship between stimulus and response. Unlike the engineer, who builds up his complicated mechanisms by the combination of parts whose properties he understands, the biologist wishing to analyze the behavioral machinery of an animal has first to identify the component parts whose properties he is intending to study in detail. In higher animals, he is fortunate that this preliminary analysis can be made on an anatomical basis. Although we are all ultimately concerned with the properties of the whole animal, we can speak meaningfully of the properties of sense organs, central nervous systems, and muscles because this particular type of anatomical differentiation has occurred in the vast majority of animal phyla. The determination of the stimulus-response relationships for these three types of major element of the behavioral machinery indeed forms a large part of the literature of neurophysiology. Since the sense organs can be considered the input elements and the muscles the output elements of the mechanism, with most of the computing machinery being concentrated in the central nervous system, the physiologist is already relating the biological system to the pattern of man-made mechanisms.

Beyond this elementary stage, the biologist has two different objectives. If he is a student of behavior in the widest sense, he will wish at once to try to resynthesize from his knowledge of the properties of the parts an understanding of the behavior of the animal as a whole. To this end, he will investigate the properties of his anatomical units in combination, as in the study of reflexes. He will, or should, try to express his results on the properties of the parts in such a way that their properties in combination can be predicted. The best example of success of this sort is provided by the muscle-control servomechanism of the mammalian limb. It is already possible to go a long way toward a prediction of the dynamic properties of the intact limb from a detailed knowledge of the properties of muscles, motor neurons, muscle spindles, and their associated reflexes (Merton, 1953; Houk and Stark, 1962; Robson, 1962).

The biophysically-minded biologist has a different aim. He wishes to pursue the analysis of the properties of the components to deeper levels, and to interpret their working in physicochemical terms. To this end he tries to set up conceptual models with constituent parts which in combination imitate or reproduce the properties of the component he is studying. He tries to perform a further experimental separation of the constituent parts, again to study their properties in isolation and their properties in known combination, and thus, by a series of stages, to work down to functional units whose properties are fully explicable in physicochemical terms.

If, at this deeper level of analysis, it is again possible to identify anatomically the component parts of the functional unit, model-making rests on a sure foundation, but unfortunately this is very often not possible with our present techniques. The component parts of organs may be resolvable only with the electron microscope, which cannot study mechanisms in action. In such cases, another sort of test is needed of the validity of a conceptual model; it is held to be satisfactory if the theoretical mathematical formulation of its performance is identical with that of the actual biological system. This test is not absolute, for there is always the danger of analogy. There may be many different types of mechanism with the same over-all relationship between input and output but no correspondence between the properties of the parts. Where structural analysis is impossible, good sense and judgment of relevance are sometimes the only guides to a choice between alternative models. Many biologists would feel happy if they could get even this far. Often the test of similarity of performance between the model and the actual system has to be judged satisfactory without the full demonstration of mathematical identity. Sometimes a mere qualitative resemblance is accepted as adequate, though this can be very misleading without at least an approximate comparison of quantitative performance.

In a recent review (Pringle, 1960) I tried to illustrate some of these analytical principles in a general discussion of models of muscle. In that paper I dealt with some of the concepts which are current in muscle physiology with particular reference to the large amount of experimental work which has been done on the properties of the frog's sartorious muscle at 0°C. I want to discuss two examples of neurophysiological analysis, one of an input element and one of an output element of the behavioral machinery, to illustrate better some of the ways in which the biologist is using the concepts of the engineer. I shall try to show how we have now got to the point at which we need help from engineers if the analysis is to be rigorously pursued. The examples I choose are, on the input side, mechanical sense organs subserving touch, pressure, and the like, and, on the output side, the peculiar type of muscle found in the flight system of higher insects. My own recent experimental work has been concerned with the second of these; a more extensive review of sense organs as input elements was presented recently at another symposium (Pringle, 1961).

Mechanoreceptors

Figure 1 is an example of the sort of information-flow diagram which the physiologist can draw as a conceptual model of the operation of a mechanical sense organ. At the first stage of analysis, each of these "black boxes" is simply assumed to perform a transformation on the incoming information, so that the central nervous system finally receives some measure of the mechanical phenomena going on in the environment of the animal. The reason why we set up this particular model and not any other is that there are methods of making measurements at these particular intermediate points in the causal chain; no useful purpose is served at present by postulating any more constituent functional parts. As we shall see, it may indeed be necessary to have more constituent parts in our model, but these have an altogether different degree of validity as analytical units, since we cannot at present get experimental evidence to establish their real identity. Figure 1 goes as far as it is possible to go in performing a structural analysis; although we cannot separate the black boxes and study their properties in isolation, we can separate them effectively by making measurements at intermediate points in the system.

Box 4 needs no discussion. Box 3 is justified by the observation that in all sense organs where it has been looked for, there is a nonpropagated depolarization at the origins of the sensory fiber (the generator potential) when a mechanical stimulus is applied to the sense organ. Figure 2 shows a recording from a mechanical sense cell, made with an internal microelectrode. The burst of five impulses is superimposed on a depolarization which is maintained for the

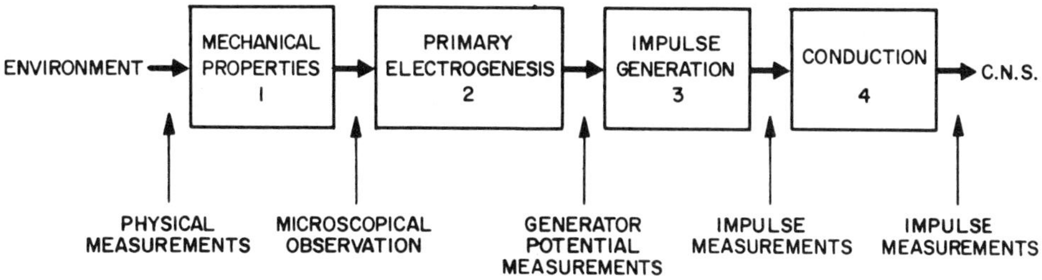

FIG. 1. Control sequence in mechanoreceptors and the types of measurement and observation that can be made.

duration of the stretch stimulus. The fact that the potential does not return to the base line between impulses shows that the two processes of generator-potential initiation and impulse initiation are taking place at different locations in the nerve, thereby giving structural validity to the conceptual analysis of Fig. 1.

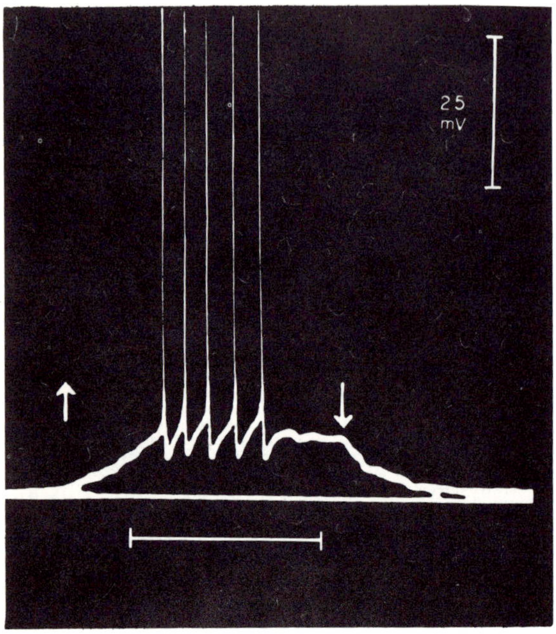

FIG. 2. Generator potential with superimposed nerve impulses from a stretch receptor of the crayfish. Arrows mark duration of stretch; time 1 sec (from Gray, 1959, after Eyzaguirre and Kuffler, 1955).

Katz (1950) tried to apply to the analysis of a mechanoreceptor the second type of test of validity, to see how far one over-all property of the sense organ, its adaptation to a constant stimulus, could be interpreted in terms of the sum of the properties of the constituent parts. The over-all property of adaptation manifests itself as a gradual decline in the frequency of nerve impulses when a constant stimulus is applied to the sense organ. Since he found a linear relationship between generator potential and impulse frequency, the derivative response in this case (the frog muscle spindle) must be a property of either box 1 or box 2 or of both. Katz discussed, in terms of more fundamental concepts about the nature of the processes occurring in boxes 1 and 2, how this might arise, but without microscopical observation to separate the effects in boxes 1 and 2, he was left with several possible explanations. Katz (1961) continued the discussion in the light of electron-microscope evidence about the fine structure of the sensory terminals responsible for the processes in box 2 (the transducer action) but was still left with uncertainty, owing to the difficulty in making observations on the living preparation. Gray (1959) discussed this problem for other types of mechanoreceptor and concluded that, although much of the adaptation is undoubtedly due to mechanical properties (box 1), as suggested by Eyzaguirre and Kuffler (1955) for the crayfish muscle receptor, the transducer processes in box 2 may have a dynamic as well as a static component contributing to the over-all effect.

It is worth noting at this point that the analytical scheme of Fig. 1 satisfies one important criterion for a good conceptual model, namely that the constituent functional parts of the model should not be mutually interacting. Processes in box 2 do not influence processes in box 1, and similarly for the other conceptual units; the action is in one direction only. This means that if the sense organ is considered a linear system, a mathematical prediction of the over-all performance can be obtained by multiplying the transfer functions for each of the constituent processes. A model which does not satisfy this criterion is a poor model because the boundaries of the conceptual units are imprecise (Pringle, 1960) and are not established with certainty; it is not then valid to multiply together the transfer functions of the postulated constituent parts in order to compute the transfer function for the whole complex.

Even this feature of model making rests on unsure foundations in the case of most biological systems, since they clearly do not perform in a linear fashion over a wide range of stimulus intensities. Until some general theoretical treatment of nonlinear systems is available, the biologist nearly always has to make use of the analytical schemes devised by the engineer for linear systems, but he may in this way be led into error. Some years ago (Pringle and Wilson, 1952) I tried to analyze the over-all property of adaptation of a mechanical sense organ from the cockroach leg to a step-function stimulus by fitting to the observed points a curve which was the sum of a constant and three exponentials. Studies with a wider range of input frequencies have now shown that this is not the correct type of relationship. Chapman and Smith (1963) have re-investigated the adaptation property of the same ending over a wider frequency range and find that the points on the curve can only be fitted by a "power-law" formula of the type $y = Bt^{-k}$, where y is the instantaneous impulse frequency after a step-function stimulus and B and k are constants. The log/log plot of frequency against time is linear, and the Nyquist plot shows a constant phase angle at various frequencies and a modulation sensitivity increasing as the kth power of the forcing frequency. Although, at the moment, no one has suggested a physical model which will give this performance, we may have to accept it as a fundamental property of certain biological systems. A similar formulation is needed to describe the decline in tension which occurs when a crab muscle is subjected to a step-function increase in length (Rayner, 1962), tension declining at a rate proportional to log t. Since part of the adaptation property of sense organs almost certainly derives from mechanical properties of the transducer, the similarity between sense organs and muscles may be more than a coincidence. A power law for stress relaxation may be a general rheological property of biological materials.

Discussion. The formal representation in a model of the behavioral machinery of an animal must include realistic input elements if the over-all performance is not to be an arbitrary simplification. It will be clear that even in the case of the relatively simple mechanical sense organs, we are still some distance from a sufficiently complete understanding for it to be possible to specify what is required.

Insect Fibrillar Flight Muscle

In the same way that an information-flow diagram can be drawn for a mechanical organ as an input element, the diagram in Fig. 3 shows the sequence of events thought to occur on the output side when a muscle is controlled by its motor nerve. This is not so good a conceptual model as that shown in Fig. 1, since only in some cases can the black boxes be established as fully valid conceptual units by identification with structural elements; experimental measurements at intermediate points can be made only between boxes 1 and 2 and between boxes 4 and 5. Indirect measurements can be made of the amount of neuromuscular transmitter liberated by a nerve impulse, but we cannot separate box 3 experimentally from boxes 2 and 4. Box 4 can be broken down into separate parts, such as the combination of transmitter with the receptor site, the end-plate potential, the active membrane response and the propagation of an impulse over the surface membrane of the muscle fiber (where this occurs), but since these (apart from the last) are not known to be noninteracting processes, this further analysis is not worthwhile for our present purpose.

In the case of vertebrate striated muscle, the all-or-nothing nature of the propagated muscle potential (box 4) provides a convenient unit in terms of which subsequent processes in the muscle are usually interpreted. Since this is followed by a single distinct twitch in the contractile mechanism (black box 6)

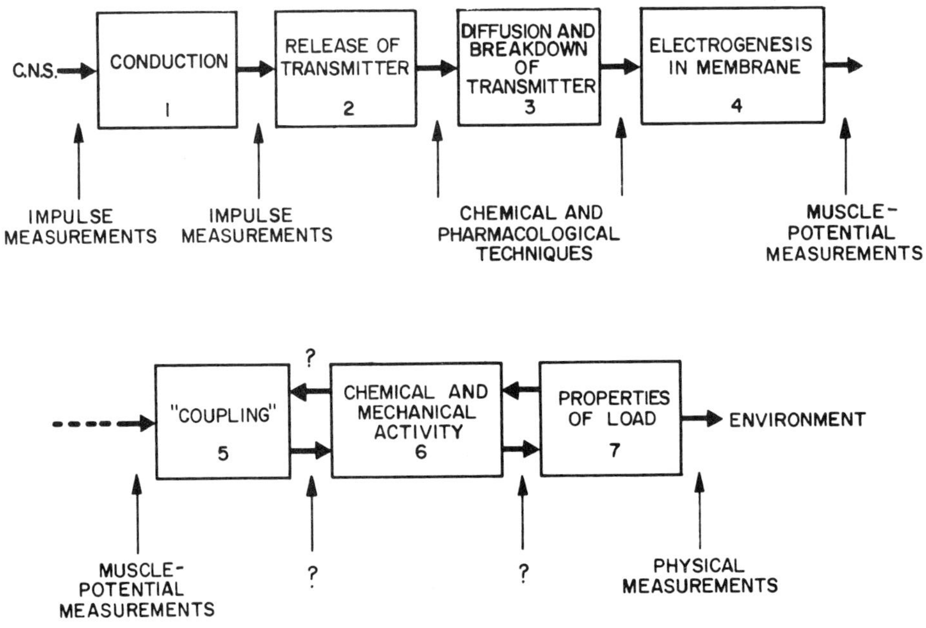

FIG. 3. Control sequence in striated muscle and the types of measurement that can be made.

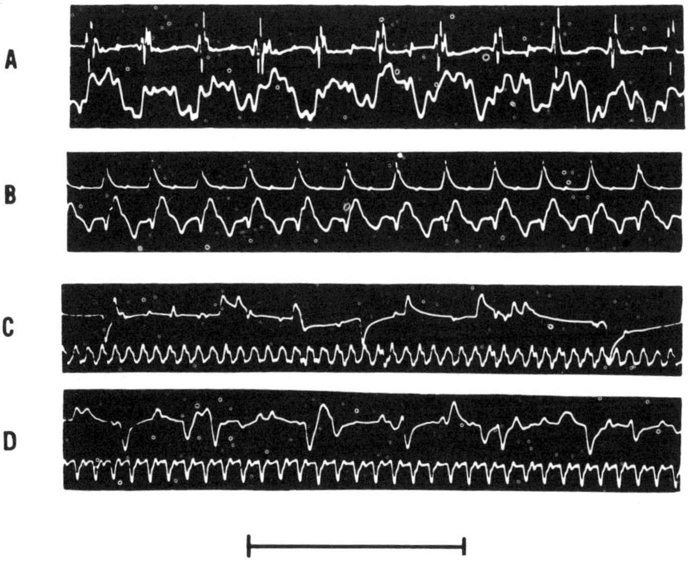

FIG. 4. Electrical (upper trace) and mechanical (lower trace) records from the thorax of flying insects. A, cockroach. B, moth. C, fly. D, wasp. Time scale, 140 msec for A, 100 msec for B, C, D. Note synchrony in A and B and lack of correspondence in C and D between electrical and mechanical events (Pringle, 1957).

evidence can be obtained about the largely unknown coupling process (black box 5) by studying the relationship between the muscle potential and the twitch under various conditions. The peculiar property of insect fibrillar flight muscle which makes it interesting and difficult to interpret is the lack of correspondence between individual events in the contractile mechanism and electrical activity in the surface membrane. Figure 4 shows this phenomenon.

Since we have here a different relationship between black boxes 4, 5, and 6 from that found in vertebrate striated muscle, the insect tissue provides an opportunity for exploring more precisely the nature of the processes going on in the internal mechanism of the muscle and helps to clarify the conceptual units in terms of which the properties of muscles have to be analyzed. Since no method has yet been found of making direct experimental measurements of the activity of the coupling process (box 5), this and boxes 4 and 6 are not conceptually distinct in the same way as the earlier elements in the control sequence. It is necessary also to take into account the properties of the mechanical load (box 7) in order to give a complete description of the performance of a muscle.

Studies of the properties of isolated fibrillar flight muscles (Machin and Pringle, 1959) have shown that the rhythmic mechanical activity of the active muscle derives from a reciprocal interaction between the contractile mechanism and the mechanical load. The frequency of oscillatory contraction is determined largely by the resonant combination of the inertia of the load with the elasticity of the muscle. In the resting condition this mechanical oscillation is damped, but when the muscle is brought into activity by a sufficiently high frequency of muscle potentials (box 4), produced by the arrival of motor nerve impulses, the tissue behaves as if the damping were negative and goes into an oscillatory condition in which mechanical work is done on the load—in the intact insect on the aerodynamic resistance of the wings. It is not easy to visualize the property of the muscle which enables it to transform itself into this oscillatory condition; for this we must consider first certain experiments done on the isolated muscle.

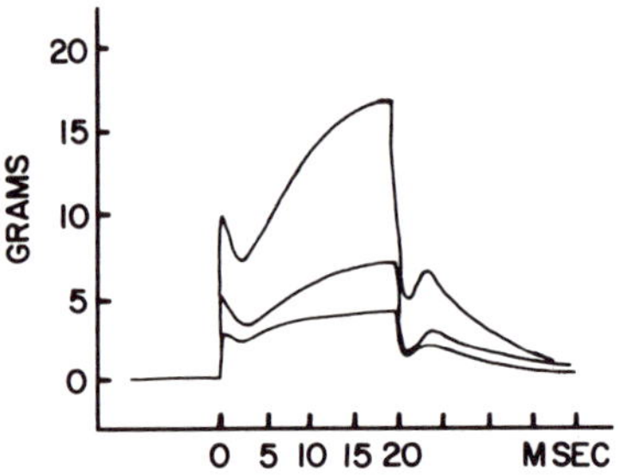

FIG. 5. Isometric tension changes in maximally stimulated basalar muscle of *Oryctes* when subjected to three different amplitudes of quick stretch and release. Ordinate shows additional tension developed in excess of about 20 g steady isometric tension (after E. G. Boettiger, unpublished record).

Open-Loop Behavior of Insect Flight Muscle. Often the performance of a closed-loop system can best be investigated by devising conditions under which the loop is effectively opened. In the case of the reciprocal relationship between a muscle and its mechanical load, the loop is opened by imposing length changes on the muscle with an apparatus sufficiently powerful to override the contractile activity of the muscle, and then measuring the resulting changes in tension. This breaks the connection (forward arrow) between boxes 6 and 7 of Fig. 3. Figure 5 shows the tension changes which occur in a fibrillar muscle of a beetle when it is subjected to a step-function increase and decrease in length.

The initial transient rise of tension (which is all that occurs in vertebrate striated muscle) is followed by a delayed rise of tension; the curves for a sudden release of length (quick release) are symmetrical for length changes of small amplitude, but the performance is nonlinear at the length changes used in the experiment of Fig. 5. This delayed rise and fall of tension after a quick stretch and quick release is the essential property in which insect fibrillar flight muscle differs from vertebrate striated muscle.

The Cicada Sound-Production Mechanism. The way in which a rhythm of mechanical activity can be generated when a fibrillar muscle is coupled to a suitable load is best illustrated not from the system found in insect flight but from the mechanism of sound production of certain cicadas, where the nature of the load imposes step-function changes of length on the muscle. Figure 6 shows an analysis of the rhythmic mechanism of the cicada tymbal muscle. Sound is produced in a manner analogous to the clicking of the lid of a tin can, and the powerful muscle is connected to the sound-producing tymbal through a thin apodeme attached at one corner of the tymbal.

Lines 1, 4, and 7 of Fig. 6 are experimentally determined. The impulse in the

FIG. 6. Diagrammatic representation of the sequence of events if a single tymbal muscle fiber is excited from its motor nerve. A, diagram of anatomy with processes numbered as in Fig. 3. B, time course of events numbered as in A (modified from Pringle, 1957).

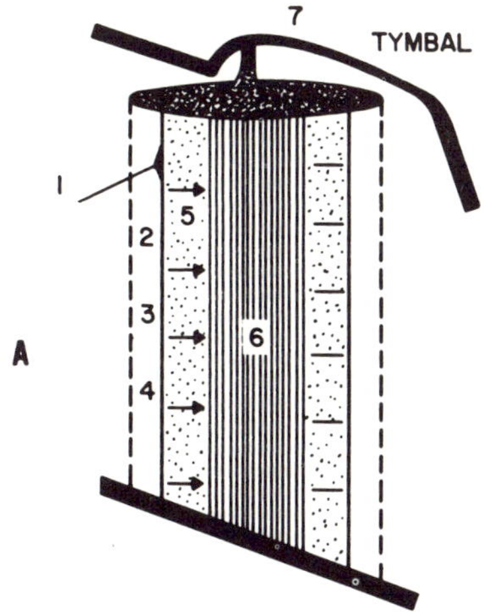

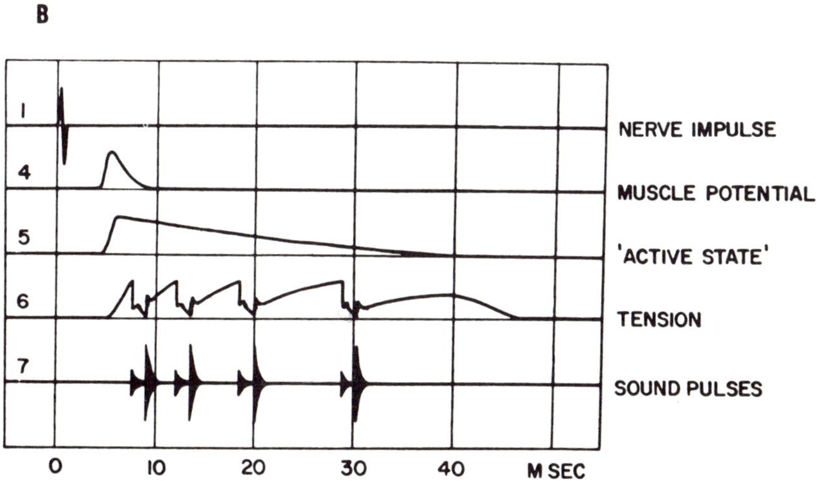

single motor-nerve fiber initiates an electrical potential in the surface membrane of the muscle fiber, and this is followed, in this species, by four double pulses of sound, each corresponding to the clicking in and out of the tymbal membrane. Lines 5 and 6 are interpolated as interpretation of the mechanism. Line 5, the "active state" of the muscle fiber, is a postulated property of the coupling process (box 5 in Fig. 3); it outlasts the muscle potential (box 4 of Fig. 3). Line 6 shows the postulated changes in tension in the tymbal muscle on the assumption that the rate of increase of tension is proportional to the magnitude of the active state and that the effect of the clicking in and out of the tymbal membrane is to impose a quick release and quick stretch on the muscle. It will be clear that the delayed changes in tension following a quick release or a quick stretch are capable of generating a rhythm of mechanical activity when the muscle is connected to a load with the properties of the tymbal membrane.

Harmonic Responses. It is possible to study the open-loop performance of a fibrillar muscle by imposing sinusoidal length changes on it and measuring the resulting sinusoidal changes of tension instead of applying quick stretches and quick releases (Machin and Pringle, 1960, 1962). Figure 7 shows a Nyquist plot of the results of such an experiment on a fibrillar muscle from a beetle.

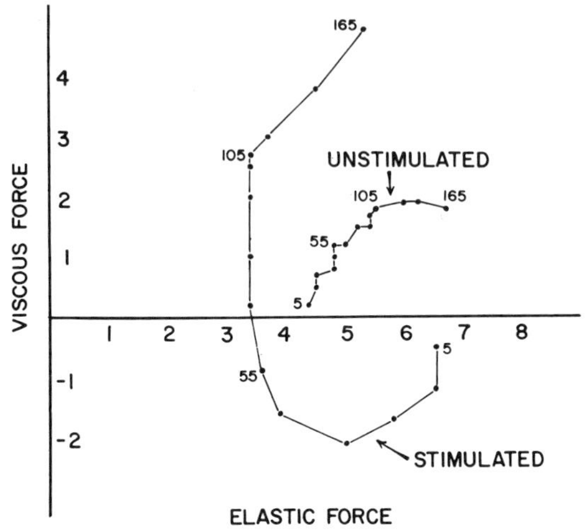

FIG. 7. In-phase ("elastic force") and quadrature ("viscous force") components of tension in basalar muscle of *Oryctes* subjected to imposed sinusoidal length changes at various frequencies (small numbers near experimentally determined points, in cps); temperature 30°C. (Based on an unpublished experiment similar to those described in Machin and Pringle, 1960.)

The plot for the resting muscle shows points entirely in the first quadrant, but the plot for the active muscle has a large loop and, over a certain range of forcing frequencies, the points lie in the fourth quadrant. At these frequencies the muscle is doing work on the driving apparatus, and it is clear that if the mechanically resonant properties of the muscle–wing system are such that it can oscillate at these frequencies, it will do so. It is also clear that the oscillation will build up to an amplitude limited by the aerodynamic damping of the wings. In the species of beetle used for these experiments, the natural frequency of wing beat is 40 beats/sec which is near the frequency at which the muscle is capable of delivering maximum power into the resonant system.

Discussion. It is thus possible to interpret the performance of insect fibrillar flight muscle according to the information flow diagram of Fig. 3, provided that a rather peculiar reciprocal relationship is introduced between the contractile mechanism (box 6) and the mechanical load (box 7). The biophysically minded biologist will wish to continue his analysis and try to explain the properties of the contractile mechanism in terms of its molecular structure, and once again it will be necessary to introduce further analytical concepts in whose terms these properties are to be described. Thus the analysis is pursued to ever deeper levels. This is not the place, however, for a discussion of hypotheses which have been put forward to explain the molecular basis of muscular contraction.

Conclusion

It will have been made clear that the concepts of the control-system engineer are indeed being followed closely by the physiologist when he explores the nature of the machinery of input and output elements of the nervous system. Only when the analysis is complete, and the whole behavior of muscles and sense organs can be understood in molecular terms, will it be possible to specify precisely the components which must be built into a model of the behavioral machinery. If attention is restricted to the computing machinery of the brain, models, however attractive, are liable to be unrealistic. The physiologist wants not merely an analogy, but something which will help him better to understand the actual mechanism.

Summary

Physiologists analyzing the neuromuscular machinery of animal behavior normally adopt a conceptual framework similar to that of the engineer and construct models in the form of block diagrams of the control systems. Mechanical sense organs and muscles form the input and output elements, respectively, of these behavioral models. Criteria are discussed for identifying the component parts of a biological system.

The operation of mechanical sense organs can be analyzed into four constituent processes, and methods are available for determining the dynamic properties of each of them.

The dynamic properties of muscles have been analyzed into a larger number of constituent processes, but only some of these are separable by experiment. The special properties of insect fibrillar flight muscle are described in some detail, and it is shown that these can be understood in terms of a mutual interaction between the load and the activity of the contractile mechanism.

Behavioral models should incorporate the properties of naturally occurring input and output elements if they are to be useful to the biologist. A model which merely performs computations on given information is unlikely to bear much resemblance even to the central nervous system and can contribute little to the understanding of natural animal behavior.

References

Boettiger, E. G., 1960. Insect flight muscles and their basic physiology, *Annual Review of Entomology*, vol. 5, Palo Alto, Calif.: Annual Reviews, pp. 1–16.

Chapman, K. M., and R. S. Smith, 1963. A linear transfer function underlying impulse frequency modulation in a cockroach mechanoreceptor, *Nature*, **197**, 699–701.

Eyzaguirre, C., and S. W. Kuffler, 1955. Processes of excitation in the dendrites and soma of single isolate sensory nerve cells of the lobster and crayfish, *J. Gen. Physiol.*, **38**, 87–119.

Gray, J. A. B., 1959. Mechanical into electrical energy in certain mechanoreceptors, *Progr. Biophys.*, **9**, 285–324.

Houk, J. C., and L. Stark, 1962. "An Analytical Model of a Muscle Spindle Receptor for Simulation of Motor Co-ordination," Quarterly Progress Report No. 66, Research Laboratory of Electronics, Mass. Inst. Techn., Cambridge, Mass., pp. 384–89.

KATZ, B., 1950. Depolarization of sensory terminals and the initiation of impulses in the muscle spindle, *J. Physiol.*, **111,** 261–82.

KATZ, B., 1961. The termination of the afferent nerve fibre in the muscle spindle of the frog, *Phil. Trans. Roy. Soc.* (*London*), **B-243,** 221–40.

MACHIN, K. E., and J. W. S. PRINGLE, 1959. The physiology of insect fibrillar muscle: II. Mechanical properties of a beetle flight muscle, *Proc. Roy. Soc.* (*London*), **B-151,** 204–5.

MACHIN, K. E., and J. W. S. PRINGLE, 1960. The physiology of insect fibrillar muscle: III. The effect of sinusoidal changes of length on a beetle flight muscle. *Proc. Roy. Soc.* (*London*), **B-152,** 311–30.

MACHIN, K. E., and J. W. S. PRINGLE, 1962. The physiology of insect fibrillar muscle: IV. The effect of temperature on a beetle flight muscle, *Proc. Roy. Soc.* (*London*), **B-155,** 493–99.

MERTON, P. A., 1953. Speculations on the servo-control of movement, in *The Spinal Cord* (CIBA Foundation Symposium), J. L. Malcolm, J. A. B. Gray, and G. E. W. Wolstenholme, eds., Boston: Little, Brown.

PRINGLE, J. W. S., 1957. *Insect Flight*, Cambridge, Eng.: Cambridge Univ. Press.

PRINGLE, J. W. S., 1960. Models of muscle, in *Models and Analogues in Biology* (Symposia of the Society for Experimental Biology, No. XIV), Cambridge, Eng.: at the University Press, pp. 41–68.

PRINGLE, J. W. S., 1962. Prologue: the input element, in *Biological Receptor Mechanisms* (Symposia of the Society for Experimental Biology, No. XVI), Cambridge, Eng.: at the University Press, pp. 1–11.

PRINGLE, J. W. S., and V. J. WILSON, 1952. The response of a sense organ to a harmonic stimulus, *J. Exp. Biol.*, **29,** 220–234.

RAYNER, M. J., 1962. Ph.D. thesis, Cambridge University.

ROBSON, J. G., 1962. Ph.D. thesis, Cambridge University.

3. The Use of Computer Technics in Neural Research

BELMONT G. FARLEY, *Massachusetts Institute of Technology*

It cannot be disputed that the theory of neural organization and behavior is at a far more primitive stage than theory in the physical sciences. However, the reasons given for this difference vary widely. Some hold that there are basic differences between the physical and behavioral sciences which prohibit a comparable approach.

The viewpoint of this paper is that the root of the difficulty lies in the extreme complexity and nonlinearity of the behaving organism as an object of study, and that there are no basic distinctions of a philosophical nature between the sciences. The consequences of the complexity and nonlinearity are so great, however, that they account for the widespread feeling that there must be a qualitative or fundamental difference. The training of the physical scientist does not ordinarily provide him with tools adequate to deal with these nonlinear complexities.

Within the past few years, however, the hope of eventually overcoming these obstacles has been increased by developments in the use of computers and computer technics for both experimental and theoretical activity.*

In addition to their use in practice, these technics can be made to serve as a conceptual framework supporting a unified philosophy of the physical and behavioral sciences and also illuminating actual differences between them. This paper attempts to present such a framework, and to discuss a few philosphical and practical problems in its terms. The practical use of a digital computer in attacking problems of nonlinear complexity in models of neural organization will also be illustrated.

The philosophical framework is intended to be an outline rather than a formalized systematic analysis. Even non-professional philosophers (a class which includes the writer) will undoubtedly perceive loose ends. Some of these are left undiscussed as not of immediate importance, and some are no doubt unintentional. However, I do not at present see any fundamental barriers to future extension and formalization, although there may be serious practical difficulties.

Based on research supported by the National Institutes of Health and the National Aeronautics and Space Administration under contract PH43-63-540.

* We use the term "technics" in the sense of computer-oriented concepts and design methodology, rather than circuitry *per se*.

In general the ideas will be recognized as stemming from the pragmatist-positivist-operationalist schools of thought, modern mathematical logic, and, of course, computer technics. However, there is also a good deal of philosophical skepticism. No claim is made for originality in details, although the synthesis may be partly new and the emphasis unusual.

Informal Models in Everyday Experience

Scientific activity as a conscious pursuit may be viewed as a natural extension of processes carried out by the nervous system of every higher organism, for each organism must continually solve problems posed by its environment. To solve them, according to our hypothesis, the organism has or acquires a sort of internal representation of those parts of its experience that repeat in some way and are therefore predictable. This internal representation enables the organism to react successfully in the present and to predict the future. We call such representations "hypotheses," "models," or "theories." The term "theory" is often used to designate the more complex representations, but we will make no sharp distinction between the terms.

We remark that the idea of an "environment" is itself a model, owing its justification to the fact that experience does exhibit certain repetitive qualities. An organism has some rather specific requirements and could not long survive if certain aspects of its experience did not repeat. However, sense impressions seldom, if ever, repeat exactly, and therefore the organism must have means for examining impressions and deciding which have the same significance. That is to say, there must be fundamental processes which classify impressions according to a scheme useful to the organism. We may say that these "perceptual" processes constitute true, if rudimentary, models of experience in that they reflect some of its repeatable aspects and, furthermore, are actually used to predict which response will be appropriate.

The higher organisms are able to improve their models as a result of experience. Thus even at the level of unconscious function, the individual human being is not merely a collector of data. He is continually formulating new models and testing them against his experience. In successful cases he produces a succession of models, each of which correctly predicts a wider range of experience than its predecessor. At some point, procedures of this general nature become a part of conscious activity, and explicit statements of models and their realm of applicability become desirable for facilitating recollection and communication. In the early stages this activity is quite informal—"red sky at night, sailor's delight"—but as the game begins to be played for its own sake, increasing precision of procedure and description is sought, and science as a specific human occupation emerges. Note that classification is a principal procedure in the early stages of science, and is as fundamental to science as to the organism.

Thus scientific activity arises naturally from certain informal activities of daily life and takes the direction of increasingly detailed specification of experimental procedures and increasingly formal statement of theoretical models, which in turn serve to predict successfully over a broadening realm of experience.

An excellent example of this increasing formality and sophistication is the evolution of the study of planetary motions through the Ptolemaic system, the re-orientation of topological viewpoint in the Copernican system, and the greater formality and progressively increasing abstraction of the Keplerian, Newtonian, and Einsteinian models. Each of these models encompasses a wider range of experience of planetary motion and, at the same time, introduces a more complex set of fundamental postulates than its predecessor.

No branch of science has reached the point of perfect specification or complete formalization, and probably none ever will, but formalization constitutes an ideal standard of comparison by which progress in a science may be judged. We will show that it is possible to establish an ideal formal "schema" of scientific experiment and theory by using theoretical properties of ideal computers, and we will then discuss some philosophical problems with the aid of the schema.

Formal Models and the Role of Mathematics as a Descriptive Language

Much emphasis has been placed on the use of mathematics in science, but it appears that discussion of its role has been somewhat distorted by an overemphasis on "number" as the central concept in scientific applications. We prefer to regard mathematics as a formal "language" which affords a means of precise expression, and to argue that it is in precision of *statement* that the "exactness" of the "exact" sciences resides. The theoretical justification for this point of view rests on the fact that mathematics as a formal discipline can be shown to depend upon formal, or symbolic, logic, which consists, in turn, of statements abstracted from ordinary discourse (Quine, 1951). With this in mind, it should occasion no surprise that mathematics is the tool of choice when precise statement is desired.

It is easy to see that even when the outcome of an experiment is a number, this number may be regarded as a shorthand notation for a series of statements. For example, the number may be expressed in binary form. It is then clear that a series of simple "go/no-go", or "yes/no," statements can be substituted for the number. Furthermore, these statements can be related directly to the operations of the experiment. Increasing the precision of the experimental determination then merely results in a longer series of statements of the same type. These considerations indicate that there exists no sharp distinction between "qualitative" and "quantitative" statements if the statements can be given any operational meaning at all. Thus we may say that to be more "quantitative" is actually to say more.

Mathematical analysis and the deductive aspects of mathematics can also be expressed by series of logical statements. The equations of a theory may be viewed as constraints which a variable must satisfy. To put it another way, equations constitute rules for selecting particular sets of points from the totality of points in the space of the variables. Such rules can always be given in the form of logical statements, and the same applies to any other rules which may be necessary to state or to apply the theory. Thus the concept of mathematics as a formal language applies also to the ordinary theoretical uses of mathematics in science.

From our viewpoint, therefore, mathematics is a means of formal expression with generality adequate for statement of any specifications, structures, assumptions, rules, equations, methods, or models required for experiment or theory in science.

The advantage of formality is that procedures and hypotheses can be made precise, i.e., can be recorded and communicated unambiguously, and that predictions can be calculated to any desired precision.

The importance to us of the foregoing considerations stems from the fact that an ordinary digital computer, if provided with an unlimited memory, can be instructed in such a way as to "simulate" any system of effectively calculable logical procedures. By simulate we mean that the output of the computer is identical with the output of the system of logical procedures under all circumstances. The metamathematical term "effectively calculable" means, roughly, capable of being defined inductively by means of explicitly stated rules. This is a very wide class of procedures. It includes the rules necessary to define or simulate all ordinary logical statements, all ordinary functions of mathematical analysis, and all existing scientific theories. It would seem that a scientific theory must be effectively calculable, since otherwise the advantages of mathematics as stated above would be lost.

An Ideal Schema Relating Experiment and Theory

We are now ready to introduce an ideal schema which will serve to express the relation between experiment and theory.

An experiment in astronomy will serve as a concrete example. We assume that a theory of planetary motion is given, and that our object is to test the theory. The function of the theory is to reproduce the measurements as precisely as possible, over as long a time as possible, and (to make the idealization complete) for as many planetary systems as possible.

However, we wish to establish a schema which is as formalized as possible. We therefore imagine that telescopes equipped with tracking devices have been trained on the planets. Attached to the telescopes are devices which transmit signals defining the position of the axes at each instant. These signals are the raw data, and the angle variables of the axes constitute the "measurables" of the experiment.

We now assume that our computer has an unlimited computing speed, in addition to an unlimited memory, and that it also has access to a physical clock whose time is specified by the theory. We will now program the computer with the statements of the theory and the necessary rules for relating its predictions to the measurables of the experiment. Furthermore, the programming will be carried out in such a way as to reproduce the measurables in "real" time, i.e., the time of the experiment itself. The measurables of the experiment and the theoretical predictions of the computer will be identically displayed by some method chosen for convenience in observation and comparison. Now we set the present initial conditions of the measurables into the computer, and wait to see how long it continues turning out predictions which agree with the display originating from the telescope axes. The more lasting the agreement, and the more numerous the initial conditions which give good agreement, the better the theory.

We may state the idea of agreement between experiment and theory in the following way. A "satisfactory" theory is one for which it is not possible by any test to determine which display is generated by the experiment, and which by a computer programmed to simulate the theory.

It is now clear that a theory provides an imitation or "model" of the experimental system under study. Perhaps the true usefulness of a theory lies in this imitative ability, since it makes possible the performance of substitute "experiments" with a model, in place of endless actual experiments. Thus a variety and depth of acquaintance with the model can be achieved which are hardly attainable by experiment on the natural system. This same function is served by the informal representation of experience acquired in everyday life.

When experience with the properties of a model becomes deep and general, a psychological state of mind may be arrived at called "understanding," and if the model is a well-established imitator, then it may be said that this understanding extends to a "natural" system.

However, it is important from a philosophical standpoint not to confuse the "model" with the "modeled." One of the principal advantages of our formal schema is that it prevents such confusion.

It is useful to describe an experiment and a theoretical model in abstract terms. Consider a multidimensional space whose coordinate axes are the measurable variables of the experiment. The result of an experiment will then be a subset of points in this "experimental" space.

In accordance with our view of mathematics as a precise language, we may now regard a theoretical model as a set of rules which serves to determine subsets of points in the experimental space. The subsets selected by an ideal theory would lie so close to those selected by experiments as to be indistinguishable from them in all cases. In the physical sciences, the rules most used are represented by differential and integral equations, always supplemented by other rules of application which may or may not be explicitly stated.

In all but the most rudimentary theories, the rules refer at least in part to abstract variables which are not measured variables, and therefore the theory makes use of a space of higher dimensionality than the experimental space. In our schema, the output variables of the computer match the variables of the experimental space, but the other variables of the theoretical model are also represented within the computer, and could be observed, if desired. The schema makes clear, however, that a theory formulated in terms of abstract, nonmeasurable variables must include rules by which the abstract variables are to be related to the measurables. Since the variables are measured by certain methods or instruments, these rules perforce include a theory of any effects of the measurement methods themselves. Without these rules, a theory cannot be applied unambiguously. For example, Newton's laws refer to "acceleration" and "mass," without specifying how they are to be measured. This deficiency has been the subject of many studies, the results of which in part led ultimately to the notions of relativity theory.

It is clear that all experiments which actually can be performed in a given experimental space can select at most a finite collection of subsets, which ordinarily will occupy only a small portion of the space. In most cases, large regions of the experimental space are actually inaccessible to experiment, owing to difficulties in measurement, or because it is not possible to force the experimental variables into these regions. If a theory is to have any predictive value, its rules must determine subsets for experiments not yet performed, and theories are usually formulated so that the subsets they can select fill the whole of the experimental space. Thus a theoretical model ordinarily represents a very great generalization from actual experiment. If we add to this the fact that many abstract, non-measurable variables can be used in a theory, we are led to inquire whether more than one generalization is possible, i.e., whether there is more than one theory which will lead to the same subsets.

As a matter of fact, there are an unlimited number of distinct theories which lead to identical subsets of experimental space, no matter how these subsets are defined. This can be seen from the following considerations. The experimental space may always be thought of as embedded in an "augmented" space of higher dimensions. The extra dimensions may correspond to measurable variables not previously included in the experiment, or they may be new abstract variables with no immediately known experimental correlates. There is no limit to the number of dimensions which may be added. The subsets to be matched may now be considered as intersections of the original space with higher-dimensional subsets in the augmented space. These "augmented" subsets in turn may be thought of as selected by a theory formulated in terms of the variables of the augmented space. Since an intersection subset does not uniquely determine an augmented subset, there are an unlimited number of augmented subsets available. Thus there are at least two ways of establishing different theories which select the same subsets in the original experimental space: by augmenting the dimensions differently and by choosing rules which select different augmented subsets containing those of the experimental space. There are an unlimited number of such theories, since, as we have seen, there is no limit to the number of ways different dimensions or different subsets may be chosen. All these theories are actually different from one another in that they predict different results in some region of experimental space, or augmented space, or both. In brief, every theory can be considered a special, or limiting, case of an unlimited number of more general theories.

In applying these ideas to experimentally selected subsets there is still more freedom, owing to the limited precision of measurements. The subsets of experimental space selected by the different theories need not be identical with one another or with the experimental subsets. The subsets need only lie so close together as to be experimentally indistinguishable. We may therefore add to the previously defined theories any others which predict results so close as to lie within the specified experimental precision. Clearly there are an unlimited number of these theories also, since there exist an unlimited number of distinct subsets arbitrarily close to any given subset.

These ideas are rather abstract, and we will illustrate them by reference to theories of planetary motion. Kepler's laws may be formulated in terms of space, time, and velocity as coordinate variables. The orbit subsets selected by the theory have elliptical projections on two-dimensional space, and Kepler's laws give rules which serve to specify the orbit uniquely, given the initial coordinates of a planet.

Newton introduced a new variable, and hence a new dimension, for each planetary body, called its "mass," and the orbits depend on the initial conditions of the mass variables as well as on the initial values of the other coordinates. Newton's inverse-square laws of planetary attraction give the new rules for calculating the orbits in the new, augmented space.

The orbits calculated from Newton's laws are very close to those of Kepler in the case of one massive central body surrounded by a few widely separated bodies of negligible mass, but in most other cases the orbits are different, since Kepler's orbits do not depend on mass variables. Kepler's theory may be regarded therefore as a limiting case of Newton's theory. Newton's theory augments considerably Kepler's space and gives new rules for choosing new subsets. These subsets coincide with the old for the case of negligible planet–sun mass ratios. Newton's orbits can be viewed similarly as limiting cases of the orbits of Einstein's theory of general relativity, although the orbits predicted by Einstein lie so close to Newton's that it is difficult to detect the difference by actual experiment.

The measurements of an experiment increase in precision and number as time goes on, since the experiment is repeated with improved measurement methods. In addition, more variables may be measured. Therefore the later experiments rule out certain theories or classes of theories admitted by the lower-precision, older experiments. Looking at this process from another point of view, we see that a greater burden is placed on the later theoretical models by the increased information which must be "absorbed," i.e., by the increased precision and greater number of measured subsets which must be matched. Hence the more recent theories tend to require additional abstract dimensions and more complex rules of subset selection. According to our interpretation, this explains the increasing complexity of theories with time.

Our discussion, however, has centered thus far on the subsets selected by a theory rather than on the selection rules. In most discussions of theory, primary attention is paid to these rules. Although we have stated the criterion for a satisfactory theory exclusively in terms of the agreement between its predictions and experiment, other criteria based on properties of the rules themselves are usually emphasized. For example, criteria such as "economy," "simplicity," or possession of certain invariance properties are often applied to the rules. Sometimes particular philosophical or even aesthetic standards are considered crucial; in many cases these criteria are not precisely defined. Some even depend on individual psychological preferences, and are therefore non-unique, being variant from person to person.

Even "simplicity" has never been completely defined, and this concept too actually depends on individual preference. We prefer to use the term "convenience" rather than "simplicity," and to recognize that it represents a pragmatic criterion rather than a philosophical principle. The convenience of a theory depends heavily on the means available for its study and application.

Strong exception is likely to be taken to this viewpoint by those who feel that particular philosophical principles are of paramount importance for theories, and we must agree that the application of such principles both has proved and will continue to prove of great value in the search for new theories. Reference to our schema and discussion, however, suggests that conclusions about the absolute philosophical status of any particular principle be drawn with caution, because there may exist satisfactory theories (in our sense) which admit contrary principles. Perhaps each philosopher can have his own set of theories. A rigorous treatment would have to specify precisely formulated sets of conditions and then discuss mathematically the nature of theories, if any, which satisfy the conditions. This is a difficult task, and to our knowledge it has seldom been attempted. Von Neumann undertook one study of this general nature in an investigation of the question of determinacy in quantum theory. He showed that no "hidden parameters" allowing a deterministic description are possible, under the assumption that certain formal mathematical characteristics of the theory are retained. That is, deterministic description and these particular characteristics are incompatible. However, this result does not prove that there does not exist a deterministic theory fitting the experiments as closely as quantum theory.

Deterministic vs. Nondeterministic Models

As it happens, determinism is one philosophical principle which can easily be discussed with the aid of our schema and the abstract frameworks of experimental and theoretical space. By "nondeterministic" we mean "not completely predictable."

It is not hard to show that there exist both deterministic and non-deterministic theoretical models which are equally satisfactory. Therefore the simple question of whether "Nature" is deterministic or non-deterministic can never be answered.

We first observe that it is an invariable fact of experience that if any measurement is made with sufficiently great precision, irreducible and unpredictable fluctuations appear in the result. These fluctuations are often called "noise." In a strict sense, then, the results of an experiment are never exactly predictable. However, we will follow common practice and consider a theory satisfactory if the statistical properties of its subsets match those of the experimental subsets.

We have imagined our model programmed in a computer, which is a deterministic device, and therefore all theoretical models discussed thus far have actually been deterministic. A computer can imitate the statistical properties of any "noise" as closely as desired by various computational processes which produce "pseudo-random" numbers. Alternatively, we may allow the computer access to a source of "actual" random numbers, such as a roulette wheel or a nuclear disintegration device. The computer then becomes a nondeterministic device, and may imitate nondeterministic models. Either the random or the pseudo-random numbers may be introduced into the procedures of the theory in any way at any point or points, just so the end result at the computer output has the required statistical properties.

Thus, insofar as experimental results can be predicted, they can be predicted equally well by either a deterministic or a nondeterministic model, and the philosopher may have his choice. If he should wish to predict the result of experiment exactly, however, he will never be satisfied.

Those who insist on an exclusive province for strict determinism in science usually pretend either that the noise doesn't exist or that it need not be considered part of the theory. A common dodge is to label the unpredictable fluctuations "experimental errors." In other words, if experiment does not produce perfectly predictable results in agreement with the tenets of determinism, then experiment is in "error"!

Of course, the noise may be segregated at some particular point of the model. In particular, it may all be added just prior to the computer outputs, and then ascribed to the measurement process. This may be very convenient. However, according to our schema, details of the measurement process cannot be separated logically from the model, and are therefore part of the theory. A corollary is that it cannot be proved by experiment that the noise is in fact lumped in this way. This is an instance of the well-known philosophical viewpoint, as formalized by our schema, that no aspect of a theoretical model can be projected "behind" the measurements. The fact that there is no unique way of making such a projection should militate against it, if nothing else.

Continuous vs. Discrete Models

It has been assumed that the computer which "contains" the theoretical model has an unlimited memory, so that it can produce continuous functions. It is of interest to inquire whether unlimited memory is really required for a satisfactory theory.

As we have seen, noise always appears in experimental measurement. Furthermore, no measuring instrument possesses an unlimited "bandwidth," and therefore, by elementary information theory, only a finite amount of information can be gathered by any experiment or set of experiments in a finite time. Thus we may say that the capacity of the "channels" of scientific information is finite. As a result, a computer of finite memory capacity can, in principle, always be chosen large enough to model any or all existing experiments to a precision small compared with the measurement uncertainties. Since restriction to such discrete models would radically alter the purely mathematical aspects of theory, theoretical scientists prefer models based on continuous analysis, and it is true that the use of continuous functions appears to have important technical and heuristic advantages, but it must be recognized that it is impossible to choose between discrete and continuous models by any experiment. Incidentally, certain mathematical advantages might perhaps be claimed for finite discrete models. Their logical systems possess properties of provable consistency, completeness, and decidability lacking in systems adequate for continuous mathematical analysis. These theoretical advantages do not seem to result in readily usable practical advantages, but it is interesting that models with such meager resources are adequate for all scientific theory which exists at any given point in history.

The Role of the Human Observer

One important condition which we have insisted on in the experimental part of the schema is that all the experimental measurables be operationally definable. That is, all the operations of the experiment must be specifiable as precise procedures in a "protocol" of the experiment. The reason for such a stipulation is clear: the experiment must be repeatable by any competent experimentalist.

In our planetary-motion example, the protocol includes the specifications of the telescope, tracking system, and data-signal takeoffs from the axes. Anyone "skilled in the art" can reconstruct the apparatus and reproduce the experimental arrangement.

The present-day trend of experiment is to use automatic equipment insofar as practicable. However, the human observer is used to some extent as a data-gathering instrument in almost every case, particularly in the behavioral and neural sciences. Unfortunately, the human being does not bring a specifiable procedure to an experiment. Of course, he is often used to perform tasks which are in fact specifiable, or so closely specifiable that the only serious obstacle to replacing him with automatic equipment would be the high cost in time and money. In such cases we may consider our ideal capable of fulfillment. However, in many situations the human observer, under present circumstances, cannot be replaced because he must use his nervous system to transform the data by processes not yet known. We do not consider such an experiment satisfactory in an ideal sense.

We will not of course propose that human observation be immediately barred as inadmissible. This would be absurdly impractical. However, we propose that human observation which makes use of unknown processes be considered an interim measure, to be replaced as soon as practicable by precisely defined procedures.

Many persons hold that the human observer is a perfectly natural choice in an ideal experimental situation, even though his data-processing methods may be entirely unknown. Precisely defined processes may be appropriate to the physical sciences, it is argued, but the nonphysical (particularly the behavioral) sciences are somehow intrinsically different and must therefore be satisfied with less stringent ideals. Our viewpoint, however, is that our ideal schema is not restricted to the physical sciences, but, on the contrary, is to be considered as a paradigm applicable to the study of science generally.

The necessity of specifiable processing as an ideal in nonphysical sciences has been denied because replacement of the human observer by a comparable "standard" observer, even in the simpler experiments, has seemed so unlikely as to be essentially impossible. If a condition is believed impossible of attainment, it will not be given the status of a basic requirement, even as an ideal. However, recent developments in information processing suggest that we need no longer be so pessimistic. For example, one of the most useful of the capabilities of the human nervous system as an information-processing device is its ability to learn to separate complex data into "meaningful" classes. We have already pointed out that ability of this nature is fundamental to the survival of any organism. In the past few years considerable effort has been directed toward the development of specifiable "recognition" processes. Existing methods do not yet compare with the human observer in performance, but they should eventually prove able to approach human capability. Thus perhaps the most important reason for denying specifiable data-processing as an experimental ideal no longer exists.

If, in spite of this consideration, the human observer were to be insisted upon as an ideal, his disadvantages would have to be accepted along with him. The differences between individual human observers, the inconstancy of the individual in time, and his tendency to do his processing in terms of his expectations are well known, as is his inability to record or communicate the processes utilized in observation.

Furthermore, there is a serious philosophical complication. We have emphasized that a complete theory must include any processes or effects of the measuring instruments themselves. To accept any unknown experimental process as an ideal is thus to give up even the hope of ever acquiring a complete theory.

The argument is sometimes advanced that, since the human being must appear somewhere, he might as well enter at the data-processing stage. Of course, we assume that the human being sets up the experiment and theory initially. However, in our ideal schema the human being is situated at the point of comparison between experiment and theory. Ideally, we may say he simply reads the output of a computer which says "yes" or "no" according to whether experiment and theory agree or fail to agree with respect to a specified criterion. He is placed in the role of creator and monitor of the experiment and the theory rather than employed as an essential process of either one. The scientific investigator is thus engaged in what would seem to be his most natural and profitable activities.

In many experiments the number of data or of observations required is so enormous that it is impracticable to use human observers. For example, in electrophysiology so many data are required for statistical significance that automatic data processing is required.

Finally, the processes used by the human being were acquired through evolution more or less specifically for his own survival. It is hardly likely that these are the optimum processes for every form of scientific experiment. On the contrary, important phenomena may be missed because the human observer lacks necessary processes comparable in complexity to, but essentially different from his own. It is even possible that supplementing brains with suitable processing tools could have an effect comparable in importance to that already obtained by supplementing the senses.

We see, then, that the human being as observer possesses advantages which make him indispensable for the present, but we need not insist that he be accepted forever as an ideal instrument, particularly with the host of practical and theoretical problems he brings with him. Soon we may even look forward to his virtual replacement or supplementation in experiments in the neural and behavioral sciences which require complex classification, including nosology and diagnosis. Eventually, complex standard processing equipment may become so common that the experimenter will no more think of using a human being as a measuring instrument than he now thinks of using a water clock.

The Behaving Animal as an Object of Experiment

It is useful to examine some of the basic differences between the experimental problems in the physical sciences and those in the behavioral sciences.

An experimenter generally has some model, formal or informal, of the "system" he is investigating. He cannot measure every conceivable variable having to do with the system. Therefore he attempts to arrange his experiment so as to study the system to the greatest advantage with the limited means of measurement at his disposal. One way of doing this is to structure the experiment so that as few variables as possible need be measured, for example, by keeping as many variables as possible constant during the experiment. Another is to "isolate" the system, i.e., to arrange things so that as few variables as possible interact strongly with the experiment. These arrangements simplify things for the experimenter, but they often seriously restrict the system under investigation. Even after the experimenter has done his best to limit variables or isolate the system, he usually cannot measure all the remaining relevant variables. Thus every experiment is, to some extent at least, an "abridged" version of a more comprehensive ideal experiment on an unrestricted system for which all conceivable variables would be measured simultaneously.

An important difference between experiment in the physical sciences and experiment in the behavioral sciences lies in the extent of this abridgment. The experimenter in physics has tended to choose experiments having comparatively few relevant variables, and thus he usually has to deal with relatively little abridgment.

On the other hand, consider an unrestricted animal in his natural environment as an idealized system for study. The variables pertaining to the animal include all measures of overt behavior such as limb and body movements, eye movements, and vocalizations. In addition the internal variables include biochemical production, autonomic responses, and neural activity in all parts of the brain. Furthermore, the environment is part of the system and all its variables must be measured. Compared with such an ideal experiment, the most ambitious experiment yet performed represents a drastic abridgment indeed.

The difficulty of carrying out anything close to such an ideal experiment has been so great that behavioral science has tended to split into fields corresponding to particular favored viewpoints and methodological techniques. Neurophysiologists have concentrated primarily on measuring variables within the nervous system or closely related to it. Psychologists, most of whom do not even consider themselves biologists, have tended to limit their measurements to variables outside the "skin" of the animal. Even cursory examination shows that both deal with extremely abridged experiments. Although there has been much argument over the supposed philosophical advantages of particular types of abridgment, from our point of view such abridgment arises from practical necessity and no one type is actually desirable. To deliberately ignore measurable data is to labor under a self-imposed handicap.

Extreme abridgment implies that the experimental subsets occupy only an extremely small portion of the ideal experimental space. Therefore it is very difficult, if not impossible, to guess at far-reaching generalizations from drastically abridged experiments.

Even in the practical sense, extremely abridged studies encounter serious difficulties. There is such a wide divergence of methodology that the abridgment differs from laboratory to laboratory, and even from experiment to experiment in the same laboratory. Needless to say, the experimental results and their models are difficult to reconcile.

A parallel may perhaps be drawn from physics. We might attempt to construct the quantum theory of the solid state by limiting experiment to bulk properties like conductivity, elasticity, and specific heat, deliberately eschewing any study of spectra, gases, and all other atomic and molecular phenomena. No one can say that it would be impossible to arrive at quantum mechanics from such an abridged study, but it would certainly be difficult. In actual practice, many of the bulk properties of matter, known from very old experiments, received a theoretical foundation only after the quantum theory had been developed, mainly through the study of atoms, molecules, and gases, and then applied to the solid state.

Considerations such as these suggest that a knowledge of neural organization, combined with behavioral measures, will provide a broad theoretical foundation of behavior long before an equally good theory can be constructed from behavioral observations alone. It has sometimes been pointed out rather emphatically by those with contrary views that neurophysiology has not yet given much help to the behavioral sciences. This is true, but it is also true that existing theories of behavior fall far short of a theory of the unrestricted animal.

As we have emphasized, experimental data never repeat exactly. This difficulty may be overcome by classifying the environmental data so that each class corresponds to some useful predictability.

Note that it is not necessary to distinguish here between those experiments in which the experimental variables are reset to some standard value before each observation and those in which repeated observations are made on the system as it performs some unrestrained natural behavior. Thus in our astronomical example, measured data on planetary motion are classified into sections of geometrically or analytically defined orbits in a suitable space. That is, it is predicted that the measured data subsets will lie near certain classes of subsets defined by a suitable theory.

In typical experiments in physics it is necessary to discard comparatively few data. After a minor statistical "reduction," necessitated by irreducible fluctuations, further classifications are usually one-to-one.

However, in the neural and behavioral sciences classification processes typically lump a very large number of data into comparatively few categories, and thus discard a great deal of information. This is an important further source of abridgment in those sciences, and an important source of difference between experimental procedures in the behavioral sciences and those in the physical sciences. Effectively, the behavioral scientist has poorer models and thus a more poorly structured notion of his experimental "environment" than the physical scientist, and he is forced to use rough many-to-one classification procedures because they can lead to a measure of predictability in a complex environment whose structure is otherwise unknown.

In such cases our computer imitates the experiment only after a many-to-one classification has taken place in the experimental procedure. In cases of very rough classification (such as separating black and white balls in an urn model), models may be so trivial as to be worthless, and since our goal is to minimize abridgment, we wish to minimize such classifications. This is possible if a theoretical model can be found to imitate the experimental measurables before the classification process, because then a suitable model in the computer can replace the many-to-one classification process in the experiment, or at least reduce the wasted information. The very fact that a particular classification process is found repeatedly useful strongly suggests that a model exists for it.

Thus many-to-one classifications are necessary in the early stages, but are replaced gradually by improved models (and improved experiments) until an irreducible minimum of "noise" is reached. How large a "noise" will remain cannot be predicted in advance.

For example, electroencephalogram (EEG) waves have been classified empirically into "normal" and "abnormal" waves of various types. If we had a model for the EEG which satisfied enough neurophysiological constraints to imitate EEG waves of various types under various conditions of the model parameters, we might hope that comparison of waves in the normal and abnormal classes with their corresponding model parameters would lead to a physiological theory of the normal and abnormal states. It is unlikely that such a success could be achieved in one jump, but it can be seen that even comparatively crude classifications can be of considerable aid to model construction. We will show some examples of such a study later.

Even abridged systems in the neural and behavioral sciences produce data in such enormous amounts, and of such complexity, that highly sophisticated processing instruments are required as integral parts of the experiments even for classification if the classes are explicitly specified. The most highly developed processing machine at present is the digital computer, but it has not been much used because it has not been designed for experimental work, but rather for business and the more routine scientific and engineering applications. Both the machines' input-output means and their modes of use strongly reflect this bias. What seems to be required is a family of computers developed around the concept of the processor as a laboratory tool. Such devices are being developed and may be expected to accelerate the introduction of advanced data processing into experimental procedures.

Some General Remarks on Brain Models

Having discussed experiment, we will now consider the nature of theoretical models.

It is most important to avoid confusing any experimental "system" with its model, and this is particularly true of brain models. For example, "the brain is a computer" is sometimes stated with such conviction that the speaker seems to believe that the brain is in fact a computer. Now, as we have seen, it is true that any measured variables of the brain can be imitated as closely as we please by models suitably expressed in mathematical language, and therefore by a computer. To say that the brain can be imitated by a computer is, however, not the same thing as saying that it *is* a computer. To do so is to confuse the system under study with its model. Of course, perhaps all that may be intended by the statement in question is a definition, which can be rephrased, "Whatever the brain is, it is a computer." This is certainly admissible, but it would appear to rob the statement of useful content, since it merely equates the concept "computer" with that of "brain."

The statement could convey precise meaning if it stated that the brain is a certain well-known kind of computer, or resembles one very closely. However, no such computer is known, and it really seems that a detailed resemblance between brains and any presently known computer is illusory.

We may ask, then, if there is any content at all to the statement. It is true that in many situations the nervous system may be viewed as an information processor, and we have already discussed it in that connection. We have also seen that a class of devices called "computers" may be used as information processors. Thus we can agree that there is a resemblance between "brains" and "computer," when viewed at a very abstract level. However, other views are possible. For example, the nervous system may be viewed as a "servo" system. Furthermore, there are devices (such as television receivers) which process information, but which are not ordinarily called "computers." Thus it would seem that both the definition of "computers" and the nature of their resemblance to "brains" remain somewhat vague. Such comparisons can often serve useful heuristic purposes, but they do not convey much meaningful information about the brain, and they can sometimes be very misleading to those who take them too literally.

Now, we have discussed models, statable in part as sets of differential equations, which serve to imitate the measurable variables of the solar system more or less successfully. We hope to have such models eventually for brains, but the brain models will bear the same relation to brains as the differential equations bear to solar systems. According to our viewpoint, as formalized by the schema, no theory can ever be more than an approximate description of certain specified measured variables. These ideas are well known to philosophy, but they are emphasized here because if they are not kept in mind constantly in the study of the brain and behavior, subtle and misleading biases can result

We will briefly discuss one other concept, "consciousness," in these terms. We do not have a definition of "consciousness," but it would appear that a child is differently "conscious" from an adult, and if an adult even loses a limb, his relation to his "experience" is changed and therefore probably his "consciousness" also. Thus "consciousness" is probably not a single entity and may be very sensitive to the biological properties and history of the organism. It is therefore premature to attribute the possibility of "consciousness" of a specifically "human" variety to any particular type of model. We have no present way of knowing how close to a human being "himself" such a model must be.

Physical Theory and Its Relation to Mathematical Analysis

So far, nothing has been said about how the theories to be tested in our schema are acquired in the first place, and of course it is not possible to say anything very precise about this, just because theories are products of unknown processes of the human brain.

We have pointed out that scientific activity may be viewed as a formalized extension of an activity of everyday life involving the construction of internal representations, or models, of experience. If, as in scientific activity, model construction is pursued in a determined way over long periods of time, it is reasonable to assume that various types of models particularly suited to the kinds of experiments under intensive investigation become fixed in the mind. These models may then serve as a store of possible models to be tried, with modifications, to fit new experiments. We note, further, that it is possible to reverse this process when a store of models becomes large. That is, the mind can sort through "experience," looking for examples which fit previously stored models. Thus it seems likely that, in the mind of the scientist, not only does an experiment seek its model, but a model seeks an experiment. These processes can reinforce one another, so that we may expect the body of experiment and the body of models in a science to be complementary. It is not unusual to find the same type of physical "law," or differential equation, appearing again and again in seemingly different contexts. For example, many of the earlier laws, such as Hooke's law, Boyle's law, Ohm's law, specific heat, and thermal expansion, are simple proportionalities or linear relations with experimentally determined coefficients. In every case these laws hold only over restricted ranges of the variables, that is, they are linear approximations to more complicated functions. One gets the impression that experiments fitting the more rudimentary models were picked out earliest for theoretical formulation. Of course this choice was probably not deliberate, since similar laws were often expressed in different ways, and to this day elementary texts in physics do not usually emphasize that such laws are all particular examples of the same general type of model.

In terms of our abstract framework, perhaps we can say that after great familiarity with the character of the subsets of space selected by various types of rules, it becomes easy to recognize which type of rule is likely to approximate a given experimentally determined subset. Conversely, it is easy to select experiments whose subsets approximate those determined by familiar types of selection rules.

Now, what types of selection rules are likely to be considered for possible models? We have seen that previously useful rules may often be drawn upon. However, the mathematical formalization of models in physics and consequent intensive study of mathematics have led to another source of great importance: theoretical models which were originally studied primarily as objects of mathematical interest. A body of knowledge has resulted which we may call "classical" mathematical analysis. From our previous discussion, it is hardly surprising that the physical scientist tends to fix on experiments amenable to imitation by the models with which he is most familiar, those suggested by classical mathematical analysis. Conversely, he tends to reject or ignore experimental results which do not easily lend themselves to such descriptions.

Since the models of theoretical physics tend to be drawn from mathematical analysis, we are led to inquire how serious such a restriction on possible models may be. The "kernel" of "classical" analysis is the mathematical theory of differential and integral equations, and related topics. It seems clear that the totality of possible theoretical models which can be formulated in logical terms greatly exceeds the number which can result from the use of differential and integral equations alone. For example, switching networks do not lend themselves to formulation in terms of analysis. One reason for this is that various requirements of single-valuedness, continuity, and differentiability must be placed on functions and transformations treated in analysis.

Furthermore, there are very serious gaps in knowledge within analysis itself owing to insurmountable difficulties in mathematical treatment. Linear equations are the only class of differential equations for which a satisfactory mathematical theory exists. Of course, the theory of linear equations comprises a considerable body of knowledge, and a student may study mathematical physics for years without realizing that there are serious restrictions on the methods he has been at such pains to acquire. Since textbooks emphasize what is known, rather than what is not, scientific mathematics may begin to appear almost synonymous with a subset of analysis.

Our familiar example of planetary motion furnishes an interesting case of a nonlinear problem in classical physics. The nonlinearity arises first in the Newtonian model because of the mass interactions between bodies. It is easy to find graduate texts in theoretical mechanics which work out the solutions to the two-body problem in Newtonian motion, but which do not even mention the fact that, after some 200 years of study, few details can be given about the orbits in the general case of multiple bodies.

We see, then, that the class of mathematical models with which the physical scientist has the most thorough familiarity, and therefore upon which he is most likely to draw, is actually small compared with the class of possible models. This is another reason for our emphasis on mathematics as a descriptive logical language which is much more general than analysis.

It is interesting that good correspondences for so many analytic models can be found, and this requires explanation. It may be that "nature prefers" these well-studied models, and since these tend to be the "simpler" models, this is more or less equivalent to the assertion that "nature is simple." This is a position common in theoretical physics. However, we must keep in mind the fact that the subsets determined by all existing experiments occupy only a tiny portion of the space of all conceivable experiments. There exists therefore the possibility that the subsets which admit of approximation by "simple" selection rules have been picked out of the total for precisely that reason, and that existing models represent merely a first approximation to a transcendently complex "nature." We have seen how the successive theories of planetary motion improve the quality of their approximations at the expense of added dimensions and complexity of selection rules. Similar examples can be given in any physical science. This succession of improved models may well be analogous to the improvement in curve fitting obtained by including more and more terms in a series expansion, except that the theorist deals with improvements in approximations to sets of curves. There is no guarantee that such a process will ever terminate, nor does it appear that any evidence of termination can be drawn from the history of science. It is not even easy to draw meaningful conclusions from the fact that certain particular types of models have proved useful, since it seems likely that a wide class of "universes" of sufficient complexity would admit the same types of simple approximate models. We will in fact conjecture that almost any reasonable model universe beyond a certain degree of "complexity" will possess at least one region within which any given "simple" model would provide approximate agreement. It may even be possible to prove a precise version of such a statement.

Perhaps enough has been said to convince engineers and physical scientists that caution is indicated when models from their fields are suggested by neurophysiological or behavioral data from experiments on severely abridged systems, because obviously the complete model for the unrestricted higher animal is far more complex than any model familiar to present science.

It is important in this connection to realize that nonlinearities of the nervous system and its functions enter at a very low level. The neuron itself appears to be intrinsically nonlinear, and cannot be approximated by a linear device over any significant range, any more than a "flip-flop" circuit can be constructed from linear components. Likewise, perceptual behavior involves an essentially nonlinear process, namely, classification.

Use of Digital Computers in Neural Modeling

We have now arrived at the crucial point of our investigation of the construction of theories of neural organization and behavior. If, as we have maintained, models from classical mathematical analysis are likely to prove inadequate, we must look to the creation of a new body of experience with complex models more suited to the needs of the science. There are essentially three ways of proceeding—exploration of new models by further mathematical analysis, exploration by means of physical "analog" computer systems, and exploration by digital-computer calculation. All three methods in fact overlap, and an optimal attack on the problem would use all three approaches in some combination. The mathematical approach—attempts to prove significant, useful theorems—can make use of computer experience to gain insight into what general theorems may be provable. In this paper, however, we wish to stress the advantages of the digital computer.

We have seen that any model whose properties can be stated precisely can be imitated as closely as desired by a digital computer. Two important practical advantages result. The first is that to use the computer is to require precision of statement. This in itself is of great value, particularly in theories of neural organization and behavior, because it forces the theorist to think his ideas through to explicit formulations. Vagueness at any step will not do. The second advantage is that, as long as they are explicitly stated, theoretical models may be considered without any constraints whatever being placed on the imagination. In particular, models need not be mathematically tractable, nor need they be realizable by particular physical devices such as are used in analogs. Both these requirements can impose severe restrictions on theoretical models, although the restrictions are, of course, different. Freedom of the imagination to concentrate solely on the forms of model suggested by experiment without distraction by extraneous considerations can hardly be overemphasized. It is particularly important in the early stages when models often must be tried and discarded in rapid succession.

Another important advantage of digital computers is the great convenience and flexibility of control over the parameters of the model. Any or all of the parameters of a highly complex model may be changed in a short time, and this can even be done while a particular state of the system is held, as it were, arrested in time. Such flexibility of parametric control is particularly difficult to achieve in physical-analog devices. Control extends of course to the precision of representation of the parameters themselves. Precision can be tailored to the needs of the particular case, from very crude specification to a precision beyond that attainable in any analog device.

John von Neumann was one of the first to recognize that the complex, nonlinear nature of the problems presented by the biological sciences (as well as very complex physical problems such as are encountered in hydrodynamics) could be attacked profitably by computer methods. The factors that led him to this realization are outlined in the following paragraph from Ulam's discussion (Ulam, 1958, p. 40) of von Neumann's work.

> Another major source from which general mathematical investigations are beginning to develop is a new kind of combinatorial analysis stimulated by the recent fundamental researches in the biological sciences. Here, the lack of general method at the present time is even more noticeable [than in modern physics]. The problems are essentially non-linear, and of an extremely complex

combinatorial character; it seems that many years of experimentation and heuristic studies will be necessary before one can hope to achieve the insight required for decisive syntheses. An awareness of this is what prompted von Neumann to devote so much of his work of the last ten years to the study and the construction of computing machines and to formulate a preliminary outline for the study of automata.

Application to a "Neuron-like" Network Model

We shall illustrate the above ideas by a short description of a dynamic model of a "neuron-like" network. This model can be made to imitate a variety of known electrophysiological phenomena by varying the parameters of the model network. The imitations are mainly qualitative at present, but there is reason to hope that further investigation can result in significant quantitative experimental comparisons.

The dynamic model being studied is a network of elements, or "cells," which have the following properties: (1) Each element has a definite "resting" threshold for incoming "excitation" below which no action occurs and above which the element "fires," (2) the thresholds and their time constants are distributed statistically over the elements of the net.

When an element fires, its threshold immediately rises to a high value and then falls exponentially back toward its resting value. Thus during this recovery the element has a "relatively refractory" period. A short time after firing, an element transmits excitation to all other elements it is connected to. In the present model all connections transmit the same amount of excitation, but the transmission delays are distributed statistically over the connections. The statistical distributions of time constants, thresholds, and delays are under the control of the operator, and their form, mean, and variance may be changed in a few seconds.

The transmitted excitation is added to any already present at the succeeding cell, and the excitation sum decays exponentially toward zero. If at any instant the excitation sum exceeds the threshold of that element, the element undergoes its own firing cycle and transmits its own excitation. During the firing cycle, the element is said to be "active." Note that in physiological terms, both "spatial" and "temporal" summation are present, and that the element is an "analog," not a purely digital, element. Note also that the elements do not, in general, fire synchronously.

There are 1296 elements, and they may be thought of as being arranged in a 36×36 array. The "operator" selects a set of candidates for connection to a typical cell, and the computer then proceeds through the net, using the same set relative to each cell, and assigns connections according to a preset probability distribution. Most nets described here have a connection probability depending only on radial distance from each cell. They are thus statistically isotropic, and also statistically homogeneous, except at the edges, where they may be thought of as having been cut out of an infinitely large homogeneous net. However, the net of Fig. 3 is nonisotropic; i.e., the connections are made predominantly in one particular direction.

After the computer program assigns connections, and other net parameters are fixed, the operator may choose the elements to be stimulated initially by means of a photoelectric device, or he may choose one of several preselected groups. The computer then proceeds to calculate the resulting activity. Firing is indicated for the experimenter by the brightening of spots on the oscilloscope corresponding to fired cells. Parameters may be changed during operation. Calculation of each time increment takes the computer 0.1 sec to 1 sec, depending on the number of cells active.

For qualitative description of observed behaviors, it is convenient to divide the networks roughly into two classes: "tightly" connected nets and "loosely" connected nets. Tightly connected nets tend to produce well-defined wavefronts, and loosely connected nets tend to oscillate diffusely.

Tightly Connected Nets

If each element of the net is connected only with its immediate neighbors, and a single stimulus is given to a few neighboring cells, a "wave" of activity (fired cells) is generally seen to progress away from the site of the stimulus. If only one stimulus is given, or if it is repeated at long intervals, the wavefronts may form expanding rings much like those caused by a pebble dropped in a pond. No reflections are observed at the edge of the net. Each wavefront pushes before it a "precursor" of excitation which has not yet fired cells and leaves behind a "trough" of refractory cells. This means that if waves interact from a rapidly repeating source or from multiple sources, the interaction will be very nonlinear. A rapidly repeating stimulus—or, under certain conditions, a single stimulus—may produce instead of rings a very complex "broken field" of wavefronts, which continues to re-excite itself. Waves produced by multiple stimuli may be quite different from waves produced by individual stimuli.

Many different "modes" of self-excited activity are possible. A particularly pervasive and striking one consists of one or more continuously rotating and expanding spirals. The arms of a spiral maintain a distance such that the refractory trough lies between them. An example is shown in Fig. 1.

Loosely Connected Nets

If there are significant numbers of connections long enough to reach back behind the refractory trough of a wave, then "backfiring" behind waves can occur. In such a case the wavefronts become indistinct, and activity may diffuse more or less evenly over the whole net. If the net constants are such that the connections just exceed the length sufficient to permit backfiring, the whole net will usually oscillate diffusely after a single stimulus. The oscillation may continue indefinitely, or it may, if the net is barely oscillating, stop spontaneously after a burst of several cycles. It is easy to understand this behavior qualitatively. The few cells initially stimulated cause several more to fire, each in turn firing others, in a numerically expanding series of events. This increase continues until a large proportion of the cells are in a refractory state; firing must then decrease progressively until the refractory cells have recovered; the cycle then recommences if there is still activity in the net. Thus, if the total number of cells that

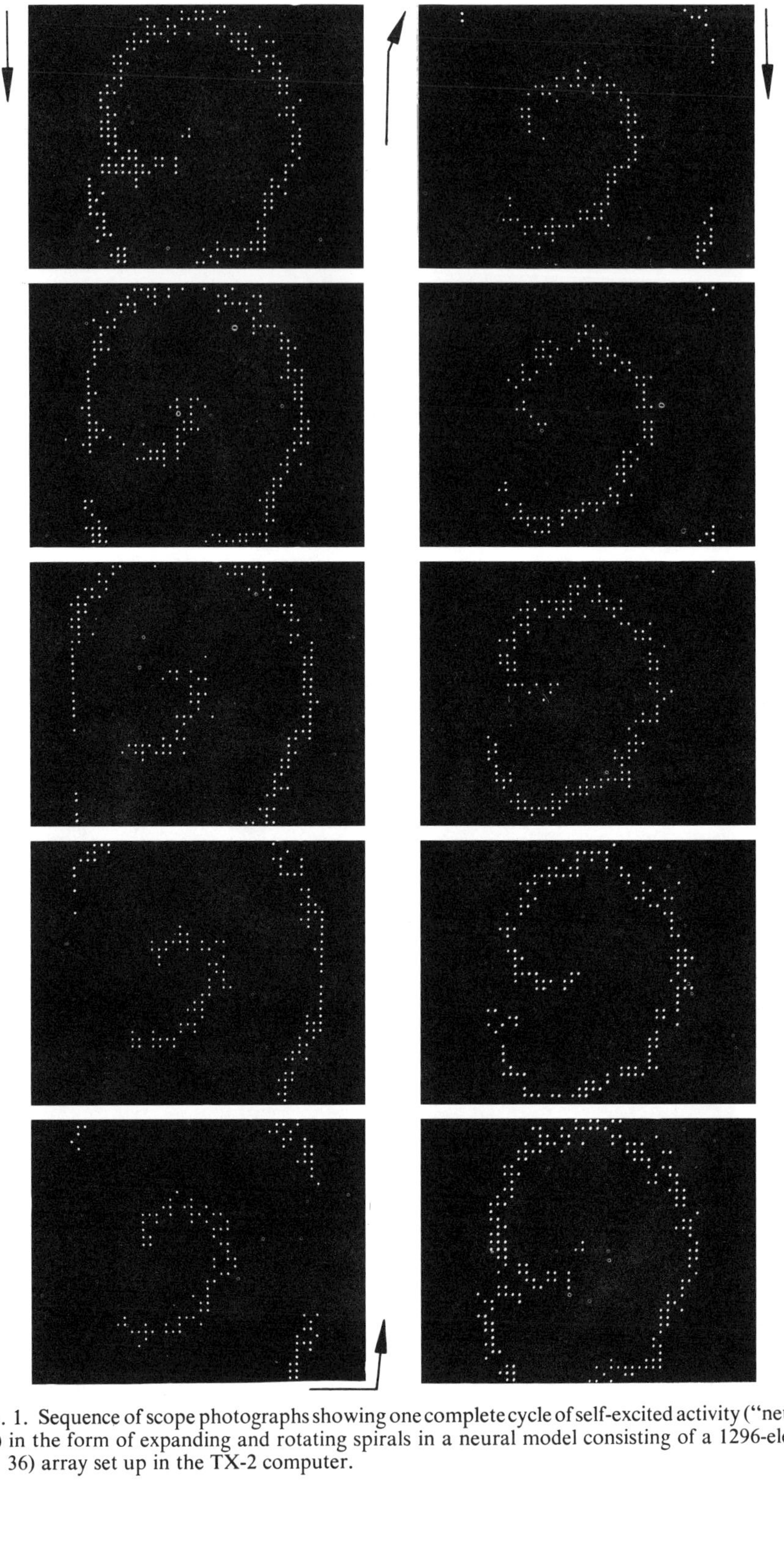

FIG. 1. Sequence of scope photographs showing one complete cycle of self-excited activity ("neuron" firing) in the form of expanding and rotating spirals in a neural model consisting of a 1296-element (36 × 36) array set up in the TX-2 computer.

fire is plotted as a function of time, a roughly periodic function results. The waveform of this rhythmic oscillation is not sinusoidal, but is likely to be considerably smoother than waveforms usually associated with "relaxation" oscillations. If the excitation at the cells is summed over the entire net, we obtain essentially the same curve, somewhat smoothed by the time constant of excitation decay. An example is shown in Fig. 2.

An important characteristic of the oscillation waveform is amplitude modulation, which occurs because multiple oscillatory modes are possible. During rhythmic oscillations, activity may be spread diffusely and evenly over the net, rising and falling more or less in phase everywhere. This "normal" mode gives rise to large-amplitude oscillations of the kind discussed above. However, modes in which activity transfers from place to place in the net are also possible. Such modes may not show large fluctuations in amplitude because cells are constantly firing. Since which cells fire at any instant depends on which cells fired at the preceding instant and also on past history, the modes are "coupled," and activity can transfer from mode to mode. The result of this "mode changing" upon the excitation-vs.-time plot is amplitude modulation of the oscillation, the oscillation period remaining the same. The net may therefore show irregular "bursts" of rhythmic oscillations, and this behavior continues indefinitely. Such effects are clearly shown in Fig. 2.

The primary determinant of the rhythmic oscillation period in the normal mode is the refractory period, because cells tend to fire whenever they return to a state near their normal resting threshold, or soon thereafter. The oscillation period is therefore dependent primarily on the dynamic properties of the elements.

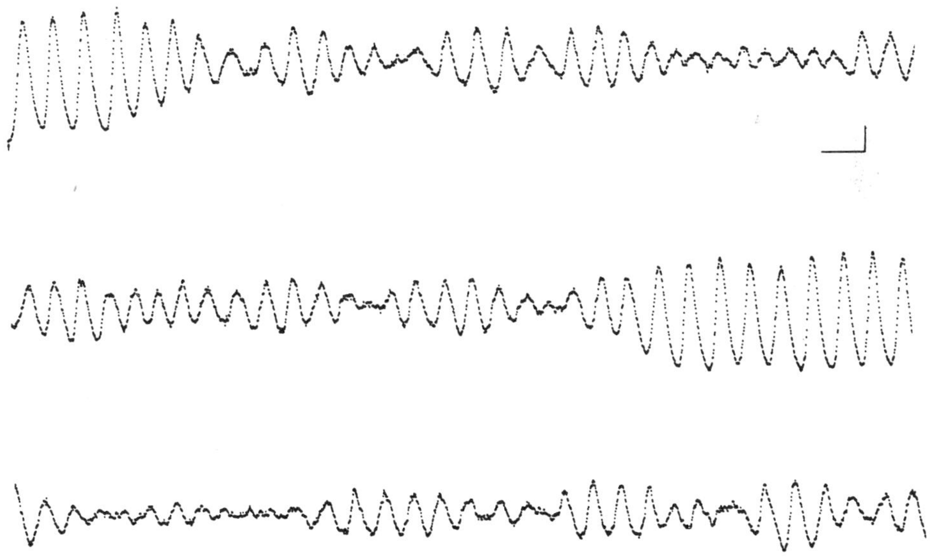

FIG. 2. Oscillations in a loosely coupled, isotropic net. Note resemblance to "alpha" rhythm.

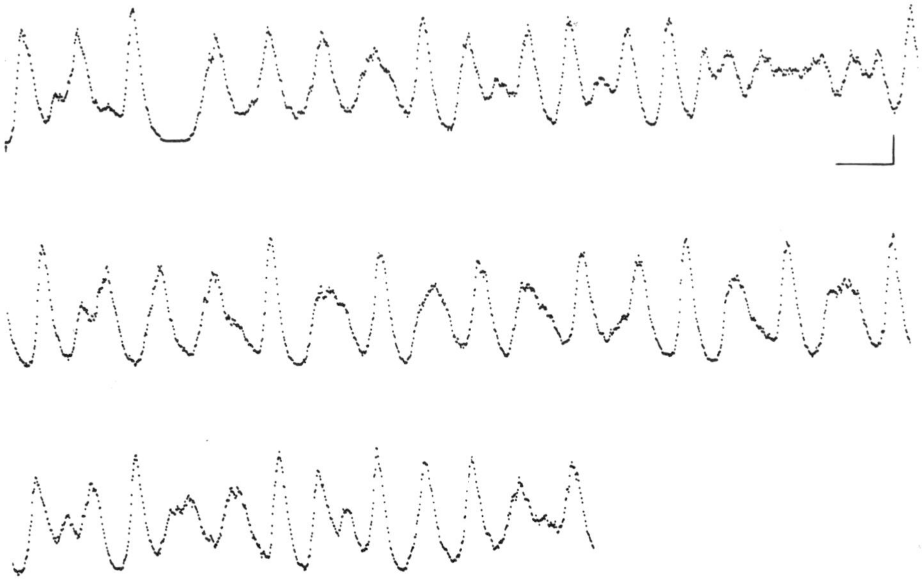

FIG. 3. Irregular waveform produced by loosely coupled, non isotropic net. This wave is reminiscent of the EEG in certain pathological conditions.

According to this explanation, the oscillation period should be roughly independent of network size and shape, amplitude of oscillation, and initial stimulus, and partly independent of the connection lengths and excitation time constant, so long as they are not extreme. These predictions are borne out by observation, to a good approximation. The oscillations do not occur if the transmission delays are longer than the refractory periods.

If the mean threshold over the net is set initially very high, so that the average firing rate is very low, the excitation output waveform is generally quite irregular. As the mean threshold is lowered, the average firing rate increases and the output becomes gradually more rhythmic. If this process is continued after a rhythmic oscillation is attained, the frequency begins to increase, because the cells fire more often, on the average, but the peak-to-peak amplitude of oscillation decreases, because the low thresholds tend to prevent the lower excursions of firing rates. Eventually, when the thresholds are very low, the oscillations may disappear into a "noisy" background of high average activity.

The type of connection may affect the output waveform markedly. If the connection patterns are highly nonisotropic, a considerably different type of oscillation can take place. An example of such an oscillation is given in Fig. 3.

Periodically Stimulated Networks

We have been discussing networks in which an initial stimulus starts a self-sustaining oscillation. Interesting effects are also observed if networks in a state of self-excited activity are given further stimuli. If a net is stimulated periodically with a diffuse constant stimulus, the average amplitude of the responses may vary with stimulus frequency in a manner superficially similar

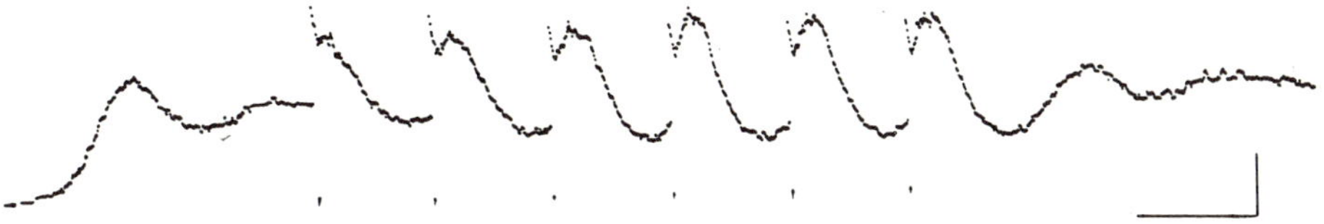

FIG. 4. Responses of active net to repetitive stimulus. Note increasing size of delayed peaks.

to ordinary resonance. At or near the "resonant" frequency, the second and several succeeding responses build up, or augment, to a much higher level than the first response; this augmentation accounts for the resonance effect. Augmentation may be understood qualitatively as follows. If a large group of cells in an active net is given a stimulus, not all cells in the group are fired because many are refractory. After a certain length of time, those cells fired by the stimulus will have recovered (except for some that may have been refired internally) and, in addition, those that were refractory at the first stimulus will be available (again, except for some that may have been refired internally). Thus, if the ratio of stimulated cells to internal average firing rate is large enough, we may expect a second stimulus, given at the proper time, to fire a larger number of cells than the first. The "transient" introduced by such a stimulus may go through several "cycles," so that, at the right period, the augmentation of response may last for several responses. The oscillations, the augmenting of responses, and the "resonance" phenomenon thus may all be viewed as consequences of the refractory properties of the units. An example of augmenting of response is shown in Fig. 4.

Physiological Interpretation

It is not unreasonable to assume, as a first approximation, that an electrode on or near a brain picks up a summation of electrical activity in nerve cells comparable to the sum of excitation over the network model. We may thus compare the electrophysiologically observed waves with the excitation-vs.-time output of the model. We shall not go into complete detail here on this question However, we have had enough discussion to make several comparisons. The similarity of the amplitude-modulated rhythmic oscillations to the alpha rhythm of the resting electroencephalogram (EEG) need hardly be emphasized. Furthermore, some of the irregular output waveforms resemble the EEG taken under other normal, or even abnormal, conditions. For example, the wave of the non-isotropic net illustrated in Fig. 3 is similar to waves sometimes observed in hepatic coma. It is also possible to synchronize both model and EEG to periodic stimuli under certain conditions. Responses which increase or augment with periodic stimuli are also known in several parts of the nervous system. Finally, high-frequency components superposed on the lower-frequency oscillations have been demonstrated both in electroencephalograms and in the model. Figure 5 shows "noisy" high frequencies observed in the model.

It seems unlikely that the cell bodies or their axons correspond in a direct way to the elements of the model. In the first place, the phenomena of similarity

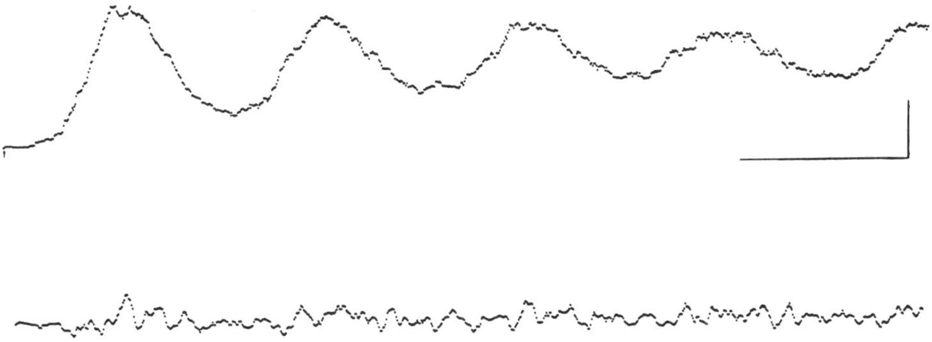

FIG. 5. Segment of oscillation at very high resolution. Note superposed high frequencies, which are filtered out and magnified (×4) at bottom.

are all of the "slow-wave" variety, and there are experimental grounds for believing that action potentials of axons or soma do not produce slow potentials directly. Furthermore, the time constants necessary for proper oscillation periods are all much longer than the 1- to 2-msec constants usually associated with soma and axons. However, time constants have been observed in the 10- to 20-msec range which may be attributable to dendritic membrane, and time constants of this order in the model produce typical oscillation periods of about 100 msec, which is in the right range for the alpha rhythm. Identification of the model elements with dendrites or parts of dendrites would seem to be attractive, but the physiology of the dendrites has not yet been well established.

If we nevertheless adopt the hypothesis of identification of the elements with dendrites (or an approximation thereto), the axons can be regarded as the transmitters of excitation, since the expected time delays fall well within the range of 10 msec or less required for oscillation (assuming, as before, 10- to 20-msec refractory time constants). In fact, many axon fibers conduct rapidly enough to transmit from one hemisphere to the other within this time, even if thalamic relays must be traversed. Thus, under proper circumstances, large portions of the brain might act together as a roughly homogeneous network in the sense of the model. This may be the condition when the resting alpha rhythm is being produced, since in this case synchrony is often observed between both sides of the head. On the other hand direct interaction between dendrites in some circumstances is considered a possibility by many neurophysiologists. In such cases, the model may be directly applicable, no axons for transmission being necessary. This would be consistent with those experiments in which poor correlation between slow potentials and action potentials has been observed.

We have presented several striking qualitative similarities between the outputs of the model and experimentally observed slow-potential phenomena of neurophysiology, including both "normal" and "abnormal" conditions. The very fact that there are several such similarities, under different conditions of parameters, gives hope that more work with the model will suggest experiments to test specific predictions, and perhaps lead eventually to an elucidation of the underlying physiology of the slow potentials.

Some Practical Considerations

The computer used in this work is the TX-2 at the Lincoln Laboratory, Massachusetts Institute of Technology. The TX-2 has a number of features that make it more useful than most digital computers in our application. For example, analog-to-digital converters are an integral part of the system. Other features are precision display oscilloscopes and a high-speed printer capable of providing graphs. In addition, the TX-2 has a number of manual controls: a photoelectric device ("light-pen") permits control of the computer directly from the oscilloscope; a typewriter and extensive sets of control switches enable the user to select or modify programs; analog/digital knobs can be used to change program parameters continuously. These controls and others can be used while the machine is operating. The computer also has under program control a precision plotting board, precision timers, and a random-number generator. It is also possible to use an automatic impulse-operated camera to photograph oscilloscopic displays of a process calculated frame-by-frame. The resulting film may then be projected at the speed best suited to observation of dynamic effects. This output medium is particularly useful in the network studies.

The central processor of TX-2 handles low-precision numbers much more rapidly than computers designed for engineering calculations. Logic features of the control element include specially designed instructions which, with the flexible arithmetic element, greatly facilitate data handling and simulation. "Multi-sequence" control makes it possible to add input-output equipment readily and use it efficiently. Finally, the TX-2 operating philosophy is such that the researcher himself can use the machine and select modes of operation appropriate to his particular experimental and theoretical problems.

The above features make the machine useful as a tool for acquiring an intuitive appreciation of the varieties of behavior exhibited by models too complex for mathematical analysis. Naturally, as the important parameters of the model and their qualitative roles are recognized, systematic studies are carried out, particularly in those cases where similarities to experimental data are observed.

The idea of the computer system as a directly used tool in both experimental and theoretical research in the neurophysiological sciences has only recently begun to be put to practical use. The widespread use of the term "general-purpose" as applied to computers is indicative of a psychological attitude which appears to have retarded development of computer systems for specific purposes. We have stressed that a digital computer is capable of simulating any precisely defined system, and the term "general-purpose" stems from this theoretical fact. However, in actual practice, computers are no different from other tools or instruments in that particular classes of applications require special features of design, construction, and methods of use.

Of course, the sizable cost of computer systems and the effort of acquiring the skills of a new methodology are among the practical factors which have slowed development and introduction of specialized instruments. Telescopes, particle accelerators, and the like have been accepted as part of the required instrumentation of astronomy and physics for many years, but the need for expensive and complex instruments in the neural and behavioral sciences is only now making itself felt. Nonetheless, computing instruments especially suited to experimental and theoretical problems in the neural sciences are now being made available, and it is likely that within the next few years their use will become widespread.

Many of the practical problems encountered in the study of a complex model by computer simulation are specific to the particular model, so that it is not profitable to discuss them in great detail. We may mention a few considerations however.

The network model just described uses the full resources of the TX-2 computer system, which is quite large and fast. The size of the network is a compromise affected by many considerations, such as memory size (69,000 words) and speed (about 6 μsec cycle time), the complexity of the elements, and the number of connections. It should be emphasized that if the network is too small, poor "statistics" may make results difficult to interpret, and the limited "dynamic range" of the output sum may make it difficult to detect small superposed effects such as the high-frequency "noise" of Fig. 5.

The use of the computer by the researcher makes it desirable to retain certain "utility" programs in the memory, or at least readily available on tape, to change parameters of the model and perform other necessary chores. Further, the type of logic used in simulation and the necessity for utilizing the computer speed and memory to the utmost preclude the use of the usual programming aids, which are primarily oriented toward problems in mathematical analysis. It is essential that the primary "loops" of the program be carried out directly in machine instructions. Even so, one "instant" of discrete time requires as long as one second of TX-2 time, and a "run" can require five to thirty minutes. Thus a considerable amount of computer time is required. This is simply a fact of life in this and other fields of comparable complexity.

The personnel required for carrying on studies of complex models are by no means excessive, given a suitable computer system. The writer and one assistant are the only personnel active in the network study described above. The effort required before obtaining the first results will vary with model and experience, of course, but it is comparable to experimental setup efforts in neurophysiology. Also, the "setup" and the model are likely to change continually during investigation, just as in "real" experiments.

Practical experience with TX-2 in the network study strongly suggests that better computer designs can produce substantially improved performance in this problem area at lower cost. It is not possible to predict at present just what computer configuration the researcher in neural organization will be using in a few years.

Summary and Discussion

We have given an outline of a philosophy of science as a consciously pursued activity analogous to, and perhaps evolved from, unconscious processes which enable the organism to survive in its ordinary environment. According to this view, the principal goals of scientific activity are to discover recurrences of some sort in experience, and to construct formal models which describe these recurrences as accurately as possible while predicting as much new experience as possible.

It is a fact that those experimental recurrences that are demonstrable are never exact or certain, but in any case we have seen that there always exist an unlimited number of theoretical models which provide equally good descriptions. Which of these models will make the best predictions in new situations can only be answered by experimental test.

We have further shown how these ideas can be made precise by a formalized schema of experiment and theory using idealized computers, and that the schema applies, ideally, to the sciences of neural organization and behavior and not only to the physical sciences.

Such a philosophy of science as we have outlined apparently appears quite radical to some, particularly since no final conclusions are possible. However, from another standpoint, it is actually a conservative philosophy, since it (ideally) recognizes only precisely specified experimental variables, requires specific definition of "recurrences," and makes no assertions which are unverifiable. This means, of course, that any assertions which are made must be regarded as approximate and provisional. We may say that only "a degree of reasonable expectation based on experience" is required of the scientist as a philosophical attitude, but we have of course made no attempt to elaborate his actual psychological motivations.

It should be noted in a philosophical discussion that studies of the functional organization of the brain and its relation to its environment can have important consequences for philosophy as well as for science. For example, resolution of some difficult problems of epistemology would seem to depend ultimately on the nature of the environmental models constructed by the brain. Investigation of such questions would require a much more formal discussion than we have given, since a complete model of an individual brain would have to include within it "nested" submodels on various levels. For example, an individual has a model of his environment, models representing other individuals within that environment, models representing models of the other individuals, and so on. In our informal discussion we have not always made these distinctions explicit. However, there does not seem to be any logical barrier to a complete formalization, either by methods analogous to the separation of "metalanguages" and "object languages" in logic, or by analogy with the idea of computers represented within computers, although it would be a very complicated process to carry out such a formalization.

We have pointed out that although our schema appears adequate for science generally, a number of practical difficulties are posed by the neural and behavioral sciences, which require instruments and methodologies different in many ways from the traditional tools of the physical sciences. There are, of course, considerable problems of experimental technique, but many of the difficulties arise principally from the unprecedented complexity both of experimental data and of theoretical models.

In particular, although we have shown that an unlimited number of models exist for any set of experimental data, there is at present no known practical rule for producing even one of them complex enough to be of interest to science. However, we have seen that these problems can be attacked by the data-processing and computational capabilities of modern digital computers. As an illustration, it was shown how simulation of a "neuron-like" network model on a large digital computer has demonstrated a number of similarities between behavior of the model and electrophysiological phenomena.

Finally, our experience suggests strongly that data-processing and computing instruments must be designed specifically for the needs of the science of neural organization and behavior. We may hope that present activity in this direction will lead eventually to a class of instruments commensurate with the problems, the challenge, and the importance of the science.

Bibliography

Rather than attempting to supply either citations for all statements in text, or a complete bibliography, I have listed (and occasionally cited) works in logic and scientific philosophy which have had the most direct bearing on the work presented in this paper.

Bridgman, P. W., 1946. *The Logic of Modern Physics*, New York: Macmillan.

Dantzig, T., 1954. *Henri Poincaré*, New York: Scribners.

Farley, B. G., 1962a. Some similarities between the behavior of a neural network model and electrophysiological experiments, *Self-Organizing Systems*, M. C. Yovits, G. T. Jacobi, and G. D. Goldstein, eds., Washington, D.C.: Spartan.

Farley, B. G., 1962b. Problems in the study of the nervous system, in *Proceedings of the 1962 A.F.I.P.S. Spring Joint Computer Conference*, vol. 21, Palo Alto, Calif.: National Press, pp. 147–52.

Kleene, S. C., 1952. *Introduction to Metamathematics*, New York: Van Nostrand.

Lindsay, R. B., and Margenau, H., 1944. *Foundations of Physics*, New York: Wiley.

Margenau, H., 1954. Advantages and disadvantages of various interpretations of the quantum theory, *Physics Today*, **7**(10), 6–13.

Quine, W. V., 1951. *Mathematical Logic*, revised edition, Cambridge, Mass.: Harvard Univ. Press.

Ulam, S., 1958. John von Neumann, 1903–1957, *Bull. Am. Math. Soc.*, **64**(3, part 2), 1–49.

Weaver, W., 1948. Science and complexity, *American Scientist*, **36**(4), 536–44.

4. Theoretical Significance of Dendritic Trees for Neuronal Input-Output Relations

WILFRID RALL, *National Institutes of Health*

Neural modelers have usually assumed that the synaptic input to a neuron can be treated as input delivered to a single point or a single summing capacitor. There has also been a tendency to assume that a combination of synaptic excitation and inhibition can be treated as a simple arithmetic sum of positive and negative input components. Both of these assumptions are oversimplifications. It is the purpose of this paper to draw attention to theoretical models which avoid these oversimplifications, and to present the results of computations designed to test the significance of what may be called spatiotemporal patterns of synaptic input.

It is a well-established anatomical fact that essentially all of the neurons in the gray matter of the vertebrate central nervous system have extensively branched dendritic receptive surfaces. Two examples of this (A and B) are illustrated in Fig. 1. The Purkinje cell of mammalian cerebellum (Fig. 1A) has a dendritic surface area that can be as much as one hundred times that of its cell body (estimate based upon the careful measurements of Fox and Barnard, 1957). Mammalian motoneurons (Fig. 1B) and pyramidal cells have dendritic surface areas that have been estimated to be ten to twenty times that of their cell bodies (Sholl, 1955; Aitken, 1955; Young, 1958; Schadé and Baxter, 1960, p. 175). Although the Purkinje cell is a special case, neural modelers should beware of schematic neurons (e.g., C, D, E, F) whose dendrites are either short and unbranched or completely missing. An excellent illustration and discussion of various dendritic branching patterns can be found in a recent paper by Ramon-Moliner (1962).

The entire dendritic surface of mammalian motoneurons is covered by a high density of synaptic connections (Wyckoff and Young, 1956; Rasmussen, 1957).* Thus there is little reason to doubt that the extensive dendritic surface of such neurons must be regarded as a receptive surface to which synaptic input is delivered from many afferent sources. In spite of differences between various

* These densities were estimated as 15 to 20 synaptic end-feet per 100 square micra of dendritic surface area. Motoneurons vary in size; the total dendritic surface of any one such neuron is of the order of 10^4 to 10^5 square micra. Such values imply that an individual motoneuron can be expected to receive from 1500 to 20,000 synaptic endings.

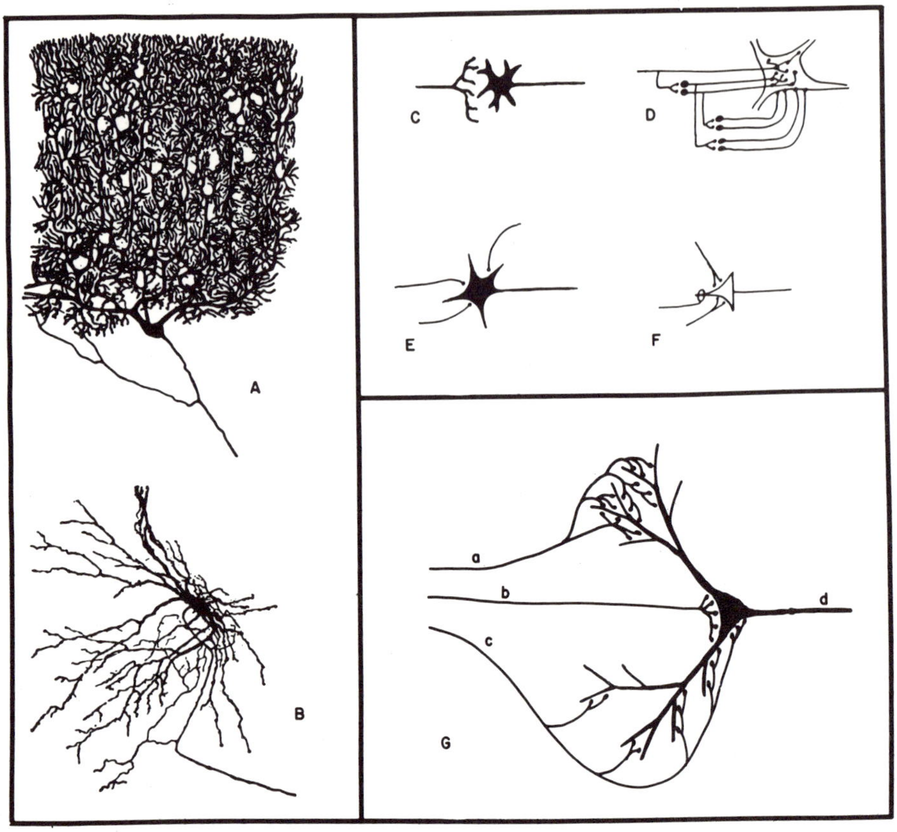

Fig. 1. Histological and schematic neurons. Parts A and B are reproduced with only slight modification from the classic histological work of Ramón y Cajal (1909); A represents a single Purkinje cell of adult-human cerebellum; B represents a single motoneuron of fetal-cat spinal cord. Parts C, D, E, F and G are schematic neurons with schematically indicated synaptic connections; they all represent neurons whose dendritic branching could be expected to resemble B. Part C represents a portion (redrawn) of a well-known diagram by Ramón y Cajal (1909) of a circumscribed reflex; D resembles a schematic drawing by Lorente de Nó (1938b); E resembles the schematic motoneurons of Lloyd (1941); F resembles the even more schematic symbols of McCulloch and Pitts (1943). Part G is also schematic; it is intended to show neither the full dendritic receptive surface of the neuron nor a full complement of synaptic connections to this surface; its purpose is to distinguish between three different patterns of synaptic connections, as discussed in the text.

types of central neurons with respect to both synaptic histology and unresolved synaptic mechanisms, it seems reasonable to assume that all of their extensive dendritic surfaces serve a receptive function; see also Bishop (1956, 1958), Grundfest (1957), and Bullock (1959). Furthermore, many questions of synaptic mechanism can be bypassed if one is willing to assume that the effects upon the receptive surface can be treated as excitatory and inhibitory membrane-conductance changes; this approach is elaborated in a later section.

The schematic dendritic neuron of Fig. 1G helps to illustrate some of the questions we wish to consider. Three sets of synaptic endings are indicated; these are not intended to represent all synaptic connections with this neuron; neither should (a), (b), and (c) be taken literally as single afferent fibers. Here, (a) is meant to represent a group of afferent fibers whose synaptic connections are made predominantly over the peripheral dendritic surface; (b) represents a different group of afferent fibers whose synaptic connections are made predominantly over the surface of the neuron soma and dendritic trunks. A third group of afferent fibers (c) makes synaptic connections first with the dendritic periphery, but also, at successively later times, with successively less peripheral portions of the dendritic system, and finally with the soma. Separate synaptic activity in (a), (b), or (c) provides a simplified example of difference in *spatial* pattern of synaptic input; (c) also implies a particular spatio-temporal pattern. Various sequences of activity in various combinations of (a), (b), and (c) can provide a variety of spatio-temporal patterns of synaptic input.

In order to gain insight into the possible functional significance of various spatio-temporal patterns, we pose the following theoretical problem: develop and explore mathematical models which permit the calculation and comparison of the effects to be expected at a neuron soma for such varied patterns of synaptic activity.

A mathematical model to aid the exploration of the physiological implications of dendritic branching has been developed over the past several years (Rall 1957, 1959, 1960, 1962a, b). The 1959 paper presents and discusses the essential simplifying assumptions and makes a detailed effort to determine, for important model parameters, the ranges of values which would be consistent with available anatomical and electrophysiological evidence. This study provided evidence for a significant dominance of dendritic properties over somatic properties in determining various whole-neuron properties of mammalian motoneurons. It was pointed out (Rall, 1959, p. 520) that "this leads naturally to a possible functional distinction between dendritic and somatic synaptic excitation: the larger and slower dendritic contribution would be well suited for fine adjustment of central excitatory states, while the relatively small number of somatic synaptic knobs would be well suited for rapid triggering of reflex discharge."

A limited aspect of this problem, the passive decay of nonuniform membrane depolarization, was dealt with first (Rall, 1960, pp. 515–16 and 527–29). This was useful for a discussion of problems in the interpretation of experimentally observed synaptic-potential decay. This limited problem has the mathematical advantage that the nonuniformity occurs in the initial condition of the boundary-value problem; the linear partial differential equation remains homogeneous, and its constant coefficients are not affected by the nonuniformity. However, this mathematical simplification is lost when one wants to treat the onset of nonuniform synaptic activity or to treat various durations of sustained nonuniform synaptic activity. As will be shown below, such problems involve a perturbation of the system; not only does the differential equation become nonhomogeneous, but the value of an important coefficient is perturbed from its passive-membrane value. For each region of soma-dendritic receptive surface which receives significantly different synaptic activity, the magnitude of the perturbed coefficient and of the nonhomogeneous term of the partial differential equation will be different. The problem becomes a multi-region boundary-value problem. The particular case of two regions has been presented, together with its formal solution and with numerical examples (Rall, 1962a). One example of those results is included below (in Fig. 5); however, the details of the two-region boundary-value problem will not be reproduced here.

Spatiotemporal synaptic patterns involving multiple regions can be more conveniently handled by using a compartmental model of a dendritic system. The term "compartmental model" is borrowed from the fields of chemical kinetics and radioactive-tracer kinetics. The essential simplifying assumption is that spatial nonuniformity *within* regions (compartments) is completely neglected; nonuniformity is represented only by differences *between* regions (compartments). For nerve models, this means that one explicitly replaces a distributed-parameter transmission line with a lumped-parameter transmission line, a procedure that is commonly used for both diagrammatic and analog representations of transmission lines. Mathematically, the multi-region (partial-differential-equation) boundary-value problem is replaced by a system of ordinary differential equations which are linear and of first order. The formal mathematics and also computational methods for dealing with such systems are well developed. My acquaintance with this body of knowledge has benefited greatly from discussions with John Z. Hearon and Mones Berman. Particularly relevant papers, which also include references to the literature, are Hearon (1963) and Berman, Shahn, and Weiss (1962a, b).

Membrane Model

Equivalent Circuit. For the present discussion, a small uniform patch of dendritic membrane is represented by the equivalent circuit shown in Fig. 2. This is only a slight generalization of the models proposed by Fatt and Katz (1953) and Coombs, Eccles, and Fatt (1955). The background for such models

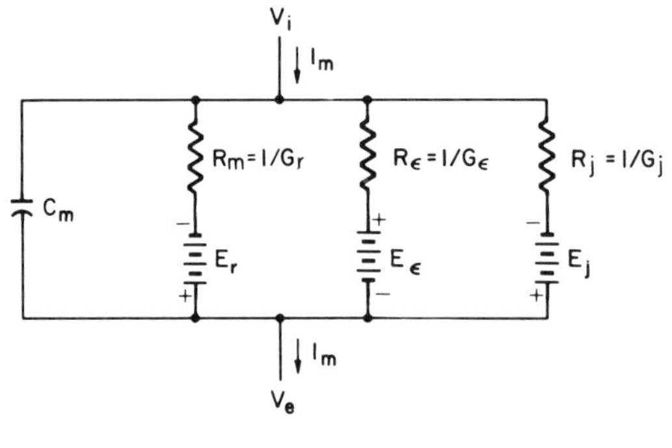

FIG. 2. Equivalent circuit for electric model of nerve membrane, where I_m is the membrane current density (i.e., current per unit area of membrane) flowing from the cell interior, at potential V_i, to the cell exterior, at potential V_e; C_m is the membrane capacitance per unit area; and G_r, G_ϵ, and G_j are separate parallel conductances per unit area, with their associated emf's designated as E_r, E_ϵ, and E_j. The subscripts r, ϵ, and j designate the conductance pathways associated with the resting membrane, with synaptic excitation, and with synaptic inhibition, respectively. Membrane potential is defined as $V_m = V_i - V_e$. The value of I_m is defined by Eq. 1 below; positive values of I_m indicate outward current.

is provided by the important papers of Hodgkin and Katz (1949), Fatt and Katz (1951), and Hodgkin and Huxley (1952); references to earlier literature can be found in these papers.

An essential feature of this model is that each of the three parallel conducting pathways contains its own series emf, which is assumed to remain constant. The resting conductance G_r is also assumed to remain constant; thus all voltage changes are treated as consequences of changes in the values of G_ϵ, G_j, and I_m. The resting state of the membrane is obtained when these three variables are all zero; then the membrane potential, defined as $V_m = V_i - V_e$, is equal simply to the resting emf E_r. When G_ϵ alone is increased from zero, this causes the value of V_m to move closer to that of E_ϵ; similarly, an increase of G_j alone causes a displacement of V_m toward E_j.

It is consistent with experimental evidence to regard E_r and E_j as around 70 to 80 mv negative (interior relative to exterior) and to regard E_ϵ as close to zero. Thus an increase in G_ϵ, assuming that V_m is initially close to E_r, makes V_m less negative; this is a positive displacement of V_m and is commonly described as depolarizing, because the absolute value of V_m is decreased. An increase in G_j, assuming that V_m is initially close to E_r and E_j, can either hyperpolarize, depolarize, or have no effect on, V_m; the direction of the change depends upon the sign of the difference, $E_j - V_m$.

Synaptic Input. Mechanisms of synaptic transmission will not be specified or discussed here. To focus attention upon spatiotemporal patterns of synaptic input, it is assumed that local changes in G_ϵ and G_j can be used as measures of local synaptic activity. Thus for a given portion of dendritic receptive surface area, ΔA, and for a given period of time, ΔT, we will think of the excitatory membrane conductance per unit area, G_ϵ, as roughly proportional to the number of synaptic excitatory events that occur over ΔA, during ΔT. Similarly, G_j shall be regarded as roughly proportional to the number of synaptic inhibitory events that occur over ΔA during ΔT. We will mention, but not elaborate on, the possible complications (1) that the elementary synaptic events might not be of equal magnitude, and (2) that a different type of inhibition could reduce G_ϵ rather than increase G_j. Here we will assume that synaptic input to a dendritic neuron can be specified as a set of G_ϵ and G_j values, each assumed to remain constant over specified ΔA for specified ΔT.

Voltage Dependence of Conductances. As presented below, the mathematical model depends upon the assumption that G_ϵ and G_j are independent of the membrane potential. This assumption confers the advantages of a differential equation having constant coefficients. The solution of the two-region boundary-value problem (Rall, 1962a) depends upon this assumption. It should be noted, however, that when one shifts emphasis from formal solutions to numerical computations with a compartmental model, it becomes practical to consider the possibility of testing the effects of mild voltage dependence. The assumption of complete voltage independence corresponds to the "electrical inexcitability" of Grundfest (1957). The assumption of a mild voltage dependence might provide a way of approximating the property of "decremental conduction" in the dendrites favored by Lorente de Nó and Condouris (1959, p. 612). Both of these cases can be distinguished from the full-blown time and voltage dependence assumed in the Hodgkin and Huxley (1952) model of the axonal nerve impulse. The approach of the present research is to explore questions of spatio-temporal pattern first with the simpler assumption (voltage-independent conductances). Later, it would be well to explore the differences obtained for various degrees of voltage dependence.

Mathematical Formulation of Membrane Model. The net current across a uniform patch of membrane is simply the algebraic sum of four component currents, one capacitive and three conductive; the density of this current (i.e., current per unit area of membrane) can be expressed as

$$(1) \qquad I_m = C_m \dot{V}_m + G_r(V_m - E_r) + G_\epsilon(V_m - E_\epsilon) + G_j(V_m - E_j),$$

where $\dot{V}_m$ represents the time derivative of membrane potential; the other symbols are all defined in Fig. 2 and its legend. To focus attention upon V_m as the dependent variable, this differential equation can be rearranged as follows:

$$(2) \quad \tau \dot{V}_m = -(1 + \mathscr{E} + \mathscr{J})(V_m - E_r) + I_m R_m + \mathscr{E}(E_\epsilon - E_r) + \mathscr{J}(E_j - E_r),$$

where the new variables $\mathscr{E}$ and $\mathscr{J}$, representing synaptic excitation and inhibition, are defined:

$$(2.1) \qquad \mathscr{E} = G_\epsilon / G_r = R_m / R_\epsilon,$$

$$(2.2) \qquad \mathscr{J} = G_j / G_r = R_m / R_j,$$

and where τ represents the membrane time constant:

$$(2.3) \qquad \tau = C_m / G_r = R_m C_m.$$

The introduction of $\mathscr{E}$ and $\mathscr{J}$ as new variables has several advantages; these quantities are dimensionless numbers which correspond to intuitively meaningful conductance ratios; they also provide a simple notation that is well suited for later compartmental assignment of subscripts.

Here we are concerned with the solution of the differential equation (2) for step changes in the values of $\mathscr{E}$, $\mathscr{J}$, and I_m. At first, I_m will be regarded as constant current applied across the patch of membrane by means of electrodes; later, I_m will include the net current supplied to one region of soma-dendritic membrane by neighboring regions or compartments. For periods of time during which $\mathscr{E}$, $\mathscr{J}$, and I_m remain constant (after a step change at time $t = 0$), the differential equation and its solution can be conveniently expressed in the reduced form

$$(3) \qquad \dot{v} = -\mu(v - v_s),$$

whose solution is

(4) $$v = v_s - (v_s - v_0)e^{-\mu t},$$

where μ, v_s, and v_0 are constants and v is a transformed variable; v_0 is the value of v at $t = 0$; and v_s is the steady-state value approached by v for large t. The following definitions complete this formulation:

(5) $$\mu = \frac{1 + \mathcal{E} + \mathcal{J}}{\tau} = \frac{G_r + G_\epsilon + G_j}{C_m},$$

(6) $$v = \frac{V_m - E_r}{E_\epsilon - E_r},$$

(7) $$v_s = \frac{\mathcal{E} + \beta\mathcal{J} + \chi}{1 + \mathcal{E} + \mathcal{J}},$$

with

(8) $$\beta = \frac{E_j - E_r}{E_\epsilon - E_r},$$

and

(9) $$\chi = \frac{I_m R_m}{E_\epsilon - E_r}.$$

This notation differs from, but remains consistent with, that of an earlier presentation (Rall, 1962a). As before, the dependent variable expresses the departure of the membrane potential from its resting value. By defining v as in Eq. (6), we normalize this variable so that $v = 1$ for $V_m = E_\epsilon$, and $v = 0$ for $V_m = E_r$; also, $v = \beta$ for $V_m = E_j$, and $\beta = -0.1$ for the particular example $E_\epsilon = 0$, $E_r = -70$ mv, and $E_j = -77$ mv.

Six examples of Eq. (4) are illustrated in Fig. 3. In all cases, the initial value, $v_0 = 0.1$, represents a preceding steady state. Curves A and D both correspond to a step increase of 0.5 in the value of $\mathcal{E}$ only, whereas curves C and F both correspond to a step increase of 0.5 in the value of $\mathcal{J}$ only (with $\beta = 0$), and curves B and E both correspond to a step increase of 0.5 in the values of both $\mathcal{E}$ and $\mathcal{J}$. In the original calculation (Rall, 1962a, Fig. 3) the initial values were $\mathcal{E} = \frac{1}{9}$, $\mathcal{J} = 0$, and $\chi = 0$ for curves A, B, and C, on the left; they were $\mathcal{E} = \frac{1}{3}$, $\mathcal{J} = 2$ (with $\beta = 0$), and $\chi = 0$ for curves D, E, and F, on the right; however, these same curves correspond equally well to other sets of initial values and to other step changes in the values of $\mathcal{E}$, $\mathcal{J}$, and χ.*

* The values of three parameters are needed to determine a particular curve from Eq. (4). All six curves in Fig. 3 have in common the value $v_0 = 0.1$; thus they differ from each other only in the values of two independent parameters. It is convenient to regard the dimensionless rate constant $\mu\tau$ (see Eq. 5 and Eq. 12.1) as one independent parameter, and the initial slope $\tau(\dot{v})_0$ (see Eq. 11) as the other. Then the values of v_s and Δv_s are not independent; they are determined by Eq. (12). A difference in the value of a single independent parameter accounts for the difference between the two curves in each of the following pairs (A, D), (B, E), (C, F), (A, C), and (D, F). One of many possible sets of conditions corresponding to curve A is given by $\mathcal{E}_0 = 0.61$, $\mathcal{J}_0 = 0$, $\chi_0 = -0.45$, with the step $\Delta\chi = 0.45$, and $\mathcal{E}$ and $\mathcal{J}$ unchanged; another set is given by $\mathcal{E}_0 = 0.61$, $\mathcal{J}_0 = 2.25$ (with $\beta = -0.1$), $\chi_0 = 0$, with the step $\Delta\mathcal{J} = -2.25$, and $\mathcal{E}$ and χ unchanged.

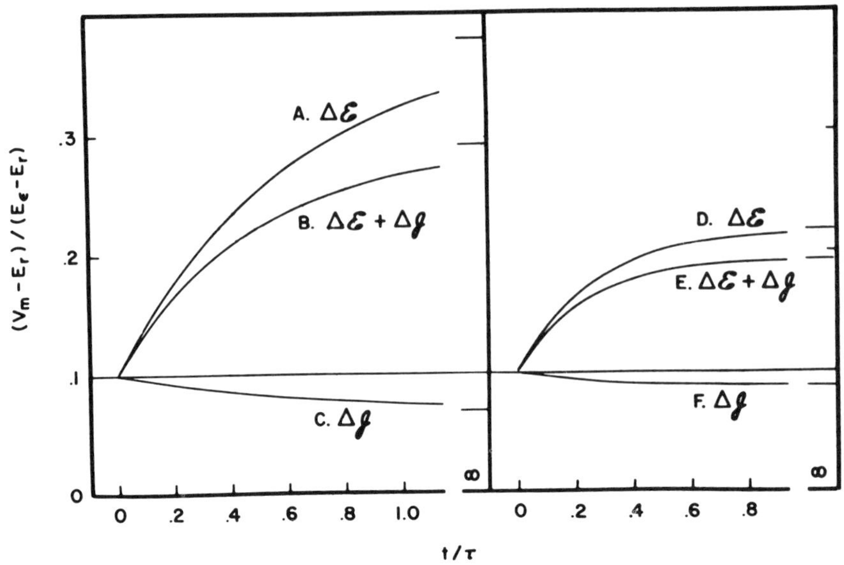

FIG. 3. Transients of v produced by a step change in the values of one or more of the variables $\mathscr{E}$, $\mathscr{J}$, and χ; see Eq. (4) and Eqs. (10)–(12). Values corresponding to each curve are listed below, where "slope" refers to the initial slope defined by Eq. (11):

Curve	$\Delta\mathscr{E}$	$\Delta\mathscr{J}$	Slope	Δv_s	$\mu\tau$
A	0.5	—	0.45	0.28	1.61
B	0.5	0.5	0.40	0.19	2.11
C	—	0.5	−0.05	−0.031	1.61
D	0.5	—	0.45	0.118	3.83
E	0.5	0.5	0.40	0.092	4.33
F	—	0.5	−0.05	−0.013	3.83

Linearity, Nonlinearity, and Related Properties. Several noteworthy properties of this membrane model are illustrated by Fig. 3; several of these are made both more explicit and more general by Eqs. (10)–(12) below. Algebraic manipulation of Eqs. (3), (5), and (7) yields for the change in slope (at $t = 0$) to be expected for any combination of step changes in the values of $\mathscr{E}$, $\mathscr{J}$, and χ at $t = 0$, the expression

$$\tau(\Delta\dot{v})_0 = (1 - v_0)\Delta\mathscr{E} - (v_0 - \beta)\Delta\mathscr{J} + \Delta\chi, \tag{10}$$

where v_0 is the value of v at $t = 0$. If we assume a preceding steady state, it follows that a zero slope existed just before $t = 0$; also, $v = v_{0s}$ at $t = 0$; then Eq. (10) simplifies to

$$\tau(\dot{v})_0 = (1 - v_{0s})\Delta\mathscr{E} - (v_{0s} - \beta)\Delta\mathscr{J} + \Delta\chi, \tag{11}$$

which expresses the initial slopes of Fig. 3. Also, it follows from Eq. (3) that the steady-state increment can be expressed as

$$\Delta v_s = v_s - v_{0s} = \tau(\dot{v})_0/\mu\tau, \tag{12}$$

where $\tau(\dot{v})_0$ is given by Eq. (11) and μ_T corresponds to constant values following the step at $t = 0$; that is,

$$\mu_T = 1 + \mathscr{E}_0 + \mathscr{J}_0 + \Delta\mathscr{E} + \Delta\mathscr{J}. \tag{12.1}$$

Linearity can be seen to hold in two restricted senses:

(i) The change of initial slope is shown by Eq. (10) to be a linear combination of the input increments, $\Delta\mathscr{E}$, $\Delta\mathscr{J}$, and $\Delta\chi$; however, the coefficients of this linear combination are different for every value of v_0. Thus in Fig. 3 curves B and E both have an initial slope equal to the sum of the separate slopes of curves A and C, or of curves D and F.

(ii) Linearity in the sense of summation of successive steps applies only to $\Delta\chi$; the contribution of $\Delta\chi$ in Eqs. (10) and (11) is independent of the value of v_0; also, the value of μ_T in Eq. (12) is left unchanged by $\Delta\chi$. Successive increments in the value of $\mathscr{E}$ and $\mathscr{J}$ do not have this property, except for the special case of simultaneous $\Delta\mathscr{E}$ and $\Delta\mathscr{J}$ with the constraint that $\Delta\mathscr{J} = -\Delta\mathscr{E}$. The fundamental requirement for linear summation of successive steps is that the value of the rate constant μ remain unchanged.

Nonlinearity is present in the following two senses.

(iii) The steady-state increment (Eq. 12) cannot be expressed as a linear combination of $\Delta\mathscr{E}$, $\Delta\mathscr{J}$, and $\Delta\chi$ because $\Delta\mathscr{E}$ and $\Delta\mathscr{J}$ also appear in the value of the denominator, μ_T. This accounts for the differences between the left and right sides of Fig. 3, in spite of the matching initial slopes.

(iv) Successive increments in the value of $\mathscr{E}$ or of $\mathscr{J}$ do not sum linearly in their effects. Successive increases in the value of $\mathscr{E}$ are decreasingly effective because the successive values of v_0 in Eq. (10) increase and because the value of μ_T in Eq. (12) increases; it follows also that successive decreases of $\mathscr{E}$ from an initially large value are increasingly effective. Successive increases in the value of $\mathscr{J}$ are decreasingly effective because the successive values of v_0 in Eq. (10) approach the value of β (i.e., V_m shifts toward E_j) and because the value of μ_T in Eq. (12) increases; it follows also that successive decreases of $\mathscr{J}$ from an initially large value are increasingly effective.

At the risk of unnecessary repetition, it is noted that these nonlinearities exist without any added complication such as voltage dependence of conductances. They result from the fact that $\Delta\mathscr{E}$ and $\Delta\mathscr{J}$ are inputs of a different kind from $\Delta\chi$. A change in χ represents a change of input current to an unperturbed system. A change in $\mathscr{E}$ or $\mathscr{J}$ is a change in a conductance which is an element of the system; the system itself is perturbed; the value of a constant coefficient in the linear differential equation is changed; hence the simple superposition rules do not hold.

Dendritic Trees

Although a compartmental model of dendritic trees proves to be more flexible for treating variety both in spatiotemporal patterns of activity and in dendritic branching pattern, it seems desirable to consider first the distributed model of dendritic branching. The latter has the advantage of more immediate appeal to intuition and thus provides a bridge from familiar concepts of anatomy and physiology to the more abstract compartmental model.

Distributed Model. A theoretical model of dendritic branching depends upon a number of simplifying assumptions. A dendritic tree is assumed to consist of cylindrical trunk and branch elements; extracellular gradients of potential are assumed to be negligible; the membrane is assumed at first to be passive and uniform throughout the dendritic tree. Also, the distribution of membrane potential along each cylindrical branch element must satisfy the same partial differential equation that is well established for passive axonal membranes (Hodgkin and Rushton, 1946; Davis and Lorente de Nó, 1947). For each cylindrical element, this equation is of the form

$$\frac{\partial V}{\partial T} = \frac{\partial^2 V}{\partial X^2} - V,$$

where $V = V_m - E_r$ represents the deviation of the membrane potential from its resting value, $T = t/\tau$ expresses the time in terms of the passive-membrane time constant, and $X = x/\lambda$ expresses axial distance in terms of the characteristic length λ, which is proportional to the square root of the cylinder diameter. We require solutions for cylinders of finite length which satisfy boundary conditions determined by the physical requirement of continuity of axial current and of membrane potential at all points of junction between branches. A systematic method for satisfying these boundary conditions under steady-state conditions for any arbitrary branching pattern has been devised (Rall, 1959).

Class of Trees Equivalent to Cylinder. The exploration of transient solutions was facilitated greatly by a consideration of the class of dendritic trees which can be shown to be mathematically equivalent to a cylinder. Figure 4 shows one example of this class. Because this particular example displays symmetric branching, each daughter branch at every bifurcation has a diameter that is 63% of its parent diameter. It should be pointed out, however, that less-symmetric trees also belong to the class under discussion. The essential requirement is that the daughter diameters at each bifurcation satisfy the following constraint: the sum of their separate diameters raised to the $\frac{3}{2}$ power is equal to the parent diameter raised to the $\frac{3}{2}$ power. The mathematical argument (Rall, 1962a) will not be reproduced here; intuitively, this constraint resembles the familiar notion of impedance-matching.

It is a particularly simple and convenient property of this class of dendritic trees that equal increments of length in the equivalent cylinder correspond to equal increments of dendritic surface area. This is indicated in Fig. 4 by the dashed lines which divide the dendritic tree into five regions of equal dendritic surface area. This property provides a convenient scale, extending from a neuron's soma to its dendritic periphery, that can be used to define and compare various soma-vs.-dendrite distributions of synaptic activity.

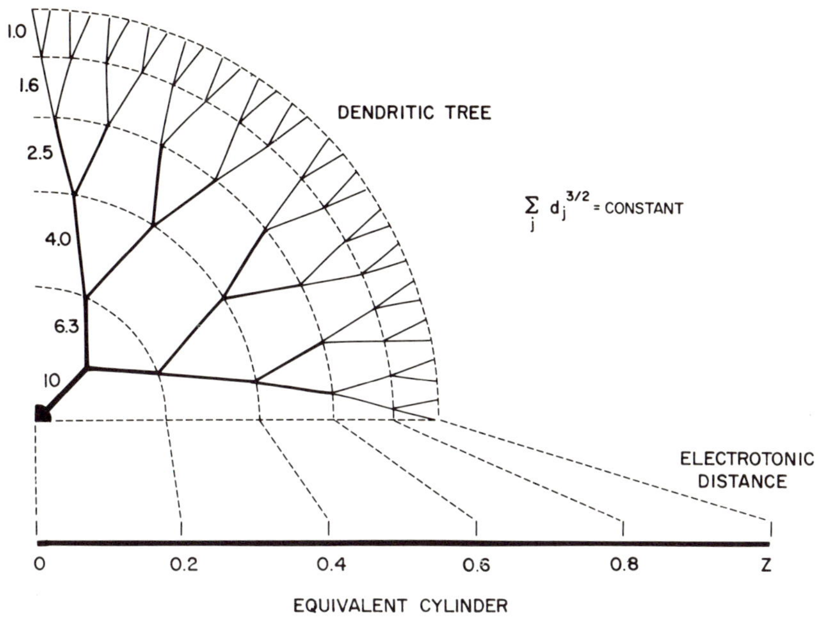

FIG. 4. A dendritic tree illustrative of a class of trees that can be mathematically transformed into an equivalent cylinder. The diagram corresponds to a specific numerical example (Rall, 1962a, Table I) of a symmetric tree with a radial extent of about 800 μ, and with cylindrical diameters decreasing from a trunk of 10 μ to peripheral branches of 1 μ. The dashed lines connect points having the same Z-value (electrotonic distance) in both tree and cylinder. Z is defined by

$$Z_2 - Z_1 = \int_{x_1}^{x_2} \frac{dx}{\lambda},$$

where x represents distance along successive cylindrical axes. λ, the characteristic length which changes with branch diameter, is defined as

$$\lambda = \sqrt{(R_m/R_i)(d/4)},$$

where R_i is the specific resistivity of the intracellular medium, d is the diameter of the cylinder, and the extracellular resistivity is assumed to be negligible.

Relation to Anatomy. The relation of such dendritic trees to anatomical evidence deserves brief comment, with respect both to the constraint upon branch diameter and to the electrotonic length of the equivalent cylinder. Inspection of the various dendritic branching patterns presented and discussed by Ramon-Moliner (1962) suggests that dendritic trees of the "tufted" type are less likely to satisfy the ($\frac{3}{2}$-power) constraint upon branch diameter than trees of the "radiate" type; however, this question has not yet been investigated in detail. Motoneurons of the spinal cord are of the radiate type; the constraint upon the $\frac{3}{2}$ power of branch diameters appears to be satisfied, to at least a rough first approximation, by some of the histological examples that have been studied (Rall, 1959; also unpublished analysis of measurements provided me by Aitken and Bridger, 1961). The same material has also been analyzed to obtain estimates of electrotonic length; when the value of Z (as defined in Fig. 4 above) is computed for all dendritic terminals, an appropriately weighted average of the values obtained for such a neuron typically lies in the range between $Z = 1$ and $Z = 2$. The order of magnitude of such estimates is important in determining that the dendritic periphery is much less functionally remote from the neuron soma than many neurophysiologists have assumed for over twenty years, from Lorente de Nó (1938a) through Eccles (1960, pp. 199–201).

Limitations of Equivalent-Cylinder Model. Although the equivalent-cylinder model has served a very useful purpose in studying dendritic trees, it has three significant limitations which can be overcome by going to a compartmental model. One limitation is the assumption of what might be called "electrotonic symmetry." Although the actual branch diameters and lengths need not be as symmetrical as the diagram in Fig. 4, all terminal branches must be assumed to end with the same Z-value (i.e., the same electrotonic distance from the soma). Furthermore, spatiotemporal disturbances in the equivalent cylinder can represent only those dendritic tree disturbances which can be assumed to be the same in all parts of the tree corresponding to any given Z-value. This assumption is readily fulfilled when a disturbance is initiated at the soma and spreads into a passive dendritic tree (e.g., Rall, 1960), or when synaptic activity is assumed to be distributed uniformly over all dendritic branches corresponding to a given range of Z-values (e.g., Rall, 1962a, and also Figs. 5, 6, and 7 below). However, this symmetry requirement is not satisfied by patterns of synaptic activity which feature significant differences between two or more branches that occupy the same range of Z-values.

An obvious limitation is the constraint upon the $\frac{3}{2}$ power of branch diameter. Some exceptions to this limitation can be represented by the more general class of dendritic trees that are equivalent to an exponential taper (see treatment of taper in Rall, 1962a); in some cases it may be useful to join two segments of different taper. However, as soon as more than two regions are required to express either complexity of taper or complexity of synaptic pattern, the rigorous solution of the boundary-value problem becomes even more complicated than in the two-region case. For practical purposes, this can be regarded as a third limitation: when more than two regions of cylinder or taper are required, it is simpler to go to a compartmental model.

Compartmental Model. A lumped-parameter transmission line is commonly used to introduce and develop the theory for the distributed-parameter case; successive subdivision of the lumps leads, in the limit, to the distributed case. Here, however, we retain the lumped-parameter approximation; each lump of membrane becomes a compartment; the rate constants governing exchange between compartments are proportional to the series conductance between them. A single compartment may correspond to a single dendritic branch, a group of branches, or just a segment of a trunk or branch element, according to the needs of a given problem. For problems compatible with the equivalent-cylinder constraints, the compartmental model is simply a straight chain of compartments of equal size; this case permits a comparison of predictions made with both kinds of models. Branching chains of compartments of unequal size can be used to approximate an unlimited variety of dendritic branching systems and of synaptic activity patterns.

The mathematical formulation of such a compartmental model consists of a system of ordinary differential equations which are linear and of first order. The ith compartment satisfies the equation

$$\dot{v}_i = \sum_j \mu_{ij} v_j + f_i, \tag{13}$$

where

$$f_i = (\mathscr{E}_i + \beta \mathscr{I}_i + \chi_i)/\tau, \tag{14}$$

and each of the coefficients μ_{ij}, represents a rate constant. When subscript j differs from subscript i, the value of the coefficient is always nonnegative, and the order of the subscripts has the meaning: from the jth compartment into the ith compartment. When subscript j equals subscript i, the coefficient μ_{ii} always has a negative value which represents the total rate of loss from the ith compartment. The values of these μ_{ii} and μ_{ij} are given by the expressions,

$$\mu_{ii} = -(1 + \mathscr{E}_i + \mathscr{I}_i)/\tau - \sum_{j \neq i} \mu_{ij}, \tag{15}$$

and

$$\mu_{ij} = g_{ij}/c_i = g_{ji}/c_i \quad \text{for} \quad i \neq j, \tag{16}$$

where c_i represents the capacitance of the ith compartment, and $g_{ij} = g_{ji}$ represents the conductance between adjacent compartments. Where i and j do not represent directly connected compartments, g_{ij} is zero. It can be seen that the net current flow from the jth compartment to the ith compartment is given by the difference

$$c_i \mu_{ij} v_j - c_j \mu_{ji} v_i = g_{ij}(v_j - v_i).$$

For the special case of a straight chain of compartments each of which corresponds to an equal increment ΔZ of an equivalent cylinder, it can be shown that the result

$$\mu_{ij} = \mu_{ji} = \tau^{-1}(\Delta Z)^{-2} \tag{17}$$

corresponds to treating all the c_i as equal to the capacitance of a ΔZ-length of equivalent cylindrical membrane and all the nonzero g_{ij} as equal to the core conductance of a ΔZ-length of equivalent cylinder. Furthermore, this result leads, in the limit, as ΔZ is made successively smaller, to the correct partial differential equation for the equivalent cylinder.

The symbols $\mathscr{E}$, $\mathscr{I}$, β, and τ have the same significance as earlier [see Eqs. (2)–(9)]; however, χ is here restricted to current that might be supplied to a compartment by means of electrodes. Subscripting of $\mathscr{E}$ and $\mathscr{I}$ is essential for synaptic patterns; β and τ could also be given subscripts, but they will be assumed, for present purposes, to be the same for all compartments.

Some Properties of Linear Compartmental Systems. For any given set of values of $\mathscr{E}_i$, $\mathscr{I}_i$, and χ_i, the linear system defined by Eqs. (13)–(16) is of the type discussed by Hearon (1963). The square matrix M, composed of the coefficients μ_{ij}, has two important properties discussed by Hearon. It has a strictly dominant diagonal composed of μ_{ii} that are all negative, and it has nonnegative off-diagonal elements whose values are so constrained by the connectivity of the system that the matrix is either symmetric to begin with (special case of Eq. 17), or symmetrizable because $c_i\mu_{ij} = c_j\mu_{ji}$. These properties of the matrix are sufficient to guarantee a number of important properties of the solutions of such a linear system. The roots of the matrix are all real and negative; this ensures that the solutions are bounded and nonoscillatory, and that the homogeneous problem (all $f_i = 0$) has no nonzero steady-state solution. For the nonhomogeneous problem, the steady-state solution is a vector which can be defined as $v(\infty) = -M^{-1}f$, where M^{-1} represents the inverse of M, and f is a constant vector composed of the f_i in Eqs. (13) and (14). As long as M and f remain unchanged, the transient behavior of the system, in passing from any given initial state toward its steady state, can be expressed as a linear combination of (decaying) exponentials, exp $(-m_i t)$, where $-m_i$ represents distinct roots of the matrix M. See Hearon (1963) for a discussion of multiple roots, separation of roots by the μ_{ii}, initial conditions permitting overshoot, and other related questions.

Whenever there is a step change in the value of $\mathscr{E}_i$ or $\mathscr{I}_i$, this has two effects upon Eq. (13). First, it perturbs the system by changing the diagonal element, μ_{ii}; this changes the matrix M and the values of its roots. Second, it changes the value of f_i; this, together with the change in M, causes a change in the steady state toward which the system now tends. In contrast, a change in only the value of χ_i would change f_i without changing the matrix M; in this case, the system itself is not perturbed.

The superposition properties of linear systems apply only to changes in f_i which are not associated with any change in the matrix M. When M remains unchanged, the system can be called a linear time-invariant system, or a constant-parameter linear system; see, for example, Mason and Zimmerman (1960, pp. 318–21). The point here is that we do not have such a system when changes occur in $\mathscr{E}_i$ or $\mathscr{I}_i$, except for the very special case where μ_{ii} remains constant because of equal and opposite changes in $\mathscr{E}_i$ and $\mathscr{I}_i$.

Method of Solution. For any given initial condition, with a constant f vector and a constant matrix M, it is possible to completely define the required transient solution. Numerical computation requires inversion of M and solution of its roots; standard procedures are available for this. However, for spatio-temporal patterns of synaptic activity which involve repeated stepwise perturbations of the system, a new transient solution with a new initial condition, a new constant f vector, and a new constant matrix M is required for each perturbation. An alternative numerical procedure replaces the set of differential equations with a corresponding set of difference equations. Beginning with the initial set of v_i, the change δv_i in each compartment is obtained from the difference equation for a short time increment, δt. As the computation progresses through successive δt's, the perturbations (changes in μ_{ii} and f_i) can be introduced at the times required by the synaptic-activity pattern being assumed. The results presented below were obtained by a numerical procedure of this kind. Use was made of a general computer program that has been developed by Berman, Shahn, and Weiss (1962a, b)* for a large variety of problems related to compartmental systems, as well as other systems; this program employs a fourth-order Runge-Kutta procedure for improving the accuracy of numerical solutions obtained for any specified time increment.

Results of Computations

Central vs. Peripheral Half of Tree. Consider a dendritic tree that is equivalent to a cylinder whose length equals one characteristic length, as in Fig. 4. We wish to compare the effect at the soma ($Z = 0$) of synaptic excitation (G_ϵ) delivered only to the central half of the receptive surface (i.e., $Z = 0$ to $Z = 0.5$) with the effect of the same amount of synaptic excitation delivered only to the peripheral half (i.e., $Z = 0.5$ to 1.0). Curve A in Fig. 5 shows the effect of a step increase (from $\mathscr{E} = 0$ to $\mathscr{E} = 2$ at $t = 0$) confined to the central half, whereas curve C shows the effect of an equivalent step increase confined to the peripheral half; curve B corresponds to a uniform step ($\mathscr{E} = 1$) over the entire receptive surface. These curves were originally computed as solutions of a two-region boundary-value problem in a distributed-parameter system (Rall, 1962a). More recently, they were recomputed for a chain of ten equal

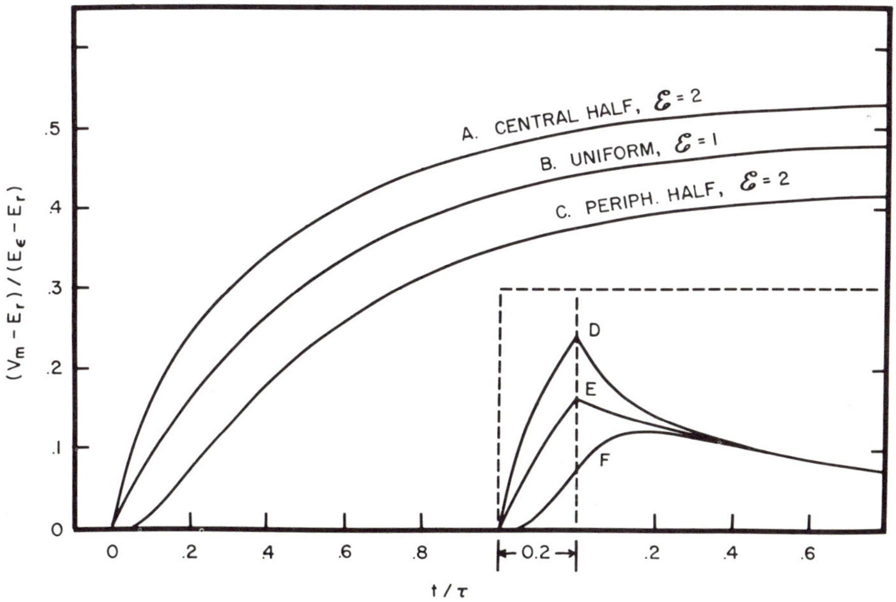

FIG. 5. Computed transients of soma membrane depolarization for step changes of G_ϵ; central vs. peripheral half of dendritic tree; see text. Curves A and D are for $\mathscr{E} = 2$ over central half of receptive surface. Curves B and E are for $\mathscr{E} = 1$ over entire surface. Curves C and F are for $\mathscr{E} = 2$ over peripheral half of surface. Curves A, B and C are for an on-step; curves D, E, and F are for a square pulse (on-step followed by off-step).

* I wish to express my appreciation to Dr. Berman and Mrs. Weiss for their advice and help in the task of achieving compatibility between their program and the idiosyncrasies of my problems. The computations were done on the IBM 7090 at the National Bureau of Standards.

compartments, using Eqs. (13)–(15) and (17). Agreement was within 2%, indicating that errors due to compartmental lumping were not serious.

The early portions of curves D, E, and F are the same as those of curves A, B and C, because they are produced by the same on-step; after a time interval, 0.2τ, each on-step is followed by an off-step. Thus curves D, E, and F can be regarded as theoretical (excitatory) synaptic potentials (epsp) for the artificial assumption of a square $\mathscr{E}$ pulse. It can be seen that curve F rises more slowly to a later peak with an amplitude half that of curve D. Because the threshold for motoneuron reflex discharge has been found (experimentally) to correspond to soma-membrane depolarizations in the range between these peak amplitudes, it is clear that the differences between curves D, E, and F could be responsible for success or failure in the initiation of a propagated impulse.

In addition to the differences that have been noted for the effects of central vs. peripheral $\mathscr{E}$ steps and pulses, it is important to note that the asymptotic value of curve C is only about 20% less than that of curve A. This means that a sustained barrage of peripheral dendritic synaptic activity can have a very significant effect on the level of soma-membrane depolarization.*

Comparison of Four $\mathscr{E}$-Pulse Locations. To demonstrate the importance of specifying the spatiotemporal patterns of synaptic input, a specific numerical example will be presented in two stages. The first stage compares four different cases in which the same magnitude of excitatory conductance pulse is delivered to four different portions of the dendritic receptive surface of a neuron. A comparison of the transient soma-membrane depolarizations calculated for these four cases provides a more detailed demonstration of the effect of central vs. peripheral dendritic location than was demonstrated above for two regions. The second stage makes use of these same four input locations to compare two different spatiotemporal sequences of $\mathscr{E}$ pulses.

Computations were carried out for a straight chain of ten equal compartments ($\Delta Z = 0.2$ per compartment). Compartment No. 1 is regarded as the soma, and compartments 2, 3, $\cdots$, 10 are regarded as equal increments of dendritic surface area, arranged in sequence from dendritic trunks to dendritic periphery; see the diagram in the upper-right corner of Fig. 6.

All four of these curves display transient depolarization of the soma membrane. The magnitude and time of occurrence of the $\mathscr{E}$ pulse is the same in each case; perturbations A, B, C, and D differ only with respect to $\mathscr{E}$-pulse location, as indicated in the figure. Curve A displays the largest and earliest peak depolarization; it results from an $\mathscr{E}$ pulse in compartments 2 and 3, a dendritic location which would correspond to the trunks and portions of primary branches. Cases B, C, and D result from progressively more peripheral $\mathscr{E}$-pulse locations; these are intended to correspond to progressively more

* The curves in Fig. 5 correspond to an over-all electrotonic length equal to one characteristic length (i.e., the Z-values extend from 0 to 1.0). For the same geometry, a reduction in the estimated value of R_m/R_i by a factor of four would reduce the estimated values of λ by a factor of two; all Z-values would be doubled. Recalculation of these curves on this assumption would increase the steady-state asymptote of curve A from a value of 0.54 to a value of 0.65 and would decrease the asymptote of curve C from 0.43 to 0.33; also, the peak of D would increase from 0.24 to about 0.28, and the peak of F would decrease from about 0.12 to about 0.07; the decaying portions of D and F would come together (within 10%) only at about $t = \tau$

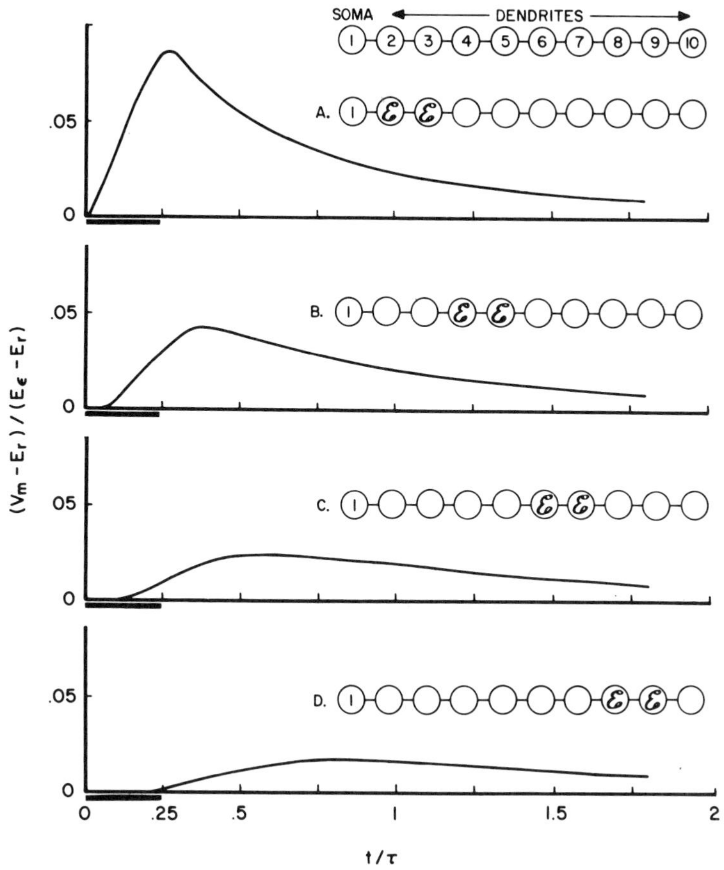

FIG. 6. Effect of $\mathscr{E}$-pulse location upon transient soma-membrane depolarization. The soma-dendritic receptive surface is divided into ten equal compartments ($\Delta Z = 0.2$) as shown, upper right. The four $\mathscr{E}$-pulse locations are shown in the diagrams beside the letters A, B, C, and D; in each case, $\mathscr{E} = 1$ in two compartments for the time interval indicated by the heavy black bars. These curves were drawn through computed values for time steps of 0.05τ. To convert to real time, a τ-value between 4 and 5 msec is appropriate for mammalian motoneurons.

peripheral distributions of restricted synaptic activity delivered to the dendritic receptive surface of a neuron. As a consequence of these locations, curves B, C, and D display progressively lower and later peaks;* these curves also display a progressive delay in their initial rise.

Each of these curves represents a theoretical synaptic potential (epsp) for the artificial assumption of a square $\mathscr{E}$ pulse. It does not seem likely that the differences between such synaptic potentials would be unimportant for neuronal input-output relations.

---

* The peak height of curve A is 0.085 at about $t = 0.25\tau$. For curve B, this is 0.042 at about $t = 0.4\tau$. Curve C has a wide peak of 0.023 centered at about 0.6τ. Curve D has a wide peak of about 0.017 centered at about $t = 0.8\tau$. At times after $t = 1.5\tau$, the decay of all four curves is approximately the same (within 10%); this is because the depolarization spreads and tends to become uniform over the soma dendritic surface as it decays.

Two Spatiotemporal Patterns. Consider next a sequence of four successive $\mathscr{E}$ pulses where each individual $\mathscr{E}$ pulse has the same magnitude and duration as in Fig. 6. What kind of difference should be expected at the soma for the following two sequences of $\mathscr{E}$-pulse locations: sequence $A \to B \to C \to D$ (trunks first, dendritic periphery last) and $D \to C \to B \to A$ (dendritic periphery first, trunks last), where these letters refer to the four input locations (not the voltage transients) of Fig. 6? The computed transient soma-membrane depolarizations for these two cases are shown in Fig. 7. Curve $D \to C \to B \to A$ has a peak value of 0.152 which occurs close to $t = \tau$. Curve $A \to B \to C \to D$ has two peaks of 0.085, one near $t = 0.25\tau$ and a second near $t = 0.55\tau$. The decay of the curves is approximately the same (within 10%) after $t = 2\tau$.

It should be emphasized that these curves were not obtained by displacing and adding curves A, B, C, and D of Fig. 6; they represent the results of separate computations for the designated sequences of $\mathscr{E}$ pulses. It is clear that the effect at the soma of these two sequences is quite different. The $\mathscr{E}$-pulse sequence $A \to B \to C \to D$ achieves no significant increase of peak soma response over that achieved by $\mathscr{E}$ pulse A alone. In contrast, the $\mathscr{E}$-pulse sequence $D \to C \to B \to A$ achieves significant summation by almost doubling the peak soma response. Both sequences produce results that differ significantly

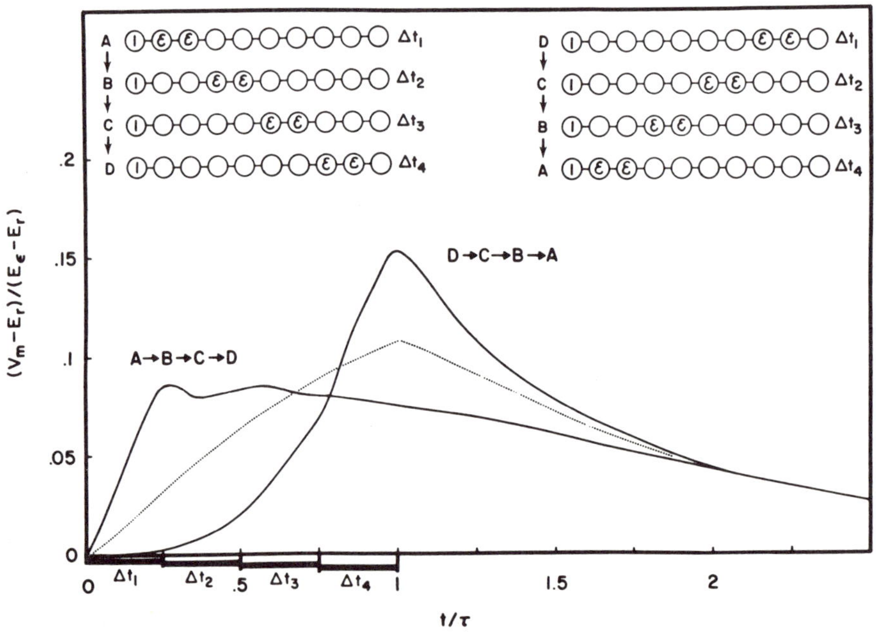

FIG. 7. Effect of two spatiotemporal sequences upon transient soma-membrane depolarization. Two $\mathscr{E}$-pulse sequences, $A \to B \to C \to D$ and $D \to C \to B \to A$, are indicated by the diagrams at upper left and upper right; the component $\mathscr{E}$-pulse locations are the same as in Fig. 6. The time sequence is indicated by means of the four successive time intervals, Δt_1, Δt_2, Δt_3, and Δt_4, each equal to 0.25 τ. The magnitude, $\mathscr{E} = 1$, in two compartments is the same as in Fig. 6. The dotted curve shows the computed effect of $\mathscr{E} = 0.25$ in eight compartments (2 through 9) for the period $t = 0$ to $t = \tau$.

from the dotted curve, obtained when the spatiotemporal pattern is removed by averaging $\mathscr{E}$ over the eight compartments and four time intervals. Three conclusions seem warranted.

(i) A temporal sequence of $\mathscr{E}$ pulses does not result in a unique output unless the dendritic locations of these $\mathscr{E}$ pulses are specified.

(ii) For maximal peak of the soma-membrane depolarization, the peripheral dendritic $\mathscr{E}$ pulses should be delivered earlier than the proximal dendritic $\mathscr{E}$ pulses.

(iii) For rapid achievement and maintainance of a steady soma-depolarization level, a proximal dendritic $\mathscr{E}$ pulse followed by a prolonged sequence of peripheral dendritic $\mathscr{E}$ pulses would be very effective.

Effect of Inhibitory-Conductance Location. Consider the presence of a sustained inhibitory conductance during the occurrence of an $\mathscr{E}$ pulse. What difference in transient soma-membrane depolarization should one expect when the inhibitory perturbation is located peripherally or proximally in relation to $\mathscr{E}$-pulse location? The results of a relevant computation are presented in Fig. 8. For A the $\mathscr{I}$ location (compartments 9 and 10) is peripheral to the $\mathscr{E}$ pulse. For B the $\mathscr{I}$ location (compartments 5 and 6) is the same as the $\mathscr{E}$-pulse location. For C the $\mathscr{I}$ location (compartments 1 and 2) includes the soma and is proximal to the $\mathscr{E}$ pulse. The peak of curve B is 93% of the control peak, while

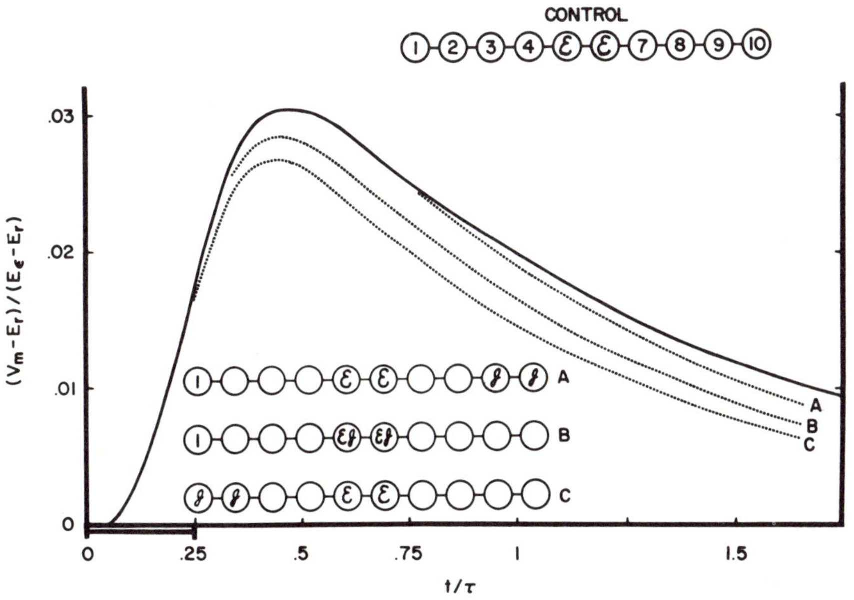

FIG. 8. Effect of inhibitory-conductance location upon transient soma-membrane depolarization. The continuous curve represents the uninhibited "control" transient in compartment 1 (i.e., the soma) for an $\mathscr{E}$ pulse in compartments 5 and 6, as shown in the diagram at upper right; $\mathscr{E} = 1$ for $\Delta t = 0.25\ \tau$, as in the two preceding figures. The dotted curves show the inhibitory effect of having $\mathscr{I} = 1$ in two compartments throughout the time shown; it was assumed for these particular computations that $E_j = E_r$. The $\mathscr{I}$ locations corresponding to curves A, B, and C are indicated by the three diagrams in the lower part of the figure.

that of curve C is about 88% of control. In another calculation with $\mathscr{I} = 10$, the peak corresponding to B dropped to 57% of control, while the peak corresponding to C dropped to 40% of control; even with $\mathscr{I} = 10$, the peak corresponding to A was 99% of control. Also, for $\mathscr{I} = 1$ in compartments 7 and 8, the computed peak was 99% of control. These results suggest a rather simple generalization. When the $\mathscr{I}$ location is peripheral to the $\mathscr{E}$-pulse location, the peak of the transient soma-membrane depolarization is not significantly reduced from the uninhibited control level; however, the declining phase of the soma transient is more rapid. When the $\mathscr{I}$ location is identical with, or proximal to, the $\mathscr{E}$-pulse location, the transient soma peak is reduced from its control level; this reduction is greatest when the $\mathscr{I}$ location includes the soma.

Some qualifications should be added immediately. These results were computed for a straight chain; other computations with branching systems have revealed situations in which a $\mathscr{I}$ location identical with a peripheral $\mathscr{E}$ location can produce more effective inhibition than an equal amount of sustained $\mathscr{I}$ at the soma. Such situations, and also the effect of E_j different from E_r, require further investigation. Furthermore, when $\mathscr{E}$ and $\mathscr{I}$ perturbations are both sustained, the steady-state results differ from those noted above for the peak of a brief transient; the effects due to peripheral $\mathscr{I}$ locations are less discriminated against. On the other hand, when $\mathscr{E}$ and $\mathscr{I}$ perturbations are both brief, their timing becomes very important and peripheral locations become less effective. Proximal $\mathscr{I}$ pulses are most effective when they are timed to center upon the peak depolarization at the $\mathscr{I}$ location; this means that for optimal effect a $\mathscr{I}$ pulse located as in C of Fig. 8 should be timed later than a $\mathscr{I}$ pulse located as in B of Fig. 8. Thus it should be clear that a strategically placed (spatiotemporally) $\mathscr{I}$ pulse can have a potent inhibitory effect.

Spatial Summation of Events in Peripheral Divisions of Tree. Here we present only one set of results obtained for a branching compartmental system. Computations were carried out for the nine-compartment system shown in Fig. 9; compartment 8 was taken to be the neuron soma, while compartments 1, 2, 3, and 4 represent four divisions of the dendritic periphery; transient membrane depolarization in compartment 8 was computed for various combinations of events in the four peripheral compartments. Curve A was computed for an instantaneous depolarization, $v = 0.5$, in any one of the four peripheral compartments; curve B was computed for a simultaneous pair of such events; curve C was computed for four simultaneous events of this kind. Except for inaccuracies in drawing these curves, the ordinates of B are twice those of A at all times; also, the ordinates of C are four times those of A at all times. Because the system is not perturbed, curves A, B, and C can be obtained equally well for other distributions of equal amounts of instantaneous depolarization in the peripheral compartments; for example, curve B would also result from $v = 1$ at $t = 0$ in any one peripheral compartment, or from some combination, such as $v = 0.1$, 0.2, 0.3, and 0.4, in the four peripheral compartments. In contrast, curves D–H on the right all involve different perturbations of the system, but all are computed for $\Delta t = 0.25\tau$. Curve D was computed for an $\mathscr{E}$ pulse

($\mathscr{E} = 2$) in any one of the four peripheral compartments. Curve E was computed for a simultaneous pair of such $\mathscr{E}$ pulses in compartments 1 and 2; the peak amplitude of E is not twice, but 1.94 times, that of D; symmetry requires the same result if compartments 3 and 4 are substituted for 1 and 2; however, for compartments 1 and 4 (or 2 and 3, or 1 and 3, or 2 and 4) a slightly different curve (not shown) is obtained with a peak amplitude that is 1.99 times that of D. If, on the other hand, the pair of $\mathscr{E}$ pulses is combined in a single peripheral compartment (i.e., $\mathscr{E} = 4$) the summation is poorer, as shown by the dotted curve, G, whose peak is 1.83 times that of D. Curve F was computed for simultaneous $\mathscr{E}$ pulses (each $\mathscr{E} = 2$) in all four peripheral compartments; the peak amplitude of F is 3.84 times that of D; in contrast, when these four $\mathscr{E}$ pulses are combined in a single peripheral compartment (i.e. $\mathscr{E} = 8$), the summation is still poorer, as shown by the dotted curve H, whose peak is only 3.12 times that of D. In these calculations all compartments were assumed to be of the same size, corresponding to $\Delta Z = 0.2$, and all μ_{ij} were taken to be equal.

Once one has accepted the fact that linearity of summation of effects should not be expected for perturbations of the system, it is interesting to reflect on the significance of the results shown at the right in Fig. 9. These results show that, although the departure from linearity can become quite large when perturbations are superimposed upon the same compartment, the departure from linearity can be surprisingly small when brief perturbations occur in separate portions of the dendritic periphery. The intuitive key to this phenomenon is the notion that separate location reduces the interference that each perturbation introduces into the effect from the other. Just as an $\mathscr{E}$ pulse in one peripheral compartment sums better (because it interferes less) with an $\mathscr{E}$ pulse in a different peripheral compartment than in the same compartment, so, also, a $\mathscr{I}$ pulse in one peripheral compartment is less effective in inhibiting the transient

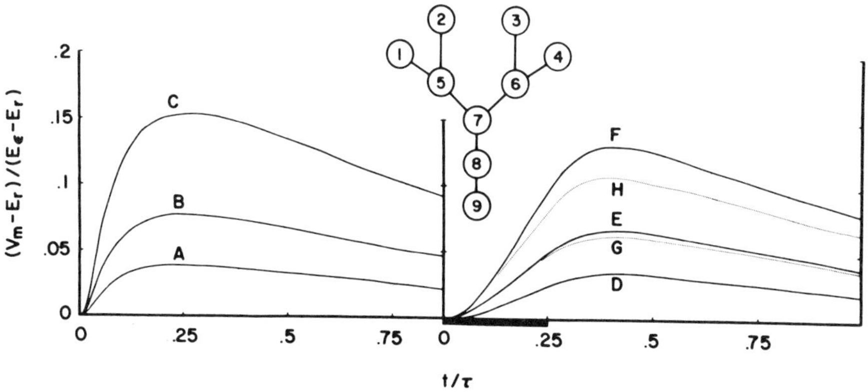

FIG. 9. Effect of peripheral dendritic summation upon transient soma-membrane depolarization. The left side of this figure illustrates linear summation for events which can be regarded either as initial conditions or as nearly instantaneous (at $t = 0$) depolarizations produced by very brief and very large applied currents. The right side of this figure illustrates various degrees of non-linearity in the summation of simultaneous $\mathscr{E}$ pulses having a duration $\Delta t = 0.25\tau$, as shown by the black bar.

due to an $\mathscr{E}$ pulse in a different peripheral compartment than in inhibiting the transient due to an $\mathscr{E}$ pulse in the same compartment.

Discussion

Although more computations will be required for exploring answers to a variety of further questions, the specific results presented and discussed above (Figs. 5–9) support the notion that spatial pattern and spatiotemporal pattern of synaptic activity must be taken into account by any comprehensive theory of neuronal input-output relations.

Dendritic Synaptic Functions. Dendritic branching provides a very large surface area which can receive synapses from many different afferent sources. Although individual brief events in the dendritic periphery suffer significant attenuation in the spread of their effects to the soma, the very large number of dendritic synapses (of the order 10^3 to 10^4) permits a ceaseless spatiotemporal spattering of synaptic events, the integration of which can provide a finely graded background level of depolarization (a biasing of neuronal excitatory state). The sluggish transient characteristic (compare curve D of Fig. 6 with curve A) provides for a smoothing of the temporal input pattern delivered to the dendritic periphery. An approximation to linear summation of $\mathscr{E}$-pulse effects is provided by separated peripheral dendritic locations (see Fig. 9); however, linear summation does not occur with common locations, whether central or peripheral. The effect of $\mathscr{I}$ location has been discussed above with Fig. 8.

Trigger Zone, Convergence Zone, and Soma. It is important to distinguish the soma from the concepts of the trigger zone and of the convergence zone, although they are often (at least implicitly) identified with each other, for convenience. In the case of motoneurons, the soma does represent the zone of dendritic convergence, and it is in this sense that the word *soma* has been used in this paper. The present theoretical results can be applied to other situations, such as receptor branches or the dendritic arborization from an apical dendrite, where the functionally important zone of convergence may be far from the soma. The situation is conceptually simplest when the trigger zone (for initiation of an impulse) is either at or near the convergence zone. This is generally assumed to be the case for motoneurons, except perhaps in those cases where the axon arises from a dendrite at some considerable distance from the soma. The possibility of two trigger zones, one for each of two different dendritic convergence zones, has been suggested for pyramidal cells (see Eccles, 1957, p. 226); particularly interesting to me is the suggestion that one of these (the more peripheral one) might serve a "booster" rather than an impulse-triggerng function (Spencer and Kandel, 1961). The importance of graded integration of synaptic activity in the dendrites, as contrasted with impulse generation and propagation in axons, has been increasingly emphasized in recent years (Bishop, 1956, 1958; Bullock, 1959; Grundfest, 1957).

Central vs. Peripheral Dendritic Function. For a neuron with a single trigger zone that can be identified with the soma and the zone of dendritic convergence, the theoretical results can be seen to suggest a general functional distinction between dendritic and somatic synaptic activity. In contrast with the relatively sluggish and finely graded depolarization that would spread to the trigger zone from the dendritic periphery, brief synchronous activity of a few synapses at or near the soma could produce precisely timed and relatively sharp depolarizations. These would be well suited to precisely timed triggering of axonal impulses; their amplitudes might well be too small to reach threshold without the help of earlier dendritic synaptic activity. Thus the dendritic activity could determine success or failure in the onward transmission of temporal synaptic patterns delivered at the soma. A different mode of behavior would result when sustained dendritic synaptic activity was sufficient to cause a rhythmic discharge of impulses (similar to the action of a generator potential in a sensory neuron); in such a case, synaptic inhibition at the soma could block the discharge of impulses over specific periods, or it could reduce the frequency of rhythmic discharge. An experimental example of this last effect is provided by the observations of Kuffler and Eyzaguirre (1955) on stretch receptor cells of crayfish. The differences in epsp time course presented in Figs. 5 and 6 are roughly similar to the differences reported by Fadiga and Brookhart (1960) for spinal motoneurons of the frog.

Implications for Theories of Nerve Networks. The incorporation of these results into a theory of nerve networks seems to lead to something more complicated than is usually assumed. In the case of the "neuronic equations" of Caianiello (1961, and this volume), the delayed-coupling coefficients (for h different from k, and r greater than zero) could be used to approximate the differences displayed here in Fig. 6; however, the assumed linear combination of inputs neglects the nonlinearities discussed above with Eqs. (10)–(12). A complete incorporation of the compartmental model would lead to four sets of equations: two sets would define all of μ_{ii} and f_i as linear combinations of input from all sources; another set would consist of systems of differential or difference equations, one for each compartmental neuron; and the fourth set would test the difference between depolarization and threshold at the trigger zone of each neuron.

Conduction Safety Factor with Branching. A question was asked regarding the electrotonic equivalence between a cylinder and a class of dendritic trees (see Fig. 4); the questioner felt, especially from the point of view of impulse propagation, that successive branching must, with the accompanying reductions of diameter, lead to a block of conduction; would this not conflict with the concept of electrotonic equivalence? The answer is that small branch diameter should not be expected to cause conduction block, when the $\frac{3}{2}$-power-of-diameter constraint (see Fig. 4, and Rall, 1962a) is satisfied at every branch point; the partial differential equation is the same for cylinder and tree, provided that distance is expressed as electrotonic distance, Z. For conditions of constant impulse-propagation velocity in a cylinder (Hodgkin and Huxley, 1952), one should expect the corresponding velocity in individual branches to decrease as the square root of branch diameter; (see Rall, 1962a, p. 1082). Furthermore, if branch diameters are smaller than required by the $\frac{3}{2}$-power constraint, the safety factor for impulse propagation into the small branches will increase, not decrease. A decrease of safety factor and an increased likelihood of conduction block would be expected when branch diameters exceeded the values required by the $\frac{3}{2}$-power constraint. For such cells (e.g., the Purkinje cell of Fig. 1A, and at least some of the "tufted" cells of Ramon-Moliner, 1962), this geometric polarity might account for a failure of impulses to invade dendritic trees, even if the membrane were electrically excitable; such cells also provide for powerful convergence from their dendritic periphery whether the spread is passive or by means of impulses. Such safety-factor considerations may be important at branched receptor terminals and at branched axonal terminals.

References

Aitken, J. T., 1955. Observations on the larger anterior horn cells in the lumbar region of the cat's spinal cord, *J. Anat. (London)*, **89,** 571.

Aitken, J. T., and J. E. Bridger, 1961. Neuron size and neuron population density in the lumbosacral region of the cat's spinal cord, *J. Anat. (London)*, **95,** 38–53.

Berman, M., E. Shahn, and M. F. Weiss, 1962a. The routine fitting of kinetic data to models: a mathematical formalism for digital computers, *Biophys. J.*, **2,** 275–87.

Berman, M., M. F. Weiss, and E. Shahn, 1962b. Some formal approaches to the analysis of kinetic data in terms of a linear compartmental system, *Biophys. J.*, **2,** 289–316.

Bishop, G. H., 1956. Natural history of the nerve impulse, *Physiol. Rev.*, **36,** 376–99.

Bishop, G. H., 1958. The dendrite: receptive pole of the neurone, *EEG Clin. Neuorophysiol.*, Suppl. No. 10, 12–21.

Bullock, T. H., 1959. Neuron doctrine and electrophysiology, *Science*, **129,** 997–1002.

Caianiello, E. R., 1961. Outline of a theory of thought processes and thinking machines, *J. Theoret. Biol.*, **1,** 204–35.

Coombs, J. S., J. C. Eccles, and P. Fatt, 1955. The inhibitory suppression of reflex discharges from motoneurones, *J. Physiol.*, **130,** 396–413.

Davis, L., Jr., and R. Lorente de Nó, 1947. Contribution to the mathematical theory of the electrotonus, *Studies from the Rockefeller Institute for Medical Research*, vol. 131, 442–96.

Eccles, J. C., 1957. *The Physiology of Nerve Cells*, Baltimore, Md.: Johns Hopkins Press.

Eccles, J. C., 1960. The properties of the dendrites, in *Structure and Function of the Cerebral Cortex*, D. B. Tower and J. P. Schadé, eds., Amsterdam: Elsevier.

Fadiga, E., and J. M. Brookhart, 1960. Monosynaptic activation of different portions of the motor neuron membrane, *Am. J. Physiol.*, **198,** 693–703.

Fatt, P., and B. Katz, 1951. An analysis of the end-plate potential recorded with an intracellular electrode, *J. Physiol.*, **115,** 320–70.

Fatt, P., and B. Katz, 1953. The effect of inhibitory nerve impulses on a crustacean muscle fibre, *J. Physiol.*, **121,** 374–89.

Fox, C. A., and J. W. Barnard, 1957. A quantitative study of the Purkinje cell dendritic branchlets and their relationship to afferent fibres, *J. Anat.*, London, **91,** 299–313.

Grundfest, H., 1957. Electrical inexcitability of synapses and some consequences in the central nervous system, *Physiol. Rev.*, **37,** 337–61.

Hearon, J. Z., 1963. Theorems on linear systems, *Ann. N.Y. Acad. Sci.*, **108,** 36–68.

Hodgkin, A. L., and A. F. Huxley, 1952. A quantitative description of membrane current and its application to conduction and excitation in nerve, *J. Physiol.*, **117,** 500–44.

Hodgkin, A. L., and B. Katz, 1949. The effect of sodium ions on the electrical activity of the giant axon of the squid, *J. Physiol.*, **108,** 37–77.

Hodgkin, A. L., and W. A. H. Rushton, 1946. The electrical constants of a crustacean nerve fibre, *Proc. Roy. Soc. (London)*, **B-133,** 444–79.

Kuffler, S. W., and C. Eyzaguirre, 1955. Synaptic inhibition in an isolated nerve cell, *J. Gen. Physiol.*, **39,** 155–84.

LLOYD, D. P. C., 1941. The spinal mechanism of the pyramidal system in cats, *J. Neurophysiol.*, **4,** 525–46.

LORENTE DE NÓ, R., 1938a. Synaptic stimulation as a local process, *J. Neurophysiol.*, **1,** 194–206.

LORENTE DE NÓ, R., 1938b. Analysis of the activity of the chains of internucial neurons, *J. Neurophysiol.*, **1,** 207–44.

LORENTE DE NÓ, R., and G. A. CONDOURIS, 1959. Decremental conduction in peripheral nerve; integration of stimuli in the neuron. *Proc. Nat. Acad. Sci.*, **45,** 592–617.

MASON, S. J., and H. J. ZIMMERMAN, 1960. *Electronic Circuits, Signals and Systems*, New York: Wiley.

MCCULLOCH, W. S., and W. PITTS, 1943. A logical calculus of the ideas immanent in nervous activity, *Bull. Math. Biophys.*, **5,** 115–33.

RALL, W., 1957. Membrane time constant of motoneurons, *Science*, **126,** 454.

RALL, W., 1959. Branching dendritic trees and motoneuron membrane resistivity, *Exptl. Neurol.*, **1,** 491–527.

RALL, W., 1960. Membrane potential transients and membrane time constant of motoneurons, *Exptl. Neurol.*, **2,** 503–32.

RALL, W., 1962a. Theory of physiological properties of dendrites, *Ann. N.Y. Acad. Sci.*, **96,** 1071–92.

RALL, W., 1962b. Electrophysiology of a dendritic neuron model, *Biophys. J.*, **2,** 145–67.

RAMON-MOLINER, E., 1962. An attempt at classifying nerve cells on the basis of their dendritic patterns, *J. Comp. Neurol.*, **119,** 211–27.

RAMÓN Y CAJAL, S., 1909. *Histologie du système nerveux de l'homme et des vertébrés*, vol. 1, Paris: Maloine.

RASMUSSEN, G. L., 1957. Selective silver impregnation of synaptic endings, in *New Research Techniques of Neuroanatomy*, W. F. Windle, ed., Springfield, Ill.: C. C Thomas.

SCHADÉ, J. P., and C. F. BAXTER, 1960. Changes during growth in the volume and surface area of cortical neurons in the rabbit, *Exptl. Neurol.*, **2,** 158–78.

SHOLL, D. A., 1955. The surface area of cortical neurons, *J. Anat. (London)*, **89,** 571–72.

SPENCER, W. A., and E. R. KANDEL, 1961. Electrophysiology of hippocampal neurons: IV. Fast prepotentials, *J. Neurophysiol.*, **24,** 272–85.

WYCKOFF, R. W. G., and J. Z. YOUNG, 1956. The motorneuron surface, *Proc. Roy. Soc. (London)*, **B-144,** 440–50.

YOUNG, J. Z., 1958. Anatomical considerations, *EEG Clin. Neurophysiol.*, Suppl. No. 10, 9–11.

5. Mathematical and Physical Problems in the Study of Brain Models

E. R. CAIANIELLO, *Università di Napoli*

The first problem that confronts a physicist entering the field of neural theory is methodology. Neuroanatomists, neurophysiologists, psychologists, logicians, mathematicians, and engineers differ widely in their attitudes toward both the scope and the methods of research; constructive discussion is often rendered impossible by differences in terminology and in mental habits rather than by the difficulty of the subject at hand.

Methodology will be, to a considerable extent, the subject of this paper; I propose to discuss the roles of mathematics and physics in this domain as seen by a theoretical physicist, and to use this discussion as a pretext to describe briefly some of the work now being done at the Istituto di Fisica Teorica in Naples.

Two dangers lie ahead: one is to see the trees but not the forest (reliance on gadgets, excessive meticulousness in simulating a given organ or function, overemphasis on some special theoretical aspect); the other is to see the forest but not the trees (belief that a large number of elements coupled together are sufficient to provide any number of functions, use of mathematical tools that are too sophisticated or—more commonly—too elementary, excessive simplifications in statistical treatment).

If and only if our goal is to obtain an efficient model of a "brain" (I shall systematically misuse this word to denote any system that contains more than one neuron), then the first question is whether we should strive for anatomical as well as physiological fidelity. This is a point at which misunderstandings arise, and the situation is further confused by the diversity of opinion regarding the role that "randomness" should play in models. The point is essential; we know that a fairly reasonable behavioral description of a sodium chloride crystal can be provided by a model composed only of ping-pong balls and rubber bands, whereas the situation is hopeless if one bases his analysis on the radiation spectra of sodium and chlorine ions. The choice made here is decisive; once it is made, the "level" of the description provided by the model is determined.

Another important question is this: how much weight should be given to information gathered by observing the gross activities of a brain (speech, pattern recognition, abstraction, and the like)? If we decide, for instance, on an "extreme physiological model" (i.e., a black-box model with arbitrary structure which can provide some of the brain's functions), then can it be hoped that this model will spontaneously provide other brain functions not taken into account in its design? We can regard such a black-box simulator as a brain model, rather than a toy or gadget, only if we are assured that its behavior and that of a real brain will "converge." Planned functions such as playing chess, solving mathematical problems, and writing poetry are important only if there is reason to expect that a model providing such behavior will converge toward that of a brain, i.e., provide many other neural functions not originally planned.

These questions are of course vague and general; I shall deal now with more specific questions concerning the role of the so-called "exact sciences" in model making. The ideas and models we are working with in Naples will serve as grounds for formulating questions of methodology.

Mathematical Problems

There are two fundamental aspects of the study and simulation of brains. One is the formulation of general principles, laws, and equations appropriate to the description (at a chosen level of approximation) of the structure and behavior of a mass of neurons (consider, for example, Navier's equations for fluids). A theory does not describe nature itself, but rather certain entities of interest which have been abstracted from nature. The theory is recognized as satisfactory only when it is cast into precise mathematical formulas which are amenable to quantitative investigation. It is thus necessary to search first for a mathematical model which preliminary analysis shows to be adequate for the analysis of neural masses. Subsequent study will lead, as is usual in science, to agreements and disagreements between the model and experimental data and then to improvements both in the theory itself and in the methods of handling the theory. All we can ask for at present is a starting point sufficiently well based and permanent to permit further elaboration.

But if we already had a "general" theory—one as good as Maxwell's and Lorentz's equations for electromagnetism—it would be no more helpful in constructing a working brain model than those equations would be in designing a television set. A brain model is a machine; we must know its *purpose* (by defining its functions and input-output relations) and find the *rules* to be followed in the construction of its organs. The detailed specification of the purpose and rules constitutes the second major aspect of the problem of model making.

In brief we may describe the two aspects as the need for general laws and the need for specific rules.

A Particular Mathematical Theory. In Naples a sustained program of quantitative neuroanatomical research has provided enough information to warrant a first estimate of the species, numbers, sizes, and mutual connectivity of the neurons in the cortex. With these data in hand we can attempt to assign realistic values to the parameters and coefficients which appear in the *neuronic* equations. These equations describe the instantaneous relations in a network of "discriminators" which simulate neurons. Memory growth and decay is accounted for by *mnemonic equations*, which are "decoupled" from the neuronal equations by an "adiabatic learning hypothesis."

The details of this theory are presented in a paper by Lauria (1963); what matters here is the general aspect of such equations:

$$(1)\ \phi_h(x, y, z, t + \tau) = \sum_k \int_{-\infty}^{\infty} dx' \int_{-\infty}^{\infty} dy' \int_{-\infty}^{\infty} dz' \int_{-\infty}^{t} dt'\ K_{hk}(x, x'; y, y'; z, z'; t, t') \cdot I[\phi_k(x', y', z', t')] - s_h(x, y, z, t),$$

where the function

$$(2)\qquad I[\phi] = \begin{cases} 1 & \text{if } \phi > 0, \\ 0 & \text{if } \phi \leqq 0. \end{cases}$$

The value of $\phi_h(x, y, z, t)$ is the "effective excitation," i.e., excitation minus threshold, which impinges upon the neuron of species h located at point x, y, z of the cortex at time t. The threshold of that neuron is $s_h(x, y, z, t)$; the kernels K_{hk} contain all of the information about density distribution, connectivity, and time decay of each neuron type in each layer of the cortex.

If ϕ_k were substituted for $I[\phi_k]$ in Eq. (1), we would obtain the ordinary Fredholm equations, which could be solved by inspection. As Eq. (1) stands, however, we are confronted with a novel and most interesting mathematical situation; this situation will be discussed below.

First, I should point out some of the possible uses of this equation. It is based upon data concerning the microstructure of the brain and upon several hypotheses; the solution could be easily compared with EEG recordings. If the parameters are adjusted to yield a solution which matches EEG records, then the values of the parameters should provide information about cortical connectivity and neural thresholds. If this mathematical model is reasonably accurate, it will be of help in the study of the effects of drugs and diseases. It might also enable us to determine which components of an EEG record are connected with brain functions and which are tied to other rhythms of the body (such as heartbeats).

However, we are interested in this equation here only because it represents a tentative description of a "neuronal medium," a dense mass of neurons. The physicist has two ways of describing the statistical behavior of a gas; à la Gibbs, or simply by writing Boyle's or van der Waals' law. Equation (1) follows the philosophy of the second approach, with this difference: exchange of information and decision making in the component elements are taken fully into account. For this reason we can say that our theory is at the "cybernetic" level of system theory, a level between the detailed (particle) and the ordinary (statistical) levels.

The Role of Mathematics in the Study of Brains. It is rather surprising to see how little consideration has been given thus far (with a few laudable exceptions) to the very real contributions mathematics can make toward the solution of our problems. No amount of skill is spared in the application of various techniques to specific problems, but no matter how elementary or complicated the required calculations may be, mathematics is assigned only the ancillary role of supplying technical aids rather than guiding ideas.

No one doubts that without the diffusion equation a nuclear reactor could not be designed or even conceived. The situation is entirely similar in the case of the nervous system: there is as yet no known and tested "equation" to help us beyond the fumbling stage at the very beginning, by offering as clear an understanding of the behavior of a neuronal medium as the diffusion equation gives of a uranium–graphite lattice.

There are times when posing a problem may be far more difficult than solving it. The diffusion equation—like all other equations of mathematical physics—followed, rather than preceded, the study of natural phenomena; when theoretical physics posed a problem, mathematics provided a solution. It is my conviction that the most urgent task at hand is the creation of concepts, equations, and problems which offer a *natural* way of dealing with the structure, function, and control of a neuronal medium, just as tensor calculus did for general relativity theory.

To make this point clearer, let us consider Eq. (1). I do not suggest that this equation is the final answer; it would be unrealistic to search for a "final answer." Emphasis should be placed on testing simple models with the aim of discovering *methods* of general applicability. From this point of view, Eq. (1) is as good a beginning as any.

A mathematical study of Eq. (1) is of interest only if it is carried out at a very general level. By this I mean that equations as such are not important; what matters is "problems," i.e., equations plus all the data (boundary values, initial conditions, and the like) necessary to specify interesting physical situations. A book on the wave equation, for example, presents an equation on the first page; several hundred subsequent pages are required to pose and solve specific problems related to the equation. In the case of Eq. (1), what additional data and hypotheses are required to pose an interesting problem?

This is a most important question. If we knew the answer to it, we would also know what types of inputs and controls are required for the normal functions of a neuronal medium, a much more complex structure than a uranium–graphite lattice. Here mathematical rigor can be of real value. A general equation, once given, implicitly contains an indication (though it is often well hidden) of the type of problem which is "natural" for the system in question. Thus, if one tries to solve a Cauchy problem with a Poisson equation or a Dirichlet problem with a wave equation, then he will most certainly be faced with unending troubles.

We are just beginning to investigate Eq. (1). One thing is evident upon inspection: the search for conditions which yield a unique solution is greatly different and far more difficult than for most equations in mathematical physics. This difficulty is due to the extreme nonlinearity of the equation, which stems largely from the function $I[\phi]$. However, the very fact that nonlinearity is such an outstanding feature of this equation may offer a clue to future research. The unit step-function $I[\phi]$ has the important property

$$I[uv] = 1 - (I[u] + I[v]) + 2I[u]I[v], \tag{3}$$

which can be generalized to any number of factors; this property is not shared by other nonlinear functions, and it may prove as useful in handling products as linearity is in handling sums.

There is perhaps some indication that a "natural problem" for Eq. (1) might have some features in common with the Darboux problem. If so, then for an equation in the two variables x (space) and t (time), the solution $\phi(x, t)$ must be determined by assigning values to the functions $\phi(x, 0)$ and $\phi(0, t)$. To predict behavior in a neuronal medium we must, then, in addition to assigning the initial excitatory states, control the state of one neuron (in this simple example) at all later times.

I believe the foregoing example suffices to illustrate my previous statement that mathematical investigations made at an appropriate level of generality can tell us things about brains and about models of brains; the value of this technique should not be underestimated. Given an acceptable mathematical theory, the list of questions we can pose can be made as long as one might wish; consider just memory growth and decay, pattern analysis, and self-organization, which can all be posed as precise mathematical problems once a theory such as ours is accepted.

I shall mention here one final example. If we wish to build a hardware model of the brain, we are immediately confronted with the necessity of knowing how many neighboring "neurons" should, on the average, be connected with any given "neuron." Suppose that our model is to be composed of 10^3 "neurons," a much smaller number than the approximately 10^{10} neurons in the brain. How must we scale down the average number of connections per neuron in our model so that the behavior of the model will be a good analog of a real brain? Problems of this type arise every day. Let us take an example from aeronautical engineering, where the weight of a wind-tunnel model is proportional to the cube, and the lifting power to the square, of the model's linear dimensions. It is possible to calculate the scale factors which will provide a good physical model of a full-size airplane. The problem is that of achieving functional similarity in a reduced-size model, and the only way the problem can be solved is through the use of mathematical equations. Here the pure mathematician may prove helpful.

Physical Problems

To a theorist, all that has been said up to this point is also physics; however, I wish to discuss here a quite different aspect of the application of the ideas and methods of physics to the problem of model-making. This aspect appears as soon as one asks the question: what kinds of organs (large sets of neurons, chunks of the "neuronal medium") are required in a brain model, and how should they be connected with each other and with the environment? I propose to discuss briefly some methodological aspects of this question.

Model Methodology. Since we assume that models are of interest only insofar as they may potentially converge with the functions of the brain itself, we have adopted the point of view of the experimenter who proposes to examine an extremely complicated machine which operates on unknown principles, and to discover those principles. One example would be a group of scientists examining a spaceship from another galaxy.

We have chosen to examine the relation between the brain and natural languages. We regard natural languages as especially important because they are the articulation of thought, the natural product of the brain. We have ignored special activities such as problem solving and theorem proving because they look to us like acrobatics rather than normal walking.

After deciding that we are interested in a physiological, rather than strictly anatomical, model, we have adopted what I shall call a "gray-box" methodology. We require that the specified input-output relations of any box in a model be of a type that can be provided by some combination of elements which behave like the "neurons" of our mathematical theory. We feel free, of course, to replace any aggregate of such "neurons" (once its function is fully understood) by any functionally equivalent piece of machinery which is technologically more convenient. With this approach we hope to stay as close as possible to nature, at the same time retaining complete mathematical control of everything we build.

Our attitude toward randomness is that, for the time being, we want as little of it as possible, although we believe it to be of the greatest importance in providing reliability and self-organizing capabilities; we cannot afford to attack such problems at present.

An essential problem in methodology is to decide what we should look for in studying a given phenomenon. Let us take, as particularly relevant to our purposes, some simple problems in visual-pattern recognition. The recognition capability of our model should be, to some extent, independent of size, translations, rotations, and deformations of one type or another. We might set out by studying the mathematical requirements for a network of neural analogs which would meet these criteria. We would quickly stumble onto the fact that *N* and *Z* are quite different patterns to the human observer, although topologically equivalent, whereas a solid line and a dotted line are essentially the same pattern to a human but radically different from a topological standpoint.

Another problem is whether we should focus our attention on models that can learn to recognize any arbitrary pattern presented, or concentrate instead on models that are to some extent pre-designed to respond (like the neurons of the retina) only to certain classes of contours and shapes from which more complex patterns are constructed.

We have been considering these and similar questions for several years, and our approach has been experimental. Whatever the value of our work or the shortcomings of my own deductions, I firmly believe that the "convergence" of models can be achieved only if we try to understand, in full humility, nature's own organizational principles. I am becoming more and more convinced that there are many "rules" for the specification and assembling of a brain model which are yet to be discovered, and that their discovery will require generations of effort.

The Procrustes Project. I shall conclude with a brief description of a particular research project which illustrates our approach to model making. The "Procrustes" program is currently being run on a large-scale computer at Pisa, and has given some satisfactory preliminary results. The subject of this study is natural language, which we regard not as something to be invented or defined, but rather as a physicist regards the universe: it is there and he proposes to unravel it.

We are convinced—primarily on the basis of anatomical evidence—that a major role in mental phenomena is played by hierarchical organizations of neurons in layers, each layer having elements with various specialized functions. "Procrustes" has the task, when presented words of a language (or sequences of polygons or other structured patterns), to find out what layers, or levels, of elements are required to analyze, memorize, and recognize those words.

If we start with Italian letters (21 symbols), Procrustes searches for syllables. (These need not be the conventional syllables of the grammarian; there are more economical choices.) A "cutting operation" (hence the name) determines when the search should be halted, and the program then specifies how the filters which select syllables are to be built (either with hardware or in the form of a computer program). Next, the groups of syllables which occur in words are examined, and a higher level of filters is formed, and so on.

As is always the case with models of this kind, Procrustes raises more problems than it solves, but at least they are raised within a clear mathematical framework. Each problem can be dissected into component parts, but even with this simple approach many problems have arisen which we could not have formulated beforehand.

Procrustes has operated correctly in cases where the "words" were letter sequences formed according to a simple mathematical law. In an initial analysis of the language it also "discovered," as expected, the five Italian vowels a, e, i, o, u; a frequency analysis of the same material yielded instead the letters a, e, i, o, t.

Summary

Questions of methodology relevant to the study of brain models are raised; they reflect the point of view of a theoretical physicist approaching this field. It is stressed that pure mathematics should play a much more significant role in the future, and in support of this position several crucial questions are presented; these are questions which can only be answered after a thorough study of unexplored regions of pure mathematics.

As to the problem of designing and building a physical model, it is stressed that experimentation, especially in neurophysiology and linguistics, is important if we wish to develop a model whose behavior converges toward that of the brain. The paper concludes with a brief description of the Procrustes project.

References

Braitenburg, V., and F. Lauria, 1960. Toward a mathematical description of the grey substance of nervous systems, *Nuovo cimento*, **18**(1, suppl.), 149–65.

Braitenburg, V., and N. Onesto, 1962. *Atti del I° Congresso Internazionale di Medicina Cibernetica*, Naples: Giannini.

Caianiello, E. R., 1961. Outline of a theory of thought processes and thinking machines, *J. Theoret. Biol.*, **1**, 204–35.

Lauria, F., 1963 (in press).

6. A Theory of Resonant Networks

RICHARD F. REISS, *General Precision, Inc.*

It must be admitted straightaway that the hypothetical networks discussed in this paper are not based on careful inductions from physiological data. The theory presented stems from two lines of research. First, during simulation studies of hypothetical neural systems involving reciprocal inhibition (Reiss, 1962), certain unexpected behavioral anomalies drew our attention to a type of "resonance" effect that can be produced by appropriate combinations of pulse frequency, axon time delays, temporal summation, and firing thresholds. Second, a general investigation of the possible roles of pulse-interval coding in computers and other kinds of machines, inspired initially by analogical reasoning from neural systems, led to the conception and study of a variety of "resonant" mechanisms; although some of these mechanisms could be physically realized only by electronic devices, others required parameter ranges that might well be within the capabilities of neurons. Thus both of these lines of research suggested an exploration, in some depth, of the possibilities of "resonant networks" in neural systems.

The aim of this paper is to outline a few basic types of hypothetical neural networks which, if they existed in nature, would exhibit resonance properties useful to the organism. Mathematical and formal logical analyses of such networks, which can be rather easily developed in some cases, have been almost entirely ignored for the sake of simplicity and brevity. From the standpoint of the biologist, to whom this paper is addressed, the central problem is whether even the simplest resonant network discussed does exist in real neural systems. Of course, this question can only be settled by observation and experiment. Therefore we have concentrated on the quantitative parameter requirements of simple networks and on their expected behavior in response to simple patterns of inputs. The examination of large systems of resonant networks is carried only far enough to suggest the potential value of such networks in neural communication and control functions.

Except to the extent that this theory was suggested by the results of a simulation study of reciprocal inhibition, it has not been developed by means of physical models, either electronic analogs or digital programs. The theory presented here is based entirely on "paper-and-pencil" analysis. However,

Much of the research this paper is based on was supported by the Air Force Office of Scientific Research, under contract AF 49(638)-1021.

further explorations of resonant networks, especially of their response in groups to complex patterns of inputs, will in most cases probably require extensive use of physical models.

Definitions and Assumptions

1:1 Neurons and *T*-Neurons. The networks to be discussed are combinations of two simple types of neuron. We define a *1:1 neuron* as any neuron which fires once and only once for each excitatory input spike. The act of "firing" is taken to be the initiation of a spike in or near the root of the neuron's axon. For simplicity we assume that a neuron has only one axon root, although the axon may branch to provide multiple terminals. If an axon terminal contacts the synaptic region of neuron N, then a spike occurring at that terminal is called an "input pulse" for neuron N. It is assumed that, for the firing frequencies to be considered, refractory effects at axon terminals and spike-initiation loci are negligible. Input pulses are excitatory or inhibitory according to the type of axon terminal involved.

The second type of neuron to be defined is the *T-neuron*. A *T*-neuron has a temporal threshold T; one excitatory pulse cannot evoke firing. If two excitatory input pulses arrive at times t_1 and t_2, then the neuron will fire (once) if and only if $t_2 - t_1 \leqq T$. The firing will occur shortly after t_2 owing to synaptic delays.

We shall generally assume that T is quite small, say less than 1 msec. Thus a *T*-neuron can serve as a "near-coincidence" detector; it is contacted by two excitatory axon terminals (at least), and if two input pulses appear at these terminals within the time interval T, then a single pulse is initiated in the neuron after a short delay. The magnitude of the synaptic delay in a *T*-neuron will not concern us, nor do we care what synaptic processes are responsible for the existence of a temporal threshold.

The value of T may, in some networks, be assumed to vary with the passage of time as a result of firing, facilitation, imposed fields, and the like. The consequences of such variation will be examined in some detail later. Although the quantity T has been defined in terms of two successive input spikes, the concept of the temporal threshold can of course be extended to cover situations in which more than two input spikes arrive at intervals slightly greater than T and yet produce firing as a result of temporal summation. In most of the networks to be examined such situations cannot arise. However, temporal thresholds for three or more input spikes will be discussed in connection with temporal pattern detection.

In network diagrams the neuron somata will be represented by circles labeled N_1, N_2, and so on. Lines originating at a circle represent axons; an arrowhead at the end of an axon denotes an excitatory terminal; an inhibitory terminal is denoted by a solid dot. For example, the network in Fig. 1 consists of three neurons, N_1, N_2, and N_3. The axon of N_1 has excitatory terminals contacting the synaptic regions of neurons N_2 and N_3 and an inhibitory terminal contacting the synaptic region of N_3. It is not necessary to assume any particular spatial or dynamic relations between axon terminals; *any* set of relations which satisfy

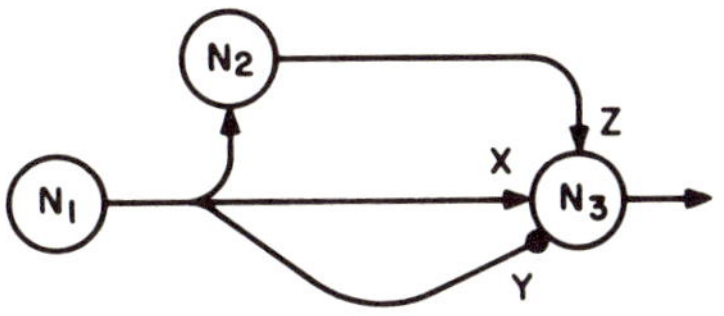

FIG. 1. A typical network. Neurons are labeled N_1, N_2, and N_3; excitatory terminals are denoted by arrowheads, and inhibitory terminals are denoted by dots.

the input-output functions specified for a 1:1 neuron or T-neuron is sufficient for our analysis. The only requirements are that a 1:1 neuron fire once for every excitatory input pulse and that a T-neuron fire once for every pair of excitatory input pulses that are separated by a time interval less than T (assuming no inhibitory terminals are active).

Frequency and Period. It is understood that *frequency* means pulse-repetition rate. The reciprocal of the time interval separating any two pulses may be considered the frequency at which *those two* pulses occurred. However, except in one special case, a pulse train composed of more than two pulses does not have a unique frequency, and therefore a statistical definition of frequency is required. Since this paper is primarily concerned with hypothetical networks whose behavior must be analyzed in terms of time intervals between individual pairs of pulses within trains, statistical notions of frequency are of little value. Consequently, pulse trains will generally be described in terms of time intervals rather than frequencies.

Unfortunately, this policy raises another problem of terminology: the double meaning of "period." A period is commonly defined as a time interval between two particular events (it is defined by the events and not by its magnitude). On the other hand a period may be defined as a quantity of time having no reference to any particular historical events. We shall use *period* to mean a *quantity* of time; "the period between the first and second pulse" refers to the amount of time separating the occurrence of the pulses, not to the historical segment of time itself. Thus a "period" will always denote a number. We shall use "interval" to refer to particular segments of historical time. Although an interval always has a magnitude (which may be used to define a period), this is only one property of the interval. Since intervals are necessarily ordered on the time dimension, it is meaningful to say that one interval succeeds or overlaps another; by contrast, periods cannot succeed or overlap one another. We can only say that one period is less than, equal to, or greater than another period.

Pulse Trains. A continuously varying signal having a complex waveform is often treated as if it were the sum of a set of perfect sinusoidal signals with various frequencies, phases, and amplitudes. We shall treat pulse trains in a similar fashion.

A pair of pulses is *successive* if there are no intervening pulses. A train of K pulses contains $K - 1$ successive pairs, and the time intervals between these pairs are called *primary* intervals of the train. Two intervals are *contiguous* if the pulse which terminates one interval also initiates the other interval; a *secondary* interval is any interval composed of two or more contiguous primary intervals.

If all of the primary intervals in a train have the same magnitude, then that train is *regular*; if two or more primary intervals have different magnitudes, then the train is *irregular*. We shall later have occasion to define certain special types of trains, but the distinction between regular and irregular will suffice for the present.

Every finite pulse train defines a finite set of *component periods*. A period p is a component of a train if and only if there is some pair of pulses in that train separated by a time interval of magnitude p. Thus, for example, the period between the second and fifth pulses is a component of that train. In the case of a regular train, all primary intervals have the same magnitude and each one can therefore be represented by the same component period; we shall call this the *fundamental* period (P_f) of a regular train. Its reciprocal is the fundamental "frequency" (F_f). Since the primary intervals of an irregular train are not all the same size, no fundamental period is defined for such a train.

A regular train of K pulses has exactly $K - 1$ components periods (hereafter often "component"). The smallest is the fundamental period, and all other components are integral multiples of the fundamental; translated into the frequency domain, the latter components define the *subharmonics* of the fundamental frequency.

In an irregular train of K pulses there are necessarily more than $K - 1$ components. However, there is a definite upper limit, namely, $\frac{1}{2}K(K - 1)$; this is true regardless of the duration of the train. By comparison, there is no upper limit to the number of sinusoidal components that may be present in a finite segment of a continuous waveform.

We shall be discussing hypothetical networks which *respond directly to secondary intervals*. So far as these networks are concerned, secondary intervals are just as "real" as the intervals between successive pairs of pulses (primary intervals). The precise position of a pulse relative to *any* other pulse in the train may be important. When one analyzes pulse trains from a statistical point of view, such considerations are usually unimportant; only the magnitudes and ordering of primary intervals are significant.

A Basic Resonant Network

In this section a simple hypothetical network, the *band detector*, is introduced. The network shown in Fig. 2 exhibits "tuned," or "resonant," behavior produced by the combined effects of time delay and pulse-coincidence detection. Variations in behavior imposed by bandwidth expansion or contraction and by neural parameters are also described.

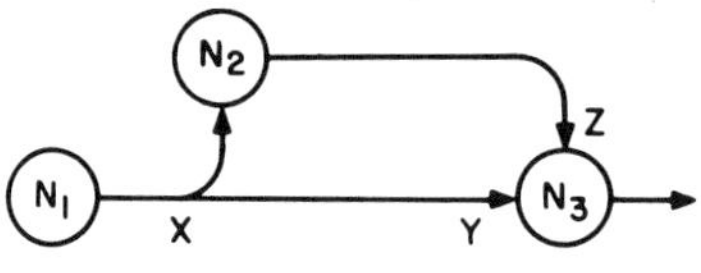

FIG. 2. The band-detector network. A pulse produced by the input neuron N_1 splits at the branch point X; one pulse travels to the excitatory terminal Y, while the other passes through N_2 to its excitatory terminal Z, arriving D msec later. Neuron N_3 is a T-neuron which fires if and only if two pulses arrive within a period T.

The Band Detector. Consider the network shown in Fig. 2. Assume that N_2 is a 1:1 neuron and that N_3 is a T-neuron. We do not care what type of neuron N_1 happens to be; it merely provides the input to the system. The axon of N_1 branches at point X, sending an excitatory terminal (named Y) to N_3 and another such terminal (not named) to N_2. Neuron N_2 in turn has an axon whose excitatory terminal (named Z) contacts the synaptic region of N_3. It is assumed here, as in most subsequent networks, that the velocity of pulse propagation is of the same order of magnitude in all axons (although this assumption is neither necessary nor ideal). The relative lengths of axons in our drawings are intended to suggest the relative propagation-time delays suffered by pulses traveling along those axons.

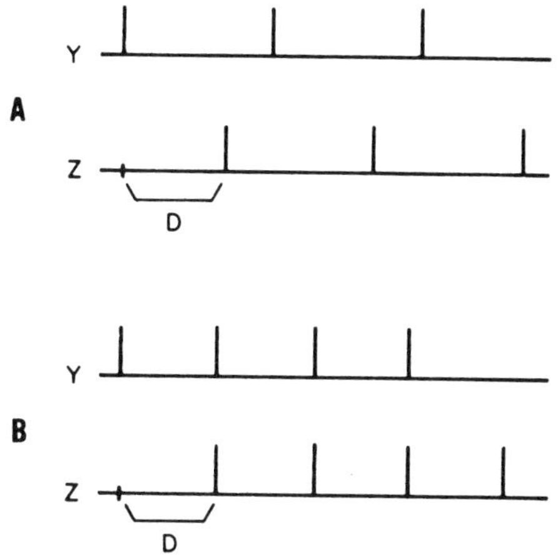

FIG. 3. Regular pulse trains produced by N_1 arriving at terminals Y and Z in the network of Fig. 2. A. The fundamental period of the trains is greater than the net delay D, and no two pulses arrive in coincidence at N_3. B. The fundamental period is equal to D, and three pairs of pulses arrive in coincidence, each pair causing N_3 to fire.

A pulse originating in N_1 splits at point X; one pulse travels directly to terminal Y, and the other pulse travels over the longer route through N_2, where it encounters a synaptic delay, to terminal Z. The propagation time from X to Y is assumed to be appreciably less than that from X to Z. We are not interested in the absolute values of these delays; only the net *difference* in propagation times is important here. The net difference in delays $(d_2 - d_1)$ is denoted by D.

In Fig. 3A a regular train of pulses arrives at Y and, delayed by time D, a similar train arrives at Z. Assuming that D does not vary appreciably as a function of time, both trains will have the same fundamental period and duration because they both originated at point X. All of these pulses have an excitatory effect on N_3, which is a T-neuron, and if any two pulses are separated in time by an interval less than T, neuron N_3 fires. Let us assume that T is small relative to D. It is clear that a regular pulse train of this fundamental period, no matter how many pulses may be involved, will never cause N_3 to fire.

But suppose N_1 generates a regular pulse train having a shorter fundamental period, a fundamental period approximately the same as D. This case is illustrated in Fig. 3B. The second pulse arriving at Y may follow the first pulse arriving at Z by an interval less than T, and N_3 fires. If so, all subsequent

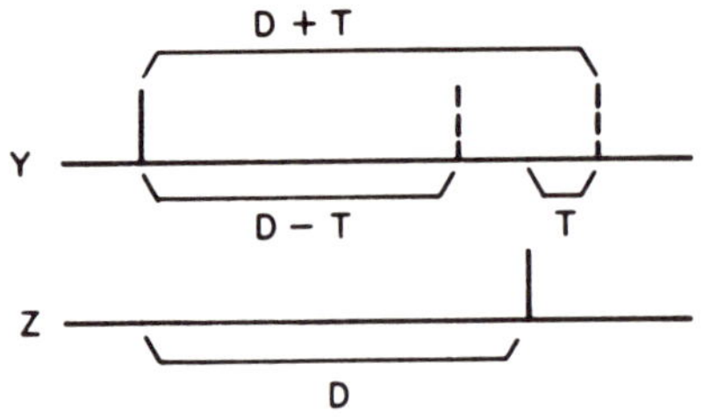

FIG. 4. The band of component periods detected by the network of Fig. 2 has the lower limit $D - T$ and the upper limit $D + T$, where T is the temporal threshold of N_3.

pulses in the train at Y will cause firing in N_3 because the train is regular and D is constant. We note first that in this case N_3 produces regular pulse train having the same fundamental period as the train generated by N_1; neuron N_3's train, of course, has one less pulse and therefore is somewhat shorter. We note further that N_3 generates a regular train if the incoming trains have a fundamental period slightly less than, or equal to, or slightly greater than, the delay difference D. To be more precise, if the fundamental period P_f of the regular input train is between $D - T$ and $D + T$, then N_3 generates a regular train having the same fundamental period but one less pulse.

Thus there is a "band" of fundamental periods which will fire N_3. The situation is illustrated in Fig. 4. The first pulses of the trains at Y and Z are represented by solid vertical lines. The second pulse arriving at Y must fall between the vertical dashed lines at $D - T$ and $D + T$ to initiate a pulse in N_3. The *width* of the band of periods is $2T$. We shall denote the lower limit of such a band by P_L and the upper limit by P_U. Since the reciprocal of a period is a frequency, a band of frequencies is defined by any band of periods; the limits of the frequency band are the reciprocals of P_L and P_U.

In the examples of Fig. 3 the pulse trains had fundamental periods either near, or appreciably greater than, D. Hence all other component periods in these trains were much greater than D. Consider now the case illustrated in Fig. 5. Here the regular trains arriving at Y and Z have a fundamental period that is approximately $\frac{1}{2}D$. Five pairs of pulses are nearly coincident in their arrival at N_3, and we may assume that N_3 is fired by each pair; hence N_3 generates a regular train of five pulses with the same fundamental period as the incoming trains. In this case N_3 is responding *not* to the fundamental period of the incoming trains, but rather to a larger component period, namely, that between the first and *third* pulses (or second and fourth, third and fifth, etc.)

The fundamental frequency F_f of a regular train is the reciprocal of the fundamental period. The network fires when it "detects" any interval whose period falls within a certain band; the "center" of that period band is D. We shall

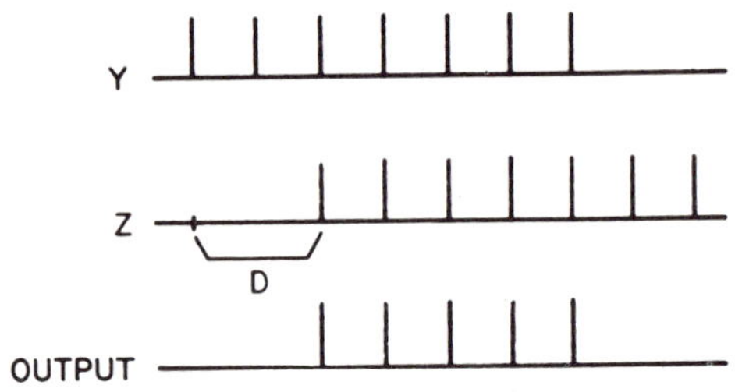

FIG. 5. The regular trains produced by N_1 have a fundamental period of $\frac{1}{2}D$ in this example; thus the trains' fundamental frequency is the first harmonic of the band detector's tuned frequency, and N_3 fires (five times). Time delays between input and output trains have been ignored in this graph.

call the reciprocal of D the *tuned frequency* of the network and denote it by F_0. The upper limit of the corresponding frequency band will be denoted by F_U and the lower limit by F_L. We have $F_U = 1/P_L$ and $F_L = 1/P_U$.

We have seen that the network of Fig. 2 will pass any regular train whose fundamental frequency satisfies the condition $F_U \geqq F_f \geqq F_L$. It is also clear from the example of Fig. 5 that a regular train will pass if its fundamental period is $\frac{1}{2}D$, i.e., if its fundamental frequency $F_f = 2F_0$. We can show that a regular train will pass through the network if its fundamental frequency is F_0, $2F_0$, $3F_0$, $4F_0$, or any other multiple of F_0. In brief, the filter will pass all *harmonics* of its tuned frequency, and about each harmonic of F_0 there is, of course, a band of frequencies that will also be passed.

To state it another way: a regular train of pulses will pass through the network if and only if its fundamental frequency or any *subharmonic* of that frequency ($\frac{1}{2}F_f$, $\frac{1}{3}F_f$, and so on) falls within the limits F_L and F_U of the network's frequency band.

This statement requires two qualifications. First, if the fundamental frequency of the incoming train is high enough, i.e., if the fundamental period is less than T, then, clearly, successive pulses within the same train can fire N_3. However, an increase in duration of the pulses, or of the refractory periods, or both, may defeat the filtering action of the coincidence-detection process even at frequencies lower than $1/T$. Second, the higher the frequency of a train, the longer the train must be to produce an output from N_3. More precisely, if the Kth subharmonic of the train falls within the frequency band of the network, then the train must contain more than K pulses, because the output train has $K + 1$ fewer pulses than the input train. This phenomenon is demonstrated in Fig. 5, where the input train contains seven pulses. The first subharmonic ($\frac{1}{2}F_f$, or $2P_f$) falls within the band determined by T; hence the train passes, but with the "loss" of two pulses.

Although it might be helpful to compare the behavior of this and related networks with that of filters and other conventional electronic devices, we shall give the networks names that indicate their basic functions rather than real or fancied resemblances to electronic devices.

Band-Detector Response to Irregular Trains. Consider the situation illustrated in Fig. 6. The irregular train of four pulses generated by N_1 has four component periods, two of which are designated P_1 and P_2. The other two have the values $P_1 + P_2$ and $P_2 - P_1$ It is assumed here that the component P_1 is equal to D. Clearly, N_3 will fire twice in response to this input, once at the coincidence of the second pulse at Y and the first at Z, and again at the coincidence of the fourth pulse at Y and the third at Z. Thus the output of N_3 will be a train composed of just two pulses. What are the components of this output train? There is only one, and it is *not* P_1 which *does* fall within the band of the network, but P_2, which *doesn't* fall within that band. Although this is in some ways an extreme, or special, type of irregular train, it should be noted that trains of this type are by no means uncommon in real nervous systems.

We can consider an irregular train of this form to be composed of a *regular* train of *regular* trains, a sequence of two-pulse trains, each having a fundamental

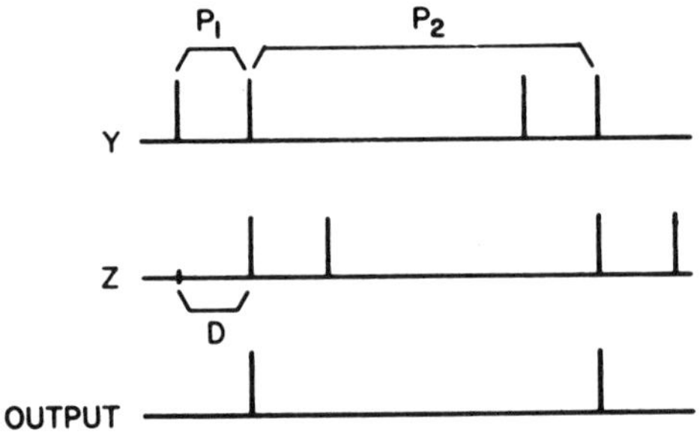

FIG. 6. Demodulation of a second-order train. The second-order train is composed of two pulse pairs and has a fundamental period P_2. Each pulse pair has a period $P_1 = D$. Pulse coincidence occurs twice, causing N_3 to produce two pulses (a first-order train) with the period P_2. Time delays between input and output trains have been ignored in this graph.

period P_1. Clearly, every two-pulse train is regular; at least three pulses are required to form an irregular train. If such regular trains appear at regular periods, in this case P_2, then—treating each train as an individual entity—we might say that there is a regular train of these trains. This will be called a *second-order train*; in Fig. 6 it is a train having a fundamental period P_2, while the *first-order* trains have a fundamental period P_1.

We have already seen that if P_1 falls within the band of a detector, each first-order train will initiate one output pulse. As a consequence, the output train will have a fundamental period P_2; the output of the band detector is a *first*-order train having the same fundamental period as the *second*-order train which constitutes the input.

Suppose now that the band detector receives an *irregular second-order* train; i.e., the periods between successive pairs of first-order trains are not all the same. The band detector's output will be an irregular first-order train having exactly the same components as the irregular second-order train fed into the detector. If the intervals of the second-order train carry information, then clearly a band detector can extract this information and recode it in the form of a first-order train. The behavior of the band detector in this case might be compared to that of a "demodulator" in continuous-wave frequency-modulated communication systems.

If information is carried by the primary intervals of a second-order, rather than first-order, train, then messages from diverse sources can be funneled through a single channel on a time-sharing basis and then distributed to appropriate destinations by simple band detectors. The fundamental periods of the regular first-order trains constituting an irregular second-order train can be used to determine which band detector will be selected and thus which destination will receive the message. This is but one possible role for systems of band detectors as will be seen later.*

Expanding-Band Detectors. We have thus far assumed that in the network of Fig. 2 the net delay D and temporal summation threshold T are constant.

* During the Ojai symposium F. Jenik brought to my attention his recent monograph (Jenik, 1962), in which he briefly discusses the behavior of a model network that is effectively identical with the network of Fig. 2. He characterizes the behavior of this network as "frequency selection by coincidence of latency" and presents statistical input-output curves (his Figures 31, 32, 33; Section 4.2) which nicely demonstrate the harmonic response of a band detector as measured on his own model and on the models of Harmon (1961) and McGrogan (1961). An excellent doctoral dissertation, "Koinzidenz-Siebschaltungen" (Schief, 1962), on the possible role of resonant networks in auditory systems, has since come to my attention. Also, D. M. MacKay (1962) has discussed resonant networks in relation to self-organizing systems.

However, fluctuations in such parameters are to be expected in a nervous system, whether due to incidental "noise" or to "purposeful" mechanisms. In the following analysis it will be assumed that D is constant but that T, although it has a normal, or "resting," value, may vary as a result of N_3 firing.

The period bandwidth will be denoted by B. Thus $B = 2T$. Suppose that when N_3 fires repeatedly, T gradually becomes larger, owing to some synaptic facilitation process. Consequently, the bandwidth B increases. In this case we shall say that the band "expands" and that the network is an *expanding-band detector*.

Consider the case where the band is initially quite narrow—i.e., B is very small—and the input to the detector is a train of pulses that have traveled an appreciable distance. Initially the train was nearly regular, but it has passed through several axons and synaptic junctions. As the first few pulses of the train passed through these axons and synapses, they introduced short-term fluctuations in propagation and transmission velocities, so that periods between later pulses were perturbed about their original values. These perturbations constitute a type of noise, and perhaps the latter part of the train has become quite "noisy," therefore irregular, by the time the band detector is reached. Perhaps originally all of the periods of primary intervals fell within the detector's band, but now many of these periods have become too large or too small. If B remains constant, then only the first few pulses might get "through" the detector, since their periods were not disturbed by the turbulence they created. Thus a large portion of the train might be "lost" at the detector, owing to noise. If, on the other hand, the detector's band expands in response to the initial pulses, then perhaps most or even all of the later pulses can get through; the detector "examines the credentials" of the leading pulses very closely, but if they pass inspection, then the detector, like a careless customs inspector, permits succeeding pulses to pass without close scrutiny.

The foregoing hypothetical case suggests one possible role of expanding-band detectors in neural communications: to "accommodate" to increasing noise. Such detectors might play a related role in processing accelerating or decelerating trains which are relatively free of noise. An *accelerating* train is one in which the successive primary intervals become progressively smaller, and a *decelerating* train is one in which they become progressively larger. Both types of train are common in nervous systems.

We must assume that there is some maximum rate at which T can increase in response to excitatory input pulses, as well as an upper limit to the magnitude of T. Consider an expanding-band detector which receives an accelerating train; suppose that the first few primary intervals are too large to excite the detector, but that later intervals fall within the band limits. At some point in the train the detector begins firing. At some later time the primary intervals become smaller than the normal lower limit of the band.

If, however, the rate of progressive decrease in the size of primary intervals is less than the rate of increase in T due to firing, then for a time the lower band limit falls more rapidly than the successive primary intervals. On the other hand, if the expansion of the detector's band cannot keep pace with diminishing sizes of primary intervals, then the detector may fire only a few times as the accelerating train's intervals "sweep across" the band. The same phenomenon occurs with a decelerating train; if the rate of deceleration is too great, the expanding band cannot "follow" it and only a few pulses are produced by the detector. In either case, even a slowly accelerating or decelerating train will be "cut off" if its primary intervals move beyond the maximum band expansion.

This analysis prompts two conclusions. First, an expanding-band detector's output indicates whether and when the intervals of an accelerating or decelerating train reach a critical magnitude (how critical depends on the narrowness of the normal band). Second, the output also indicates by its duration whether the *rate* of acceleration or deceleration is above or below a certain threshold (the rate of increase of T). If this rate threshold is exceeded, only a few pulses are produced; if it is not exceeded, then a large segment of the input train may be passed on. Assuming that the rate of increase of T is a function of the rate at which excitatory pulses arrive at the T-neuron, the rate threshold for accelerating trains might be appreciably greater than that for decelerating trains. Perhaps more importantly, if the increase of T is a synaptic phenomenon independent of firing in the T-neuron, then an accelerating or decelerating train could cause the band to expand appreciably before firing occurs and thus hasten the onset of firing, increasing the duration of the output train.

The assumption that T can be increased by input pulses regardless of firing in the T-neuron leads to an important conclusion about the expanding-band detector's behavior in response to *regular* input trains. Suppose the detector receives a regular train whose fundamental period is somewhat smaller than the normal lower band limit. The early pulses in the train do not cause firing in N_3, but they do cause T to increase. If the train is long enough, the lower band limit may drop below the fundamental period of the train, so that N_3 begins firing; the detector will reproduce the remainder of the input train. The detector will respond in a similar way to a regular train whose fundamental period is somewhat greater than the normal upper limit of the band, although here one would expect the delay to the onset of firing to be appreciably greater. Thus a regular train with a fundamental outside the normal band may, if it is long enough, "force" its way through an expanding-band detector.

Contracting-Band Detectors. Suppose that N_3 in Fig. 2 is a T-neuron in which no appreciable facilitation occurs or that, in any event, T cannot increase in value; instead, firing causes a *decrease* of T by inducing "fatigue" or equivalent processes. Clearly, a reduction in T means a decrease in the bandwidth of the detector. Hence we shall call this a *contracting*-band detector. It will be assumed that when N_3 is not in the act of firing, recuperative processes are able to increase the neuron's sensitivity, eventually restoring T to its normal value.

Consider a contracting-band detector receiving a regular train whose fundamental period falls within the normal band. Neuron N_3 will repeatedly fire, reproducing the input train, but in so doing it will cause the band to contract. If the train's fundamental is at or near the center of the band, it may not be affected by this contraction, regardless of the train's duration. But if the fundamental is sufficiently near the normal upper or lower band limit, then the contraction may carry the band limit past the fundamental and thus put an end to firing. Upon cessation of firing, the bandwidth begins its return to normal, and if the input train still endures, the band limit may shift beyond the train's fundamental period, thus permitting the detector to renew firing.

The periodic contractions of bandwidth thus in effect, "chop" the input train into a sequence of short trains, each containing only a few pulses of the same fundamental period as the input. The durations of these shorter trains and the intervals between them are determined by the relative values of P_U, P_L, and P_f and by the rates of band contraction and recuperative expansion.

The output of this contracting-band detector is a regular second-order train composed of regular first-order trains. The fundamental period of the second-order train may vary appreciably with variations in the fundamental period of the input train; the rate of decrease of T is presumably a function of the period between output pulses, and therefore the rate near P_L is different from the rate near P_U. Since the rate of decrease of T is chiefly responsible for determining the exact value of the fundamental period of the second-order train (assuming that the rate of *increase* of T is effectively a constant function of recuperation time), it follows that the fundamental period (or frequency) of the input can modulate the fundamental period (or frequency) of the second-order output train. Of course, the fundamental period of the second-order train is much greater than that of the input train; in terms of frequency, the second-order output train has a fundamental frequency which is much smaller than, but a function of, the fundamental frequency of the input.

Suppose the output of the contracting-band detector drives a bundle of muscle fibers. In this case the detector provides a means for converting a long train of pulses into rhythmic contractions of the muscle bundle at a much lower frequency; the exact value of the input's fundamental frequency would determine the extent and periodicity of contractions. It is noteworthy that a muscle of appreciable size and inertia may require a burst of pulses (a first-order train) in order to perform a single contraction having behavioral significance. Periodic contractions in such a muscle would correspond to the primary intervals of the *second*-order input train. Hence a muscle might function as an ideal "demodulator" of second-order trains; the information represented by second-order periods would be translated into the periodicity of contractions, whereas the frequency and duration of the first-order trains determined the strength of the individual contractions.

The "chopping" effect produced by a contracting-band detector makes it a more interesting device than an expanding-band detector. However, we shall not pursue the subject further here. By way of preparation for the next topic, it should be noted that the "resting" bandwidth of a contracting-band detector must be rather large if the chopping effect is to be useful as a frequency-modulating system (for second-order trains). For maximum effectiveness an expanding-band detector, on the other hand, must have a rather small resting bandwidth.

Quantitative Limits Imposed by Neural Parameters. Having argued that a simple network such as that shown in Fig. 2 can in principle exhibit resonant behavior and thereby respond selectively to a "band" of periodic components in a pulse train, we are faced with a serious quantitative problem: What resonant periods and bandwidths are possible in a band detector composed of real neurons? More precisely, what values of D and T are reasonable for real neural systems? This is a very difficult question. So little is known about the *exact* topology of neural systems at the ganglion level and above (where I assume band detection would be most useful and therefore likely) that we have little evidence for estimating "reasonable" ranges for D. Similarly, reasonable ranges for T are difficult to estimate because only a few types of neuron have been closely observed in action.

Rather than attempting to deduce these ranges from documented experimental evidence on network topology and temporal summation, I shall try to reason "backward" from the requirements of an ideal band detector operating within the rather well-established frequency spectrum of neural systems.

It will be assumed that the "neural spectrum" spans periods from approximately 1 msec to 1 sec, i.e., "frequencies" from 1 to 1000 pulses per second (pps). Granted that sensors or pacemakers may on occasion produce meaningful pulses at frequencies below one per second and that in mammals some neurons may fire at frequencies in the range of 1000 to 2000 pps, the assumed limits of the neural spectrum certainly include the vast majority of pulse-producing neurons. Thus we are concerned here with the parameters of band detectors "tuned" to frequencies between 1 and 1000 pps.

We shall say that a band detector is "tuned" to a period p (or frequency $1/p$) if $D = p$. A detector tuned to the high-frequency end of the neural spectrum must have $D = 1$ msec. A net delay of this magnitude could easily be provided by real neurons. But at the low-frequency end of the spectrum it is necessary that $D = 1$ sec. This is a very large delay and would presumably require a long chain of neurons in place of N_2 in Fig. 2. Even if 1:1 neurons were available having large synaptic delays and long, slow axons providing a total delay per neuron of, say, 25 msec, we would need a chain of 40 such neurons to provide a net delay of 1 sec. If we do not expect a band detector to be tuned to frequencies below 10 pps, the delay problem is considerably reduced; D need only be 100 msec at 10 pps, and this would not require absurdly long delay paths. At 100 pps D need be no greater than 10 msec. Thus if band detectors were restricted to the spectrum of 100 to 1000 pps, net delays would range from 1 to 10 msec. Clearly, the higher frequencies are to be favored from the standpoint of net delays.

Selectivity is another matter. Loosely speaking, the smaller the bandwidth of a detector, the greater its selectivity. However, the bandwidth, as such, is a poor criterion of selectivity. It is important to know the size of the bandwidth relative to the frequency to which the detector is tuned. We shall define selectivity in terms of frequencies: Let F_0 denote the frequency to which the detector is tuned (i.e., $F_0 = 1/D$), and let F_L and F_U denote the lower and upper limits of the frequency band. Then the bandwidth is defined as $W = F_U - F_L$. In terms of the basic parameters D and T, we have

$$W = \frac{2T}{D^2 - T^2}. \tag{1}$$

But this does not provide a good measure of selectivity. We want to know the ratio of the tuned frequency F_0 to the bandwidth W. We shall take as our measure of selectivity the ratio F_0/W. In terms of F_0 and T we have

$$\frac{F_0}{W} = \frac{1}{2}\left(\frac{1}{TF_0} - TF_0\right). \tag{2}$$

Examination of this equation shows that if high selectivity is to be obtained, either T or F_0, or both, must be small. If there is some practical lower limit to the temporal summation period T, then the band detector should operate at the lowest possible frequencies in order to achieve the greatest selectivity. Thus the desire for good selectivity conflicts with the desire to keep D small; the former favors detectors operating near the lower end of the frequency spectrum whereas the latter favors detectors operating at the higher end of the frequency spectrum. The crucial factor is the value of T. Let us look at some particular cases.

Assume that if the bandwidth W is 10% of the tuned frequency F_0, i.e., if $F_0/W = 10$, the selectivity is acceptable. Consider first a detector tuned at 100 pps. By Eq. (2) we see that T should be approximately 0.5 msec if we are to obtain the desired selectivity. If it were possible to achieve a delay $D =$ 100 msec, then the detector could be tuned to 10 pps and the T-neuron could be very sloppy, with $T = 5$ msec, and would still have a selectivity of 10. At the low end of the neural spectrum it would in fact be difficult to *avoid* rather high selectivities.

But suppose that with $T = 0.5$ msec the detector is tuned to the high end of the spectrum, 1000 pps, so that the delay is reduced to the small value of 1 msec. We find that the selectivity ratio has plummeted from 10 to 0.75 and that the band now extends from 666 pps, near the middle of the neural spectrum, all the way to 2000 pps. To obtain the selectivity of 10, we must have $T = 0.05$ msec. This means the T-neuron is so insensitive that two pulses must arrive within an extremely short time span in order to cause firing; it must be on the borderline of total refractoriness, a situation that is not very encouraging from the standpoint of reliability.

Perhaps the best way to summarize the parameter problem is simply to specify a frequency spectrum which does not require extremely large or small values for either D or T. We assume again that a moderate selectivity of 10 is desired; the tuned frequency is ten times the bandwidth. The spectrum from 50 to 500 pps appears to be the compromise we seek. A detector tuned to 50 pps must have $D = 20$ msec and $T = 1$ msec; to be tuned to 500 pps it must have $D = 2$ msec and $T = 0.1$ msec. We note that if at all possible, the lower limit of the spectrum should be reduced to a value below 50 pps because of the greater selectivities that may be easily and reliably achieved at low frequencies; reducing this value is contingent upon the availability of pulse paths providing delays greater than 20 msec. In a later section we shall show that a single, long, delay path can simultaneously serve an entire set of band detectors, thereby making large delays more reasonable and economical.

A few additional remarks on parameters are necessary. First, for the sake of completeness, we note that the tuned frequency $F_0 = 1/D$ is *not* at the "center" of the frequency band unless $T = 0$, and we throw out that special case on the grounds of unreliability in the T-neuron. When $T > 0$, frequency F_0 is below the center of the frequency band; the fractional part of the band that is below F_0 is less than one-half. This fraction is defined as the ratio $(F_0 - F_{\mathrm{L}})/W$, and its value is given by

$$\frac{F_0 - F_{\mathrm{L}}}{W} = \frac{1}{2}\frac{D - T}{D}. \tag{3}$$

For example, if $D = 2$ msec and $T = 1$ msec, then only $\frac{1}{4}$ of the frequency band is below F_0. Of course, this is a rather extreme case, with a very large band and low selectivity. If the selectivity is 10, as was assumed in most examples above, then F_0 will be very near the center of the band: it can be shown that approximately 0.47 of the band will be below F_0, regardless of the value of F_0, when the selectivity is 10. The greater the selectivity, the more closely F_0 approaches the exact center of the band.

Another important point is the independence of D and T. The value of D is determined by synaptic delays and rates of axon propagation preceding the T-neuron. The value of T, on the other hand, is determined by synaptic and other parameters of the T-neuron itself. Thus any agency which can effect changes in the value of T independently of the value of D can alter the detector's bandwidth without changing its tuned frequency.

Perhaps the most important aspect of the independent variability of D and T is that the selectivity of a band detector can be adjusted without affecting the tuned frequency. Weak dc fields might be effective agencies for altering temporal summation behavior and therefore the value of T. If the T-neuron N_3 in Fig. 2 possessed a dendritic structure in addition to the soma synapses indicated, then other neurons could control T by biasing N_3 toward depolarization via the dendritic synapses. Similarly, if N_2 has a dendritic input, other neurons might be able to alter the synaptic delay, hence D, hence the tuning of the detector.

In expanding- and contracting-band detectors, the quantitative relations between frequency bandwidths W and the parameters D and T are of special interest. Consider a band detector with a reasonably good selectivity of around 10 or greater. T will be small relative to D ($T \approx 0.05D$), and the term T^2 can be discarded in Eq. (1) without introducing serious error. This gives us the simple approximation equation

$$W = \left(\frac{2}{D^2}\right) T. \tag{4}$$

With D constant, the bandwidth is seen to be linearly proportional to T. However, the constant of proportionality is twice the reciprocal of D^2; hence if D is small, the *absolute* change in W produced by a change in T is quite large. This suggests that expanding- and contracting-band detectors would be most effective at the higher end of the frequency spectrum. At the lower end of the spectrum, T would have to alter considerably to effect a given change in bandwidth.

As a final note we should draw attention to the "sharpness" of a detector's band. A band detector's output, when plotted against frequency, is a very square "curve." This is due to the all-or-none output, which has only two values: firing and nonfiring. Of course, moment-to-moment random fluctuations in T and D would make the upper and lower limits of the band a bit fuzzy from a statistical point of view. However, for any given value of T, the band of a detector is very sharply defined, much more so than in the typical sinusoidal bandpass filter.

Mutations of the Band Detector

The basic band detector has been discussed at some length. Before proceeding to systems of band detectors, we shall briefly describe three other simple "resonant" networks; they are essentially variations on the network of Fig. 2 but are quite distinctive in their behavior or possible roles in neural communications. The first two involve inhibitory synapses, and the third network involves the use of "external" delays.

The Band Detector with Harmonic Suppression. In the previous section it was shown that a band detector will respond to regular trains whose fundamental frequencies are integral multiples, or "harmonics," of any frequency in the detector's band. In such cases the detector is actually responding to secondary intervals of the train, intervals which are the sum of two or more primary intervals.

Strictly speaking, this characteristic of band detectors cannot be considered a defect. The detector's function is presumably to detect the presence of *component* periods falling within its band regardless of whether those periods are produced by successive or nonsuccessive pulse pairs. However, if we assume that in some parts of a nervous system the lengths of primary intervals are the carriers of information, then the problem of harmonic responses could be quite serious, especially if long trains are involved. Is there a way to make the detector respond only to primary intervals, a way to "suppress" harmonic responses?

The answer is yes, and the simplest solution is represented by the network in Fig. 7. This is merely the network of Fig. 2 with an inhibitory terminal added to neuron N_1. The axon branch with inhibitory terminal Y is longer than the

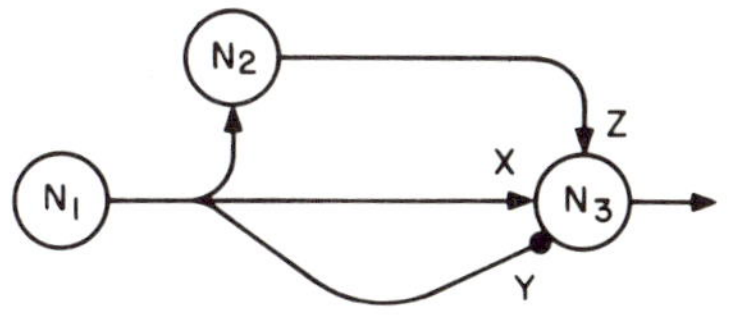

FIG. 7. Band detector with harmonic suppression, a mutation of the network in Fig. 2. The dot at Y represents an inhibitory terminal.

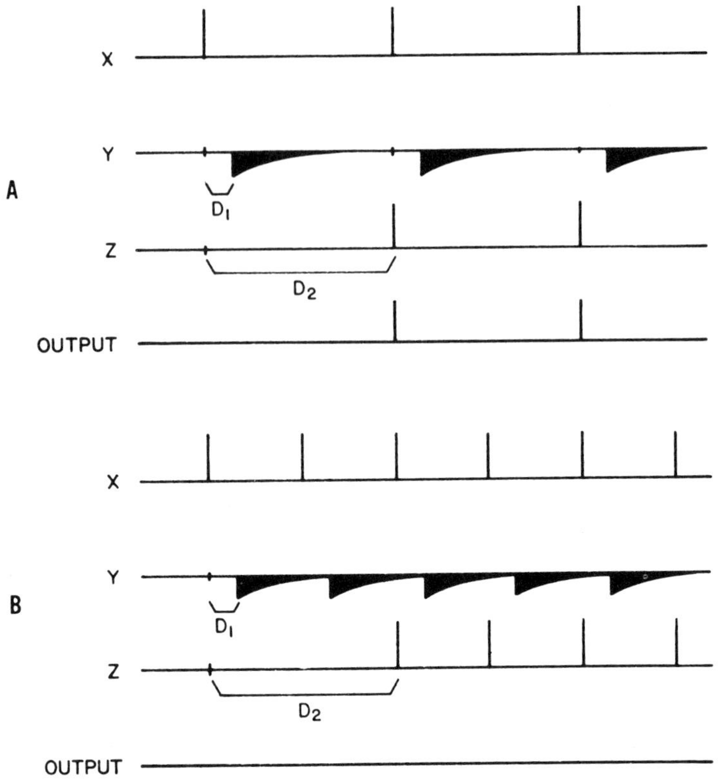

FIG. 8. Time relations for the network of Fig. 7. Solid black areas represent the inhibitory action at terminal Y; while Y is active, N_3 is unable to fire even if pulses arrive at X and Z in perfect coincidence. A. The input train has a fundamental period equal to D and passes through the detector. B. The input train's fundamental period equals $\frac{1}{2}D$ (first harmonic); although excitatory pulses arrive in coincidence, inhibitory terminal Y is always active, and prevents firing in N_3. Time delays between input and output trains have been ignored in these graphs.

branch with excitatory terminal X; thus a pulse arriving at X is followed shortly by a pulse at Y. This short delay between activity at X and Y will be called D_1. The delay between X and Z, which is the tuned period of the detector, is D_2. The behavior of this network is illustrated in Fig. 8.

First, consider an input train with a fundamental period approximately equal to D_2. In Fig. 8A the first few pulses are shown as they arrive at neuron N_3. The inhibiting effect of terminal Y has a sharp rise and a gradual decay, as indicated by the solid-black negative sawtooth waves. It is assumed that during these intervals N_3 will not fire even if two excitatory pulses arrive simultaneously. But here the inhibition has died away before the near-coincidence of pulses at X and Z, so that N_3 fires.

Let I be the inhibition period. If $D_2 - D_1 + I < T$, then the lower limit of the period band is determined by the inhibitory terminal rather than by T, and the band may be quite asymmetrical about D_2. Thus the selectivity of the detector can to some extent be controlled by the inhibitory terminal.

If the fundamental period of a regular train falls within a certain band about D_2, then the train will be passed, as in the case of the simple band detector. But suppose the input train has a fundamental period that is half of D_2, i.e., its fundamental frequency is the first harmonic of the detector's tuned frequency. The effect of such a train is shown in Fig. 8B. Neuron N_3 is effectively inhibited continuously and is unable to fire. Hence this train is "suppressed"—it does not pass through the detector. Clearly, higher harmonics will produce even stronger inhibition, and therefore *no* harmonic trains can fire the detector. We have then a band detector with harmonics suppressed.

If the inhibited periods are very short, then one or two of the lowest harmonics might pass through the detector. For example, suppose that $D_1 + I < \frac{1}{2}D_2$. Then the first harmonic could pass, although all higher harmonics might be suppressed. Consequently, the strength of inhibitory terminal Y may determine how many, if any, harmonics are passed.

It is not necessary that inhibition be provided directly by a branch axon of N_1 as shown. A fourth neuron similar to N_2 could be interposed between N_1 and N_3. This would be a 1:1 neuron with an inhibitory terminal on N_3; it would be excited by N_1 and would presumably have a smaller synaptic delay than N_2. Such a neuron would provide an opportunity for other neurons outside the network to determine whether the detector would suppress harmonics at any given time, for if this neuron were itself inhibited, then no pulses would reach Y.

If band detectors were used to control the routing or decoding, or both, of "second-order" trains (as defined above), and if the first-order trains involved were only pulse pairs whose intervals determined which detectors were activated, then there would be no harmonic problem. It will be recalled that if a regular train has a fundamental frequency which is the Kth harmonic of a detector's tuned frequency, then in effect the first $K + 1$ pulses are "lost"—the train must have at least $K + 2$ pulses to cause firing in the detector. If a harmonic train consisting of two pulses is received by an ordinary band detector, that train will not cause the detector to fire even if its fundamental frequency is only the first harmonic of the detector's tuned frequency. Thus if information is carried by intervals between such pulse pairs (these intervals being long compared with those between a pair of pulses, as in Fig. 6), there is no danger that the information will be decoded by the wrong detector owing to harmonic effects. To provide a maximum range of variability for the intervals between pulse pairs, the period between a pair of pulses should be as short as possible; consequently, detectors tuned at frequencies near the high end of the neural frequency spectrum are the most desirable. Whether such communication methods are used by any real nervous systems can only be settled by experiment.

The Band Suppressor. Consider the network of Fig. 9. Assume that neuron N_3 will fire once for each pulse arriving at the excitatory terminal from N_1 if the

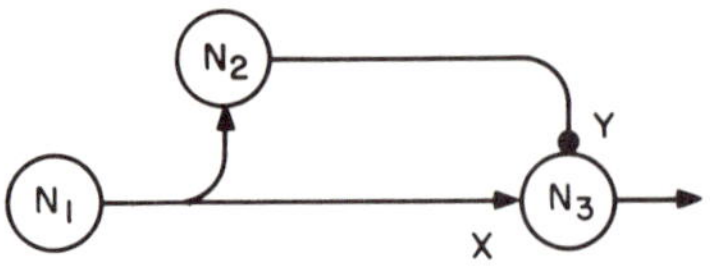

FIG. 9. A band suppressor. A pulse arriving at excitatory terminal X causes N_3 to fire unless the inhibitory terminal Y is concurrently active.

inhibitory terminal from N_2 is not active; in this respect N_3 behaves like a 1:1 neuron rather than a *T*-neuron. However, each time a pulse arrives at the inhibitory terminal, N_3 is inhibited for a period *I*. Since we assume that, as in previous networks, N_2 is a 1:1 neuron, each pulse arriving at the excitatory terminal X is followed, after a net delay *D*, by a pulse arriving at the inhibitory terminal Y. The situation is summarized in Fig. 10. If a pulse reaches X at time *t*, then N_3 is inhibited over the interval between $t + D$ and $t + D + I$. A second pulse arriving at X during this interval cannot cause N_3 to fire and is not, therefore, represented in the output train from N_3; it is "suppressed." Clearly, there is a band of periods following the first pulse at X, delimited by the dashed vertical marks, which is suppressed in any input train. Thus we call this network a *band suppressor.*

In terms of frequencies, the lower band limit $F_L = 1/(D + I)$ and the upper band limit $F_U = 1/D$. However, it should be noted that all harmonics of frequencies falling within this band are also suppressed; thus the band suppressor's response to harmonics complements that of a band detector.

If *any* two pulses in an input train, whether the train is regular or irregular, are separated by a period falling within the suppressor's band, then the second pulse will be suppressed. Thus this network responds to *component* periods, not just to the durations of primary intervals. For example, Fig. 11A shows the response to an irregular input train. The magnitudes of *D* and *I* are indicated. All of the suppression here is due to secondary intervals; each suppressed pulse is affected not by the immediately preceding pulse, but rather by an earlier pulse. In Fig. 11B the input is a leading pulse followed after some delay by a burst of pulses. Some of the component periods of this train fall within the network's band, and the corresponding pulses are suppressed; the burst is chopped into two shorter bursts. All of the suppression here is due to the isolated leading pulse. However, if the burst were a bit longer than *D*, then the early pulses of of the burst would cause some of the trailing pulses to be suppressed.

This effect could be used in neural communications in the following way. Assume that the input to a band suppressor is a burst of pulses, the duration of the burst being greater than *D* and the primary intervals being less than *I*.

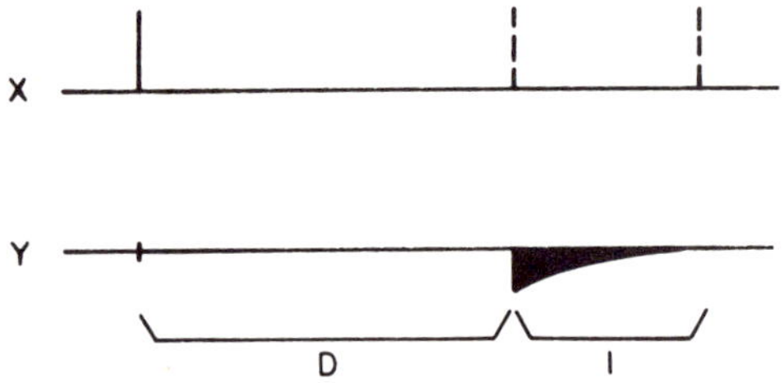

FIG. 10. Time relations for the network of Fig. 9. Component periods falling in the band from *D* to $D + I$ are "suppressed"; i.e., the output train does not contain any components in this band. Period *I* is the effective duration of inhibitory activity in terminal Y after the arrival of a pulse.

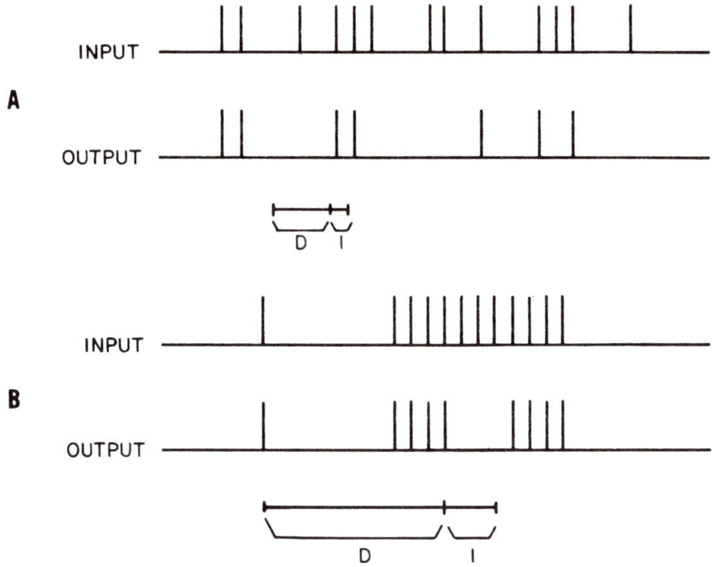

FIG. 11. Behavior of a band suppressor. A. Six pulses in the irregular input train are suppressed; the output train has no component periods in the band D to $D + I$. B. An isolated leading pulse causes the suppression of a group of pulses in the following train. Time delays between input and output trains have been ignored in these graphs.

Whether the burst is regular or irregular, it is uniformly "chopped" to duration D. Thus a band suppressor could transform a series of bursts (separated by long intervals) having many different fine structures into a series of bursts of uniform length; over a wide range the "average frequency" of a given burst is not important, although it will be preserved in the output (within limits). By contrast, if bursts of uniform duration are to be produced by a combination of temporal summation and negative feedback processes (in a network having the same topology as in Fig. 9), then the average frequency within a burst must be approximately the same for all bursts. The higher the burst frequency, the shorter the burst, and conversely. The "impact" of such bursts on statistically sensitive devices downstream tends to be constant because burst duration cancels out information carried by the frequency of pulses within the burst. A band suppressor, on the other hand, makes it possible to transmit information by average frequency within a burst. It can do this because it chops bursts to uniform length regardless of this frequency, so long as the primary intervals in the burst are less than I. We have here a curious case: a non-statistical device—i.e., a band suppressor—serving the ends of a statistical communication system. Of course band detectors can similarly be used in statistical systems.

In the network of Fig. 9, the first pulse of an input train always gets through if there have been no previous inputs for a period $D + I$. Exchanging terminals, i.e., making X inhibitory and Y excitatory, alters only this property. The network remains a band suppressor, but the first pulse may not get through. More generally, this property may be characterized by saying that if any two pulses in a train are separated by a period between D and $D + I$, then the *first* pulse of the pair is suppressed. In the network of Fig. 9, as noted perviously, it is the *second* pulse that is suppressed. This has an important effect on burst processing; if an input burst has a duration greater than D and primary intervals shorter than I, then the burst will again be chopped to a duration D. However, the output burst will copy the *last* D sec of the input burst rather than the *initial* D sec. If we assume that the input bursts are like those commonly seen in real neurons—i.e., with primary intervals growing successively larger—then the last D sec of the burst may contain primary intervals larger than I; some of the early pulses may get through, and the burst is not uniformly chopped to duration D. Thus if band suppression is used in neural systems to produce bursts of uniform length, then one would expect the network of Fig. 9, rather than that produced by exchanging terminals, to be preferred. It passes only the first D sec, during which the primary intervals are the shortest and it is most likely that all are less than I.

Space does not permit further discussion of the possible roles of band suppressors. It is clear that suppressors working in conjunction with band detectors could improve the reliability, flexibility, and selectivity of frequency-sensitive systems.

From a quantitative standpoint the band suppressor is similar to the band detector. The major difference is that $1/D$ is not near the center of the frequency band but instead defines the upper limit of the band. The lower-frequency limit is of course $1/(D + I)$. The most important similarity is that the bandwidth of the suppressor can be varied by changing I while holding D constant. However, the relation between bandwidth and I is more nonlinear than the relation between W and T in a band detector. We shall not bother to set down the band-suppressor equations paralleling Eqs. (1) through (4); they are simple functions as easily derived.

It should be noted that by introducing processes such as synaptic facilitation, or fatigue into neuron N_3, one can obtain either expanding- or contracting-band suppressors. As in the case of detectors, such suppressors can, for example, transform a long regular train into a series of shorter trains of the sort that might produce rhythmic muscle contractions.

As a final remark, we draw attention to the effect a band suppressor has on an irregular train. In "passing through" the suppressor, those pulses responsible for component periods falling within the band are "removed"; the output train contains no components in the band. However, when a pulse is removed, many other components falling well outside the band *may* also be eliminated from the train. Thus it is not correct to say that a suppressor necessarily eliminates *only* components in its band. Both larger and smaller components may be suppressed.

The Phase-Band Detector. All of the networks previously considered can be characterized as single-input devices; they operate upon pulse trains initiated at a single neuron. There is, however, a simple mutation of the band detector which operates upon two input trains. This double-input device is shown in Fig. 12. The synapses or spike-initiation loci of N_1 and N_2 provide the inputs for this network, and the pulse trains produced by the T-neuron N_3 are the output.

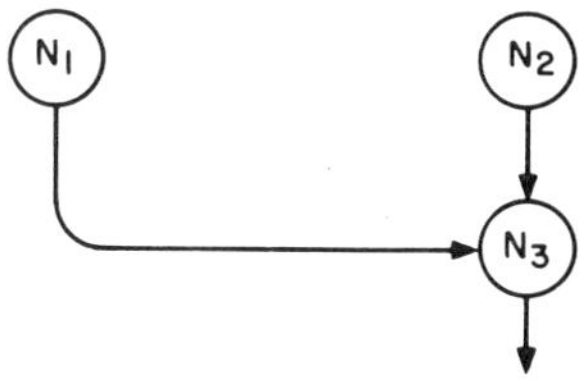

FIG. 12. Phase-band detector. Neuron N_3 is a T-neuron; coincidence of pulses arriving at N_3 is contingent on the proper phase relation between firings of N_1 and N_3.

Assumptions about the pulse speeds, net delay difference D, synaptic delays, and firing requirements are just those of the basic band detector. As before, if the period between pulse initiations at N_1 and N_2 is no less than $D - T$ and no greater than $D + T$, then N_3 will fire.

Although the term "phase" is normally used to refer to time relations between two repetitive processes, we shall extend its meaning to cover the relation between single pulses initiated at different neurons. If a pulse is initiated at N_2 *after* a pulse is initiated at N_1, then we shall say that the phase of the N_2 pulse is *positive* relative to the N_1 pulse. The magnitude of the phase relation will be denoted by ϕ. Thus if $D - T \leqq \phi \leqq D + T$, then N_3 will fire. In the special case where $D = 0$, the network becomes a simple near-coincidence detector.

To economize discussion a few special terms must be introduced. The *structure* of a train is the specific set of periods between successive pulses in that train. If two trains have the "same" structure, then there is a one-to-one correspondence between successive pulses and the periods separating these pulses. In comparing two trains, the kth *pair* of pulses refers to the kth pulse of one train and the kth pulse of the other. Thus the "first pair" of pulses consists of the first pulse of each train. More generally, these are called *matching* pairs. The train produced by a particular neuron N_i will be called the N_i train. For example, in Fig. 12 the N_1 train is one of the input trains of the network. We shall use the "phase of the kth pair" as a shorthand expression for "the phase of the N_2 pulse relative to the N_1 pulse in the kth pair."

If the N_1 and N_2 trains have the same structure (whether regular or irregular), then it is obvious that the phase of every matching pair will be the same; in this case it is meaningful to speak of the phase of the N_2 *train* relative to the N_1 *train*. If two trains do not have the same structure, then we cannot generally speak of the phase between them.

The behavior of the network in Fig. 12 can be characterized as follows: If the N_1 and N_2 trains have the same structure and the phase ϕ between them falls between $D - T$ and $D + T$, then N_3 will produce a train with the same structure. Since the network responds to a band of phases, we call it a *phase-band detector*. The condition that the N_1 and N_2 trains have the same structure may be considered a special case; there are obviously many other cases of interest, but space does not permit their analysis here.

Unlike the previous networks, a phase-band detector does not selectively respond to period components within a train (except in the degenerate case where primary intervals are shorter than the period T). So long as the input trains have the same structure and proper phase relation, every matching pair of pulses will cause N_3 to fire regardless of the period components in the trains. From this standpoint it is perhaps best to view the detector as a "phase switch." If N_1 and N_2 produce trains that have the same structure, it is reasonable to suppose that there is some single stimulating source somewhere upstream from N_1 and N_2. For example, assume that there is a single neuron N_0 which generates a pulse train; this train is subsequently split by a branching axon into two trains (with identical structure), which then proceed along parallel paths to N_1 and N_2. If the path to N_2 involves a greater total delay than the path to N_1, then the N_1 and N_2 trains will be "out of phase," i.e., $\phi > 0$. Any agencies, such as other neurons or electric fields, which can differentially modify the delays of the two paths (most probably by acting on synaptic delays) may therefore control ϕ. Suppose that D and T are constant in the phase-band detector; then these agencies can determine whether or not any pulse train originating at N_0 will be reproduced by N_3. If ϕ is adjusted to fall within the detector's phase band, the N_3 train is a copy of the N_0 train, regardless of its structure. If ϕ is moved outside the band, then N_3 produces nothing. This constitutes a very "clean" switching action. The magnitude of T determines how much "noise," in the form of random variations in pulse-to-pulse periods, may be introduced along the parallel paths without interfering with the switching action.

Since the phase of each matching pair can be independently measured and may differ from pair to pair, it is meaningful to speak of the average value of all these phases as the "average phase" of the *train*. Presumably noise would affect the variance of matching-pair phases about this average. If the variance were of the same order of magnitude as the width ($2T$) of the phase band, then many pulses might be "lost" at N_3. For example, if the average phase were near the upper or lower limit of the band, it is likely that approximately half of the pulses in the N_0 train would be missing from the N_3 train. Thus noise, of the form considered here, could quite seriously interfere with switching action.

On the other hand, noise in the form of "spontaneous" pulses appearing at points in either of the parallel paths would have little or no effect on the system, especially if refractory periods within each path were of the same order of magnitude as T. If the two trains' average phase is in the detector's band, then every matching pair fires N_3, and "extra" firing can only occur if *two* pulses spontaneously appear on the parallel paths with just the correct phase relation, an unlikely event. The same is true if there are no trains traveling down the paths. However, if there are trains with an average phase outside the detector's band (the "switch" is "off"), then spontaneous pulses can cause N_3 to fire. Even in this case the spontaneous pulses must occur with just the correct phase relations to pulses in the train on the other input path. The probability of this occurring would be proportional to T. Thus the parallel paths from N_0 to the phase-band detector should be as free as possible of moment-to-moment fluctuations in propagation velocities in order to minimize the first kind of noise, but the system could function well in the presence of extensive "spontaneous-firing" noise. The absolute lengths of the parallel paths are not important, spatially or temporally, because only the net difference in delays determines the phase between trains reaching the detector. Since the paths can be quite long, spatially, and can be composed of many neurons in series, a diffuse and weak agency could, by its *cumulative* effects on many synapses, produce relatively large changes in ϕ and thereby turn the "phase switch" on and off. Of course, the structure of the output train from N_3 is determined by N_0 rather than the phase-control agency.

Perhaps a more interesting role for a phase-band detector is the sensing of "velocities." Suppose that N_1 and N_2 are receptors sensitive, say, to light. They are spatially separated as suggested in Fig. 12. Consider a stimulus object moving from left to right in front of the receptors, exciting first N_1 then N_2. If each cell produces one pulse when excited, then the period between excitation of N_1 and N_2 must fall within the phase band if it is to fire N_3. Let the distance between N_1 and N_2 be A; then the velocity V of the stimulus object must satisfy the condition $D - T \leqq A/V \leqq D + T$ in order to fire N_3. The smaller the value of T, the more selective the network for velocities. There is a "band" of velocities proportional to the phase band, and any velocity within that band fires N_3; any velocity outside the band produces no output whatever. Furthermore, a *single* stimulus object moving from right to left cannot fire N_3 at *any* velocity.

However, a *sequence* of stimulus objects moving from right to left can cause N_3 to fire repeatedly. There are many combinations of velocity and spatial separation which will be effective, but we shall consider only one case. Suppose that we have a long series of stimulus objects separated by equal intervals of magnitude S, and that they can pass in front of N_1 and N_2 at a velocity V. As before, N_1 and N_2 are separated by a distance A. Suppose further that S is less than A but greater than $A/2$. It can be shown that there is a *lowest* velocity V_{LR} from left to right that stimulates N_3, and that similarly there is a lowest velocity V_{RL} from right to left. Actually, there is a band of effective velocities about each of these "thresholds," the width determined by T at N_3, but we shall ignore these. In general V_{LR} and V_{RL} do *not* have the same value, because $V_{LR} = (A - S)/D$ whereas $V_{RL} = (2S - A)/D$. This would be experimentally important if one were attempting to determine the values of both A and D in, a real network of this type. If one could detect the occurrence of firing in N_3 then the lowest effective velocities from left to right and from right to left could be measured. Substituting these for V_{LR} and V_{RL} in the equations above and solving simultaneously, we would obtain the values of both A and D. If it happens that $V_{LR} = V_{RL}$ experimentally, then we can conclude that $A = 3S/2$, and either equation yields the value of D. Should the velocity bands be wide enough for measurement, then of course the center values of each band are V_{LR} and V_{RL}.

It should be noted that if N_1 or N_2, or both, produce a short train of several pulses when stimulated, then several bands of velocities, perhaps overlapping to form one large band, would become effective; for example, if the period between the first firings of N_1 and N_2 does not fall within the detector's phase band, then perhaps the period between the first firing of N_1 and the second firing of N_2 does fall within the phase band. In general, the greater the number of pulses generated by each receptor, per stimulus, the greater the chance that some combination of pulses will have the correct phase relation and fire N_3. It follows that the selectivity of velocity detection is greatly reduced if the receptors produce bursts of pulses in response to a single stimulus.

As in the networks discussed earlier, if T varies appreciably as a result of excitatory inputs or firing at N_3, or both, then an expanding- or contracting-phase-band detector becomes possible. The behavioral consequences parallel those outlined for the period-band detector, and their significance will depend upon the role of the phase detector in a larger neural system.

A System of Detectors

In discussing the basic band detector and simple mutations on that network, we have put forward some suggestions about the possible role of such devices in larger communication systems. Since it is not known whether resonant networks of this type do in fact exist in any nervous system, it may appear absurd to push speculation still further by examining hypothetical ensembles of hypothetical networks. However, a theory can suffer from underdevelopment as well as from overelaboration. This is particularly true in regard to experimental verification, for it may be difficult or impossible, at a given time, to observe the elementary mechanisms postulated in a theory, whereas the gross effects of an ensemble of these mechanisms may be easily observed. The theory of Brownian motion is but one example of the importance of extrapolating from the behavior of a single element to the behavior of a system of such elements.

The small size of the basic resonant networks here proposed, not to mention their sensitivity to small variations in temporal relations between pulses, may make it very difficult to determine experimentally whether or not such devices are used in real nervous systems. If we assume (as I think we must) that these devices are most likely to occur in ganglia or at higher levels, then a search for them will be exceedingly difficult because of the dense masses of neurons surrounding them. Under these circumstances it would be useful to have in hand some ideas about the behavior of coordinated groups of band detectors, for such behavior may be easily observed. Space does not permit the exploration of more than one such system of band detectors, and even that system can only be sketched.

The Period Analyzer. As noted in a previous section, the net delay D must be quite large (10 msec to 1 sec) if a band detector is to be tuned to frequencies at the lower end of the neural spectrum, the region in which most neural activity appears to be concentrated. Such delays could be achieved by many neurons in series so that their synaptic delays would be summed, or by a few neurons with unusually large synaptic delays or unusually long and slow axons, or both. In either case the nervous system would have to evolve special structures. Although organisms have in fact evolved special, even exotic, nervous mechanisms, it would seem desirable to minimize the necessity for specialization, wherever possible, in hypothetical networks. We therefore assume that the fewer the special structures required, the greater the probability that band detectors have been evolved for use in neural communications.

A second major problem is that of coordinated parameters: how can a set

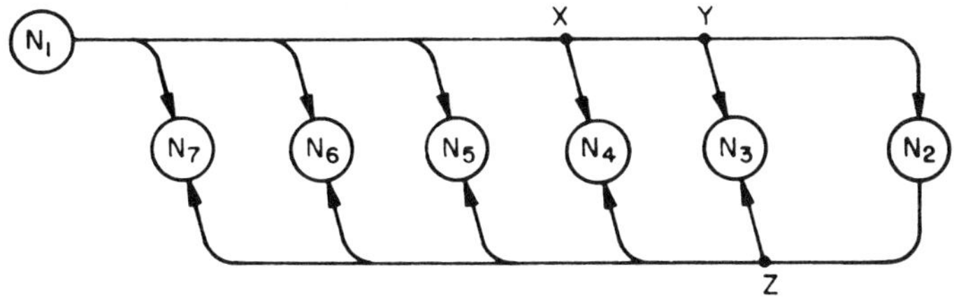

FIG. 13. A period analyzer composed of five band detectors. N_2 is a 1:1 neuron; N_3 through N_7 are *T*-neurons. The N_1 and N_2 axons and the synaptic delay of N_2 provide the requisite delays for all five band detectors. Output axons of N_3 through N_7 are not shown.

of band detectors be arranged so that their bandwidths and tuned frequencies have an orderly relation to one another?

The problem of specialized structures and that of coordinated parameters would both be alleviated, if not solved, by the network of Fig. 13. Here a single delay path serves five band detectors (N_3 through N_7). For the moment we assume that the axons of N_1 and N_2 are unusually slow and that the synaptic delay of N_2 is unusually large. Thus these two neurons represent a specialized structure. Since this special delay path serves not one but five (and in principle many more) detectors, the specialization cost *per detector* is greatly reduced. The detectors themselves need only be spatially ordered in a simple linear array (the axons of N_3 through N_7 are not shown).

The behavior of this *period analyzer* is easily predicted. Neurons N_1 and N_2 are 1:1 neurons (firing of N_1 constitutes input to the system), whereas N_3 through N_7 are *T*-neurons. In the simplest case we assume that the distances between axon branch points, such as X and Y, and between these points and the terminals on the *T*-neurons are all the same; the axon propagation velocity is also assumed to be uniform throughout the system. Thus the delay between any two branch points, denoted by *d*, is everywhere the same. We do not care about the delay from a branch point to a *T*-neuron so long as it is the same in all cases. Finally, we assume that *T* has the same value for all five *T*-neurons.

Let the basic delay between branches Y and Z be D_3; this includes the synaptic delay of N_2 and becomes the net delay for the band detector N_3. The net delay for N_4 denoted by D_4, will be $D_3 + 2d$, and so on. Figure 14 shows one possible set of relations between the period bands of the five detectors. The net delays D_3 through D_7 are marked off on the abscissa (as measured by any unit of time). As noted, each successive pair of net delays is separated by $2d$. The bandwidth is of course $2T$; the corresponding *T*-neuron is indicated by symbol in each band. The ordinate scale has just two values: Fires and Doesn't Fire. If and only if two pulses in the input train are separated by a period falling within the band of a given *T*-neuron, they will cause it to fire. This graph characterizes the response of the system to all component periods in any input train of pulses produced by N_1.

It is clear that the network selectively responds to five ordered bands of component periods (or frequencies), and has five spatially separated output channels corresponding to these bands. Thus a single input train may produce up to five distinct output trains whose structures depend upon the structure of the input train. The relations between these bands are determined by T and d. Three classes of period analyzer are possible (when T and d are uniform). The first class, as shown in Fig. 14, has *separated* bands. This results when $d < T$. A second class has *contiguous* bands, the upper limit of one band being identical with the lower limit of the next band. This results when $d = T$. The third class has *overlapping* bands; this results when $d > T$. It is obvious that if d and T are not uniform throughout the net, then some neighboring bands may overlap whereas others are contiguous or separate.

A glance at physiological parameter restrictions is in order here. Suppose that the axons of N_1 and N_2 are very slow; let us say they propagate pulses at 1 m/sec. Suppose further that successive branch points (and T-neurons) are separated by 1 mm. Consequently, $d = 1$ msec, and each period band has a width of 2 msec. Assign a modest value to the "basic" delay D_3 of say 10 msec (produced primarily by the synaptic delay of N_2). Finally, let $T = d = 1$ msec so that the bands are contiguous. How will the network behave from the standpoint of the neural frequency spectrum?

We find that the network as a whole will respond to all frequency components from 53 through 110 pps. The T-neuron N_3 is tuned to the highest frequency, 100 pps, and has a bandwidth of 20 pps. At the lower extreme, N_7 is tuned to 56 pps with a bandwidth of only 6 pps. Taking the ratio of tuned frequency to bandwidth as a measure of selectivity, we find that the detectors become progressively more selective from N_3 through N_7, the selectivity of N_3 being 5.00 while the selectivity of N_7 is 9.25, almost twice as great.

Thus the neural parameters assumed above provide a period analyzer that spans, with contiguous bands, about half of the lowest tenth of the neural spectrum. The most disturbing parametric assumption was that the axons have a propagation rate of only 1 m/sec; this is probably a lower extreme for interneuron conduction velocities in higher animals. Furthermore, we assumed that

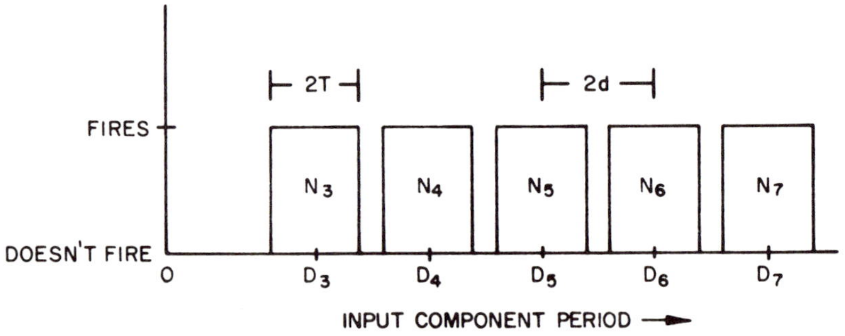

FIG. 14. Period bands of detectors in the period analyzer of Fig. 13. In this example the bands are separated, and all have the same width (T assumed the same in all five T-neurons). The successive band-center periods D_3 through D_7 are separated by $2d$, where d is the time delay between successive axon branch points such as X and Y in Fig. 13. The period from 0 to the center of the lowest period band, D_3, is the delay between points Y and Z. If an input train contains a component period falling within one of these bands, the corresponding T-neuron will be fired.

the *T*-neurons were spaced at intervals of 1 mm, a large distance in ganglia, and to reduce this spacing without altering the spectral location of the bands it would be necessary to assume axon velocities *below* 1 m/sec. Of course, one way to avoid these parametric difficulties is to assume that analyzers are used only at the high end of the neural frequency spectrum. However, this is a rather unsatisfactory solution; most neural communications appear to involve frequencies at the lower end of the spectrum, and therefore analyzers would be of marginal utility at best.

There is a way to satisfy all three parametric demands, i.e., close spacing of *T*-neurons, reasonably high axon velocities, and low-frequency coverage. If a synapse is introduced between each pair of branch points (such as X and Y) in the network of Fig. 13, then the value of *d* will be determined primarily by synaptic delays rather than axon delays. A chain of four 1:1 neurons would be substituted for the axon running from N_1 to N_2, and a similar chain interposed between N_2 and N_7. The spacing between *T*-neurons N_3 through N_7 would in this case be unimportant and could be quite small. By *adding* eight neurons, the total length of the analyzer could be *reduced* well below 1 mm. This seeming paradox is characteristic of all of the resonant networks thus far examined; by increasing the number of neurons used (thus the number of synaptic delays), the over-all dimensions of a network can be reduced, because synaptic delays require much smaller linear spaces than axon delays. This suggests that the natural habitat of resonant networks is a dense mass of neurons where chains of neurons can easily be formed. If the extremely fine axon fibers in neuropiles should also tend to have unusually low propagation velocities, so much the better. From the standpoint of resonant networks, the trouble with axons is not that they are too slow, but rather that they are *too fast*.

This may seem to be a rather startling proposition; one usually assumes that since the main business of neural systems is communication, axon conduction could not be too fast. Our analysis suggests that this assumption may be reasonable only in the case of point-to-point communication channels. The encoding and decoding of complex messages would create a definite demand for slow axons if resonant networks were employed. It is interesting that Kennedy (this volume) has also concluded that slow axons would be desirable, and that his conclusion is based on an investigation of existing networks which do *not* display resonant properties (of the sort we have considered here).

We also note that since resonant networks would enable a nervous system to utilize complex codes, much more information could be packed into a pulse train of given duration than would be possible if only statistical decoding were available; consequently, the evolution of slow axons for resonant networks would *reduce* the need for fast conduction between encoding and decoding sites.

Harmonic Behavior of Analyzers. Another difficulty encountered in period analyzers stems from harmonic responses. In the analyzer described above, a regular input train with a fundamental frequency $F_f = 110$ pps would pass through N_3 as well as N_7 because $\frac{1}{2}F_f$ falls within the band of N_3. Unless harmonics are suppressed in the lower bands by inhibitory terminals, as discussed in the section on the Band Suppressor, a regular input train may produce two or more output trains simultaneously at widely separated detectors in an analyzer. This phenomenon is not necessarily a defect; if the outputs of an analyzer are used to activate a variety of mechanisms, then harmonic effects could be utilized in the control and recruitment of, for example, a group of synergistic muscles.

However, if the outputs of the analyzer are to be mutually exclusive, and harmonic suppression is too costly, then it is necessary to restrict the total span to one octave of the neural spectrum. Under these circumstances the number of possible bands is limited by the widths of the bands (assuming that overlapping bands are to be avoided). The minimum practical value of T then becomes very important.

Owing to the reciprocal relation between frequency and period, an analyzer with frequency bands of either uniform width or uniform selectivity must have nonuniform values of T. In the network of Fig. 13, for example, if N_3 through N_7 are to have equal selectivities, then it is necessary that the corresponding T's have the relation $T_3 < T_4 < \cdots < T_7$. If, furthermore, the bands are to be contiguous, then the differences between the successive net delays D_3, D_4, $\cdots$, D_7 must be progressively larger; the distance between branch points of the N_1 and N_2 axons will not be uniform.

The net delay between points Y and Z, denoted by D_3, may be a controlled variable if the synaptic delay of N_2 can be appreciably affected by external agencies. If the synaptic delay of N_2 is increased, then all bands of the analyzer are uniformly shifted to lower frequencies; conversely, a decrease in the synaptic delay shifts all bands to higher frequencies. Such shifts do not alter the relations between bands with respect to separation, contiguity, and overlapping. Thus the synaptic region of N_2 provides a single point through which external agencies such as other neurons and electrical fields can control the "position" of the frequency analyzer in the neural spectrum. If a major portion of the net delay D_3 is due to this synaptic delay, then the range of variability can be quite large.

As in individual band detectors, variations in T due to input or output pulse trains, or both, can markedly affect the over-all behavior of an analyzer. For example, if the analyzer's bands expand in response to a long input train, then they may initially be separate but later overlap. This effect could have important consequences in many applications. However, we cannot pursue the subject further here.

Summary and Conclusions

It is possible, in principle, for small networks of neurons to respond selectively to bands of component periods in any pulse train. The behavior of such a network is similar, at an abstract level, to that of resonant physical systems; thus we have called these networks "resonant." The resonant behavior is produced by certain simple combinations of time delays and near-coincidence detection.

The primary theoretical question is whether or not existing neurons could meet the parametric requirements of resonant networks. Since the limits of the neural frequency spectrum are rather well known, it is possible to deduce with quantitative precision the corresponding parameter ranges of neurons in resonant networks. At the lower end of the neural spectrum, say from 1 to 100 pps, the most critical parameter appears to be the net delay; it is necessary that one or more neurons provide reliable and constant delays in the range of 10 msec to 1 sec. Although empirical data are fragmentary and by no means conclusive, they suggest that such delays could not be economically provided by axons (see, e.g., Prosser and Brown, 1961). Since it is very difficult to measure propagation velocities in short axons of very small diameter, these data may be misleading because they reflect instrumental rather than neural limitations. If small axons with velocities of approximately 0.01 m/sec do in fact exist, then the case for resonant networks is materially improved. However, one or more synapses in series, as would be provided by a chain of small neurons, offer an alternative to slow axons; a small neuron with somatic synapses and a short axon requires little space and may provide a synaptic delay of several milliseconds.

Over the upper region of the neural spectrum, say from 100 to 1000 pps, the most critical parameter is T, the maximum period that separates two input pulses capable, by temporal summation, of causing a T-neuron to fire. To achieve reasonable selectivity at high frequencies—for example, a bandwidth of 50 pps in a detector tuned to 500 pps—T must have a value in the neighborhood of 0.1 msec; the T-neuron must reliably fire if two input pulses arrive within a period of 0.1 msec and *reliably fail to fire* if the period is greater. This calls for very stable combinations of firing threshold, synaptic properties, and axon terminal activity. The magnitude of net delays must also be very stable in this frequency region.

Since small fluctuations in net delays and in T have negligible effects at low frequencies, and since the majority of neural messages appear to involve pulse frequencies in the lower part of the neural spectrum, we conclude that if resonant networks have been evolved in neural systems, then they are most likely to be found operating at frequencies between 1 and 100 pps. As a corollary we conclude that net delays greater than 10 msec constitute the primary parametric problem. Can such delays be provided by one or a few small neurons packed into a small volume? Only the experimental biologist can answer this question; in this paper we simply assume that such delays are not only possible but reasonable.

A special delay path becomes more economical, and presumably more reasonable, if it services a group of neurons rather than just one. We have described a "period analyzer" in which a single, long, delay path provides the net delays for two or more band detectors; in such a network the cost of providing long delays for each detector *decreases* as the number of detectors is increased. We suggest, therefore, that if resonant networks (of the type considered in this paper) exist in neural systems, they are most likely to be found in groups rather than isolated, and to be functionally related to each other. One might object that networks such as the proposed analyzer are so complicated and require so many precise connections that they are unlikely to be found in nature. In answer we point out that the period analyzer actually involves only a simple and regular pattern of connections between neighboring neurons; such structures may be easily controlled genetically. Furthermore, intricately organized special neural structures of many types have already been discovered and may indeed be the rule rather than the exception in higher animals.

The most important feature of a resonant network would be its ability to respond to pairs of pulses separated by one or more intervening pulses. A neural system which responds only to "average frequencies", whether they are long- or short-time averages, cannot behave in this way. The ability to respond selectively to *any* component period in a pulse train, not just those defined by primary intervals, would enable resonant networks to decode rather complicated messages; much more information can be packed into a pulse train of given duration if that train is to be decoded by resonant rather than average-frequency detectors. And even from a statistical point of view, a resonant detector has a great advantage in that it can quickly respond to sudden small changes in average frequency; it does not have the "inertia" necessary in averaging detectors. Another advantage is that a resonant network responds to a band of periods (or frequencies) with upper and lower thresholds. Averaging detectors are generally assumed to respond to all frequencies above, or all below, a single threshold; they cannot provide sharply selective band responses.

Since resonant networks would enable a nervous system to take full advantage of the information capacity of pulse-interval coding, the question must be raised how neurons could *encode* (as well as decode) information in secondary intervals of complex pulse trains. This problem is at least as important and interesting as the decoding problem, but even an equally sketchy discussion would require a paper on decoding systems as long as the present one. Hence the subject has been avoided here. However, we venture to point out, without explanation, that networks having essentially the *same* structure as those described above, but with T-neurons replaced by 1:1 neurons, provide basic mechanisms for encoding information in both primary and secondary intervals of complex trains. The chief problem is the prevention or control of "spurious" components in irregular trains; such components appear as the sums or differences of component periods representing information. Possible solutions to this problem have not been adequately explored by this author, and it is therefore impossible to make any summary estimate of the ease or difficulty with which a nervous system might reliably encode information in irregular trains. We can only say that "second-order" trains of "double" pulses seem to offer the simplest basis for nonstatistical coding schemes. This suggests that experimentalists should pay close attention to closely spaced pairs of pulses if they occur frequently in a record. Each pair should be treated as a single entity, and pairs classified by the period separating the two pulses. The intervals between pairs of the same period should then be examined for correlations with the behavior of relevant mechanisms. If such correlations were found, our theory would suggest that information from several sources was being funneled through a single axon or chain of neurons to be later sorted out (decoded) and sent to various mechanisms by a group of band detectors. If the double pulses are actually produced by two parallel axons, one pulse on each, then the presence of phase-sensing networks is suggested, especially if the axons in question carry signals produced by receptors.

The structure of a pulse train is often called a "temporal pattern." Two pulses separated by an interval of time constitute the simplest kind of temporal pattern, and a band detector can be viewed as a primitive "pattern-recognition" device. The problem of detecting specific temporal patterns involving three or more pulses has not been discussed in this paper because it is such a large subject. We note, however, that resonant networks would provide ideal devices for detecting complex temporal patterns; in principle there is no limit to the size and complexity of patterns that can be selectively detected by appropriate combinations of delay paths and coincidence detectors.

In the simplest cases—say, of patterns of three to five or six pulses—the most direct means for detection would be a mutation of the basic band detector. If we add one or more delay paths between the input neurons and *T*-neurons, and raise the firing threshold of the *T*-neuron so that the near-coincidence of three or more excitatory pulses is required for firing, the band detector becomes a "pattern detector" in the general sense of the word. The magnitude of *T* determines a "band of patterns" that can produce an output. As in the case of compotent periods, a pulse train may contain many "component patterns," and the modified band detector could respond to those it is "tuned" to, regardless of the presence of pulses—intervening or otherwise—which produce other pattern components. Furthermore, if the number of different delay paths exceeds the number of coincident pulses required to fire the *T*-neuron, then the network could detect several distinct "bands" of patterns. Finally, if one considers small groups of pattern detectors, perhaps sharing some delay paths as in the period analyzer proposed above, then the immense range of patterns that could be differentiated becomes apparent.

The provision of more than four or five different delay paths to a single *T*-neuron and the high firing threshold thereby required of the *T*-neuron may present serious difficulties for a nervous system. Fortunately, an alternative method of detecting specific patterns of three or more pulses is possible; by connecting the outputs of two or more simple band detectors to the input of another simple band detector the need for special *T*-neurons and many-input synapses is avoided. Even two- or three-level hierarchies of simple band detectors could respond selectively to rather large and complex temporal patterns. We note that the output pulses from the various detectors in the period analyzer indicate not only which period components are present in the input train, but also *when* the corresponding intervals occur. Since any temporal pattern can be defined, and therefore recognized, as a certain sequence of time intervals, a period analyzer could provide the first level of analysis for a number of pattern recognition networks. The outputs of the analyzer, in various combinations and with various delays, could be fed into a second set of band detectors.

One of the most striking features of the neurophysiological literature, from the theoretical standpoint, is the absence of an adequate terminology for describing (and therefore thinking about) pulse patterns. In this paper we have more or less formally defined a few simple concepts such as regular vs. irregular and accelerating vs. decelerating trains, primary vs. secondary intervals, and component periods. However, these terms are still far too primitive to permit economical discussion or thought about pulse-interval-coded systems. There is an urgent theoretical need for a rich and flexible classification scheme that can be applied to pulse trains, a "taxonomy of pulse patterns" as it were. It seems reasonable to suppose that such a taxonomy would not only facilitate the development of nerve-net theory, but would also sharpen the observational powers of the experimental biologist. Even if it turns out that the types of resonant networks discussed here do not exist in nature, they will have served a useful purpose if they stimulate biologists to take some steps toward developing a pulse-pattern taxonomy.

The more direct experimental implications of our resonant-network theory are rather obvious. Stimuli, whether natural or artificial, which produce accelerating or decelerating pulse trains, are most likely to elicit responses indicating the presence of resonant networks, singly or in cooperative groups. An input train sweeping across a wide frequency spectrum should clearly be more effective than constant-frequency trains in evoking selective responses of band detectors. Simultaneous recording from two or more points in a ganglion is clearly called for. Groups of double pulses deserve special attention, whether they occur on single or adjacent fibers. More generally, the theory suggests that biologists should carefully look for correlations between behavioral output and the durations of *secondary* intervals (not just primary intervals, as in the case of statistical-averaging systems). The special parameter requirements of resonant networks suggest that the experimentalist keep an eye open for unusually *slow* axons and *long* synaptic delays, as well as for cells which behave like our hypothetical T-neurons.

Although we have assumed throughout this paper that pulse-coincidence detection, inhibition, and pulse-train structures are all produced by axon–synapse interactions, many (and perhaps all) of the resonant behavior patterns discussed could be produced by axon–axon interactions. Crane has examined several axon configurations that could be utilized in resonant systems (Crane, this volume). We cannot pursue the subject here, but it is clear that from the standpoint of resonant-network theory, axon–axon interactions deserve special attention wherever they may be discovered.

The most general conclusion we draw from our study of hypothetical resonant networks is that nervous systems are able, in principle, to utilize the information capacities of pulse-interval coding to a much higher degree than would be possible if they were restricted to statistical averaging mechanisms of the sort most often assumed. Although the theory outlined here can rather easily be quantitatively expanded and elaborated either by paper-and-pencil analysis or by simulation models, the most desirable development would be the discovery of one or more such resonant networks in existing nervous systems; the experimental biologist is the key agent in further development of resonant-network theory.

References

Harmon, L. D., 1961. Studies with artificial neurons: I. Properties and functions of an artificial neuron, *Kybernetik*, **1**, 89–101.

Jenik, F., 1962. Electronic neuron models as an aid to neurophysiological research, *Ergebnisse der Biologie*, **25**, 206–45.

MacKay, D. M., 1962. Self-organization in the time domain, in *Self-Organizing Systems*, M. C. Yovits, G. T. Jacobi, and G. D. Goldstein, eds., Washington, D.C.: Spartan, pp. 37–48.

McGrogan, E. P., 1961. Improved transistor neuron models, in *Proceedings of the National Electronics Conference*, vol. 17, Chicago, Ill.: National Electronics Conference, Inc., pp. 302–10.

Prosser, C. Ladd, and Frank A. Brown, Jr., 1961. *Comparative Animal Physiology*, 2nd ed., Philadelphia: Saunders.

Reiss, R. F., 1962. A theory and simulation of rhythmic behavior due to reciprocal inhibition in small nerve nets, in *Proceedings of the 1962 A.F.I.P.S. Spring Joint Computer Conference*, vol. 21, Palo Alto, Calif.: National Press, pp. 171–94.

Schief, R., 1962. Koinzidenz-Siebschaltungen, Doctoral Dissertation, Technische Hochschule, Hannover.

7. Possibilities for Signal Processing in Axon Systems

H. D. CRANE, *Stanford Research Institute*

Neurophysiologists generally view the axon as a clever evolutionary trick for sending pulses from one place to another without attenuation. But for signal processing itself, it is generally suggested that one look elsewhere, in particular at the synapses in the dendritic and cell-body regions. Although this is reasonable, treating the axon merely as "interconnecting wire" may lead one to overlook some important interactions and processing capabilities in the axon networks themselves.

This paper consists of two parts. First, some aspects of "neuristor techniques" are introduced to show how this material evolved. Second, a number of examples are given to show what is meant by processing in axon systems.

Neuristor Systems

The general concept of a neuristor evolved from consideration of problems in the miniaturization and mass fabrication of electronic computer systems (Crane, 1960a). The most pertinent problem was that of signal attenuation along very small wires. (Techniques are being developed for depositing "wires" having cross-sectional dimensions in the 1000-Å to 1-μ range, and even a 1-μ copper wire has a resistance of the order of 10^3 ohms/in.) One potential method of overcoming such high attenuation is the use of an "active wire." By "active wire" we mean devices such as the firecracker fuse, which permits unattenuated propagation of a burning zone, and the axon, which permits the unattenuated propagation of the synaptic discharge. The fuse does not recover after such a propagation; it can be used only once. The axon, however, recovers in a few milliseconds and can be used an indefinite number of times. We are especially interested in using electronic principles to synthesize continuous-line analogs of the axon. Neuristor is the name suggested for the whole class of lines that exhibit unattenuated propagation with recovery. The axon is a member of this class.

A Relay-Line Model

Consider the simple relay line of Fig. 1. A line such as this is useful because one need not know organic or membrane chemistry to appreciate its operation,

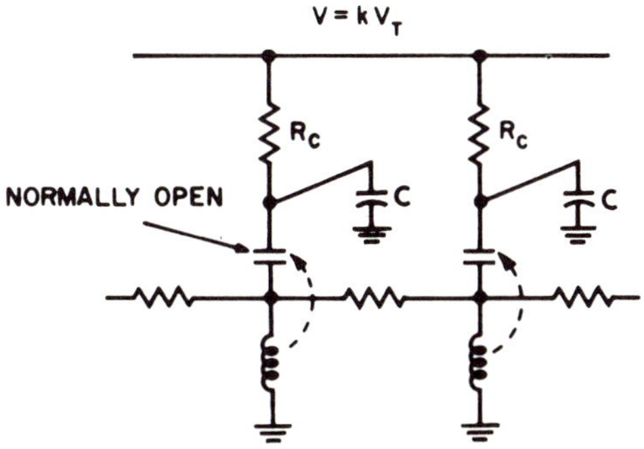

FIG. 1. Two stages of a relay line that exhibits all-or-none propagation.

nor is it necessary to understand electronics or solid-state physics. Briefly, the line operates as follows: In the resting state, all relays are de-energized (contacts open) and the capacitors (C) are charged to the line voltage V ($V = kV_T$). There exists a resting energy $\frac{1}{2}C(kV_T)^2$ for each stage, where k is a constant and V_T represents the energizing voltage of the relay. If any node voltage (voltage across the relay coil) reaches the value V_T, the corresponding relay closes, applying voltage kV_T to the node. For $k > 1$, this represents a signal gain when the contacts close. With k sufficiently large, voltage kV_T causes the voltage of the following node to exceed V_T, so that the contacts of that stage close, and so on. Once the contacts in any stage close, the relay remains energized until the corresponding capacitor is fully discharged (each section of line "burns" to completion), after which the contacts open and the capacitor recharges through resistor R_C. The length of the propagating active zone, i.e., the number of adjacent relays energized at any instant, depends on the exact properties of the line. [In experiments with lines of this type, the active zone was typically six or seven stages (Crane and Rosengreen, 1961).]

Note that if the recharge paths were removed ($R_C = \infty$), the line could not be re-used after a single discharge; it would then provide a good electric analogy for an ordinary chemical (firecracker) fuse.

A Second Relay Line

In the line described above, only one transductive process (relay) controls the flow of energy, and energy flow ceases when available energy is depleted. However, lines can be arranged in which more than one process is sensitive to the trigger parameter. For example, in a second relay line, Fig. 2, there are two antagonistic processes (relays) that respond at different rates. Briefly, this line operates as follows:

There are two relays per stage—a fast one, F (normally open), and a slower one, S (normally closed). Assume that both relays have the same pickup voltage, V_T. Whenever a node voltage reaches V_T, both relays are activated, but F picks up first. As a result, source voltage kV_T is applied directly to the node. If the value of k is sufficiently large, the adjacent node voltage surpasses V_T, and the propagation continues. Note, however, that when the S relay in

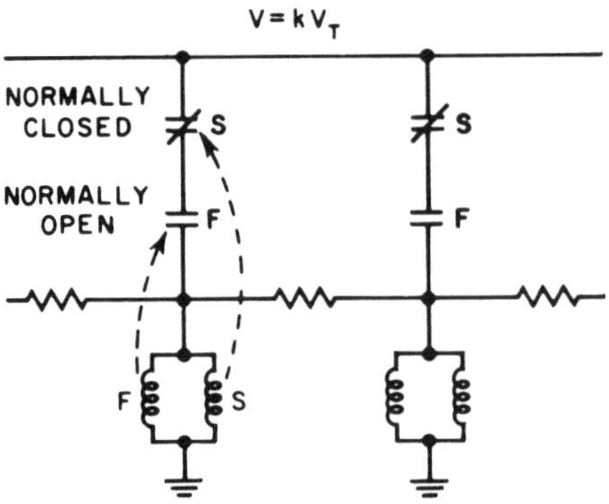

Fig. 2. Two stages of a relay line utilizing a fast relay and a slow relay in each node. This is much like the (relatively fast) sodium activation and (slower) sodium inactivation processes.

any stage is energized, the opening of its normally closed contacts removes the supply voltage from the node in that stage. Thus both relays of that stage tend to release, but because *F* releases before *S*, the supply voltage is not again applied to the node during release, and the stage recovers. It is thus clear that an unattenuated, recoverable discharge can propagate along this relay line.

In terms of Hodgkin-Huxley models, the *F* and *S* relays are somewhat analogous to sodium activation and inactivation, respectively. For convenience, the type of relay line shown in Fig. 1 is sometimes referred to as the *F*-type, for fuse, and that of Fig. 2 as the *N*-type, for nerve (Crane, 1960b).

Some Basic Properties

Four properties of neuristor lines, all well known from axon theory, are described below.

Symmetry. All of the lines discussed are longitudinally symmetrical. They can be triggered from either end or from the middle. When they are triggered in the middle, identical discharges propagate in both directions.

Refractoriness. Each portion of line becomes refractory for some period after the passage of a discharge. A primary effect of the refractory period is that some minimum spacing must exist between pulses propagating in the same direction along the line. The existence of a refractory period also has great significance for pulses propagating toward each other on a collision course.

Destructive Collisions. A basic property of neuristor lines, of utmost importance in the development of logic systems, is that two discharges propagating toward each other are annihilated upon collision. This is similar to the way in which two fires burning toward each other are annihilated when they meet—the basis for lighting "backfires."

Reflectionless Collisions. It is well known that signal reflections occur at discontinuities in linear transmission lines. A neuristor line is inherently free of such reflections. When a burning zone reaches the (open) end of a fuse line, the burning terminates; it cannot propagate back through its own refractory zone. This property of reflectionless collisions (whether with another pulse or at discontinuities in a line) is also important in the logical design of neuristor networks.

Basic Interconnections

Thus far we have considered these lines only in terms of unattenuated pulse propagation between two points. Let us now consider the additional properties that may be achieved by interconnecting such lines.

The two main attributes of a signal propagating on a neuristor line are (1) the ability to trigger further energy discharge, and (2) the presence of a refractory period. These attributes provide the basis for two types of coupling, a *T*-junction (*T* for trigger) and an *R*-junction (*R* for refractory).

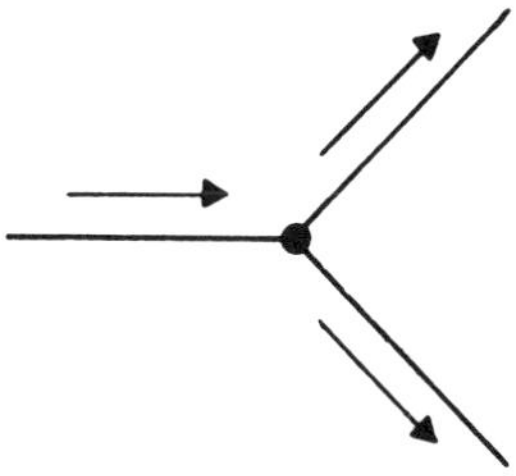

FIG. 3. A *T*-junction. (A pulse coming in on any branch will trigger a pulse on each connected branch.)

The *T*-junction. In a *T*-junction, symbolized in Fig. 3, lines are interconnected so that each line can trigger the others. That is, a signal propagating toward the *T*-junction on any one line initiates a signal on all connected lines. A symmetrical axon branch is an example of a three-legged *T*-junction.

The *R*-junction. In an *R*-junction two or more lines are coupled so that they are mutually refractory, but a signal on one will not trigger a signal on the other. The symbol used for this junction is shown in Fig. 4A, where the shaded area represents the region of coupling. Such a junction between *F*-type relay lines is indicated in Fig. 4B. In such a relay line, refractoriness is due to energy depletion in the storage capacitor. Hence an *R*-connection (i.e., mutual refractoriness) is achieved by sharing the energy storage element, as indicated. Note

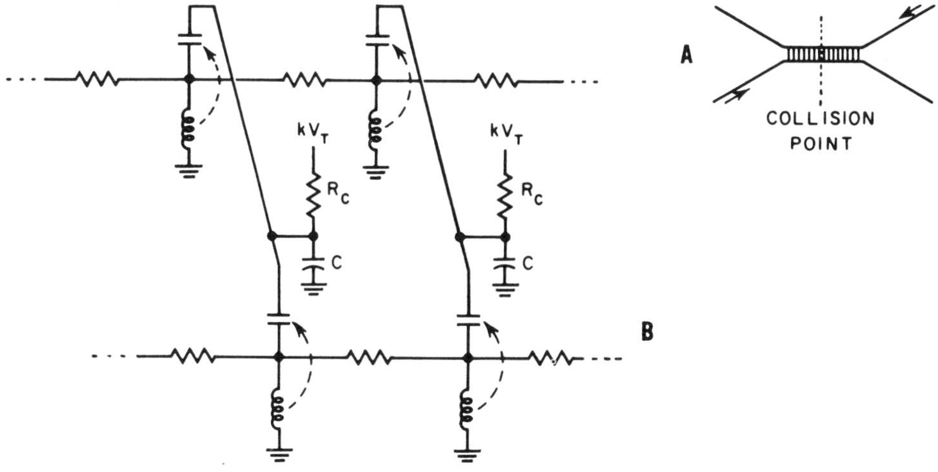

FIG. 4. An *R*-junction. A. Two pulses tending to pass in the region of the junction are annihilated as if they had collided on a single line. B. An *R*-connection between relay lines.

that with such a connection, a pulse passing along either line cannot initiate a pulse on the line coupled to it. But two pulses, one on each line, attempting to pass at the junction are destroyed just as if they had collided on a single line. This type of junction therefore provides a logical inhibition function, in the sense that a pulse on one line can inhibit, or control, the passage of a pulse on the coupled line without initiating any pulse on that line.

A refractory junction can exist between two *N*-type lines such as the one in Fig. 2 if they share the slow process but not the fast process (Crane and Rosengreen, 1961). The pulse-interaction properties for this junction would be the same as those described above.

The *T*-junction, as well as the *R*-junction, may be considered "strongly coupled" in the sense that a single pulse arriving at a *T*-junction actually causes a pulse to be initiated on every other connected line. We shall also discuss weak coupling, in which a pulse is generated on a *T*-connected branch only if several pulses arrive simultaneously at the junction and their effects are summed. (This, of course, is the situation at synapses of a highly convergent field, where a single pulse generally has little effect.) Such weak coupling schemes offer still further properties upon which logical design can be based. The fatigue junction to be discussed later is an example of a structure of this kind.

Complete Logic Capability

Many different kinds of junctions can be developed. It has been shown, however, that the *T*- and *R*-junctions can perform any desired logical function for digital-computer systems (Crane, Minnick, and Short, 1962). Such digital-computer systems take the form of networks of neuristor lines interconnected by *T*- and *R*-junctions. (For a brief introduction to logic synthesis with neuristors, the reader is directed to Crane, 1962.)

Pursuing the use of axon-like lines for realizing conventional computer systems has revealed properties that, although important for computers, have no apparent relevance for neural systems. Two such properties are the following.

(1) It is possible to show, at least in principle, that digital-computer systems can be designed without a single break in longitudinal connectivity. That is, there are no "synaptic breaks." The signal paths, even in the regions of the junctions, can be made continuous. Suppose we could visualize the use of "firecracker" fuses for such a physical realization (assuming they were rechargeable). A burning zone, once entered into the system, would continually burn along smooth lines, through and past junction regions, and be annihilated if and only if it collided with another pulse or reached the open end of a line, both possibilities being important logically. This longitudinal continuity may be of importance in electronic systems because a potentially severe manufacturing problem in highly miniaturized systems (using conventional device techniques) will be the dense set of "solder connections" required in the signal paths between heterogeneous materials. Longitudinal connectivity in a sense eliminates this problem, or at least makes it a different type of problem.

(2) Nonplanar logic systems can be physically synthesized in a plane. Consider two lines that cross in three dimensions. If the "lines" are conventional

wires, they must not touch upon crossing or a short circuit will result. If two similarly crossing neuristor lines touch at the crossing, a *T*-junction results, and the functional independence of the two lines is again lost. However, it is relatively easy to alter the crossing structure so that a signal on one line passes only straight through the junction (Crane, 1962); in this way the independence of the crossing lines is re-established. Such a crossing structure is considered to be planar.

The ability to build an entire computer system (or at least parts of one) in a single plane, in the sense indicated above, is of potential interest to the computer designer because, although he has certain capabilities for mass depositions on a plane, three-dimensional wiring is difficult. In biological systems, the Designer seems to have been able to cope nicely with three-dimensional wiring.

Axon Systems

We noted above the background for the interest in axon-like lines that exhibit all-or-none propagation. We saw that it is possible to build digital systems in which propagation is everywhere all-or-none. That is, there is no basic need for "graded response" (although it is clear that neural systems make extensive use of graded responses). The study of such systems, in particular pulse-train interactions in such systems, has revealed that the axon has information-processing possibilities beyond unattenuated signal propagation. The examples below suggest the possibilities that evolve from taking this point of view.

Pulse Trapping

Consider again the *R*-junction of Fig. 4B; assume that the two lines have different velocities of propagation and that the pulses travel in the same direction rather than on a collision course.

The velocity of propagation along a line depends on the values of all its parameters. For example, if the tension on all relay springs is weakened, then the velocity will be higher. Velocity will also depend on the supply voltage; a reasonable curve for this relation is suggested in Fig. 5A. The curve shows that there is a minimum value of supply voltage, E_{min}, below which the line cannot propagate at all.

Assume now that the coupled lines are supplied from a voltage E_0, and assume that the velocity difference in the two lines is due to a difference in spring tension in the relays of the two lines. Assume also that the curve of Fig. 5A is for the fast line, so that at voltage E_0 the velocity of the fast line is v_F. Suppose now that a pulse "*a*" moves to the right on the slower line, as in Fig. 5B. A pulse "*b*" on the faster line also moves to the right, and at some instant is a distance *d* behind. Then in a time interval of about $d/(v_F - v_S)$ the faster pulse will catch up with the first pulse. What will be the future course of the two pulses?

At any particular time, the profile of voltage vs. distance behind the slower, first pulse is suggested by Fig. 5C. (We are considering the voltage across the capacitors.) Thus E_R represents the minimum voltage reached during discharge. As the faster pulse approaches the slower pulse, the former propagates into a zone of continually lower energy levels, and therefore slows down. From the curve of Fig. 5A we see that at a voltage level E_S the fast pulse would propagate at the velocity of the slow pulse, i.e., at v_S. If $E_S > E_R$, then the second pulse approaches the first pulse until the level E_S is reached, after which both pulses propagate together. (Note that we can say only that the second pulse stabilizes in *the vicinity of* the point marked in Fig. 5C, since the curve of Fig. 5A was assumed for propagation along a resting line.)

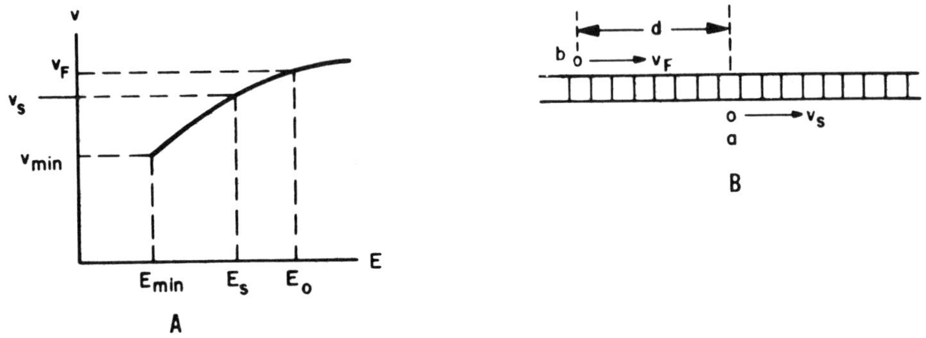

FIG. 5. Side-by-side interaction between lines of different velocity. A. Relation between velocity of propagation and (resting) energy level. B. *R*-connected lines of differing velocity. C. Instantaneous profile of energy level vs. distance behind a propagating impulse.

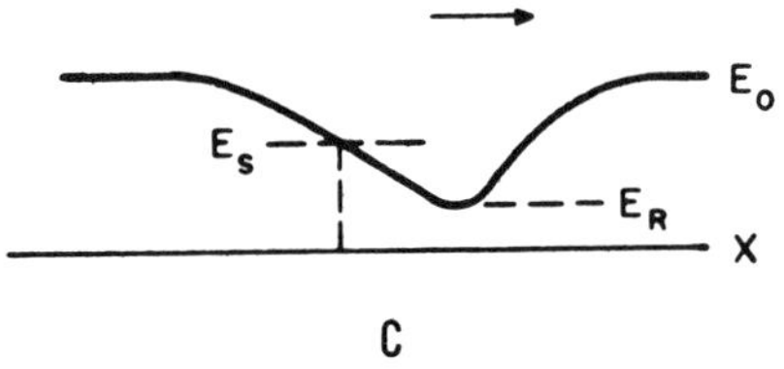

Different results can be obtained, depending on the initial velocity difference, the strength of the refractory coupling between the lines, and so on. For example, if $v_S < v_{min}$, where v_{min} is the minimum velocity at which the faster line can propagate, and if $E_R < E_{min}$, then there is no stable position for the second pulse and it is destroyed as it drives deeper into the refractory zone of the first. With still other energy and velocity relations, the faster pulse can actually pass, and in passing destroy, the first pulse.

Side-by-Side Axons

To what extent do the arguments above hold for side-by-side axons? First, we must note that whereas the relay line was treated as a one-dimensional lineal structure (i.e., a single chain of nodes), the nerve membrane is really a two-dimensional sheet folded on itself. To represent such a membrane with discrete equivalent circuits would require a two-dimensional array of nodes arranged to form a cylinder. Of course, when the "wavefront" is exactly transverse to the line, there are no lateral currents in a cross-sectional plane, and for calcula-

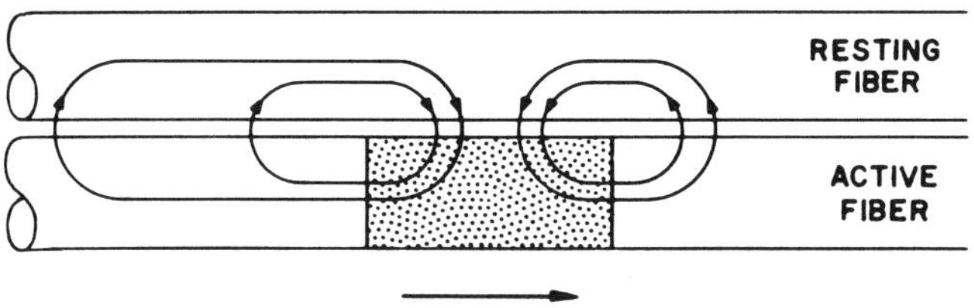

FIG. 6. Illustration of the tri-phasic electrical coupling between two adjacent fibers.

tion purposes the line can again be considered as lineal. But for the interaction of two side-by-side axons, the two-dimensional reality is important because the lateral coupling varies between different portions of the membranes, and one has to think in terms of some average coupling.

There is still another important difference. With relay lines, the lateral coupling was such as to enforce the same voltage on adjacent nodes in the two lines. The result was "pure" refractory coupling, i.e., there was no tendency whatever for triggering between the coupled lines. With axons, however, the nature of the coupling is considerably different. As a pulse propagates on one line, there exist both a short-term electrical coupling and a longer-term ionic coupling. As for electrical coupling, Fig. 6 shows that the current penetrating the adjacent fiber has essentially the same temporal and spatial distribution as the membrane current of the fiber carrying the pulse. Since this current is proportional to the second derivative of the "node voltage," the electrical coupling is basically tri-phasic. Thus, as a pulse propagates past a point on the coupled fiber, there is first a refractory (i.e., inhibitory) phase, then a trigger (i.e., excitatory) phase, and finally another refractory phase. Hence, with the coupling proportional to the second derivative of voltage, the coupling cannot be said to be simply refractory or excitatory, but is a combination of both.

This interaction of single pulses on side-by-side axons has been discussed in the literature. (For an interesting discussion see, for example, Katz and Schmitt, 1940.) We shall not pursue the details further here except to note that, with two axons that do not differ too much in velocity, pulse trapping has actually been demonstrated experimentally. With axons, however, the locked pair travel at a velocity closer to that of the faster line than to that of the slower line, as was the case with the *R*-coupled relay lines of Fig. 4B. This can be seen to be a result of the central trigger phase in the tri-phasic coupling.

Pulse Bunching

An interesting consequence of pulse trapping is that if the two coupled lines carry pulse trains, then the interaction between lines can shift the distribution of pulses along the lines. For example, in the relay-line case, if the slower line carries no pulses, then a uniform pulse train on the faster line is essentially unaffected by the presence of the slower line. If, however, the slower line also

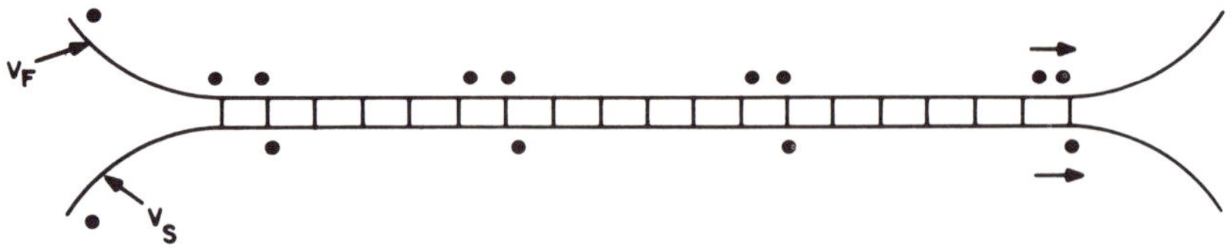

FIG. 7. Pulse bunching along two *R*-connected lines. (For the relative pulse frequencies shown each pulse on the slow line traps and bunches two pulses on the fast line.)

carries a train, bunching of pulses on the fast line can occur, as indicated in Fig. 7, where it is assumed that the pulse frequency on the faster line is just twice that of the slower line. With a three-to-one frequency ratio, the pulses would be bunched in triplets. With nonintegral ratios, the bunching would be uneven.

Although in axons locking occurs on the faster rather than the slower pulse, results similar to those observed in relay lines can be expected. The extent to which such effects actually occur depends on many factors: the strength of coupling, the difference in velocities, the activity in all axons adjacent to the pair in the nerve bundle, and so on. Study of such interaction phenomena appears important because effects such as these might actually be used in certain places as a means for having one signal act on another. But even if this were nowhere a "purposeful design" arrangement, bunching results of this type could be expected experimentally.

In our discussion of relay lines, we indicated that with certain energy and velocity relations, a fast pulse could pass and destroy a slower one. In this respect, it is interesting to speculate that high-speed pulses on giant axons might kill activity in nearby fibers in the same way.

In this section we have considered the effects of transient electric coupling between adjacent fibers. There are also longer-term effects that are due to shifts in ion concentration as a result of high-frequency pulse-train propagation. But first, a few brief comments on rear-end collisions are in order.

Rear-End Collisions

Consider a line that is uniform along its length, except for a section that has a reduced velocity of propagation (for example, the dashed section of line in Fig. 8). If two closely spaced pulses impinge on this section, and if the dif-

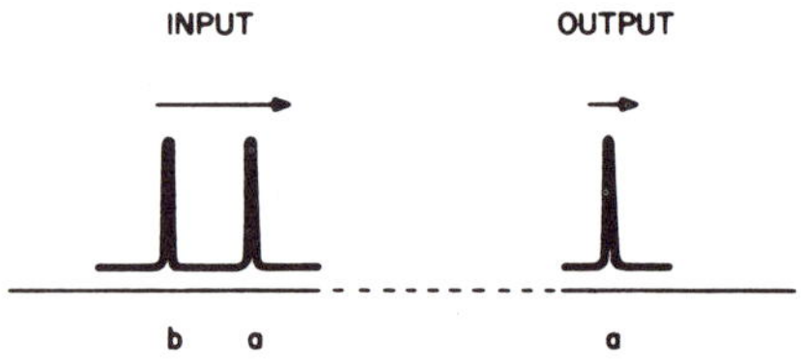

FIG. 8. Rear-end collision. Two closely spaced pulses impinging on a region of slow propagation (shown dashed); the second pulse can be destroyed by arriving in the refractory period of the first pulse.

ference in velocities is sufficiently large, then the second pulse will find the section still not completely recovered from the passage of the first pulse, and will be annihilated. Thus only the first pulse will emerge from the other side (Adrian, 1959, p. 18). If a high-frequency train is applied, then a greatly reduced number of pulses may appear at the other end. Since the velocity of propagation decreases whenever propagation occurs along a line of expanding diameter, we can expect to find such slowdown regions wherever the axon expands in diameter, such as at the hillock region, or at axon branches. In such regions we can therefore expect destructive rear-end collisions of closely spaced pulses. (For a discussion of such collisions at *T*-junctions of relay lines, see Crane and Rosengreen, 1961.) This property is considered further in a subsequent section.

Fatigue Connection

The property of fatigue can be incorporated into the relay line of Fig. 1 by arranging a multiple-level recovery system, as shown in Fig. 9A. Assume first that resistor R_R is eliminated, and that reservoir capacitor C_R is greater than C. After each propagated discharge, line capacitor C in each stage is recharged from its corresponding reservoir capacitor C_R. The line can therefore sustain a finite number of propagations before showing fatigue, with each propagating signal having slightly less amplitude than the preceding one. With a finite recharge resistor R_R, however, pulses can be propagated indefinitely, and a train of pulses initiated along a recovered line shows normal decay in amplitude to a steady-state value that depends on the pulse frequency.

If a refractory coupling is now attempted at the "higher level," i.e., by means of a common reservoir capacitor between corresponding stages of the coupled lines as in Fig. 9B, then single pulses (one on each line) can pass each other on the junction without being annihilated. Only if at least one line carries a relatively high-frequency train will the common energy be sufficiently depleted so that pulse interaction can occur between the two lines. What types of interaction can be expected?

If a pulse train propagates along one of the lines of Fig. 9B, the average potential across the reservoir capacitor, at each stage of the junction, depends on the frequency of that train. Hence, in the fatigue connection symbolized in Fig. 9C, if one line is conducting a pulse train, PT, then Pulse A arriving on the other line will enter a region of reduced velocity, the magnitude of reduction depending on the frequency of the pulse train. (The general effects would be the same if the pulse train were traveling in the opposite direction.) Hence the time of propagation of Pulse A through this section can be controlled by the frequency of the pulse train. If, instead of a single pulse, pairs of closely spaced pulses appear on the upper line, then rear-end collision in this slowdown region results in only a single pulse of each pair being able to emerge from the other end. (It is clear that there are many possibilities of interaction.) Can this same type of fatigue coupling be expected between axons?

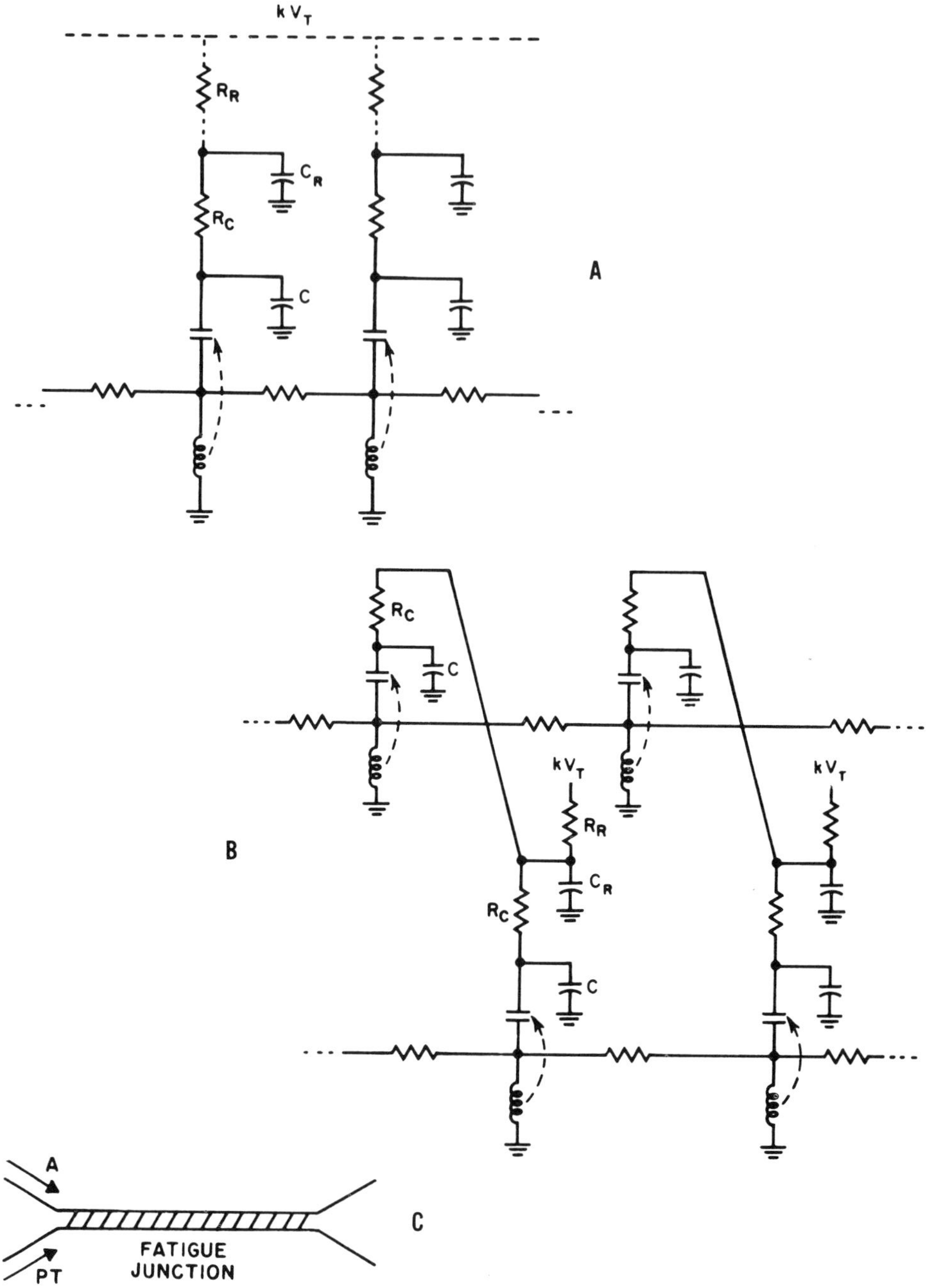

FIG. 9. Fatigue junction. A. With a "second layer" of recovery elements added to the line of Fig. 1, typical fatigue properties will be exhibited. B. A fatigue junction between lines of the type shown in part A. C. Effective interaction is achieved in this case only if at least one line carries a relatively high-frequency pulse train.

Neglecting complexities due to sheathing layers, we note that there is some net ion exchange through the membrane as a result of a single pulse propagation. If a second pulse propagates before the "metabolic pump" has restored the resting state, then there can be a cumulative change in ion concentrations, and hence also in diffusion potentials in terms of the Hodgkin-Huxley model. (This results in reduced responses, or the usual "fatigue.") As additional pulses are propagated, the cumulative change increases toward a steady-state value that depends on the frequency. To the extent that an adjacent fiber can experience this changed ionic environment (increased potassium and decreased sodium), this is inhibitory to the second fiber. Using the sketch of Fig. 10, with a high-frequency pulse train applied to the right-hand fiber, we can think of the effective zone of ionic change as enlarging, with each additional pulse, toward a steady-state radius. Hence we can expect results with axons similar to those of the "fatigue coupling" in Fig. 9; in particular, we can expect increased interaction with higher pulse-train frequencies on one or both lines.

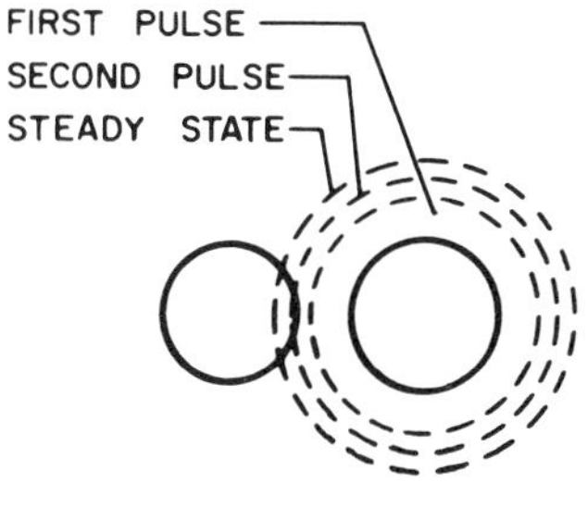

FIG. 10. Increase in the range of interaction, by summation of net ion exchange when one axon carries a relatively high-frequency pulse train. Two adjacent axons are shown in cross section.

Spiral Interconnection

It was implied in the section on neuristor logic that a strongly coupled *R*-junction, in which pulses can be annihilated by collision, is an important logical structure. In the previous section, we noted that time summation of the net ion exchange due to a high-frequency pulse train on one line might significantly affect pulse propagation along an adjacent fiber. In this section let us take an additional step wherein, by proper geometry, we may increase still further the degree of coupling between adjacent lines and actually obtain destructive collision.

For this purpose it is important to note again that the axon membrane is actually a two-dimensional sheet. Now, one cannot stop a brush fire by cutting the grass in some confined region; the fire will burn around the cut area and rejoin on the other side. Only by cutting a swath that prevents burning along every path can the fire be stopped. Similarly, with two side-by-side axons, conduction along one fiber tends to produce some limited interaction, in the sense of Fig. 10. If this range includes only a portion of the membrane of the second fiber, then the second fiber can still "burn" along the outer portion of the membrane. To eliminate even this possibility, it is therefore of interest to consider wrapping the first fiber around the second fiber to form a relatively tight spiral. This might form an effective block around the entire periphery of the second fiber. The spiral connections of axon collaterals in Mauthner cells (Retzlaff, 1957) are interesting in that they too appear to be inhibitory couplings, although the arguments presented above are wholly conjectural.

The Fan-in Junction

Consider the "fan-in" or *T-R* interconnection of Fig. 11A, in which a *T*-junction appears at the right end of the *R*-junction. Suppose a pulse arrives along line K. As this pulse reaches the *T*-junction, a pulse tries to pass backward along line L but is annihilated because this section of line has not recovered from the forward passage of pulse K. Consequently, a pulse arriving on line K (or line L) will propagate only along "output" M, and input lines K and L do not cross-talk. It is then clear that if a pulse on K arrives at the same time as one on L, or at nearly the same time, only a single pulse will emerge along M.

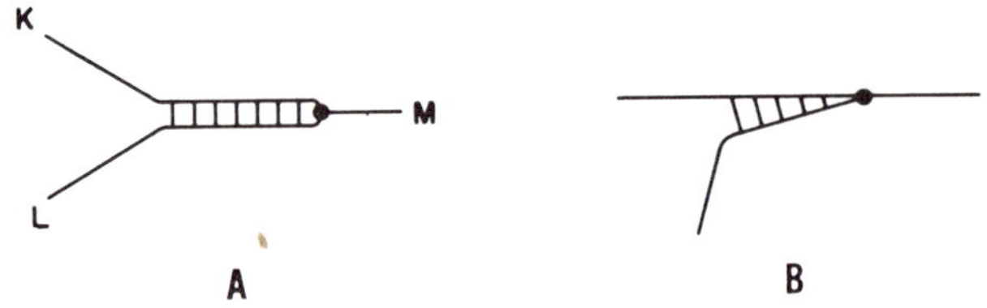

FIG. 11. Fan-in junction. A. An input pulse on Line K or Line L propagates out along line M only. B. An acute-angle axon branch that could exhibit the same properties.

To the extent that side-by-side axons can have an inhibitory effect on each other, by means of the net ion exchange or the final inhibitory phase of the tri-phasic electrical coupling, one might expect fan-in properties at an axon branch if two of the branches had a very acute angle, as in Fig. 11B. This might be most apparent in very small fibers, where "safety factors" are relatively low, and where the net ion exchange from one or a few pulses might be relatively large. I have not seen such a structure described in the literature, however, and so there is no point in speculating further here, although the hypothesized structure is used in the final example of the next section. (Physiological evidence that such modes may actually be found is suggested by the results of Tauc and Hughes, 1963.)

Analog of the Spiral Innervation of the Cochlea

The title of this section is the same as that of a paper by van Bergeijk (1961). It is introduced here to indicate how the properties of fan-in junctions, together with the properties of rear-end collision, could also be used to explain his hypothesis of increased dynamic range.

The following is from his summary: "On the basis of the experimental observations the proposition is advanced that the spiral innervation serves to extend the dynamic range of the ear. Dynamic range extension follows from elementary considerations of the refractory and summing properties of nerve fibers."

Van Bergeijk's experimental model is similar to that shown in Fig. 12A, where a set of inputs labeled f_i join into a converging arbor. The inputs are driven from different output taps along an electrical model of the cochlear membrane. For a steady tone applied to the cochlear model a steady pattern of excitation exists along the membrane. The outputs from the model are uniform pulse trains whose frequencies depend upon the level of excitation at each point of the membrane model. Evidence is cited that the spiral nerves of the cochlea actually arborize in this fashion, connecting via branches to an extended region of the cochlear membrane.

In van Bergeijk's experimental model for this anatomical configuration, a neuromime (Harmon, 1961) is inserted at each junction point in the converging arbor. Typical results obtained by van Bergeijk for a network with ten inputs are shown in Fig. 12B. The ten plotted lines indicate the value of f_i (for $1 \leqq i \leqq 10$) as a function of the intensity of a tone applied to the cochlea. Note that at low intensities the output f_{out} is higher than the individual input frequencies. At high intensity the output frequency is lower than the individual input frequencies. In other words, dynamic range is increased.

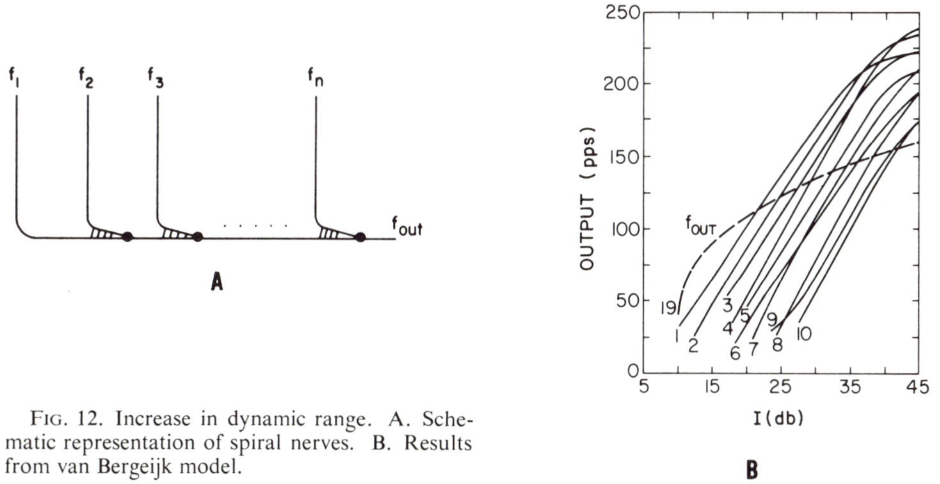

FIG. 12. Increase in dynamic range. A. Schematic representation of spiral nerves. B. Results from van Bergeijk model.

Although the results from this experimental arrangement tend to support the hypothesis of increased dynamic range, it is of interest to consider whether such results are also possible on the basis of models that may be a bit closer to the anatomical structure. (Note that introduction of a neuromime at each junction point endows an axon branch structure with synaptic properties.) It is suggested below that rear-end collision and fan-in properties could produce similar results.

Consider again the convergent arbor of Fig. 12A. It is clear that if each junction point were a simple T-junction, in the sense of Fig. 3, the structure could not possibly work as predicted. The highest frequency among the inputs f_1 through f_n would permeate the whole network; that is, the output frequency would exactly equal the maximum input frequency. This can easily be verified by noting that if two pulse trains collide head-on in an axon, the pulses are destroyed one-for-one and the collision point continually moves in the direction in which the higher-frequency train is traveling.

Suppose, however, that instead of being simple T-junctions, the arbor points were fan-in junctions as shown, so that pulses from any input could propagate only in the direction of the output. With low-intensity audio input, the discharge frequencies from the hair cells are low, and pulses would be widely spaced along the input lines. It is reasonable to suppose therefore that the output in this case would approximate a summation of the inputs. At high intensity, however, pulses would be closely packed. Two properties must now be recalled. If two pulses arrive at a fan-in junction at essentially the same time, only a single pulse will emerge. This is in itself a source of compression, but this alone would not be adequate to produce an output frequency that is actually lower than the input frequencies. However, here we could expect significant reductions caused by destructive rear-end collisions. At high intensities there would be high probabilities of tightly spaced bursts of pulses along the lines. As we pointed out in connection with Fig. 8, when such a burst arrived at a junction, only one or just a few pulses of each burst might actually survive to reach the other side. The other pulses would have been destroyed by rear-end collisions at this discontinuity in the line. Hence van Bergeijk's data may be explained by different arguments altogether.

Summary and Discussion

The material of this paper relates primarily to axon systems. The reason for such a narrow choice was discussed in the introduction. It resulted from an earlier study of logic synthesis using a more generalized class of device termed the neuristor, a class to which the axon belongs. It was indicated that such devices may offer significant advantages in the achievement of highly miniaturized electronic computer systems.

The second part of the paper is concerned with those properties evolved in the course of neuristor logic studies that might be pertinent to biological systems. Of special interest are structures that mediate the interaction of pulse trains. Since the equations for nerve membranes are relatively complex, as are the equations for continuous electronic structures, a simple relay-line model was introduced to simplify the details of discussion.

A sampling of basic axon properties was discussed, as well as a sampling of basic means of interconnection; in particular, head-on and rear-end collision of pulses, and T-, R-, fatigue, and fan-in junctions were discussed. Qualitative arguments were advanced to suggest the possible role of spiral interconnections between axons. Some expected properties of pulse-bunching, pulse-annihilation, and velocity control due to interaction between adjacent axons were also discussed; and fan-in junction and rear-end collision properties were used together in an arrangement in which increased dynamic range was produced. Other important properties can also be achieved in axon networks—e.g., multiple-level memory, detection of the front or back edge of a pulse train (on/off detection), and frequency filtering (Crane, 1960).

The discussion has been primarily qualitative and descriptive in nature. Although there has been considerable computer simulation of the Hodgkin-Huxley equations (e.g., Fitzhugh, 1961), to the author's knowledge there has as yet been no extension of these studies to include geometry "in the small." Considering just a single axon, for example, it would be interesting to try to compute the effects of certain nonuniformities of structure and membrane properties. We know that a pulse may be destroyed when propagating into an expanding region, such as the hillock region. Further, wherever there is a decrease in propagation velocity, due perhaps to variation in membrane properties, or to propagation into physically expanding regions, one might expect destructive rear-end collisions (destruction of a second pulse as it rams into a first). It would be helpful to know the critical parameters for such interactions. It would also be interesting to be able to compute the effects of interaction between adjacent membranes under different conditions. It seems important therefore to try to extend computer simulations of the Hodgkin-Huxley equations to these and other similar structural arrangements.

An objective of this paper has been to hypothesize certain properties of axon networks. For the proper interpretation of experimental data it seems important to understand to what extent these properties may actually be exhibited by axon networks. In addition, results like those of Tauc and Hughes (1963) suggest that axon trees might actually have some processing significance.

References

Adrian, E. D., 1959. *The Mechanism of Nervous Action*, Philadelphia, Pa.: Univ. Pennsylvania Press.

Bergeijk, W. A. van, 1961. Studies with artificial neurons: II. Analog of the external spiral innervation of the cochlea, *Kybernetik*, **1**(3), 102–7.

Crane, H. D., 1960a. "Neuristor Studies," Stanford Electronics Laboratories Technical Report No. 1506–2, Stanford, Calif.: The Laboratories.

Crane, H. D., 1960b. "The Axon as a Neuristor, An Analysis of Nerve Transmission," Stanford Research Institute Interim Report 1, Project 3286, Menlo Park, Calif.: The Institute.

Crane, H. D., and A. Rosengreen, 1961. "Results from Experimental Relay Neuristor Lines," Stanford Research Institute Interim Report 2, Project 3286, Menlo Park, Calif.: The Institute.

Crane, H. D., R. C. Minnick, and R. A. Short, 1962. "Neuristor Logic," Stanford Research Institute Interim Report 5, Final Report on Project 3594, Menlo Park, Calif.: The Institute.

Crane, H. D., 1962. Neuristor—A novel device and system concept, *Proc. IRE*, **50**(10), 2048–60.

Fitzhugh, R., 1961. Impulses and physiological states in theoretical models of nerve membrane, *Biophys. J.*, **1**(6), 445–66.

Harmon, L. D., 1961. Studies with artificial neurons: I. Properties and functions of an artificial neuron, *Kybernetik*, **1**(3), 89–101.

Katz, Bernhard, and Otto H. Schmitt, 1940. Electric interaction between two adjacent nerve fibres, *J. Physiol.*, **97**, 471–88.

Retzlaff, E., 1957. A mechanism for excitation and inhibition of the Mauthner's cells in teleosts: a histological and neurophysiological study, *J. Comp. Neurol.*, **107**, 209–25.

Tauc, L., and G. M. Hughes, 1963. Modes of initiation and propagation of spikes in the branching axons of molluscan central neurons, *J. Gen. Physiol.*, **46** (3), 533–49.

8. An Electronic Model of the Neuron Based on the Dynamics of Potassium and Sodium Ion Fluxes

E. R. LEWIS, *General Precision, Inc.*

In recent years T. H. Bullock and others have proposed a new functional concept of the nerve cell. In this view a single neuron is composed of many functionally distinct regions, all interacting, and is thus endowed with a high capability for information processing. In this respect it is far removed from the simple two-state device envisioned by many early modelers. This capability may well be fundamental to the ability of small groups of neurons to perform, with great reliability, complex control functions such as those performed by the nine-neuron cardiac ganglia of *Homarus* and *Panulirus* (Maynard, 1953, 1955) or perhaps the walking control in crabs (Cohen, *this volume*, p. 273) or the flight control in locusts (Wilson, 1961, and *this volume*, p. 331). Some physiologists seriously doubt the importance of many of the potential information-processing capabilities of a single cell (such as response to a pulse pattern) in a system containing thousands or even millions of cells. Nevertheless, there is strong evidence that these capabilities exist, and for that reason I feel they are well worth a thorough examination by neural modelers, particularly those interested in studying small nerve nets and ganglia.

The Locus Concept

Direct measurement of postsynaptic potentials has shown that both excitatory and inhibitory synapses have several possible modes of behavior (Bullock, 1957, 1958, 1959; Hagiwara and Bullock, 1957; Otani and Bullock, 1959). Responses to a series of presynaptic pulses may be facilitated, anti-facilitated or neither. The magnitude and time course of facilitation or anti-facilitation are parameters which may vary from synapse to synapse, even within a single neuron. In addition, rebound phenomena can occur on the termination of a burst of presynaptic pulses, as evidenced by postinhibitory firing and post-excitatory depression. This behavior can be oscillatory with successive cycles of depression and excitation, and can vary widely from synapse to synapse. Complicating matters even more, there is considerable evidence that individual neurons can be functionally much more complex than the classical concepts of a synaptic region coupled more or less directly to a pulse-generating region (Bullock, 1957, 1958, 1959; Hagiwara and Bullock, 1957; Bullock and Terzuolo, 1957; Otani and Bullock, 1959). Certain regions of the somatic or dendritic membrane may exhibit regular spontaneous activity observable as low-frequency sinusoidal or sawtooth variations in the transmembrane potential, producing periodic changes in excitability and even periodic pulses or pulse bursts. Other regions may have integrating and amplifying functions, responding in a nonlinear but graded manner to potentials impinging upon them. These regions may also be capable of pulse generation, but generally with a much higher threshold than is found in the axon. Furthermore, the generation of a pulse and its subsequent propagation along the axon do not necessarily affect the potentials in the somatic or dendritic structure; a spike may reset the soma potential completely, or partially reset it, or not affect it at all.

We are now confronted with neurons possessing, in addition to the all-or-none pulse-generation properties, several forms of graded, subthreshold activity—each presumably confined to a distinct region, or locus, in the cell membrane. Thus we have the various synaptic loci: excitatory and inhibitory, facilitating and anti-facilitating, with or without rebound. We have spontaneous "pacemaker" loci. We have regions of nonlinear graded (local) response; and, finally, we have pulse-generation loci—presumably located in or near the axon. The spatial distribution of these regions or loci could critically influence the characteristics of an individual neuron.

The locus concept suggests that simulation with conventional electronic components requires special circuits for each type of locus; the whole neuron might then require dozens of circuits, all different. One might reasonably ask whether there exists some simple approach, perhaps a single circuit which might, with minor parameter changes, accomplish any of these functions. In this paper I shall discuss such a circuit, as well as the implications that can be drawn from it concerning the origin of some of the various forms of graded activity.

A Locus Model. If as a first approximation we ignore the dendritic tree, a neuron might be represented by a structure similar to that in Fig. 1A. All of the loci are located in the somatic membrane, except the spike-initiating locus, which resides in the axon root. It is generally assumed that the large volumes of highly conductive intracellular and extracellular fluids effectively short-circuit the various regions of the high-impedance somatic membrane, whereas the small diameter of the axon cylinder produces a moderate resistance between the soma and regions of the axon membrane (Rall, 1959a, b; Fatt, 1957; Eccles, 1957). If this is true, the neuron of Fig. 1A may be viewed functionally as a number of graded-response loci in parallel, connected through a lossy network to a pulse generator as in Fig. 1B. In reality many loci are probably located in the dendritic structure and thus are electrically isolated from one another to some extent, owing to the small diameters of the dendritic cylinders. The simplified model of Fig. 1A is useful, however, in that it suggests a new approach to the problem of electronic simulation. Although the conventional approach would be to build an analog in a serial manner with complex feedback and feedforward loops, the model suggests a parallel arrangement in which

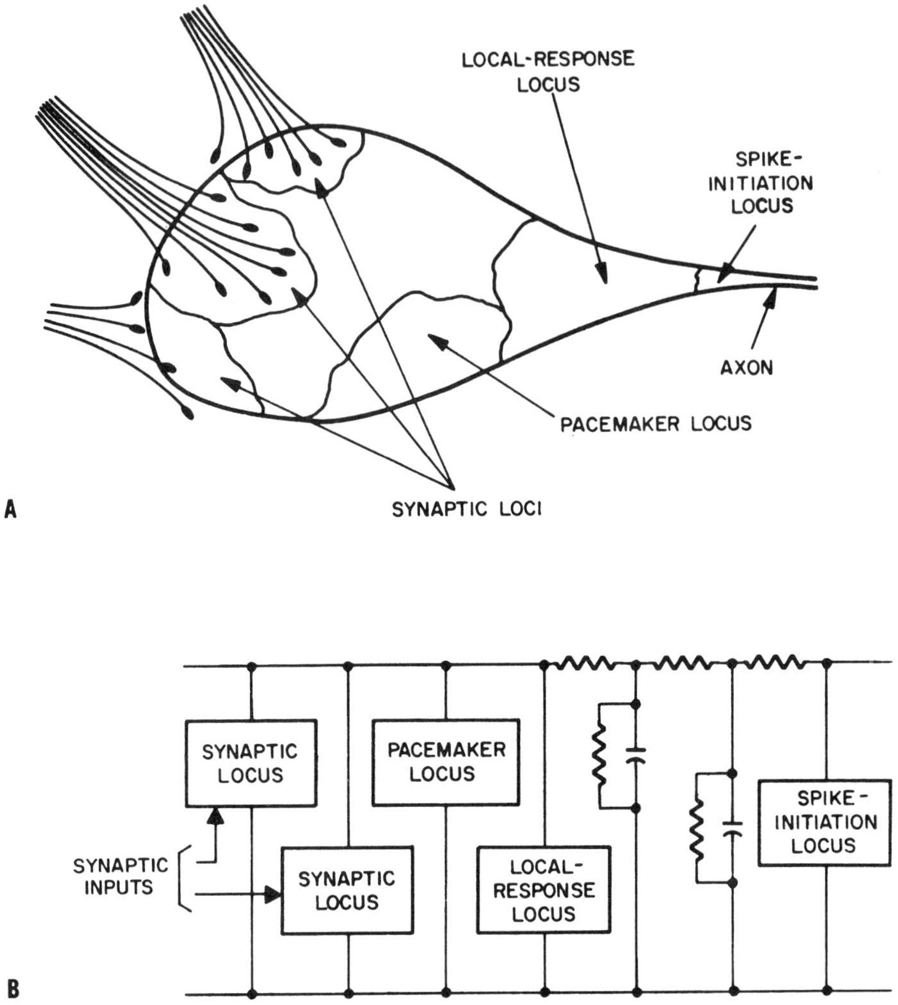

FIG. 1. A. Idealized neuron showing possible arrangement of various graded-response loci on cell body. B. A functional equivalent of the neuron in A.

one would merely supply the components of Fig. 1B. These components are in themselves, however, quite different from the usual unit building blocks of electronic systems. Aside from the probable chemical inputs to the synaptic loci (which are isolated), the basic units of this system are two-terminal devices: the input terminal pair is also the output terminal pair. Since the potential across the membrane capacitance cannot change instantaneously, it is convenient to think of each unit as supplying a current either spontaneously or in response to the potential across it, with the sum of all such currents determining the rate of change of transmembrane potential at any instant.

The synaptic loci may be viewed as transducers supplying a current which is related in some complex manner to the presynaptic input; the pacemaker loci are equivalent to either sinusoidal or relaxation oscillators; and the local-response loci are equivalent to active two-terminal devices which are capable of amplifying the potentials appearing across their terminals. Both the active properties of the local-response loci and the oscillatory behavior of the pace-makers imply that negative resistance is a crucial element in these two-terminal units. Interestingly enough, negative resistance is also essential to spike, or pulse, generation and is an important ingredient of the modern ionic hypothesis, where it is assumed to result from the nonlinear dependence of sodium permeability on the transmembrane potential (Hodgkin and Huxley, 1952a–d).

This leads to an interesting question: Would it be possible to simulate the various types of graded somatic and dendritic activity with a system based on the hypothesis often used to explain the all-or-none spike-generation properties of an axon? In the present paper this activity is extended to include synaptic as well as pacemaker and nonlinear graded-response activity, and the question becomes in essence: Can such behavior as rebound, facilitation, anti-facilitation, spontaneous low-frequency oscillations, and nonlinear graded response all be explained in terms of a dynamic balance between two opposing fluxes such as those of potassium and sodium?

To explore this possibility, an electronic circuit was built which simulated most of the important aspects of sodium and potassium currents as stipulated in the ionic hypothesis and outlined in the papers of Hodgkin and Huxley. Ten seemingly important parameters were made variable, and the circuit behavior has been examined over a fairly large volume of the resulting parameter space. With minor parameter changes, this circuit was able to produce most of the modes of graded behavior attributed to neurons. The remainder of this paper is devoted to a discussion of the circuit, the dynamic system on which it was based, and the results obtained from it.

Parameters of the Hodgkin-Huxley Model of the Giant Axon of the Squid

In a series of four papers, Hodgkin, Huxley, and Katz (1952) and Hodgkin and Huxley (1952a–c) describe detailed measurements of voltage, current, and time relationships in the giant axon of *Loligo*. In a fifth paper, Hodgkin and Huxley (1952d) formalized these data into a set of simultaneous differential equations describing the hypothetical time course of events during spike generation and propagation, thereby laying the foundation for the modern ionic hypothesis. The system they describe is basically one of dynamic opposition of two ionic fluxes (potassium and sodium) across the membrane surrounding the axon, as shown in Fig. 2. The net flux of either ion species across the membrane may be thought of as the sum of two components, *diffusion* down the concentration gradient and *drift* down the electric potential gradient, both components being limited by the resistance of the membrane. The potassium ion concentration is greater on the inside, with the result that potassium ions

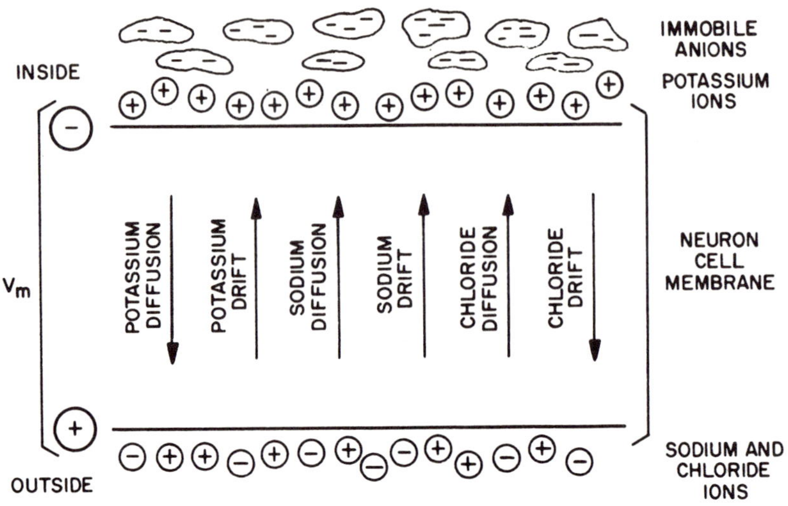

FIG. 2. Ionic fluxes across the neural membrane.

tend to diffuse outward, carrying positive charges with them. Sodium ions, on the other hand, are in greater concentration on the outside and tend to diffuse inward, again carrying positive charges. The potential across the membrane at any instant is related to the net difference in charge from inside to outside; the rate of change of this potential is related to the relative magnitudes of the ionic fluxes, which are in turn related to the relative permeabilities of the membrane to potassium and sodium ions. The Hodgkin-Huxley equations describe mathematically the dependence of these permeabilities on the transmembrane potential and on time, as they were found to hold for the giant squid axon.

At equilibrium (or rest), the potential across the axon membrane is constant; hence the net charge transfer across the membrane must be zero. This does not imply, however, that the ionic fluxes across the membrane are all independently zero, but only that the net rate of charge transfer outward is balanced by an equal inward rate. According to the hypothesis, the membrane at equilibrium is nearly impermeable to sodium ions, and the sodium flux is effectively zero. The equilibrium permeabilities to potassium and a few other ions (such as chloride) are small but finite, and equilibrium is established when the electrical potential across the membrane is sufficient to just balance the inward and outward charge fluxes carried by these ions. Owing to the presence of the other ions, this potential is slightly less than that required to balance the outward diffusion and inward drift of potassium ions alone. If, however, the potassium permeability were increased, the equilibrium potential would more closely approach that of potassium (i.e., that of a potassium concentration cell).

Although the physical properties of the membrane are not well understood, it is generally not to be expected that the diffusion and drift components of ionic flux are linearly related to the concentration and potential differences across the membrane. This is not particularly detrimental to the Hodgkin-Huxley model, however, since any nonlinearities are lumped together with membrane-permeability changes into a single variable, defined as membrane conductance. This leads to the electrical analog shown in Fig. 3, which represents a cross section of axon membrane. The batteries represent effective ion-concentration differences across the membrane, each voltage being the equilibrium potential for the particular ion species, with the effects of all ions other than potassium and sodium lumped into a single "leakage" potential (V_ℓ) in series with a constant conductance (G_ℓ). The effective transmembrane capacitance has been found to be essentially constant and is represented by a single capacitor C in series with a small resistor. The potassium and sodium conductances (G_{K} and G_{Na}, respectively) are the critical variables in the circuit and depend on the transmembrane potential V_{m}, its time derivatives, and time. The electronic analog discussed in this paper is based on the circuit in Fig. 3, with active transistor networks being used to simulate the variable potassium and sodium conductances; all important parameters of the analog are related to these conductances. Before discussing the analog circuit in detail, therefore, I should like to review the parameters of sodium and potasssium conductances as developed in the papers of Hodgkin and Huxley.

Most of the conductance data of Hodgkin and Huxley were inferred from experiments in which the transmembrane potential was suddenly changed and held at a new value, and the time course of the resulting transmembrane current was observed. For that reason the parameters are discussed below in terms of conductance changes in response to a suddenly applied voltage step. It must be remembered that these parameters apply specifically to the giant axon of the squid; they were evaluated at 6°C and may change radically with temperature. They may also vary widely from animal to animal, from neuron to neuron, or even from axon to soma to dendrite in a single neuron.

With the membrane at rest, the sodium conductance G_{Na} is extremely low, a few micromhos per square centimeter. (If Fig. 3 represents one square centimeter of membrane, this corresponds to a resistance of several hundred thousand ohms in series with V_{Na}.) If the transmembrane potential is suddenly reduced

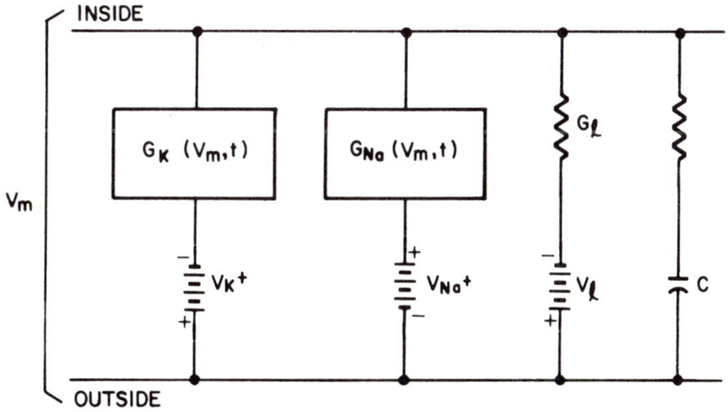

FIG. 3. The Hodgkin-Huxley electrical analog.

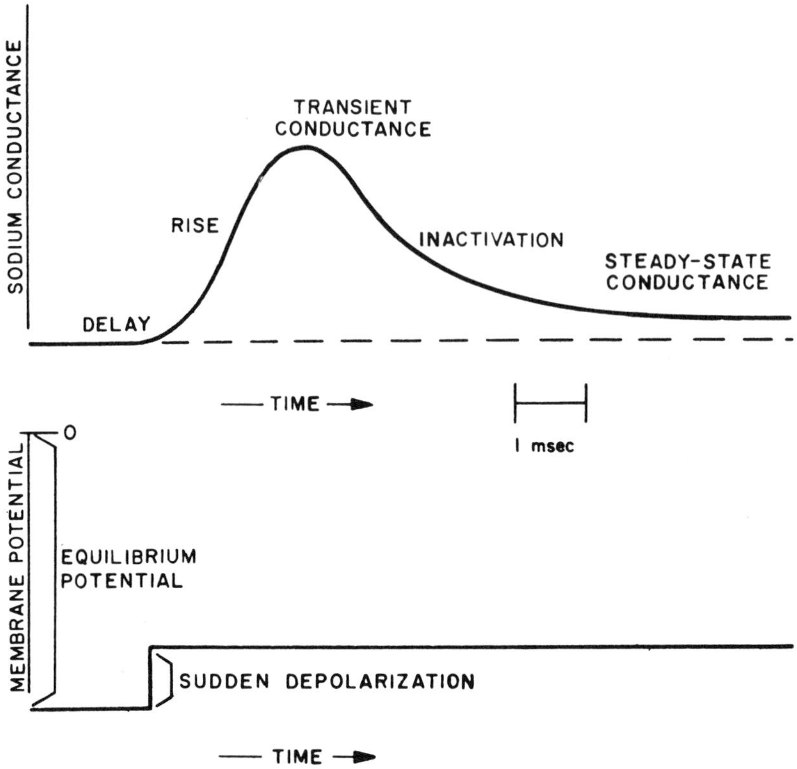

FIG. 4. Time course of sodium conductance in response to a sudden depolarization (from Hodgkin and Huxley, 1952d).

(depolarized) and held at a value less than the equilibrium potential, two components of change are observable in the sodium conductance, as shown in Fig. 4. One is a transient change in which the sodium conductance rises with a finite time constant and some delay to a peak value, only to fall again with a different time constant. The second component is steady and persists as long as the membrane is held at the nonequilibrium potential. The decline of the transient portion is generally thought to be due either to inactivation of sodium carriers or to clogging of the passages through which the sodium ions flow* (Hodgkin and Huxley, 1952d; Mullins, 1956). Once restored to the equilibrium potential, the membrane recovers from this inactivation or clogging exponentially, with time constants ranging from less than 5 to more than 20 msec. If the equilibrium potential is restored instantaneously, both components of sodium conductance fall rapidly to the equilibrium value, with time constants of generally less than 1 msec. Thus, for a stepwise depolarization and subsequent repolarization to

* It may be that the membrane is capable of storing a certain amount of sodium ions in some manner, and that these ions are available for immediate conduction. Inactivation would then correspond to depletion of these stored ions, and recovery from inactivation would correspond to their replenishment from the extracellular fluids.

the resting potential, the change in sodium conductance can be characterized by the following seven parameters.

1. *Delay time* is much less than 1 msec and depends to some extent on the magnitude of depolarization.
2. *Rise time* is 1 msec or less and depends on the magnitude of depolarization.
3. *Inactivation time constant* decreases monotonically with increasing depolarization, varying from approximately 10 msec to less than 1 msec (see Fig. 5).
4. *Time constant of recovery from inactivation* was specifically measured for only one case, a 44-mV depolarization. Here the inactivation time constant was about 1.8 msec, and that for the recovery from inactivation was approximately 12 msec. Other values can be inferred, however, from the Hodgkin-Huxley data (1952c, p. 500).
5. *Peak value of sodium conductance* increases monotonically with increasing depolarization, as shown in Fig. 6.
6. *Steady-state sodium conductance* apparently increases monotonically with depolarization, but the dependence is obscured at higher levels by the much larger transient conductance (see Fig. 6).
7. *Fall time on sudden repolarization* is less than 1 msec.

Although these parameters are reasonably inclusive, two important aspects of sodium conductance have been omitted: its behavior when sudden depolarization takes place from a nonequilibrium potential, and its behavior when the membrane potential is varied slowly about the equilibrium value. In an experiment described by Hodgkin and Huxley (1952c), the membrane potential was preset for some time at a nonequilibrium value and then depolarized to a level 44 mV below the equilibrium potential. This was repeated for several values of preset potential, both above and below the equilibrium potential, and the peak magnitude and form of the sodium conductance were observed in each case during the subsequent depolarization. It was found that the peak magnitude

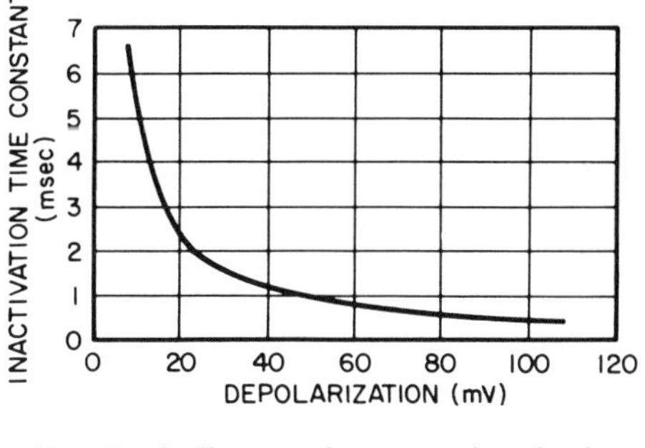

FIG. 5. Sodium-conductance inactivation time constant as a function of depolarization (from Hodgkin and Huxley, 1952d).

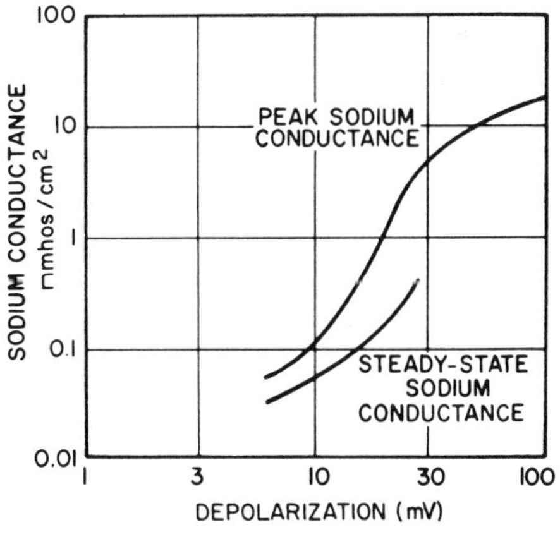

FIG. 6. Magnitude of sodium conductance as a function of depolarization (from Hodgkin and Huxley, 1952d).

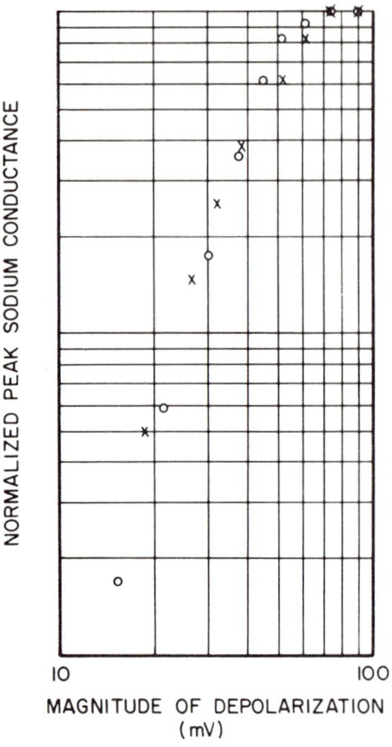

FIG. 7. Comparison of peak transient sodium conductance on depolarization *from* a constant level (equilibrium) and depolarization *to* a constant level (44 mV below equilibrium). Crosses denote depolarization from equilibrium; circles denote depolarization to 44 mV (data from Hodgkin and Huxley, 1952c, d).

of the sodium conductance decreased as the preset potential was decreased. In other words, if the membrane had originally been hyperpolarized (set at a potential greater than the equilibrium value), the peak conductance was greater than that reached from equilibrium, which in turn was greater than that reached from a slightly depolarized state. Hodgkin and Huxley interpreted these results in terms of a steady-state "inactivation" as a function of preset potential (1952c, p. 501).

Another way of viewing these results is in terms of accommodation of the transient sodium conductance to any sustained level of membrane potential. In this view, the magnitude and form of the transient sodium conductance depend very little on the absolute magnitude of membrane potential, but are determined almost entirely by the rate and magnitude of any change in that potential and on the history of recent membrane activity. If the results of this experiment are displayed in a plot of the peak sodium conductance as a function of the magnitude of depolarization —disregarding the preset potential from which the depolarization took place—the curve is essentially identical to that obtained by plotting peak conductance as a function of depolarization from the equilibrium potential. Such a plot is shown in Fig. 7.

The transient sodium conductance requires a certain amount of time to accommodate to any preset, or conditioning, potential; in fact, the time constants are simply those of inactivation or recovery from inactivation, depending on whether the conditioning involves a decrease or an increase in membrane potential. Thus, if the membrane potential is slowly varying, accommodation lags the potential somewhat, and the transient conductance depends on the the recent history of membrane potential variation. In view of Fig. 7, it seems valid as a first approximation to characterize the two components of sodium conductance as follows.

1. A steady-state sodium conductance which depends solely on the absolute magnitude of transmembrane potential.
2. A transient sodium conductance which is independent of the magnitude of membrane potential, but which depends on the rate and magnitude of any change in that potential, and on the recent history of potential fluctuations.

The potassium conductance behaves very much like the steady portion of sodium conductance. On sudden depolarization, it rises with considerable delay to a steady value, which apparently depends only on the absolute magnitude of membrane potential. An important distinction is that the delay and rise times of potassium conductance are much greater than those of sodium. The magnitude of the steady-state potassium conductance increases nonlinearly with decreasing membrane potential (see Fig. 8), whereas the delay and rise times decrease (Fig. 9). The important parameters of potassium conductance can be listed as follows:

1. *Delay time* decreases monotonically with increasing depolarization, varying from approximately 2 to less than 1 msec.
2. *Rise time* decreases monotonically with increasing depolarization, varying from more than 10 to approximately 1 msec.
3. *Steady-state potassium conductance* increases monotonically with increasing depolarization, as shown in Fig. 8.
4. *Fall time on sudden repolarization may be* 8 msec or more, and apparently decreases slightly with increasing depolarization.

Aside from the parameters of sodium and potassium conductance, several other details must be supplied to complete the circuit analog of Fig. 3. They are listed below, and they too apply to the giant axon of the squid at 6°C.

1. *Potassium potential* (V_K): 80 to 85 mV, inside negative with respect to outside.
2. *Sodium potential* (V_{Na}): 45 to 50 mV, inside positive with respect to outside.
3. *Leakage potential* (V_ℓ): 38 to 43 mV, inside negative with respect to outside.
4. *Leakage conductance* (G_ℓ): approximately 0.23 mmhos/cm^2.
5. *Membrane capacitance* (C): approximately 1 μF/cm^2.
6. *Resistance in series with membrane capacitance*: approximately 7 ohms for one square centimeter of membrane.

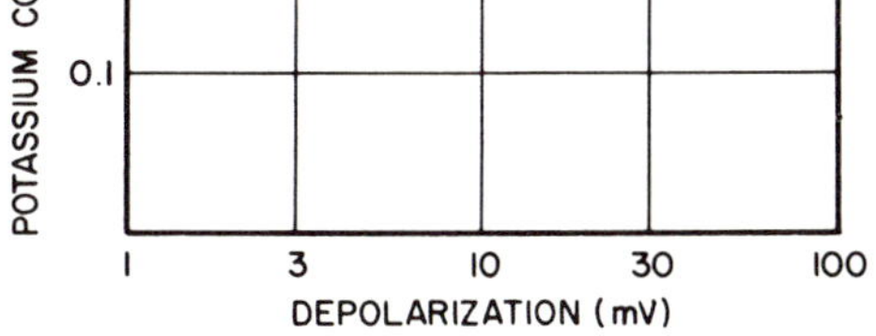

FIG. 8. Magnitude of potassium conductance as a function of depolarization (from Hodgkin and Huxley, 1952d).

FIG. 9. Rise and delay times of potassium conductance change as a function of depolarization (from Hodgkin and Huxley, 1952d).

7. *Resting potential*: 60 to 65 mV, inside negative with respect to outside.
8. *Spike amplitude*: approximately 100 mV, inside positive with respect to outside.

Possible Application of the Ionic Model to Subthreshold Phenomena

The experiments of Hodgkin and Huxley were designed to disclose the electrochemical events responsible for the generation of an all-or-none spike and its subsequent unattenuated propagation along the axon. Since the spike involves complete depolarization and is over in one or two milliseconds, they were primarily interested in rapid, large-magnitude excursions from the equilibrium potential. Subthreshold potentials, on the other hand, generally involve smaller and slower changes in membrane potential. Interestingly enough, this is quite consistent with their data, where the observed changes in sodium and potassium conductances take place much more slowly for small depolarizations and repolarizations than for large ones. Unfortunately, the data are scant in the range of depolarizations applicable to subthreshold phenomena such as postsynaptic potentials and pacemaker potentials, which are generally less than 10 mV and may persist for tens or even hundreds of milliseconds. In addition, phenomena which may be very important in slow potential fluctuations, such as the slow recovery from inactivation, were not thoroughly explored, since they do not directly affect the single spike. It would be interesting, however, to extrapolate the Hodgkin-Huxley data to low depolarization levels and see whether the resulting model could account for the various modes of subthreshold behavior. This seems especially pertinent, since the region of the cell membrane in which the various subthreshold phenomena are found is probably merely an extension of the membrane which surrounds the axon. It seems likely, therefore, that the membrane in this region is similar in structure to that of the axon; yet some modifications probably exist, since this region of the membrane is often electrically inexcitable, being incapable of producing the all-or-none spike. Thus, as well as extrapolating the Hodgkin-Huxley data, one must look for reasonable over-all parameter changes which would account for inexcitability. As yet too little is known of the membrane to relate these changes to actual structural differences; a more tenable approach is to relate them to changes in the measurable electrical properties of the membrane, such as the relative magnitudes and time constants of the potassium and sodium conductances as they relate to changes of membrane potential. This can be done by means of an electronic analog with variable parameters.

An Electronic Analog of the Ionic Model of the Membrane

If one attempts to build an electronic analog of the Hodgkin-Huxley model, the real problem he faces is that of simulating the variable sodium and potassium conductances. Probably the most direct approach is to build a network whose transfer function is the same as the desired relationship between conductance

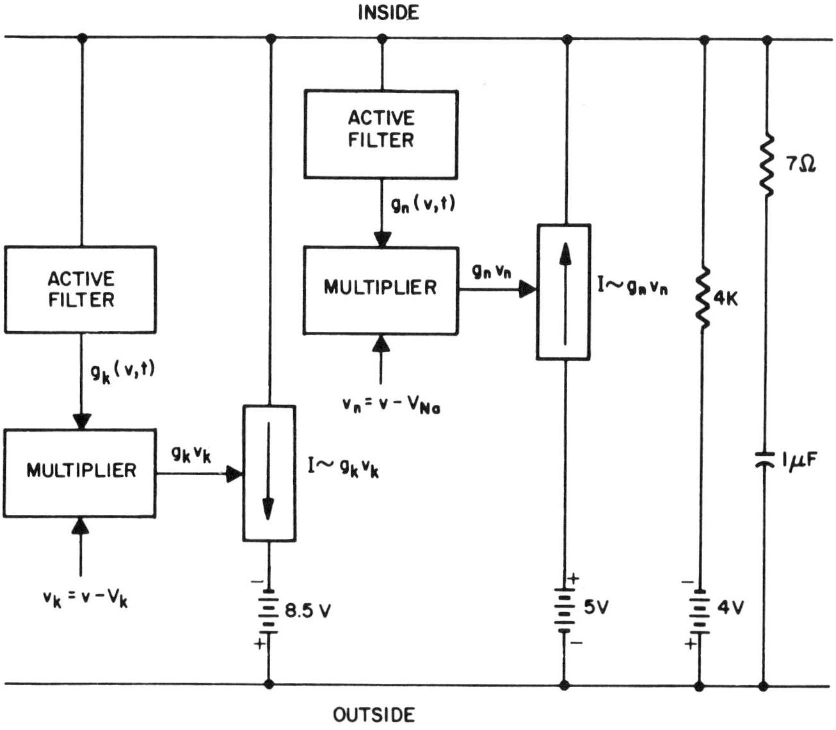

FIG. 10. System diagram for electronic simulation of the Hodgkin-Huxley model.

and membrane potential (and its time derivatives and so on). This only solves half of the problem, however, since the output of such a network is a current or voltage which must in turn be transformed into conductance. This can be accomplished in either of two ways. One method is to apply the output directly to a voltage- or current-controllable conductance, such as a field-effect transistor. The second method, the one used in this model, is to multiply the output voltage by the difference between the ionic potential (e.g., V_K) and the instantaneous simulated membrane potential v, producing a current proportional to the product. This is obviously equivalent to a voltage controllable conductance, since the current through any conductor is proportional to the product of the conductance and the voltage across it. The resulting electronic equivalent of the ionic model is outlined in Fig. 10. This network represents one square centimeter of membrane, with conductances and capacitance equal to the values obtained by Hodgkin and Huxley. For compatability with electronic circuitry, however, the potentials have been increased a hundredfold, 10 V representing 100 mV. The networks representing potassium and sodium conductances are described below.

The Potassium Conductance. The circuit used to simulate potassium conductance is shown in Fig. 11. The conductance function $g_K(v, t)$ is generated

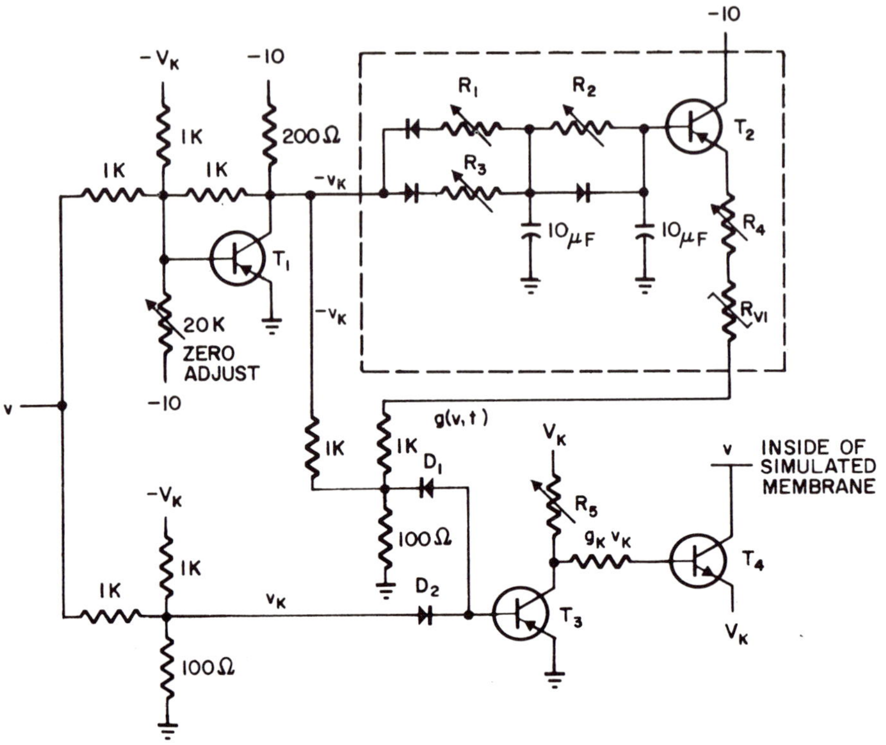

R_{V1} = VECO 023L4	R_3 = 10K MAX	T_1, T_2 = 2N1379
R_1 = 5K MAX	R_4 = 10K MAX	T_3 = 2N1381
R_2 = 5K MAX	R_5 = 5K MAX	T_4 = 2N1304

FIG. 11. Electronic circuit for simulation of potassium conductance. The diodes D_1 and D_2 are each four 1N100's in series; a feedback loop may be employed to reduce the input impedance of T_3. The active filter within the dashed line generates the conductance function, $g(v, t)$.

from the simulated membrane potential through a nonlinear active filter (enclosed in the dashed box), with variable resistors providing control over parameters as follows:

1. *Delay time* and *rise time* are dependent on the combination of R_1 and R_2, the delay time being controllable from 0 to about 10 msec, rise time from 0 to over 100 msec.
2. *Fall time* can be varied independently with R_3 over a range from 0 to 200 msec.
3. The *nonlinearity* in the dependence of steady conductance on simulated membrane potential is controlled by R_4 in series with the varistor R_{v1}, the dependence being nearly linear for maximum settings and nearly cubic for minimum settings.
4. The *magnitude* of the conductance is controlled by R_5, which is the amplitude control of the multiplier.

Multiplication is accomplished by means of a diode circuit. Since the diode is essentially a square-law device, the current I_b into the base of T_3 can be shown to be proportional to $g_K(v, t)v_K$, where v_K is the difference between the simulated potassium potential V_K and the simulated membrane potential v. Thus

$$I = A[v_k + g_k(v, t)]^2 - A(v_k)^2 = 2A\, g_k(v, t)v_k \qquad (g_k \ll v_k),$$

where A is a constant of proportionality. Finally, the current through transistor T_4, which is the simulated conductance, is proportional to $g_k(v, t)v_k$. Transistor T_1 is simply an inverter stage. The input to this stage is $v - V_K$, and the output is $V_K - v$, which is $-v_k$.

The Sodium Conductance. The difference v_n between the simulated sodium potential V_{Na} and the simulated membrane potential v is 12 V at equilibrium (corresponding to 120 mV in the ionic model). Since we are primarily interested in small deviations in v, generally less than 1 or 2 V (10 to 20 mV in the real membrane), v_n can be considered to be approximately constant, and thus the multiplier can be eliminated in the simulated sodium conductance. That is to say, the changes of sodium flux for small deviations in membrane potential depend almost completely on changes in sodium conductance, and very little

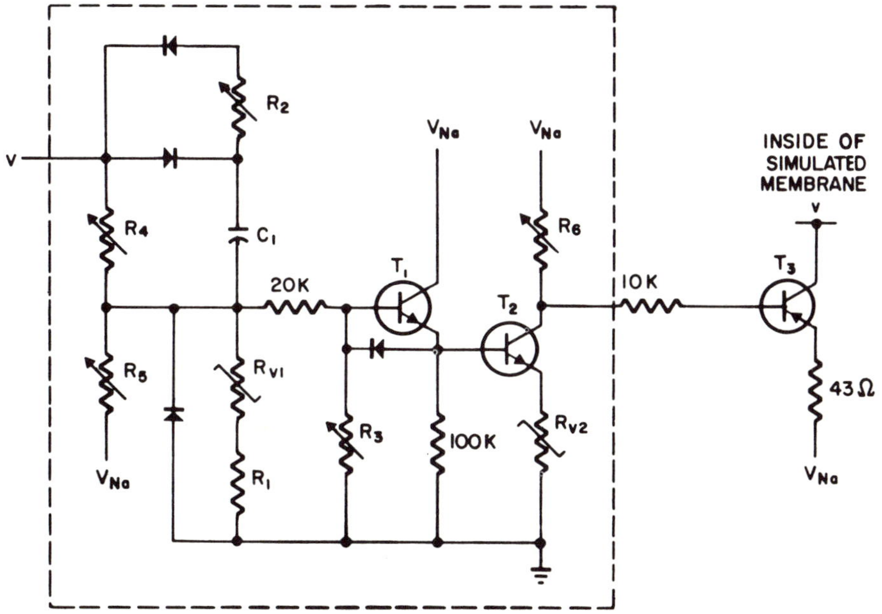

R_1 = 500Ω	R_4 = 50 K MAX	R_{V1}, R_{V2} = VECO 023L4
R_2 = 100K MAX	R_5 = 100K MAX	T_1, T_2 = 2N1304
R_3 = 50 K MAX	R_6 = 50K MAX	T_3 = 2N1373

FIG. 12. Electronic circuit for simulation of sodium conductance. The active filter for generation of g_n is enclosed in the dashed box.

on the variations in membrane potential. The circuit for simulation of sodium conductance is shown in Fig. 12. Here we apply the output, $g_n(v, dv/dt, t)$, from the active filter directly to the base of the conductance transistor T_3. Again, the parameters of conductance (given below) are established by resistors in the filter. In this case, however, some of the parameters have been fixed at the values proposed by Hodgkin and Huxley.

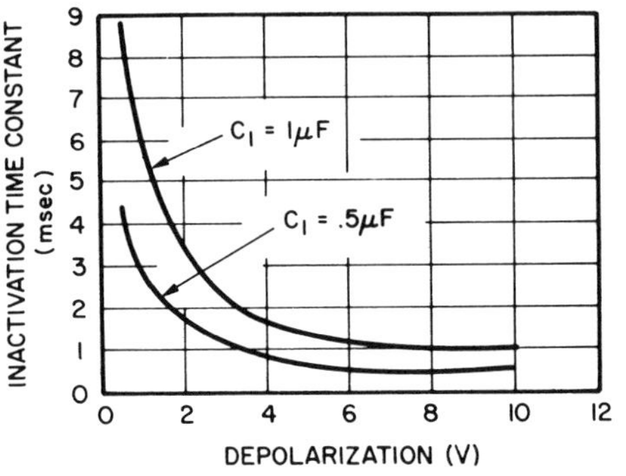

FIG. 13. Initial-inactivation time constant as a function of depolarization in the simulated sodium conductance, shown for two values of C_1.

1. The *inactivation time constant* is fixed by varistor R_{v1}, in series with resistor R_1, decreasing monotonically with depolarization in approximately the manner described by Hodgkin and Huxley (see Fig. 13).
2. The *time constant of recovery from inactivation* is dependent on R_2, and is controllable from 0 to 30 msec.
3. The *delay time* and *rise time* are fixed at less than 1 msec.
4. The *fall time* is also fixed at less than 1 msec.
5. The *nonlinearity* in the dependence of transient conductance on simulated membrane potential is controlled by R_3 and may be varied from quadratic to nearly quartic. This is accomplished by shifting along the voltage-current characteristics of varistor R_{v2}.
6. The *steady-state value of conductance* is controlled by R_4 and can be balanced to any desired equilibrium value with R_5. The *nonlinearity* of the steady conductance is dependent on the relative settings of R_3 and R_4 and may be varied from a square-root dependence to a cubic or even quartic relationship.
7. The *magnitude of total conductance* is controlled by R_6.

Capacitor C_1 simulates the accommodation of the transient sodium conductance to persistent potentials, the accommodation time constants being the same as those for inactivation and recovery from inactivation, as in the ionic model. For very large depolarizations of the simulated membrane potential, the sodium-conductance transistor becomes saturated, and the conductance becomes constant at a little more than 20 mmhos, as does that of the ionic model. Thus the circuit is a good approximation to the ionic model at both low levels and high levels of depolarization. It should be adequate, therefore for simulation of spikes as well as low-level potential fluctuations, since, during the spike, the potential falls through the intermediate range very rapidly.

The Electronic Analog as a Spike Initiator

The first test of the analog was to determine whether it was capable of producing a spike similar in form to that predicted by the ionic model and observed in real axons. For this test the outside of the simulated membrane was grounded, and a depolarizing current pulse of 1-msec duration was applied to the inside (see Fig. 14). The variable conductance parameters were set at values closely corresponding to the Hodgkin-Huxley model, as follows:

Simulated Potassium Conductance

Delay time:	approximately .5 msec.
Rise time:	approximately 3 msec.
Nonlinearity:	between quadratic and linear
Magnitude:	adjusted to give approximately 20 mhos at peak depolarization.
Fall time:	approximately 10 msec.

Simulated Sodium Conductance

Time constant of recovery from inactivation:	approximately 20 msec.
Nonlinearity of steady conductance:	approximately cubic, saturating at 20 mmhos.
Nonlinearity of steady conductance:	approximately quadratic.
Equilibrium conductance:	balanced to essentially zero.

With these parameter settings, the analog produced spikes of approximately 2-msec duration with a threshold of approximately 5 to 10% of the spike amplitude. The form of the spike can be seen in the oscillograms of Fig. 15. In Fig. 15A a minimal stimulus is applied during the first millisecond and is followed by a "prepotential" of approximately 3 msec. The spike then rises to peak amplitude in less than 1 msec. This peak is followed by "active" recovery due

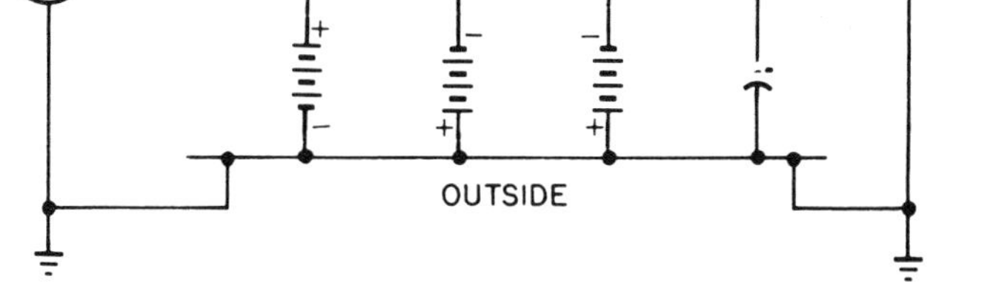

FIG. 14. Analog test configuration.

to increased conductance in the simulated potassium circuit. The after-depolarization lasts about 6 msec, followed by after-hyperpolarization, which lasts about 50 msec. Strength-duration data were not obtained, but the effect of increasing stimulus repetition rate was examined. Figure 15B shows the maximum rate at which spikes could be produced with the same minimal stimulus as in Fig. 15A (about 10 per second). As the stimulus rate was increased slightly above this frequency, the analog responded with one spike for every two stimuli, as shown in Fig. 15C. With further increase, the analog responded to fewer and fewer of the stimuli until, at approximately 100 stimuli per second, it responded to only two and was then inactive until the excitatory stimulus was removed (Fig. 15D). This effect is analogous to Wedensky inhibition.

The ability of the analog to produce spikes was maintained over a wide range of parameter variations, the threshold depending on the relative magnitudes and nonlinearities of the simulated sodium and potassium conductances. The Wedensky type of inhibition depended on the time constant of recovery from inactivation of the simulated sodium conductance and on the fall time of the potassium conductance.

The analog was capable of spike generation and recovery in the complete

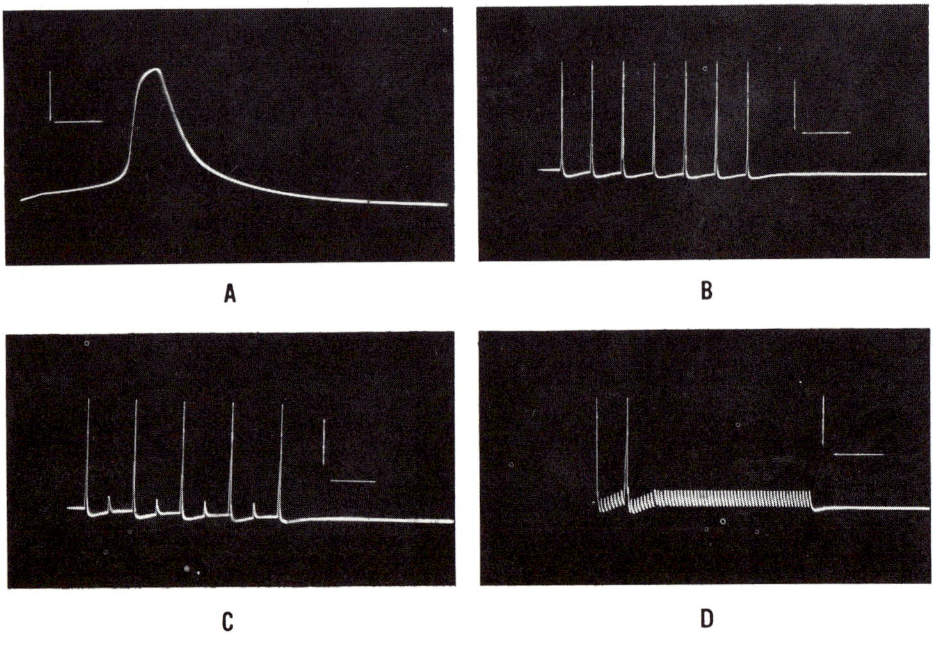

FIG. 15. A. Single spike produced by analog in response to minimal stimulus of 1 msec duration. Vertical line, 5 V; horizontal line, 2 msec. B. Series of spikes produced by analog, showing maximum spike repetition rate for stimulus of A. Vertical line 5 V; horizontal line 100 msec. C. Series of spikes produced by analog, showing firing on alternate stimuli. Vertical line 5 V; horizontal line 100 msec. D. Response of analog to high stimulus-repetition rate, showing effect analogous to Wedensky inhibition. Vertical line 5 V; horizontal line 100 msec.

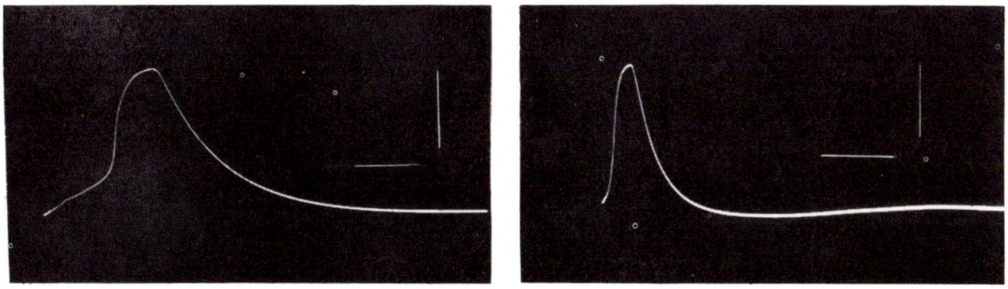

A B

FIG. 16. A. Spike generated by analog without simulated potassium conductance, resting potential established by leakage circuit alone. B. Spike generated by analog without simulated potassium conductance, resting potential established by balance between leakage and sodium currents.

absence of the simulated potassium conductance. In this case the resting potential was established by the leakage current, which produced a passive-recovery time constant of about 4 msec. Recovery was made possible by the inactivation of the simulated sodium conductance. The form of the spike can be seen in Fig. 16A; as would be expected, no after-hyperpolarization occurred. In Fig. 16B, the resting potential was established by a balance between the simulated leakage current and a small simulated sodium current that was allowed to flow at equilibrium. In this case after-hyperpolarization was effected by allowing the spike to inactivate the simulated sodium resting conductance, and the period of hyperpolarization was determined by the time constant of recovery from inactivation. This last case may well be artificial, but it serves to demonstrate the power of such an analog in testing possible modifications of the ionic hypothesis.

Simulated Synaptic Potentials with the Electronic Analog

In many neurons the postsynaptic membrane has been found to be electrically inexcitable. It is generally believed that postsynaptic potentials in such neurons are generated by chemically induced changes in specific ion permeabilities of the postsynaptic membrane (Fatt, 1958; Florey, 1962). Thus, in the ionic model, excitatory postsynaptic potentials would result from chemically induced increases in the sodium conductance, inhibitory potentials from similar increases either in the potassium conductance or in some aspect of the leakage conductance (such as chloride conductance). The chemical transmitter substances are thought to be released into the synaptic region during the presynaptic spike, and having acted upon the postsynaptic membrane, are thought to either diffuse away or be inactivated by enzymes in the synaptic region. If the enzymatic inactivation is a first-order reaction, then either diffusion or inactivation will result in an exponential decrease in transmitter concentration with time after the presynaptic spike. This can be simulated with a simple resistance-capacitance network, and the time constant can be controlled by making the resistance variable. As far as the present electronic analog is concerned, inhibitory inputs are assumed to

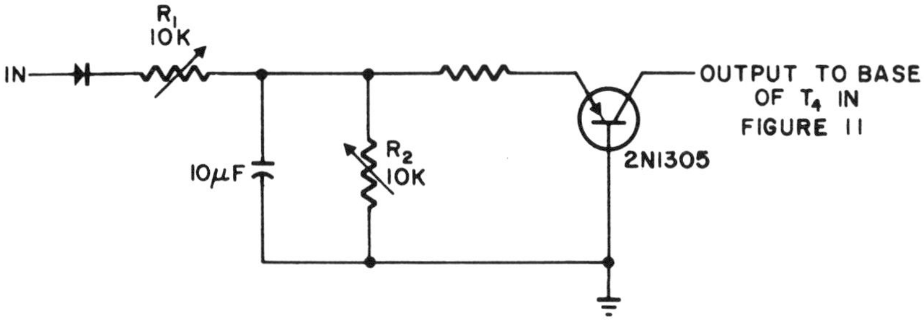

FIG. 17. Electronic simulation of inhibitory synapse: R_1 controls the quantum of simulated transmitter substance per presynaptic spike; R_2 controls the transmitter decay or inactivation rate.

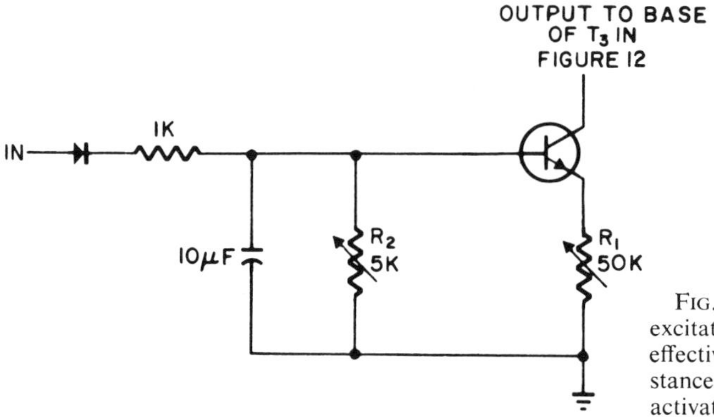

FIG. 18. Electronic simulation of excitatory synapse: R_1 controls the effective quantity of transmitter substance per spike; R_2 controls the inactivation rate.

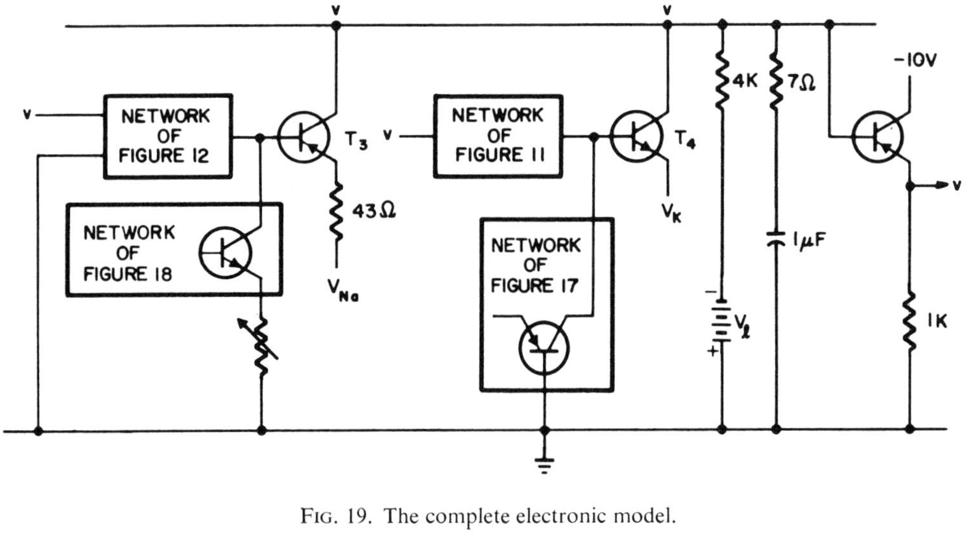

FIG. 19. The complete electronic model.

increase the potassium conductance, excitatory inputs to increase the sodium conductance. Inhibitory pulses are fed into the network of Fig. 17, whose output represents the effective concentration of transmitter substance. This output is applied directly to the conductance transistor in the potassium circuit. The effective quantity of simulated transmitter substance injected per inhibitory pulse is controlled by variable resistor R_1, while the rate of transmitter inactivation is controlled by R_2. A similar network (Fig. 18) is used to simulate excitatory synaptic transmission, again with simulated transmitter quantity per pulse and inactivation rate controlled by R_1 and R_2, respectively. The entire electronic model, with simulated synaptic inputs, is shown in Fig. 19.

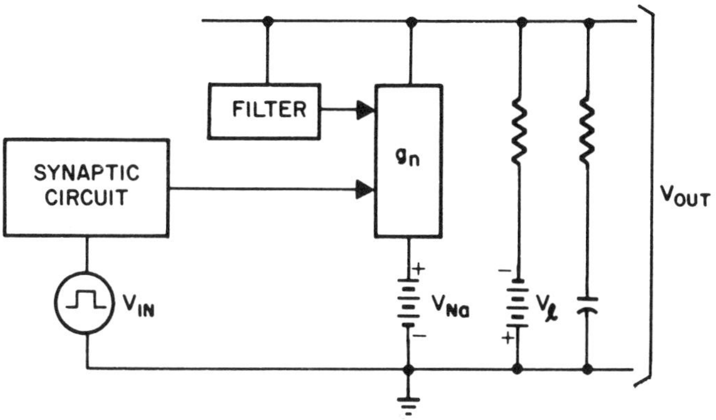

FIG. 20. Analog test configuration for simulation of epsp's.

Simulated Excitatory Postsynaptic Potentials. The first attempts to simulate epsp's with the electronic analog were made without the simulated potassium conductance, as shown in Fig. 20. Here two independent pathways control the simulated sodium conductance. One represents the synaptic input; the other represents the dependence of sodium conductance on membrane potential and is essentially a positive-feedback loop. A positive pulse applied to the simulated synaptic input causes a corresponding increase in sodium conductance, resulting in a depolarization of the membrane potential. This depolarization, through the positive-feedback loop, causes a further increase in conductance: the process is regenerative. It is limited, however, by the increased current through the leakage circuit, by inactivation of the sodium conductance, and by the transmitter inactivation. Figure 21 shows simulated epsp's resulting from a burst of 11 positive pulses applied to the synaptic network. Each pulse was 10 V in amplitude and 1 msec in duration. In Fig. 21A the positive-feedback loop was open, so that any changes in simulated sodium conductance were due entirely to the voltage representing transmitter-substance concentration. In this case successive simulated epsp's summed, but without facilitation or anti-facilitation. Each potential showed the typical ballistic form, continuing to rise after the 1-msec pulse was terminated, and falling slowly, with a time constant of 30 msec (see

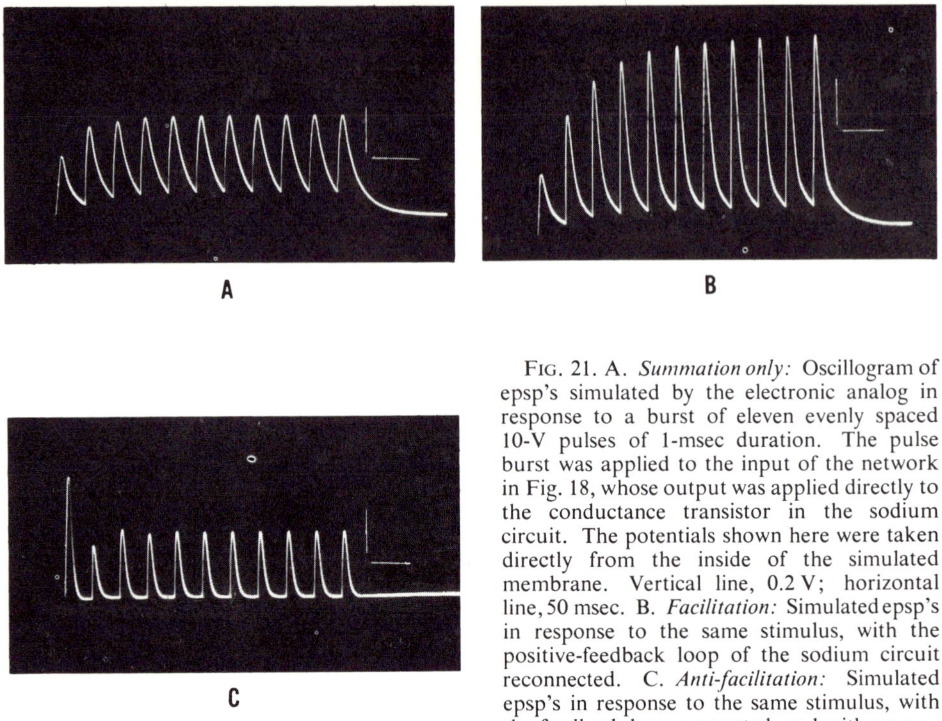

FIG. 21. A. *Summation only:* Oscillogram of epsp's simulated by the electronic analog in response to a burst of eleven evenly spaced 10-V pulses of 1-msec duration. The pulse burst was applied to the input of the network in Fig. 18, whose output was applied directly to the conductance transistor in the sodium circuit. The potentials shown here were taken directly from the inside of the simulated membrane. Vertical line, 0.2 V; horizontal line, 50 msec. B. *Facilitation:* Simulated epsp's in response to the same stimulus, with the positive-feedback loop of the sodium circuit reconnected. C. *Anti-facilitation:* Simulated epsp's in response to the same stimulus, with the feedback loop connected, and with parameter changes as described in the text.

Arvanitaki and Chalazonitis, 1961b). The slow rise was due to the large simulated transmembrane capacitance, whereas the prolonged fall was due to the slow inactivation of simulated transmitter substance, whose time constant was set at approximately 30 msec. Figure 21B shows the simulated epsp's under identical conditions except that here the positive-feedback loop was closed. Again the successive potentials sum, but with marked facilitation. With one exception, the parameters of simulated sodium conductance were identical to those outlined in the section "The Electronic Analog as a Spike Initiator," thus being very close to those outlined by Hodgkin and Huxley. The exception was the time constant of recovery from inactivation, which was set at approximately 10 msec. Figure 21C shows a dramatic change in the epsp's, yet only three parameters were changed: two of the parameters of the simulated synaptic input and one involving the sodium conductance. The effective quantity of simulated transmitter substance per presynaptic pulse was doubled, whereas the time constant for its decay was reduced to 10 msec. The time constant for recovery from inactivation was increased to 50 msec. Again the positive-feedback loop was closed. Thus by changing three parameters we have gone from marked facilitation to a form of anti-facilitation almost identical to that shown by Hagiwara and Bullock (1957). Further examination showed that the transition could be accomplished by changing the two synaptic parameters alone, setting the time constant for recovery from inactivation of the simulated sodium conductance at an intermediate value (e.g., 20 msec).

Thus, with the parameters of simulated sodium conductance essentially the same as those of the Hodgkin-Huxley model, the occurrence of facilitation or anti-facilitation depended on only the two parameters of the simulated synaptic-transmitter concentration: the quantity injected per presynaptic pulse, and the decay or inactivation rate.

Inclusion of the simulated potassium conductance had two important effects on the form of the potentials. First, it brought in an "active" component of recovery, increasing the rate of fall of the simulated epsp's and making the rate of fall dependent on epsp amplitude. The second and perhaps the more interesting effect was a rebound hyperpolarization on termination of an excitatory pulse burst, as shown in Fig. 22. This would correspond to a postexcitatory period of depression lasting for 50 msec or more. The parameters of the simulated potassium conductance were set at values consistent with the Hodgkin-Huxley data for low depolarizations, as follows:

Rise time: 10 msec. *Nonlinearity*: quadratic.
Delay time: 1 msec. *Fall time*: 6 msec.
Magnitude: adjusted to give 20 mmhos at complete depolarization.

The potassium conductance, because of its delayed rise and relatively rapid fall, had little or no effect on the amplitude of individual simulated epsp's in the experiments of Fig. 21. It did not interfere with facilitation or anti-facilitation or the change from one to the other.

Three mechanisms are primarily responsible for facilitation in this analog. Chief among these is a residue of simulated synaptic transmitter substance persisting from one pulse to the next. Other parameters remaining fixed, the quantity of this residue determines the degree of facilitation. The second important mechanism is the extreme nonlinearity in the dependence of the transient portion of the simulated sodium conductance on changes in membrane potential. A small initial change in conductance results in a proportional initial rate of depolarization. If inactivation were absent, the simulated transient sodium conductance would increase nearly as the cube of depolarization, lagging it by less than a millisecond. This would cause a corresponding increase in the rate of depolarization, and the process would become regenerative. The initial rate of

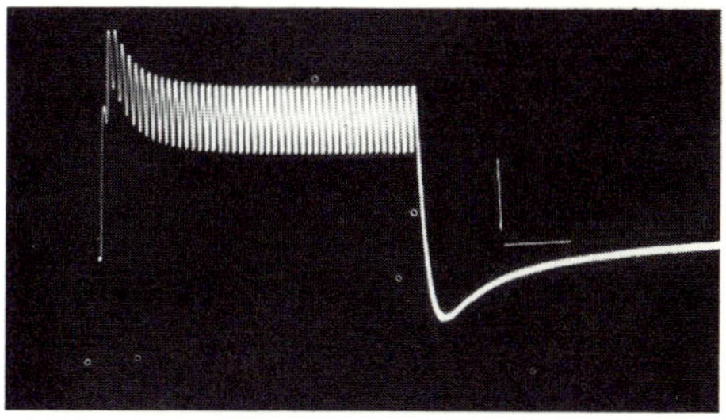

FIG. 22. Simulated epsp's with both the potassium circuit and the sodium circuit connected. Note the rebound hyperpolarization on termination of the stimulus. Vertical line, 0.2 V; horizontal line, 50 msec.

rise would thus be amplified with time in an extremely nonlinear manner. Since the initial rate of rise is finite, however, this nonlinear regeneration is limited by the third important mechanism, inactivation of the simulated transient sodium conductance. Thus the peak magnitude which a given simulated epsp reaches is determined by a balance between the rate of rise of the potential itself and the rate of inactivation of the transient sodium conductance. The dynamic relationships of this system can be expressed in terms of simultaneous differential equations. Unfortunately these equations are insoluble by normal means, owing to the complex interdependence of the variables, e.g., the dependence of rate of inactivation on depolarization. However, a highly simplified, concrete example may show how such a system is capable of facilitation. Let us ignore the simulated leakage and potassium currents, as well as the simulated steady-state sodium conductance and the inactivation of the transient sodium conductance. If we now assume a cubic relationship between the simulated transient sodium conductance and the depolarization, the over-all system dynamics are given by

$$\frac{dv}{dt} = \frac{I}{C} = \frac{v_n(g_0 + av^3)}{C}, \tag{1}$$

where C is the simulated transmembrane capacitance, v is the simulated transmembrane potential, v_n is the difference between the simulated sodium potential and v, g_0 is the initial conductance due to the simulated synaptic input, and a is the constant of proportionality between simulated sodium conductance and membrane potential. One means of testing this highly simplified system for possible facilitatory mechanisms is to assume a set of values for g_0 applied at $t = 0$ and solve the equation for each of these, evaluating v at a fixed time, e.g., 5 msec after the application of g_0. Even this is extremely difficult, however, since the solution of the differential equation gives v as a complicated implicit function of time. A simpler approach is to assume that after a short time, e.g., 1 msec, g_0 results in a small initial depolarization, v_0, and that further depolarization depends entirely on the simulated transient sodium conductance. The differential equation may now be written

$$\frac{dv}{dt} = \frac{v_n}{C}\,a(v^3),$$

with the boundary condition $v(t) = v_0$ at $t = 1$ msec. In the Hodgkin-Huxley model v_n is approximately 100 mV, C is approximately 1 μF, and a is approximately 10^{-4} mmho/mV^3. Translating these to corresponding values for the analog, expressing time in milliseconds and potential in volts, and recalling that 1 V in the model represents 10 mV in the neuron, we obtain $a = 10^{-7}$ (coul/msec · volt^4). Then

$$\frac{dv}{dt} = \frac{(10\ \text{volts} \cdot 10^{-7}\ \text{coul/msec} \cdot \text{volt}^4)}{10^{-6}\ \text{coul/volt}} \cdot v^3,$$

or

$$dv/dt = v^3.$$

Applying the boundary conditions and solving for $t = 5$ msec, we obtain

$$(2) \qquad v(t) = \frac{v_0}{(1 - 8v_0^2)^{1/2}} \qquad (t = 5 \text{ msec}).$$

Solutions of this equation are shown in Table 1 for various values of v_0.

Equation (2) exhibits a vertical asymptote at $v_0 = (\frac{1}{8})^{1/2}$. As the values of v_0 approach this asymptote, very small increments will produce extremely large changes in $v(t)$ ($t = 5$ msec). Thus if an excitatory stimulus produced an initial potential of 0.32 V, the response at 5 msec would be 0.755 V. If at the time of arrival of this stimulus, however, there existed a residual potential of only 0.02 V, the response at 5 msec would be 1.242 V. Thus a 3% residue would result in facilitation of more than 50%. Other examples from Table 1 will show clearly that in this system we have a very powerful mechanism for facilitation. It must be remembered, however, that Eq. (2) does not describe the entire analog, but is only an isolated facilitatory mechanism. If we take into account the rate of inactivation of the simulated transient sodium conductance, which increases monotonically with v, and also the simulated steady-state sodium current, the simulated potassium current, the simulated leakage current, and the time and voltage dependence of each of these, the resulting system is far too complex for ordinary mathematical treatment. Probably the best method for examining the response of such a system is precisely the method discussed in this paper: the building of an accurate analog of the system and the examination of its responses. It is then enough to say that the system probably is capable of facilitation because the analog exhibits facilitation. It must be remembered that the analog may contain mechanisms which were not intentionally built into it, and which may not be part of the ionic hypothesis. The observer must constantly be on the alert for such contingencies and must remember their potential effects on analog behavior.

TABLE 1

POTENTIAL AT 5 MSEC AS A FUNCTION OF THE POTENTIAL AT 1 MSEC (VOLTS)

v_0	$v(t)$ ($t = 5$ msec)
0.1	0.104
0.2	0.242
0.25	0.353
0.30	0.568
0.32	0.755
0.34	1.242

The mechanism for anti-facilitation in the analog is much more easily seen than that for facilitation. After a reasonably large simulated epsp, the transient sodium conductance remains for some time inactive. Recovery from inactivation takes some time, and if during this time a second excitatory stimulus arrives, the response to that stimulus will be diminished by the inactivation. Thus facilitation or anti-facilitation, as far as the analog is concerned, depends on whether or not the residual simulated transmitter concentration is sufficient to overcome the residual inactivation of transient sodium conductance. The criteria for facilitation and anti-facilitation in the analog are shown qualitatively in Fig. 23. With other analog parameters remaining fixed, and both the potassium and sodium circuits connected, the occurrence of facilitation, anti-facilitation, or neither depends only on the magnitude and rate of decay of the

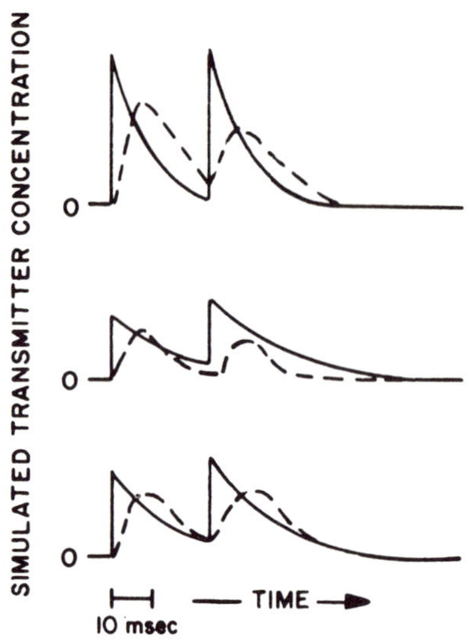

FIG. 23. Qualitative criteria for facilitation, anti-facilitation, and neither in the electronic analog.

simulated transmitter concentration. In the top trace the quantum of transmitter substance per presynaptic impulse is large and would result in a correspondingly large initial epsp, which in turn would result in a relatively high residual inactivation of sodium conductance (dashed line). Since the rate of transmitter decay is also high, the residual transmitter concentration is unable to overcome the residual sodium inactivation, and anti-facilitation occurs. In the center trace the quantum of simulated transmitter per presynaptic impulse is reduced. The resulting initial epsp would be correspondingly reduced—as would the residual sodium inactivation. In this case the transmitter decay is slow enough to allow the residual transmitter concentration to overcome the residual sodium inactivation, and facilitation occurs. In the bottom trace, the residual transmitter concentration just balances the residual sodium inactivation, and neither facilitation nor anti-facilitation occurs.

Simulated Inhibitory Postsynaptic Potentials. Figure 24 shows two examples of simulated ipsp's from the electronic analog. In Fig. 24A the simulated sodium conductance was removed so that the effects of potassium conductance alone might be observed, and the resulting potentials are due to simulated synaptic-transmitter concentration alone. The potentials simply sum; neither facilitation nor anti-facilitation occurs. Again the simulated synaptic potentials show the typical ballistic form, rising in several milliseconds to a peak, then tending to fall slowly back to the equilibrium potential. In Fig. 24B the simulated sodium conductance was reintroduced. The time constant of recovery from inactivation was set at 14 msec, and the equilibrium conductance was set at a low, nonzero value so that the simulated sodium flux contributed to the equilibrium potential, making it approximately $\frac{1}{2}$V less negative.* Thus the simulated sodium conductance would not only increase nonlinearly with depolarization, but would also decrease nonlinearly with hyperpolarization. The

* This corresponds to approximately 5 mV in the real neuron.

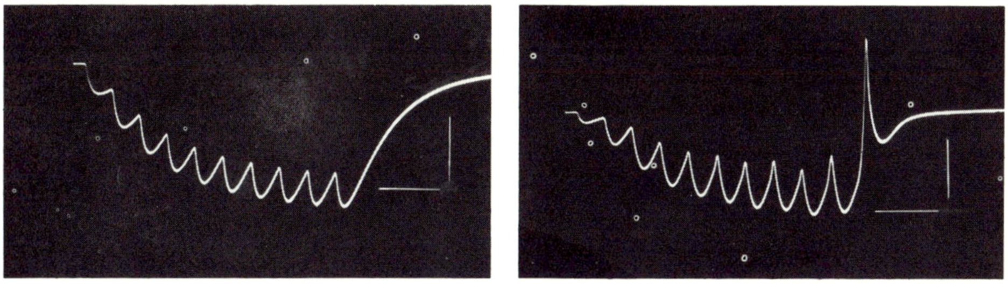

A B

FIG. 24. A. Oscillogram of ipsp's simulated by the electronic analog in response to a burst of 10 evenly spaced pulses, each having an amplitude of 10 V and a duration of 1 msec. The pulse burst was applied to the network of Fig. 17, whose output was applied directly to the conductance transistor of the potassium circuit. In this case the simulated sodium conductance was removed from the analog. The potentials were taken directly from the inside of the simulated membrane. Vertical line, 0.5 V; horizontal line, 20 msec. B. Simulated ipsp's resulting from the same stimulus, with the sodium conductance returned to the analog. In this case the resting potential was established by a small equilibrium sodium flux, as well as by potassium and leakage fluxes.

remainder of the parameters in both the simulated sodium and the simulated potassium conductances were set at the values outlined in the section "The Electronic Analog as a Spike Initiator." It can be seen from Fig. 24B that with these settings the simulated ipsp's showed not only marked facilitation, but also a bi-phasic rebound on termination of the inhibitory pulse burst. As the equilibrium value of simulated sodium conductance was returned to zero, the form of the potentials returned to that of Fig. 24A with no facilitation and no rebound.

Pacemaker Potentials with the Electronic Analog

With the analog parameters adjusted to the values listed in the section on spike initiation, the equilibrium value of simulated sodium conductance was increased slowly from zero. As a result the simulated membrane potential was gradually decreased until, at a moderate value of conductance (approximately 0.08 mmho), threshold conditions were reached. At this point the analog commenced regular, spontaneous spike generation at a rate of approximately one spike per second. Further increase in the equilibrium conductance resulted in a gradual increase in spike frequency up to about 50 spikes per second. This spontaneous activity continued until the equilibrium value of simulated sodium conductance was reduced below the threshold level. The form of the individual spikes was quite similar to that of spikes produced from direct stimulation of the simulated membrane, the primary distinction being that, following the normal prolonged after-hyperpolarization, the membrane potential did not decay simply to the equilibrium value, but continued to decrease slowly until threshold conditions again obtained and a new spike was generated. This slow decline allowed some accommodation in the simulated sodium conductance, so the threshold potential was considerably higher than that for a sudden depolarization. At higher spontaneous frequencies, residual inactivation carried over

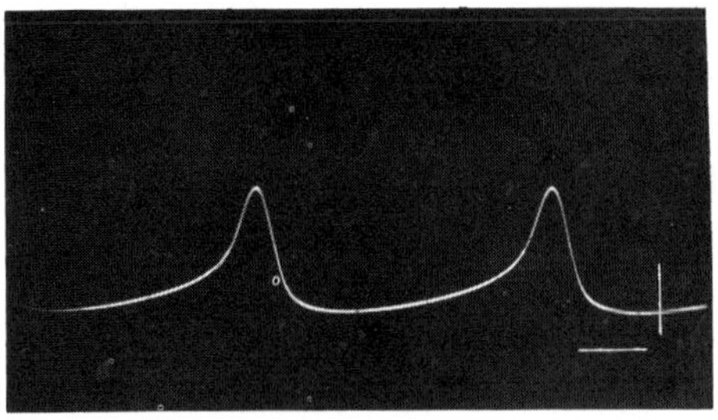

A

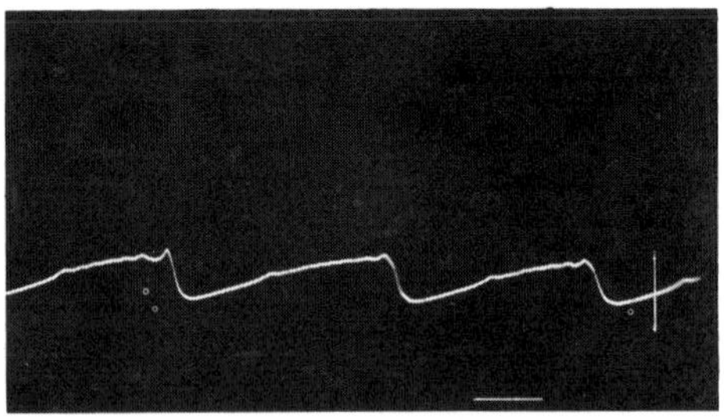

B

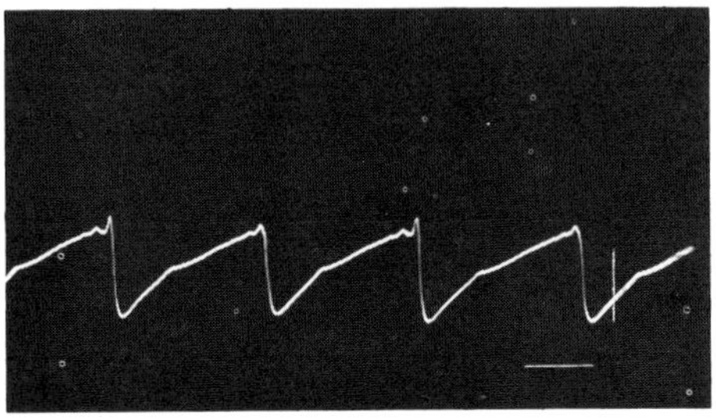

C

FIG. 25. A. Simulated pacemaker potentials on the electronic analog. Vertical line, 0.5 V; horizontal line, 20 msec. B. Simulated pacemaker potentials with parameter changes as described in the text. Vertical line, 1.0 V; horizontal line, 100 msec. C. Simulated pacemaker potentials with further parameter changes. Vertical line, 1.0 V; horizontal line, 200 msec.

from one spike to the next, and the threshold was further increased. This limited the maximum frequency of spontaneous spike generation.

Thus spontaneous spike generation could be obtained from the analog with a slight leakage of simulated sodium flux at equilibrium, and with the remaining parameters set at essentially the values given by Hodgkin and Huxley. With these settings, however, it was found that a spike was always required to terminate the depolarizing phase and reset the simulated membrane potential. In order to produce the low-frequency subthreshold oscillations described by Bullock *et al.*, a more general change in parameter values was necessary. Principally, the excitability of the analog had to be reduced. In order to reset the potential without the all-or-none spike, it was necessary that the simulated potassium flux be able to overtake the sodium flux on the slow depolarizing phase. Although there are a number of ways to accomplish this in the analog, the most straightforward approach seemed to be simply to decrease the nonlinearity in the dependence of sodium conductance on depolarization while increasing that of potassium. Throughout the experiments shown in Fig. 25 the nonlinearity for transient sodium conductance was approximately quadratic and that for the steady-state conductance was approximately linear. To compensate for the reduced nonlinearity, the magnitude of the total simulated sodium conductance was increased to maintain the original over-all range of conductance variation. At the same time, the nonlinearity in the potassium conductance was increased to a quadratic dependence on depolarization, which is more consistent with the Hodgkin-Huxley data for low values of depolarization (see Fig. 8). In Fig. 25A the remaining parameters of simulated potassium and sodium conductances were set close to the Hodgkin-Huxley values. The simulated equilibrium sodium conductance was then increased until spontaneous oscillations developed. It can be seen that the resulting potentials had a form somewhere between the sinusoidal and the sawtooth types of pacemaker potentials. The peak amplitude of these potentials was approximately 0.8 V, corresponding to 8 mV in the real membrane. This agrees well with the amplitude of pacemaker potentials in real neurons. The frequency was determined by the simulated equilibrium sodium conductance, with a minimum frequency of approximately 5 cps. In Fig. 25B the time constant for recovery from inactivation of the simulated sodium conductance was increased to 50 msec, and the magnitude of the dependence of potassium conductance on membrane potential was increased twofold. Spontaneous potentials in this case commenced with approximately one-half the simulated equilibrium sodium conductance required in Fig. 25A. Here the form of the potentials was very nearly a sawtooth wave. The peak amplitude was approximately 1 V, and the minimum frequency was approximately 2 cps. In Fig. 25C the magnitude of the voltage dependence of simulated potassium conductance was further increased. Again the minimum equilibrium value of simulated sodium conducance necessary for spontaneous oscillations was reduced. Here again the potentials exhibited a sawtooth form with a peak amplitude of approximately 1 V. The minimum frequency was now approximately $\frac{1}{2}$ cps.

Graded Response with the Electronic Analog

With the parameters of the analog reset to the values listed in the section "The Electronic Analog as a Spike Initiator," subthreshold stimuli of 1-msec duration were applied to the inside of the simulated membrane as in Fig. 14. For the weakest of these stimuli the simulated membrane potential responded in direct proportion to the stimulus magnitude, depolarizing exponentially during the 1 msec of stimulus application and commencing to repolarize exponentially immediately on termination of the stimulus. As the magnitude of the stimulus was increased, the simulated membrane potential responded in an increasingly disproportionate manner, rising with a positive inflection during the application of the stimulus and continuing to rise for some time after termination of the stimulus. Figure 26 shows a plot of the maximum depolarization as a function of stimulus amplitude. It can be seen that the resulting curve is very similar to the curves of graded or local response found in the literature (e.g., Bullock, 1957). The section of the response curve for very weak stimuli is linear; as the stimulus strength is increased, the response curve becomes increasingly nonlinear, approaching a vertical asymptote at the spike-threshold level. This is essentially the same type of behavior as that ascribed by Bullock to the graded-response regions of the soma (Bullock, 1957, 1958, 1959). The graded-response loci differ in at least one respect, however, in that the threshold for spike initiation is generally much higher than that found in the axon. One means of raising the threshold in the analog is simply to decrease the total magnitude of the dependence of simulated sodium conductance on depolarization. The effect of this is to move the vertical asymptote of the graded-response curve to the right, as shown in Fig. 26.

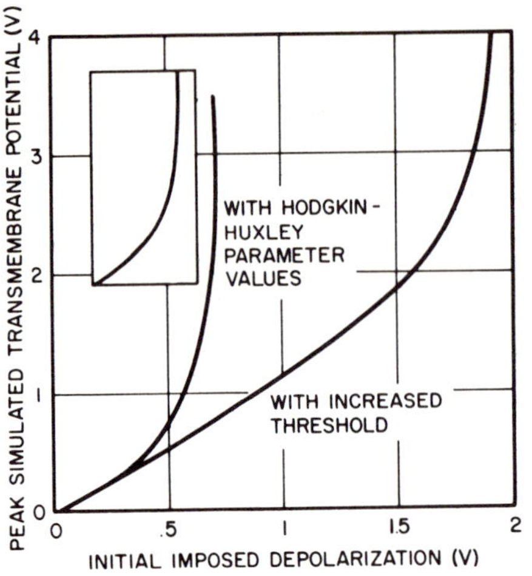

FIG. 26. Graded-response curve for the analog. The curve on the left was obtained with parameters set very close to those found by Hodgkin and Huxley. The curve on the right was obtained by increasing the threshold. The insert shows a graded-response curve from a real neuron (Bullock, 1957).

In the analog, the nonlinear region of the graded-response curve is a result of the nonlinear transient increase in the simulated sodium conductance. Since this conductance will accommodate to slowly varying potentials, the degree of nonlinearity in the graded-response curve should depend to some extent on the time derivative of the stimulus potential. To test the analog for such dependence, the following experiment was performed. With all other parameters set close to the Hodgkin-Huxley values, the magnitude of the dependence of total simulated sodium conductance on membrane potential was reduced until the spike threshold was increased to about 2 V for sudden depolarizations (an approximately threefold increase). A sawtooth stimulus potential with a peak amplitude of 4 V was then applied across the simulated membrane in the same manner as shown in Fig. 14. The graded response was then observed while the slope of the sawtooth potential was increased. The oscillograms in Fig. 27 show the results of this experiment. It can be seen that for gradually increasing stimulus potentials the depolarization was directly proportional to the stimulus, with no nonlinearities even a 4 V depolarization, which was twice the threshold value for sudden depolarizations. As the imposed rate of depolarization was increased above approximately 0.1 V/msec, a small graded potential appeared. This can be seen in the oscillograms as a small hump superimposed on the ramp of the sawtooth stimulus. As the rate was increased further, the graded response became more and more pronounced until, at very high rates, the graded potential developed into a full all-or-none spike. This type of behavior corresponds roughly to what Bullock found* in the graded- or local-response loci. The primary distinction seems to be that a threshold rate of imposed depolarization exists in the real loci. For rates below threshold no graded response occurs, whereas for rates above threshold the local response depends on the magnitude of depolarization and not on the rate. In the analog, on the other hand,

* Personal communication.

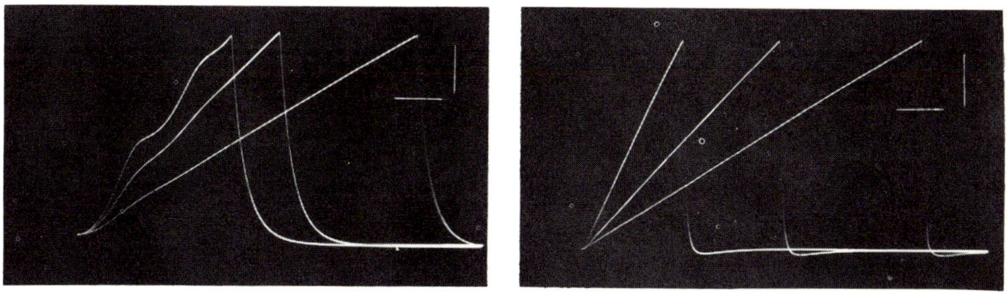

A B

FIG. 27. A. Response of analog to imposed sawtooth depolarization. Vertical line, 1 V; horizontal line, 50 msec. B. Response of analog as imposed rate of depolarization is increased. Vertical line, 1 V; horizontal line, 10 msec.

although a threshold rate seems to exist, the magnitude of graded responses for superthreshold rates seems to depend on both the magnitude and the rate of imposed depolarization. Further experiments will be necessary, however, to determine these dependencies more precisely in the analog. The important finding for the moment is that the analog does exhibit accommodation to slowly varying potentials. This is really not surprising, however, since Hodgkin and Huxley predicted just such accommodation with their mathematical model. (See Hodgkin and Huxley, 1952d.)

Summary of Behavioral Patterns Available with the Electronic Analog

The electronic analog which has been discussed in this paper is essentially equivalent to the ionic model of Hodgkin and Huxley. The various conductances, electrochemical potentials, and transmembrane capacitances were simulated with electronic circuits which were placed between two conductors representing the inside and the outside of the neural membrane. Under direct electrical stimulation of the conductor representing the inside of the membrane, the analog exhibited both graded response and, above a certain threshold, all-or-none spike generation. The form of the spike was very similar to that found in real neurons. It included the 1- to 2-msec spike itself, the 5- to 6-msec after-depolarization, and the 40- to 50-msec after-hyperpolarization. The analog also exhibited accommodation to slowly varying stimuli, and, in addition, a phenomenon similar to Wedensky inhibition for high stimulus-repetition rates. The magnitude of the graded response, when plotted against stimulus amplitude, followed essentially the same nonlinear curves that have been determined for real neurons. In addition, when a stimulus was applied directly to the simulated sodium or potassium conductance, the analog responded with potentials very similar to the epsp's and ipsp's found by intracellular recording in real neurons, exhibiting the typical ballistic form and summing the response to a given stimulus with the residual response from all previous stimuli. These potentials also exhibited facilitation, anti-facilitation, or neither, and could be shifted from one of these modes to another merely by changing the values of the simulated synaptic parameters (i.e., the simulated transmitter substance concentration per presynaptic impulse and the decay rate of this concentration). The simulated membrane potential also exhibited both mono-phasic and bi-phasic rebound on termination of a burst of either excitatory or inhibitory simulated presynaptic impulses. The analog was capable of spontaneous spike generation over a wide range of frequencies and, with slight parameter modifications, was capable of generating low-frequency, subthreshold potential fluctuations very similar to the pacemaker potentials in real neurons. These were produced in either a relaxation (sawtooth) mode or a nearly sinusoidal mode, depending on the settings of specific parameters. Finally, with the parameters of simulated sodium conductance adjusted to increase the analog spike threshold, behavior very similar to that found in the local response loci of real neurons was observed.

All of the various modes of analog behavior were found to occur within a small volume of the parameter space. Many other modes of behavior were observed as the parameters were varied outside this region, and vast regions of the parameter space are still totally unexplored. Although it might be interesting to examine these unexplored regions and catalog the observed behavior, it seems more reasonable to limit such examination to searching for known neural phenomena, particularly since our principal interest here is to simulate just such phenomena. As far as the modes of neural behavior discussed in this paper were concerned, the search of the analog parameter space for corresponding behavior was reasonably straightforward. Once the dynamics of the analog system were fairly well understood on a qualitative basis, a specific phenomenon could be singled out, and the search of parameter space could be carried out in a logical manner.

Discussion

The animal nervous system performs at least three basic functions: gathering, processing, and applying information. This much we know; but what are the mechanisms and interactions underlying these functions? The state of physiological knowledge has been greatly advanced since the time of Borelli and Willis and other seventeenth-century biological pioneers. The final experiment of Galvani on frog nerve–muscle preparations (1791) and the subsequent work by Du Bois-Reymond (1848) demonstrated that an electrical current at least accompanied nerve action; but was it, as they both declared, proved to be the essence of nervous transmission? Most modern physiologists believe that these currents are merely part of a complicated chain of electrochemical changes in the nerve membrane—changes such as those postulated in the ionic hypothesis. Very few present-day scientists, on the other hand, would presume that the ionic hypothesis is the final solution to the problem of understanding nervous activity. Sherrington (1906) concluded, after a long series of experiments on mammalian reflex arcs, that conduction was only one of many functions performed by individual neurons. He suggested that much slower, integrative processes were carried out at the junctions between cells, which he called the synapses. He postulated that with synaptic integration a single neuron responded not only to a single event, but also to the temporal summation of a large number of events or stimuli. In addition, he postulated that many conduction pathways converged on a single cell and that the resulting synapses were capable of interacting. Excitation in a single pathway would not only stimulate the receiving cell directly, but would also enhance or diminish the effect of excitation in the other paths.

The work of Sherrington was one of the first attempts to understand the information-processing capabilities of individual neurons. Workers since that time have accumulated large bodies of data concerning individual neuron properties and, as the data mount, the information-processing capabilities of a single cell seem to become more and more obscure. Bullock and his co-workers have found a large number of neural properties (pacemaking, local-response, and others) which may well contribute to these capabilities, but in what manner? How do the various synaptic, pacemaker, and graded-response regions interact to produce the one end result: an output which is a complex but relevant function of a large number of spatially and temporally distributed inputs? The problem seems to be one of limiting, rather than expanding, the possible behavioral modes of such a complex system. As Bullock puts it: "We know a lot about what it can do; what we need to know now is what it can't do." One thing is certain: very little is yet known about information processing in the single cell—let alone large groups of these cells. When such groups are studied, they are generally viewed as "black-boxes" with partially controllable inputs and partially observable outputs; little or nothing is known of the processes going on within the box itself. When it comes to the brain, the neural system is so complex and the scientific data so sparse that it seems impossible to apply even the black-box approach. Some may even wonder whether modern models of the brain (such as those composed of randomly connected two-state devices) are really much closer to the truth than the pneumatic or explosive models of Willis or the fermenting-fluid model of Borelli. The information processing in single neurons or in very small groups of neurons seems much more amenable to study, and such a study is one of the primary goals of our laboratory.

In order to explore the information-processing capabilities of neurons and small groups of neurons, we are attempting to develop physical analogs whose behavior closely matches that of real neurons. Each analog neuron is to be based on the locus concept, being composed of some combination of simulated synaptic, graded-response, pacemaker, and spike-initiating loci. Thus each model neuron is in itself a system, a system of model loci.

Each model locus has several variable parameters. For example, we wish to control the time course of facilitation or anti-facilitation, rebound, and the amplitude and time course of individual postsynaptic potentials. The characteristics of an entire simulated neuron will depend not only on the number and distribution of its various model loci, but also on the parameter settings of each locus. It can readily be seen that the number of variable parameters in an individual model neuron may become extremely large, and the number of parameters in a small system of model neurons may rapidly become unmanageable. In designing and constructing these model neurons we are faced with two problems of economics—economy of effort in adjusting parameters and economy of construction in building such a large number of system components (the loci). It has been hoped that as the system was developed, logical limitations

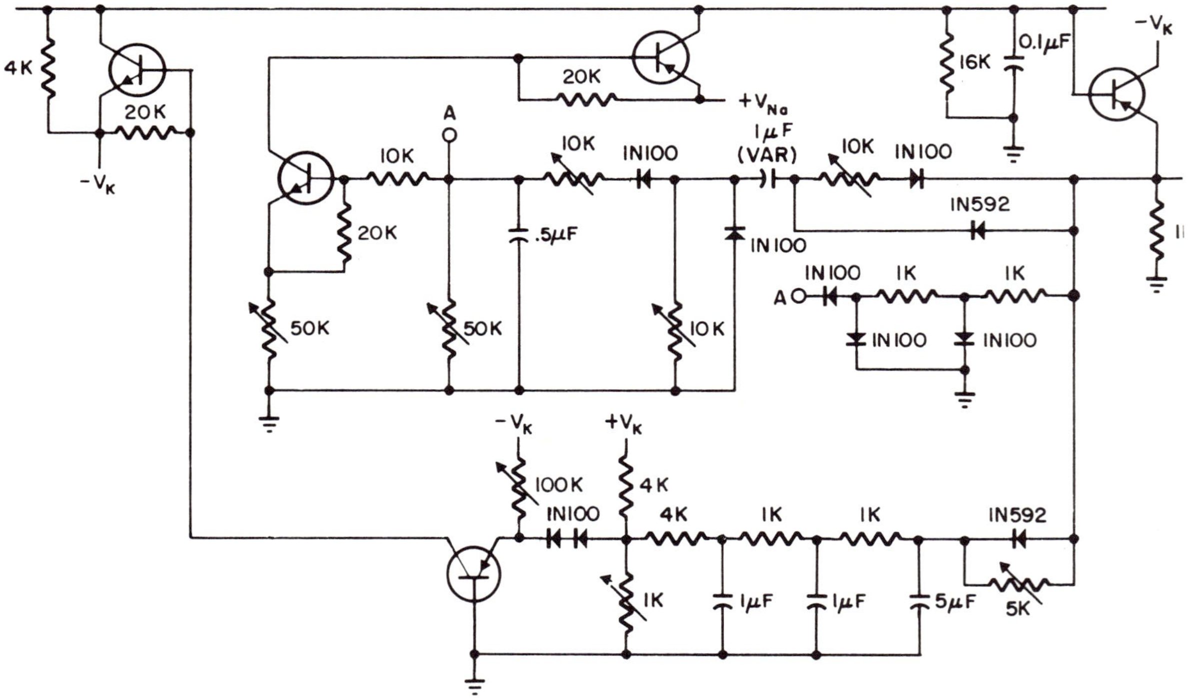

FIG. 28. A simple analog of the ionic model. With slight parameter variation this model provides all of the various modes of behavior ascribed to neural loci.

on the parameter space would evolve, thus at least diminishing the problem of economy of effort. This leaves us for the moment with the second problem, that of limiting the number of electronic components in the system.

In our first attempts to simulate the neuron on the basis of the locus concept, we decided on a serial approach. Each locus was to be represented by a single two-port electronic network. The output from one of these loci would then be applied through lossy connecting networks to the inputs of other loci. The resulting analog promised to be a complex system of distinct circuits, with a large number of feedback and feedforward connections. A review of the neurophysiological literature, particularly the papers of Rall (1959a, b, 1960), induced a shift from the serial to a parallel approach. The simulated neuron is now a parallel combination of simulated loci, each being represented by a two-terminal, or one-port, network. It was also found that a basic ingredient of the various neural loci seemed to be an effective negative resistance. Since this is also a basic ingredient of the Hodgkin-Huxley model, it was thought that a single type of network based on that model but with variable parameters might suffice for simulation of all of the various loci. To explore this possibility an early, relatively simple, six-transistor network was developed which simulated most of the important aspects of the ionic model. Eight parameters were made variable, each representing a combination of parameters in the Hodgkin-Huxley model. Subsequent experimentation proved that this simple circuit (Fig. 28) could indeed produce all of the modes of behavior found in the various neural loci. In fact, one of these circuits, with parameters adjusted appropriately, could provide several locus functions (e.g., pacemaker and local-response) simultaneously.

The over-all neural analog now promises to be a comparatively simple system of a few of these circuits connected in parallel, with the resulting potential fed through lossy networks to simulated spike initiators. In order to determine precisely what parameter changes were necessary to go from one mode of behavior to another, the simple 6-transistor circuit was expanded to the network described in this paper, permitting more control over the specific parameters of conductance. The new analog was constructed with 11 transistors, 3 varistors, and 15 diodes. Accurate parameter control and repeatability were provided by 10-turn, direct-readout potentiometers. With this new network it was possible, as described in the paper, to relate the various modes of analog behavior to specific parameters of the ionic model. In addition, experiments with the expanded analog suggested several possible approaches to limiting the parameter space of the single simulated neuron. For example, if we assume that because the membrane of a neuron is continuous its properties are nearly constant, then we might logically limit the parameters of the synaptic loci to the quantum of simulated synaptic transmitter per presynaptic impulse and the subsequent transmitter decay or inactivation rate, particularly since variation of these parameters alone in the expanded analog produced facilitation, anti-facilitation, or neither, as well as the various rebound phenomena. This condenses the synaptic parameter space from several dimensions to only two. Other similar parameter limitations may be quite reasonable, particularly since the expanded analog produced all of the various locus properties with settings quite close to those proposed by Hodgkin and Huxley. Thus we at least have a glimmering of a solution to the problem of economy of effort. Any limitations on the parameter space, however, will at best be hypothetical, and operation of the complete analog under these limitations will be merely one test of the hypotheses.

References

Arvanitaki, A., and N. Chalazonitis, 1961a. Slow waves and associated spiking in nerve cells of *Aplysia*, *Bull. Inst. Océanogr. Monaco* (1224), 1–15.

Arvanitaki, A., and N. Chalazonitis, 1961b. Slow changes during and following repetitive synaptic activation in ganglion nerve cells, *Bull. Inst. Océanogr. Monaco* (1225), 1–23.

Bullock, T. H., 1957. Neuronal integrative mechanisms, in *Recent Advances in Invertebrate Physiology*, B. T. Scheer, ed., Eugene, Ore.: Univ. Oregon Pubs., pp. 1–20.

Bullock, T. H., 1958. Parameters of integrative action of the nervous system at the neuronal level, *Exptl. Cell Res.* Suppl. 5, 323–37.

Bullock, T. H., 1959. Neuron doctrine and electrophysiology, *Science*, **129**(3355), 997–1002.

Bullock, T. H., and S. Hagiwara, 1957. Intracellular recording from the giant synapse of the squid, *J. Gen. Physiol.*, **40,** 565–77.

Bullock, T. H., and C. A. Terzuolo, 1957. Diverse forms of activity in the somata of spontaneous and integrating ganglion cells, *J. Physiol.*, **138,** 343–64.

Du Bois-Reymond, E., 1848. *Untersuchungen über thierische Elektricitat*, vol. I, Berlin: G. Reimer.

Eccles, J. C., 1957. *The Physiology of Nerve Cells*, Baltimore, Md.: Johns Hopkins Press.

Fatt, P., 1957. Sequence of events in synaptic activation of a motoneurone, *J. Neurophysiol.* **20,** 61–80.

Fatt, P., 1958. Physics of nerve processes, *Repts. Prog. in Physics*, **21,** 112–43.

Florey, E., 1962. Recent studies on synaptic transmitters, *American Zoologist*, **2,** 45–54.

Galvani, L., 1791. *Commentary on the Effects of Electricity on Muscular Motion*, Burndy Library edition, 1953, Norwalk, Conn.: Burndy.

Hagiwara, S., and T. H. Bullock, 1957. Intracellular potentials in pacemaker and integrative neurons of the lobster cardiac ganglion, *J. Cell. Comp. Physiol.*, **50**(1), 25–48.

Hodgkin, A. L., and A. F. Huxley, 1952a. Currents carried by sodium and potassium ions through the membrane of the giant axon of *Loligo*, *J. Physiol.*, **116,** 449–72.

Hodgkin, A. L., and A. F. Huxley, 1952b. The components of membrane conductance in the giant axon of *Loligo*, *J. Physiol.*, **116,** 473–96.

Hodgkin, A. L., and A. F. Huxley, 1952c. The dual effect of membrane potential on sodium conductance in the giant axon of *Loligo*, *J. Physiol.*, **116,** 497–506.

Hodgkin, A. L., and A. F. Huxley, 1952d. A quantitative description of membrane current and its application to conduction and excitation in nerve, *J. Physiol.*, **117,** 500–44.

Hodgkin, A. L., A. F. Huxley, and B. Katz, 1952. Measurement of current-voltage relations in the membrane of the giant axon of *Loligo*, *J. Physiol.*, **116,** 424–48.

Hoyle, G., 1962. Comparative physiology of conduction in nerve and muscle, *American Zoologist*, **2,** 5–25.

Maynard, D. M., 1953. Activity in a crustacean ganglion: I. Cardio-inhibition and acceleration in *Panulirus argus*, *Biol. Bull.*, **104**(2), 156–70.

Maynard, D. M., 1955. Activity in a crustacean ganglion: II. Pattern and interaction in burst formation, *Biol. Bull.*, **109**(3), 420–36.

Mullins, L. J., 1956. Structure of nerve cell membranes, in *Molecular Structure and Functional Activity of Nerve Cells*, R. G. Grenell and L. J. Mullins, eds., Washington, D.C.: A.I.B.S.

Otani, T., and T. H. Bullock, 1959. Effects of presetting the membrane potential of the soma of spontaneous and integrating ganglion cells, *Physiol. Zool.*, **32**(2), 104–14.

Rall, W., 1959a. "Dendritic Current Distribution and Whole Neuron Properties," Naval Medical Research Institute Report No. NM 01 05 00.01.02, National Institutes of Health, Bethesda, Md.: Naval Medical Research Institute.

Rall, W., 1959b. Branching dendritic trees and motoneuron membrane resistivity, *Exptl. Neurol.*, **1**, 491–527.

Rall, W., 1960. Membrane potential transients and membrane time constant of motoneurons, *Exptl. Neurol.*, **2**, 503–32.

Sherrington, C. S., 1906. *The Integrative Action of the Nervous System*, New Haven: Yale Univ. Press.

Wilson, D. M., 1961. The central nervous control of flight in a locust, *J. Exptl. Biol.*, **38**, 471–90.

9. Pulse Processing by Neuron Models

F. JENIK, *Technische Hochschule, Darmstadt*

A mathematical model of the axon membrane of *Loligo*, developed by Hodgkin and Huxley, has been extended to include an excitatory and an inhibitory input channel. These added channels permit the alteration of the sodium and potassium conductances of the postsynaptic membrane by presynaptic impulses. Electronic models of mammalian motoneurons, suggested by Dr. K. Küpfmüller, have been developed and used to examine input characteristics and input-output relations. The principal objective of this modeling is the investigation of pulse processing in nervous systems. Examples of such studies are given.

Two Neuron Models

The Extended Model of Hodgkin and Huxley. It has been demonstrated that potentials and currents in excitable cell membranes may be simulated by potentials and currents in an electrical circuit (Cole and Curtis, 1939; Hodgkin and Huxley, 1952; Eccles, 1957). A representative diagram of such a circuit is shown in Fig. 1, where C_M is the membrane capacitance; g_L is the constant leakage conductance; g_{Na} and g_K are the variable sodium and potassium conductances, respectively, and where ϕ_{Na}, ϕ_K, and ϕ_L represent the equilibrium potentials at which the sodium current i_{Na}, the potassium current i_K, and the "leakage current" i_L (due to chloride and other ions), respectively, are zero. We use φ_M to represent the membrane potential; the membrane voltage u

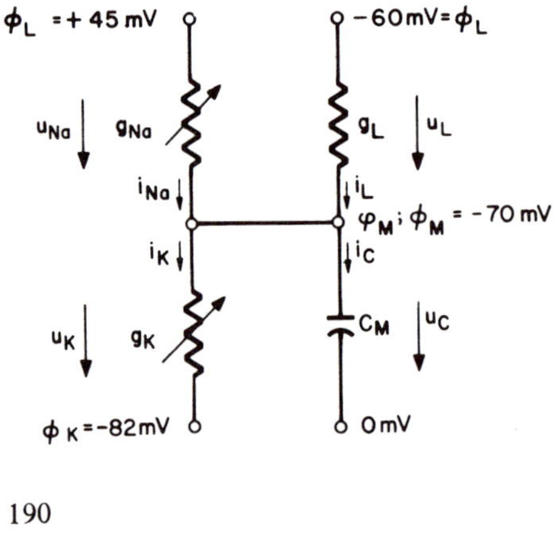

FIG. 1. Diagram of an electrical circuit representing a cell membrane.

indicates the displacement of the membrane potential φ_M from the resting value $\phi_M = -70$ mV. Also, u_{Na}, u_K, and u_L are the driving voltages for the ionic currents i_{Na}, i_K, and i_L; i_C represents the capacitive current, u_C the driving voltage for i_C.

The sodium and the potassium conductances may be influenced in two ways: (1) from the inside by feedback of the membrane potential φ_M (Hodgkin and Huxley, 1952), and (2) from the outside (via input channels) by presynaptic pulses (Eccles, 1957).

A rather complex system of nonlinear differential equations describing the feedback system of the giant axon of *Loligo* was elaborated by Hodgkin and Huxley (1952). Relying on findings of Eccles (1957), Küpfmüller developed simple linear differential equations describing approximately the transformation of presynaptic pulses into corresponding changes of membrane conductance (Küpfmüller, 1959; Küpfmüller and Jenik, 1961).

Presynaptic pulses in neurons cause changes of membrane conductances at the synaptic sites, whereas the feedback system influences the conductances at the synapse-free parts of the membrane. In the investigation discussed below

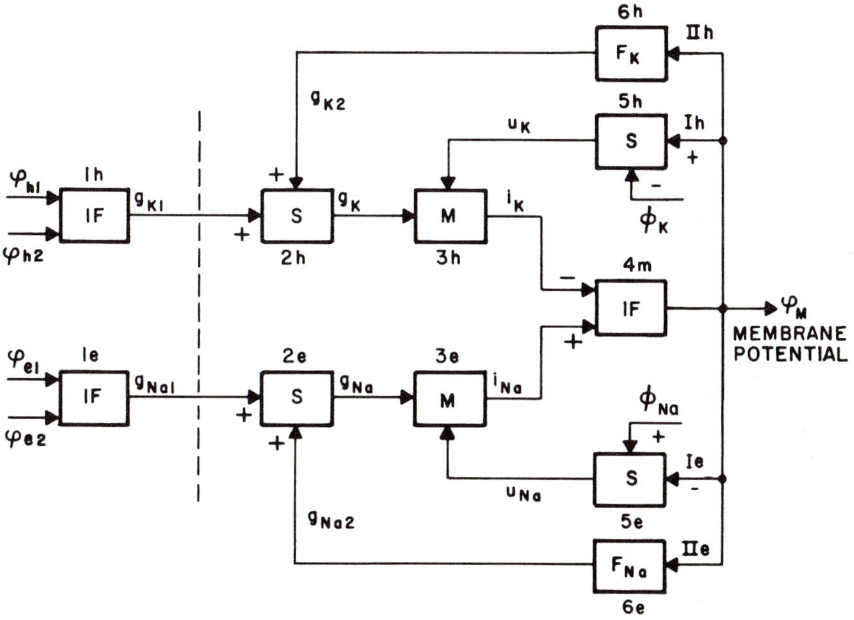

FIG. 2. Simplified block diagram of the extended Hodgkin-Huxley system.

Left of the dotted line are inhibitory (h) and excitatory (e) input channels, transforming presynaptic pulses (φ_{h1}, φ_{h2}, φ_{e1}, φ_{e2}) into corresponding changes of membrane conductance (IF = integrating pulse shaper). Right of the dotted line (except for blocks 2h and 2e, which are simple summing devices to couple the input channels into the system) is the feedback system of Hodgkin and Huxley. The membrane potential φ_M is generated at the central membrane block (block 4m, representing a parallel circuit of C_M and g_L) by the sodium and potassium currents, which are induced by the sodium and potassium feedback loops, respectively. Each feedback loop consists of a voltage path and a conductance path. The letter S indicates summation or subtraction, M multiplication, F_K and F_{Na} the complicated conductance functions.

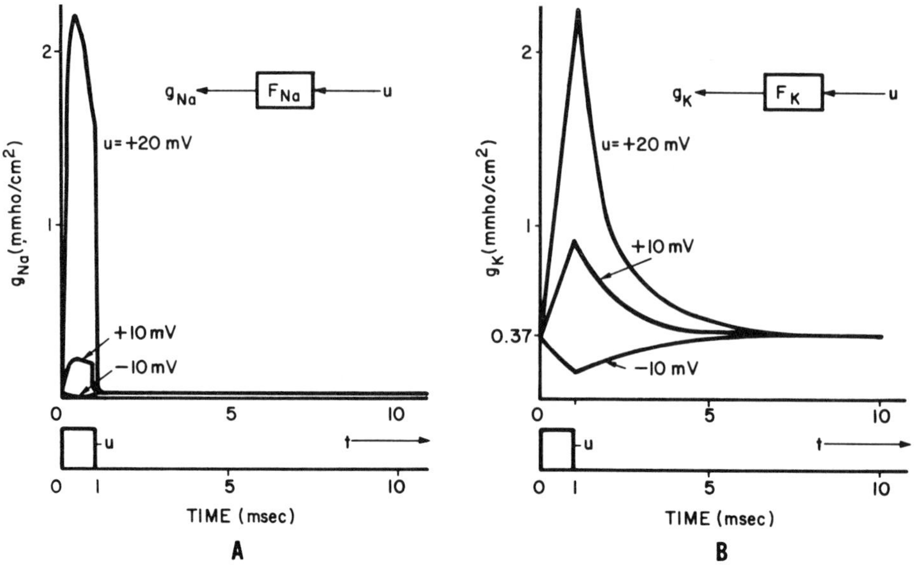

FIG. 3. Transient responses of conductance feedback paths, calculated from the Hodgkin-Huxley equations by digital computer. A. Sodium conductance path stimulated by different amplitudes of rectangular voltage pulses (u) of 1 msec duration. Each of the three responses is for a different stimulus intensity, as indicated. Depolarization is positive. B. Potassium conductance path; same stimulation as in sodium path.

the spatial separation of synapses and synapse-free areas has been neglected, and it has been assumed that the conductance changes caused by input and by feedback are additive. This allows us to combine the input and the feedback systems. The combined system may be regarded as a one-point neuron model; the expression "one-point" indicates that spatial extension and propagation have been neglected. Figure 2 is the simplified block diagram of this model.

The configuration of this model enables us to see clearly the influence of the nonlinear feedback loops on the input-output relations; this will be shown in detail in the section "Characteristic Curves and Input-Output Relations."

The properties of the feedback loops which are important for biological pulse processing and which are particularly relevant to electronic neuron modeling can be summarized as follows.

Both feedback loops are always active, above and below threshold. Each loop consists of a linear voltage path and a nonlinear conductance path. The sodium conductance path exhibits positive (self-exciting) feedback. The driving voltage u_{Na} decreases with increasing depolarization. Below threshold, alterations of u_{Na} are relatively small (less than 10 %); therefore, to a first approximation, the multiplication at block 3e (Fig. 2) may be neglected. With hyperpolarization and small depolarizations ($u \leq 10$ mV) the sodium conductance path behaves like an integrating pulse shaper (Fig. 3A). With increasing depolarization the transient response of the sodium conductance path shows a differentiating component.

The potassium loop exhibits negative (in general, stabilizing) feedback. The driving voltage u_K increases with increasing depolarization. The alterations of u_K below threshold are great, and the multiplication at block 3h (Fig. 2) may not be neglected. The potassium conductance path also behaves like an integrating pulse shaper (Fig. 3B). The resting value of potassium conductance (0.37 mmho/cm^2) is high compared with that of sodium conductance (0.01 mmho/cm^2). This permits a substantial decrease of potassium conductance during hyperpolarization, which leads to the well-known phenomenon of anode-break excitation (Hodgkin and Huxley, 1952).

The Electronic Neuron Model of Küpfmüller and Jenik. Following a proposal of Küpfmüller (1959), I developed several electronic one-point neuron models (Küpfmüller and Jenik, 1961; Jenik, 1962a). The main object of these models was the investigation of pulse processing in small nerve nets. The block diagram of a recent model is shown in Fig. 4. The designations are the same as in Fig. 2, with the following additions: P for proportional transmission, Th for threshold, and F_e and F_h for transfer functions of parts of the sodium- and potassium-conductance feedback paths, respectively. In blocks 3e and 3h transistor collector currents (i_{Na}, i_K) result from collector-base conductance (g_{Na}, g_K) and collector-base voltage (u_{Na}, u_K). Over almost the complete operation range the currents are independent of voltage and proportional to conductance (proportional transmission). Only in a very narrow range are

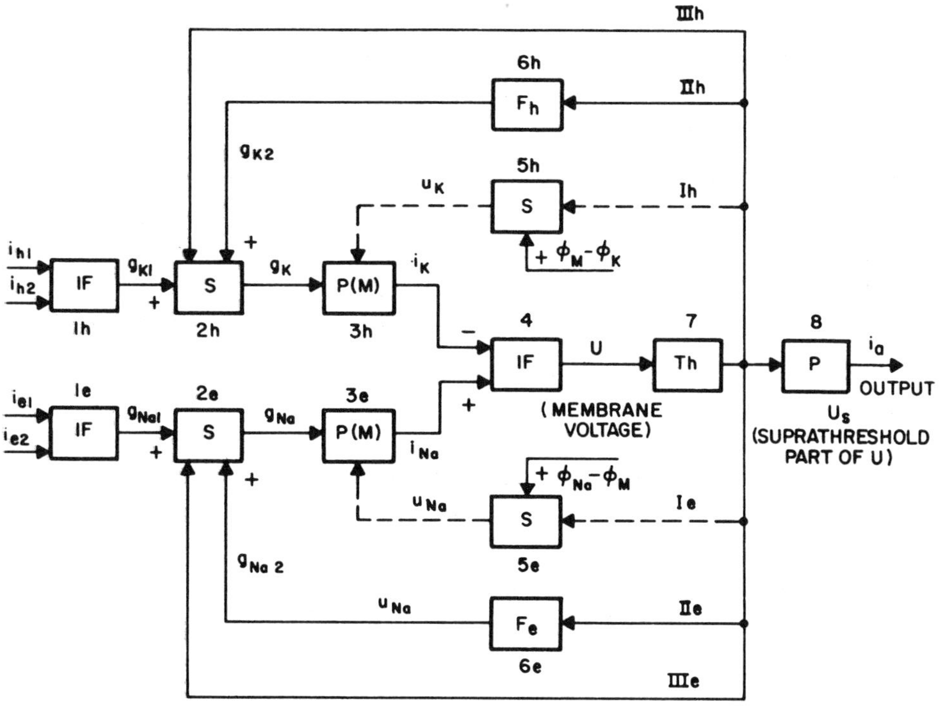

FIG. 4. Block diagram of Küpfmüller and Jenik's neuron model.

currents influenced by both voltage and conductance, and in this range multiplication holds. Block 8 represents an output amplifier.

The structure of the electronic model is similar to that of the extended Huxley-Hodgkin system (Fig. 2). It includes the central membrane block (time constant $T_m = 3$ msec), the positive and negative feedback paths, and two input channels, one for inhibitory and one for excitatory input pulses (time constant $T = 1$ msec). The input-output relations of the membrane block (block 4) and the input channels (blocks 1e, 1h) are the same as those of the extended Hodgkin-Huxley system.

As a consequence of the results cited in the previous section, blocks 6e and 6h (the sodium and potassium conductances) were designed to be differentiating and integrating pulse shapers, respectively. In spite of these similarities, major differences exist between the two systems, owing partly to the lack of appropriate semiconductors and partly to our simulating a motoneuron rather than a squid axon. These differences are as follows:

1. There is a sharp voltage threshold (block 7). As long as the membrane voltage is below threshold, both feedback paths are inactive. As soon as the threshold is exceeded, both feedback paths are activated and an action potential is generated.
2. In blocks 3e and 3h multiplication occurs only in a very narrow voltage range. In most cases the separate voltage paths (block 5e and 5h) have little influence. This restricted multiplication is a major disadvantage, particularly for the potassium path.
3. To compensate for the lack of multiplication, at least to some extent, two feedback paths (IIIe and IIIh) have been introduced which diminish the influence of the input channels during spike potential, thus providing an absolute refractory period and a rough simulation of the sodium inactivation.
4. The parameters of all blocks have been chosen to simulate motoneurons of mammals. Some of the major functional differences between motoneurons of mammals and the giant axon of *Loligo* will be elaborated on in later sections.

The model contains 18 transistors. It may be excited and inhibited by dc or by positive-going pulses. In general, spike-shaped pulses of 10-V amplitude are used. The output frequency range is from 0 to approximately 800 pps, for orthodromic dc and pulse stimulation. All voltages are 100 times those found in biological neurons. A circuit diagram of the model is given in the Appendix.

Characteristic Curves and Input-Output Relations

Characteristic Input Curves. Ordinarily, a mathematical model constitutes the most efficient description of a system. However, it may be rather difficult to visualize the behavior of an extensive or complicated system activated by complex stimuli. In such cases graphical representation is a valuable aid to description. Working models, used as computers, readily provide these representations.

For systems as intricate as neurons (or their models) a great many characteristic curves can be specified. One very important class is that of the input functions, representing the relations between the stimulus amplitudes and the resultant postsynaptic potentials. In models the input intensity is represented by the amplitudes of input pulses; in actual neurons an input intensity may be defined by the number of synchronously activated synaptic knobs (Eccles, 1953, p. 111) and by the amplitude of presynaptic pulses (Hagiwara and Tasaki, 1958; Takeuchi and Takeuchi, 1962). These characteristic curves readily allow determination of the amplitude of the postsynaptic potential evoked by an input pulse of given intensity (for a given state of activation*).

Typical curves are shown for the giant synapse of *Loligo* (Hagiwara and Tasaki, 1958; Takeuchi and Takeuchi, 1962), for the extended Hodgkin-Huxley system (calculation by digital computer), for motoneurons of cats (Eccles, 1961, Fig. 5) and for the electronic neuron model. They have been derived by stimulating the particular cells by afferent volleys or pulses of different intensities and observing the amplitudes of the stimuli and of the postsynaptic potentials.

The Excitatory Input. Figure 5 shows excitatory postsynaptic potentials (epsp) and characteristic input curves. Part A shows the epsp at the giant synapse of *Loligo* (Takeuchi and Takeuchi, 1962, Fig. 7) at a temperature of 10–15°C; $\hat{u}_{el}$ is the voltage amplitude of the presynaptic spike. In part B curve 1 is the input characteristic of the giant synapse as derived by Takeuchi and Takeuchi (1962, Fig. 8) from experimental data. Curve 2 is the input characteristic of the extended Hodgkin-Huxley system with a 33-mV threshold added to the input channel. The curve was obtained by shifting the curve in part D to the right by 33 mV. Part C is the calculated epsp of the extended Hodgkin-Huxley system, stimulated by rectangular input pulses of 0.2-msec duration at a temperature of 18°C; $\hat{i}_{el}$ is the current amplitude of the input pulse. The threshold is at $u = 15$ mV. Part D is the calculated input characteristic curve of the extended Hodgkin-Huxley system; D is derived from data partly shown in C. The abscissa has a standardized scale factor; unity represents 0.38 μA. Parts E and F are the epsp and the characteristic curve of the biceps-semitendinosus motoneuron of a cat (Eccles, 1961, Fig. 5); the input intensity indicates the amplitude of the afferent volley recorded near the entry of the dorsal nerve roots into the spinal cord; an arbitrary scale factor is used. Parts G and H are the epsp and the characteristic input curve of the electronic neuron model, which was activated by a spike-shaped pulse of 1-msec duration. The threshold is at 1.5 V.

The giant synapse of *Loligo* (Fig. 5A, B) and the extended Hodgkin-Huxley system appear to exhibit nonlinearity of amplitude response, distortion of the shape of the epsp with increasing amplitude, and considerable hyperpolarizing undershoot. The hyperpolarizing undershoot and the subsequent oscillations are chiefly caused by the activation of the potassium-conductance feedback path (Hodgkin and Huxley, 1952). The nonlinearity and the distortion of shape are caused by the increase of activity in the sodium-conductance feedback

---

* A particular state of activation may be described by membrane potential, membrane currents, and membrane conductances, by the preceding activity of the cell, and by the given and the preceding stimuli.

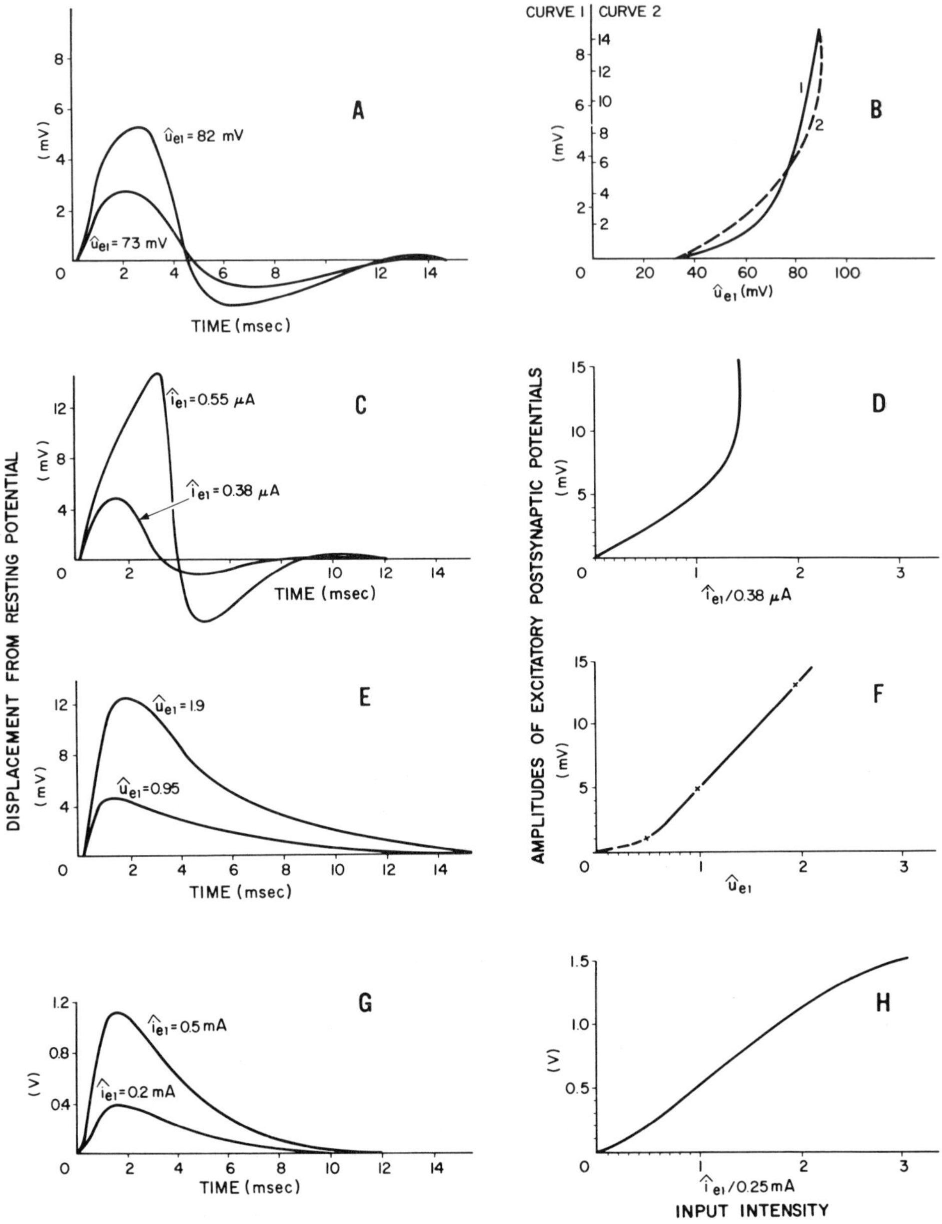

FIG. 5. Excitatory postsynaptic potentials (A, C, E, G) and characteristic input curves (B, D, F, H) of a giant synapse of *Loligo* (A, B), of the extended Hodgkin-Huxley system (C, D), of the motoneuron of cats (E, F), and of the electronic neuron model (G, H).

path. With increasing depolarization this path (which supports excitation) becomes more efficient and provides a sort of "potentiation" of excitation. Owing to this potentiation, slight variations in the amplitude of presynaptic spike amplitude might have great influence on the amplitude of the postsynaptic potential (Hagiwara and Tasaki, 1958, Takeuchi and Takeuchi, 1962). The linear input channels of the extended Hodgkin-Huxley system cause a linear increase of the input curve at low intensities (Fig. 5D). The giant synapse of *Loligo*, however, shows no output with lower input intensity (Fig. 5B, curve 1). This means that it cannot be simulated by a combination of linear input channels and the Hodgkin-Huxley system.

Better agreement between the giant synapse and the mathematical model may be obtained if a threshold of 33 mV is added to the linear input channels of the model. In this case the characteristic input curve of the extended Hodgkin-Huxley system is shown by curve 2 of Fig. 5B. By introducing some additional nonlinearity, which may be easily derived from Fig. 5B, the input characteristic of the extended Hodgkin-Huxley system could be brought into close agreement with that of the giant synapse. Another input channel of the stellar giant axon of *Loligo* is that of proximal synapses (Bryant, 1959). The input characteristic curve of this channel, which represents the relation between the amplitude of the epsp and the number of synchronously activated synapses, would be of great interest.

Compared with this nonlinear action, distortion and overshoot are small in motoneurons (Fig. 5E, F). These differences may be explained in one of two ways: either the synaptic input channels are nonlinear to such an extent as to compensate for the influence of the feedback path, or else there is little subthreshold activity of the conductance feedback paths in motoneurons. If the second assumption holds, the neglect of subthreshold activity of the conductance feedback paths in the electronic model is justified. And in fact, the excitatory postsynaptic potentials (Fig. 5G) of this model are more similar to those of motoneurons than to those of the extended Hodgkin-Huxley system.

The characteristic curves of the motoneuron (F) and of the electronic model (H) may be regarded as approximately linear over a wide range of the epsp.

The Inhibitory Input. Figure 6 shows the inhibitory postsynaptic potentials and the inhibitory input characteristic curves. The values of $\hat{i}_{h1}$ are the input intensities. With the biceps semitendinosus motoneuron (C, D) the input intensity indicates the amplitude of an afferent volley recorded near the entry of the dorsal nerve roots into the spinal cord; an arbitrary scale factor is used (Eccels, 1961, Fig. 25). The input intensities of B and F have been standardized; in B, 6.3 μA equals unity; in F, 190 μA equals unity.

The most striking feature of this figure is the marked nonlinearity of the input characteristic curves. The amplitude of the ipsp increases less than the input intensity and approaches a limiting value. This may be explained by the influence of the potassium voltage; u_K diminishes with increasing hyperpolarization and becomes zero when the membrane potential reaches the value of the potassium equilibrium potential (-82 mV).

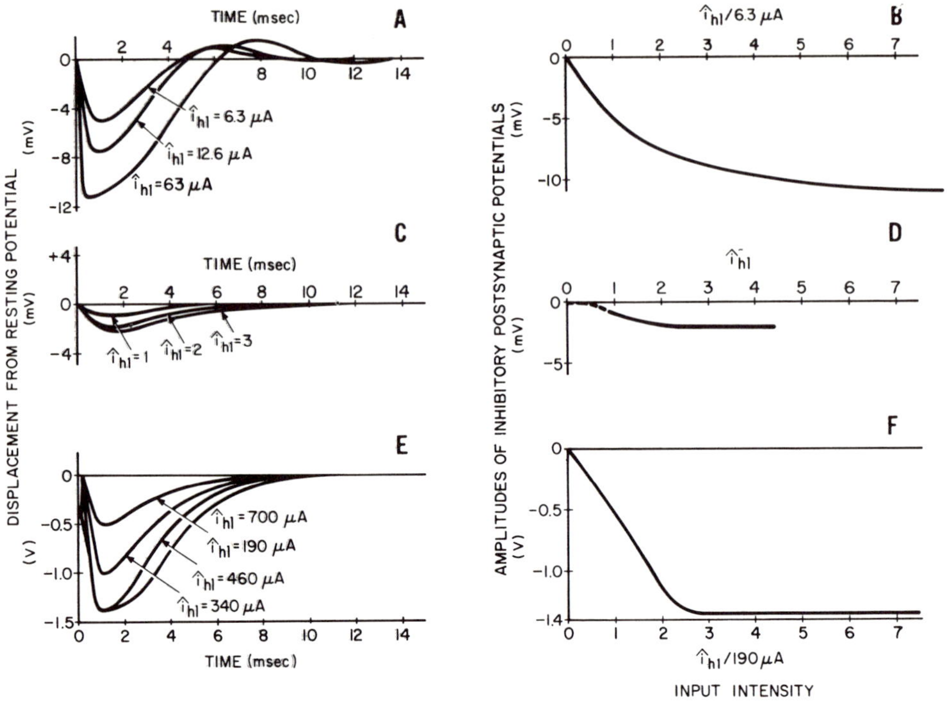

FIG. 6. Inhibitory postsynaptic potentials (A, C, E) and input characteristic curves (B, D, F) of the extended Hodgkin-Huxley system (A, B), a mammalian motoneuron (C, D), and a model (E, F).

Like the epsp, the ipsp of the extended Hodgkin-Huxley system shows considerable overshoot, which is due chiefly to the potassium-conductance path. With particular states of activation this may lead to the initiation of an action potential. Compared with the extended Hodgkin-Huxley system, the motoneuron exhibits very little overshoot, and the model exhibits none. The characteristic curve for the model is linear to -1.3 V, where it is abruptly limited. This is due to the small multiplication range of block 3h (Fig. 4), which runs from -1.3 V to -1.4 V (sodium equilibrium voltage). Only in this range does the diminished potassium voltage u_K influence the potassium current and thus the amplitude of the ipsp.

The Relative Refractory Period. The relative refractory period may have a great influence on pulse processing in a nervous system. It is characterized by decreased excitability and spike amplitude. Both effects have been investigated for the extended Hodgkin-Huxley system and for the model. The results are shown in Fig. 7. The threshold values (e_2) represent the amplitude of the smallest excitatory input pulse which evokes an action potential (values plotted at the time of stimulus onset). The asymptotic value $e_2 = 1$ is the threshold amplitude for an input pulse applied in the resting state. Input pulses to the extended Hodgkin-Huxley system were rectangular pulses of 0.2-msec duration, and those to the model were exponential pulses of 1-msec duration. The data of B and C were derived from two slightly different configurations of the model. In D a motoneuron of cat was stimulated by electrical pulses applied directly to the penetrated cell. Curve e_2 represents second-spike threshold

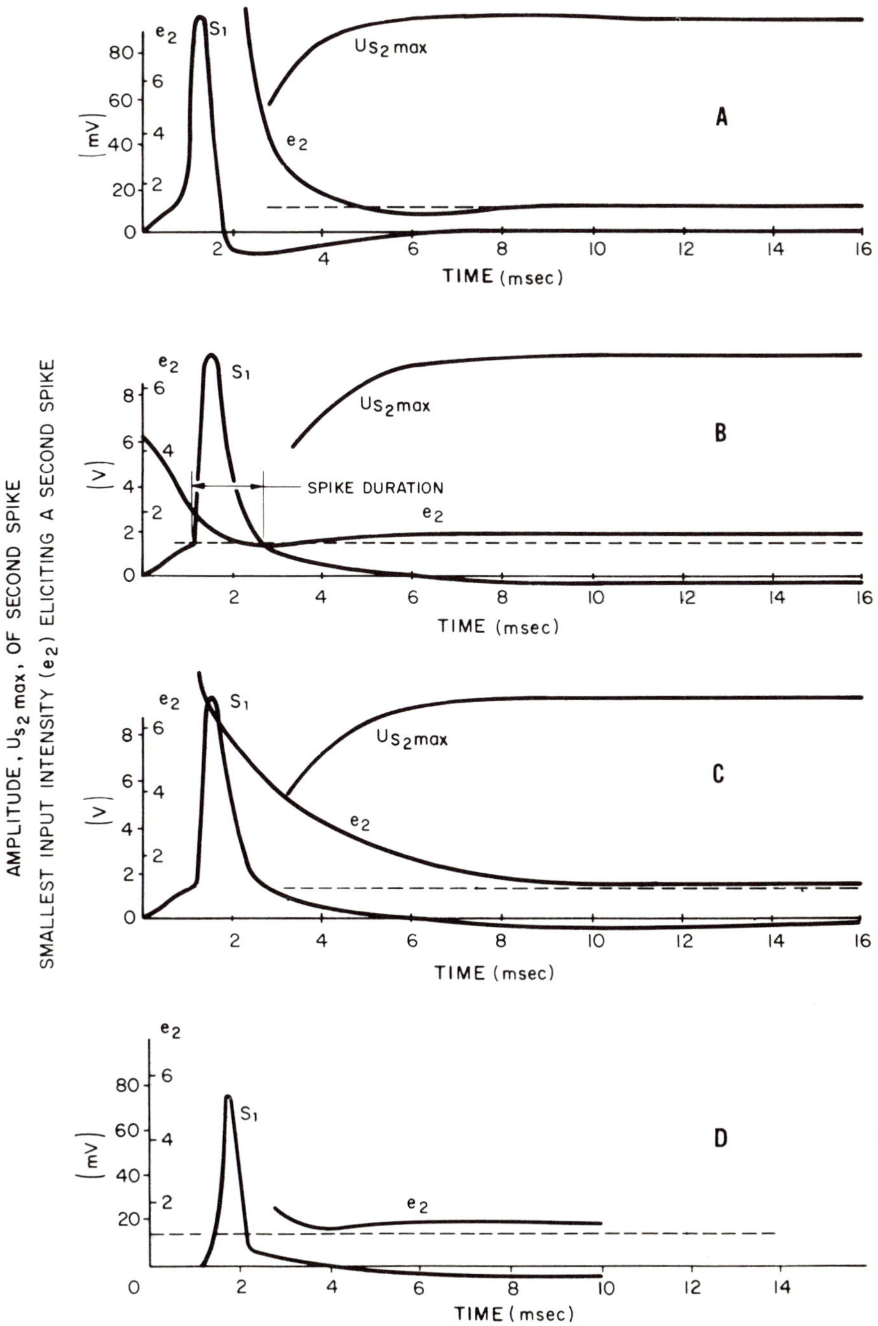

FIG. 7. First spike (S_1), second-spike threshold (e_2), and second-spike peak amplitude (u_{S_2max}) during the relative refractory period for the extended Hodgkin-Huxley system (A), for the model (B, C), and for motoneurons of cats (D).

following a full spike of antidromic origin (Fuortes, Frank, and Becker, 1957, Fig. 8).

The action potential of the extended Hodgkin-Huxley system is short. The durations of spike, after-depolarization, and after-hyperpolarization are 1.3 msec, 0.2 msec, and 4 msec, respectively. The same values for motoneurons and for the model are much longer: 1.7 msec, 3–5 msec, and 50–100 msec. The input intensity (e_2) necessary to evoke a second spike with the extended Hodgkin-Huxley system is very high immediately after the first spike, then it declines, enters a subnormal phase, and approaches its normal value ($e_2 = 1$). The high values of e_2 just after the spike may be explained by the combined effect of an increase of the potassium conductance and a decrease of the sodium conductance (Hodgkin and Huxley, 1952). The subnormal phase must be attributed to a potassium-conductance decrease evoked by the after-hyperpolarization. It should be kept in mind that e_2 indicates threshold at the *onset* of the second stimulus; the corresponding spike sometimes appears only after several milliseconds of latency.

We have been unable to find precise physiological data on the threshold variation of motoneurons; consequently, this parameter has been made flexible for the model. Two curves are shown in Fig. 7 (B and C); others are possible. In Fig. 7B, the e_2 extends to $t = 0$, indicating that the model fires two spikes for a sufficiently large single input pulse. In this case absolute refractoriness is caused by strong diminution of the gain of input channels by feedback paths IIIe, IIIh (Fig. 4) during the spike (1.5 msec). Immediately after the spike, the gain of the input channels assumes its normal value. Because of the storage in the input channels (time constant $T_e = 1$ msec), some "excitatory transmitter substance" may outlast the first spike, and a second one may be evoked after the absolute refractory period. This phenomenon has been observed for neurons in Clarke's column (Curtis, Eccles, and Lundberg, 1958). Starting from the value 4.2, the threshold (e_2) declines, passing through a subnormal phase ($e_2 < 1$), then rises to a supernormal ($e_2 > 1$) value, from which it slowly approaches the resting value. The resting state is reached after 50–100 msec. The subnormal phase is caused by the after-depolarization, the supernormal phase by the after-hyperpolarization.

The threshold variation in the model which produced the data of Fig. 7C is to some extent similar to that of the extended Hodgkin-Huxley system. This model differs from the one which gave the results of Fig. 7B by a long-lasting, graded diminution of gain of the input channels proportional to the time course of e_2 in Fig. 7C.

The threshold variation of motoneurons (Fig. 7D) is similar partly to that of the model used in Fig. 7B and partly to that of the Hodgkin-Huxley system (Fig. 7A). However, this curve (D) is not fully comparable to the upper ones because the conditioning spike is of antidromic origin, and stimulation is performed by electrical pulses applied directly to the penetrated cell rather than by presynaptic pulses. With orthodromic activation the excitability of motoneurons immediately after the spike probably will be increased if it is accepted that some depolarizing action, which is caused by presynaptic pulses, may outlast the first spike. In this case curve e_2 of Fig. 7D would approach that of the electronic model shown in Fig. 7B.

The amplitudes of spikes triggered during the relative refractory period are diminished as shown by the u_{S_2max} curves in Fig. 7. A wide variety of such curves is obtainable with the model. The effective extent of such reduced-amplitude spikes on succeeding neurons (i.e., whether they evoke normal or reduced postsynaptic potentials), can be critically important to determination of pulse processing, particularly with cells showing input characteristic curves similar to those of Fig. 5B and D.

The ν-diagram. Three kinds of input variables can be specified for a particular input fiber of a neuron or a neuron model: (1) The input intensity, (2) the average frequency or pulse rate, and (3) the temporal pattern (Bullock, 1961). Pulse trains in nervous systems show a great variety of patterns (e.g., periodic, cluster, or beat periodic (Fig. 9), nonperiodic, and statistic pulse trains). Definition of detailed patterns of nonperiodic pulse trains is rather difficult and can be accomplished only in particular cases.

A particular input situation is characterized by a set of values of all input variables. The input-operation range of a cell is characterized by all possible sets.

Three output variables may also be specified, as follows:

1. The output intensity, which indicates the amplitude of the output spike (decreasing during the relative refractory period).
2. The output frequency for periodic pulse trains or the output rate for nonperiodic pulse trains.
3. The "pattern" of the output pulse train.

A particular output situation is characterized by a set of values of all output variables; the output operation range of a cell is characterized by all possible sets. The relations between these and the input-dependent variables are termed the input-output relations. Clearly, input-output relations of neuron models should agree with those of neurons as far as possible. Because there is obviously such a great number of situations, all possible ones cannot be tested, but a number of revealing test situations can be chosen.

In communication engineering, where similar problems occur, periodic waveforms turn out to be very suitable test functions. For example, from frequency-response measurements, the input-output relations of a system for many classes of inputs can be estimated.

We find that for neuron models a periodic pulse train provides a revealing test function. Two input variables and one output variable have been considered. They are the excitatory input intensity e, the input frequency f_e, and the output rate r_a. (The output pulse train is expressed as an average rate (pps) since it is not always periodic.) The graphical representation of the input-output relations between e, f_e, and r_a may be either a set of characteristic curves (Harmon, 1961) or a geometrical locus plot which has been called the ν-diagram (Küpfmüller and Jenik, 1961). Three examples of ν-diagrams are shown in Fig. 8.

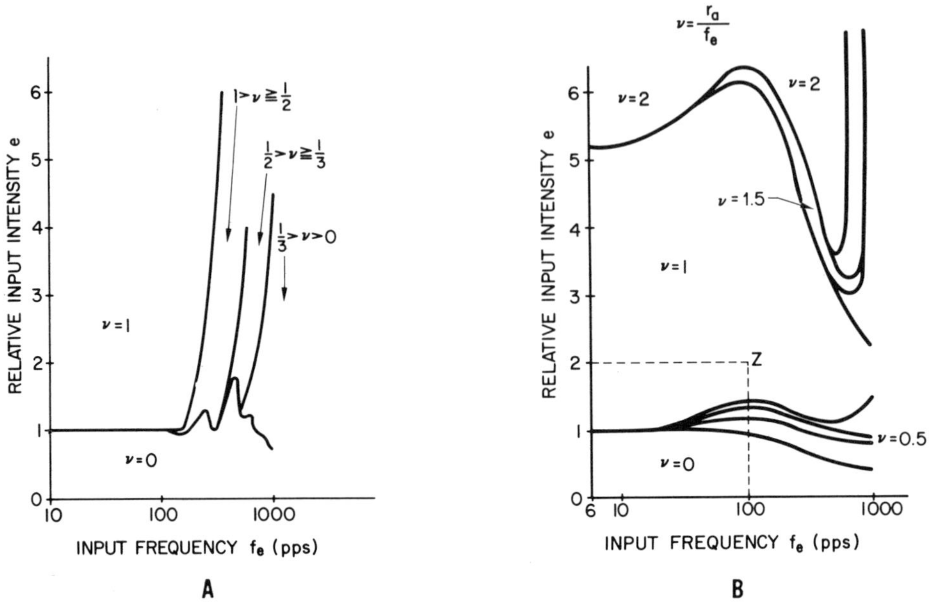

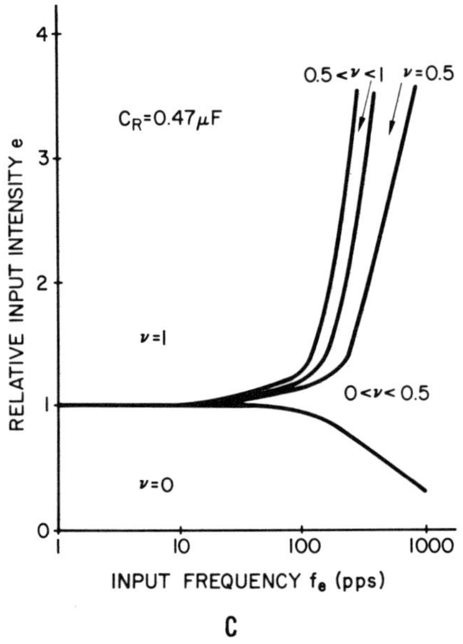

FIG. 8. The ν-diagrams of the extended Hodgkin-Huxley system (A) and of the electronic neuron model (B, C). The diagrams in B and C apply to the units represented in B and C, respectively, of Fig. 7.

The ν-diagram represents the input-output relations of a cell excited orthodromically by a periodic pulse train with frequency f_e and an intensity e that indicates the amplitude of an input pulse relative to threshold amplitude. The ν is the ratio between the output rate r_a and the input frequency f_e, and depends on e and on f_e. The shape of the ν-areas characterizes a particular model.

Diagram A was obtained by observing the pulse rate r_a of the output pulse train, which is evoked by application of an input pulse train of constant frequency f_e and intensity e. In B and C input frequency f_e has been held constant, and input intensity has been increased slowly until a change in the output rate occurs.

The input frequency f_e is plotted on the horizontal axis and the input intensity e on the vertical. The range of input frequencies extends from 0 to 1000 pps. Input intensity is seen to be of considerable importance for pulse processing in nervous systems and neuron models. The intensity range in Fig. 8 was arbitrarily chosen, since no exact data have been obtained for actual neurons.

The area encompassed by the diagram represents the input operating range. Each point represents one particular input situation; e.g., point z represents an input intensity $e = 2$ and an input frequency $f_e = 100$ pps.

The ratio of the resultant output rate to the input frequency, ν, may be assigned to each point. It turns out that there are a number of distinctly bounded areas which have the same ν-value. Figure 8B shows areas of $\nu = 0, 0.5, 1, 1.5$, and 2 (the subthreshold range is represented by $\nu = 0$). In the area $\nu = 1$, each input pulse evokes one output pulse; $\nu > 1$ indicates that (average) frequency multiplication occurs. Frequency division occurs in the region $0 < \nu < 1$ [many stable values of $0 < \nu < 1$ can exist (Harmon, 1961)]. The shapes of the boundary curves between the ν-areas are characteristic for a particular model. The absolute and relative refractory periods, the shape of postsynaptic potentials, and the after-hyperpolarization all influence the ν-areas. The models represented in 8B and 8C differ only in their excitability during the relative refractory period (7B and 7C, respectively). Note that in 8A and 8C no average-frequency multiplication occurs (i.e., $\nu \leqq 1$), but the area of frequency division is enlarged. A rise in the boundary curves indicates a decrease of excitability, whereas a fall indicates an increase of excitability. At low frequencies—up to 15 pps with the electronic models and up to 100 pps with the extended Hodgkin-Huxley system—all boundary curves are horizontal because there is no interaction between consecutive pulses. Above $f_e = 15$ pps, at least every second input pulse arrives during the after-hyperpolarization of the previous action potential and encounters less excitability with the electronic models. All boundary curves in 8B and 8C, except the lowest, therefore increase up to $f_e = 100$ pps, at which point coincidence between maximum after-hyperpolarization and input pulses occurs. Above 100 pps, temporal summation of succeeding epsp's occurs; this produces a drop in the boundary curve between the areas $\nu = 0$ and $\nu > 0$ of 8B and 8C and an "oscillatory" pattern of this boundary curve in 8A. The drops in the other boundary curves of 8B are caused partly by the temporal summation of epsp's and partly by the increase of excitability (decrease of e_2 in 7B). The steep rise of the curves near 1000 pps marks the upper firing-frequency limit of the model. The boundary curves in the areas $\nu > 0$ of 8A and 8C show, instead of a decline above 100 pps, a steeper rise; this is due to the decrease of excitability (increase of e_2 in 7A and 7C).

All ν-diagrams show the great influence of input frequency f_e and input intensity e on the input-output relation, particularly above 100 pps. This range, in which the time interval between consecutive pulses drops below 10 msec and in which summation of succeeding epsp's occurs, seems to be of great importance for pulse processing. There must be close agreement between the ν-diagrams of neurons and models if these are to be an aid in quantitative investigation of pulse processing in nervous systems.

Pulse Processing by Electronic Neuron Models

The main object of electronic neuron models (as developed by Taylor, Harmon, Derksen, McGrogan, Hiltz, Lewis, and Küpfmüller and Jenik) is the investigation of information processing in small nerve nets containing 1 to 100 cells (Harmon and Wolfe, 1959; Taylor, 1960; Derksen, 1961; Hiltz, 1962; Küpfmüller and Jenik, 1961; McGrogan, 1961; Jenik, 1962a; Lewis, *this volume*). The investigations may be carried out by systematic examination of well-known general structures or by simulation of particular nets as suggested by Eccles (1953) and by Fessard (1961). Examples of both kinds of approach may be found in recent literature (Taylor, 1956; Tischner, 1960; van Bergeijk, 1961; Harmon, 1961; Levinson and Harmon, 1961; Küpfmüller, 1961; Martin, 1961; Putzrath and Martin, 1961; Jenik, 1962a; Schief, 1962).

Information in a nervous system is usually transmitted and processed by nonperiodic pulse trains. The description and handling of nonperiodic pulse trains is very difficult; therefore we shall start with the investigation of pulse processing by periodic pulse trains. Here the distinction between coherent and noncoherent pulse trains is important, the laws of pulse processing being quite different for these two classes. The laws of pulse processing of coherent pulse trains are well known; almost all electronic digital computers and data-handling systems use coherent pulse trains (synchronized by a clock generator). In physiology little attention has been paid to the question whether pulse trains are coherent or noncoherent. Noncoherent pulse trains are probably prevalent in nervous systems. This supposition has stimulated the investigation of the little-known rules of noncoherent pulse processing (Jones, Li, Meyer, and Pinter, 1961; Jenik, 1962b). Two examples of processing noncoherent pulse trains by electronic neuron models will be given in the next two subsections.

Multiplication and Pulse Pattern. Let a cell be activated by two noncoherent trains satisfying the following conditions.

1. Consecutive pulses of the same train shall not influence one another either at the presynaptic fiber or at the postsynaptic membrane.
2. The ratio of the input frequencies shall not be expressible by a ratio of small whole numbers (e.g., 1:2, 2:3),

For two excitatory subthreshold pulse trains with intensities e_1, e_2 and frequencies f_{e1}, f_{e2}, the output rate r_a may be approximately calculated by the equation

$$r_a = k_e \cdot f_{e1} \cdot f_{e2}$$

(Küpfmüller and Jenik, 1961; Jenik, 1962a, b). If there is one excitatory (e, f_e) and one inhibitory (h, f_h) pulse train, then the relation

$$r_a = f_e - k_h \cdot f_e \cdot f_h$$

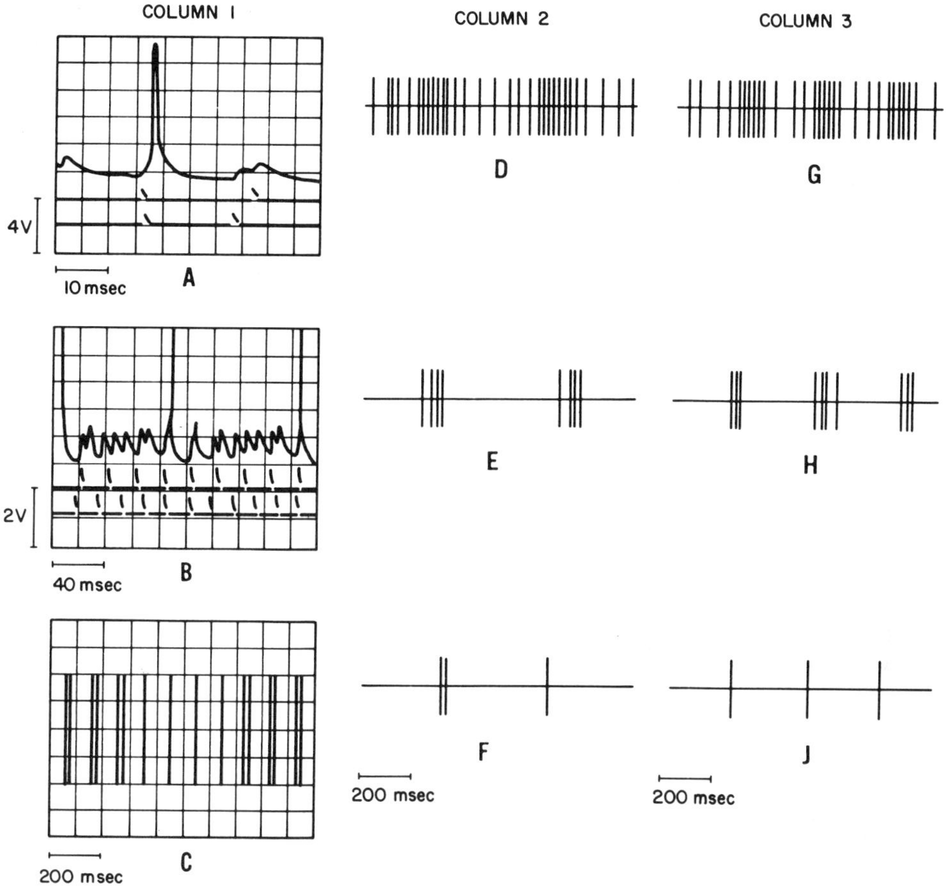

FIG. 9. Output pulse trains (rate r_a) of the model excited by two noncoherent periodic pulse trains of frequencies f_{e1}, f_{e2} and subthreshold intensities e_1, e_2.

holds approximately. The constants k_e and k_h depend on the input intensities. The output pulse train is strongly patterned (owing to beat frequencies). The patterns depend on the input frequencies and intensities. Some examples of these output pulse trains are shown in Fig. 9. In A, B, and C, $f_{e1} = 50$ pps, $f_{e2} = 57.3$ pps, $e_1 = e_2 = 0.8$, and $r_a = 11.2$ pps. Parts A and B show input pulse trains (lower traces) and membrane potential (upper trace); all other figures represent the output pulse train only. In D, E, and F, $f_{e1} = 50$ pps; $f_{e2} = 48.0$ pps; $r_a = 30.3$, 9.1, and 3.1 pps. In G, H, and J, $f_{e1} = 50$ pps; $f_{e2} = 47.0$ pps; $r_a = 29.7$, 8.9, 3.0 pps. In D and G, $e_1 = e_2 = 0.96$. In E and H, $e_1 = e_2 = 0.8$. In F and J, $e_1 = e_2 = 0.6$.

The first column illustrates the generation of the output pulse trains from the two independent subthreshold input trains. In A the shapes of epsp's and of the spike may be seen. The compressed time scales of B and C reveal temporal summation of the epsp (B) and details of the output pulse train (C), which consists of a sequence of single and double spikes. The output rate is smaller than either input frequency. The influence of input intensity on the output pattern may be seen in D–J. In column 2 the input frequencies were $f_{e1} = 50$ pps and $f_{e2} = 48.0$ pps, and in column 3 they were $f_{e1} = 50$ pps, $f_{e2} = 47.0$ pps. The slight change in f_{e2} produces a noticeable change of pattern. Input intensity decreases from top to bottom in steps of 20% of the threshold value. This decrease of amplitude causes a decrease in output rate of 60% and 90%, and a radical change of output pattern. Superposition of subthreshold periodic pulse trains at cells turns out to be an effective means of intensity and frequency contrast enhancement, provided that the resultant output patterns can be adequately discriminated.

Lateral Inhibition. Another way to obtain contrast enhancement is by lateral inhibition (von Békésy, 1960; Hartline and Ratliff, 1958; Fessard, 1961; Reichardt and MacGinitie, 1962). Figure 10A shows a simple network consisting of two model cells connected for lateral inhibition.

The cells are excited by noncoherent periodic pulse trains of frequencies f_{e1}, f_{e2}; f_{e1} is kept constant at 30 pps; f_{e2} is varied between 0 and 200 pps. Each cell is inhibited by the output of the other; excitatory and inhibitory input intensities are the same ($e_1 = e_2 = 1.45$ and $h_{21} = h_{12} = 1.25$) for both cells. For $f_{e2} = 0$, then, clearly $r_{a1} = f_{e1}$, and $r_{a2} = 0$. With increasing f_{e2} the output rate r_{a2} increases, and r_{a1} decreases to zero (Fig. 10B). At those frequencies f_{e2}, at which $f_{e2}:f_{e1}$ may be expressed approximately by a ratio of small whole numbers, beats occur, and the observation time must be large compared with the beat period. The curves may be calculated by the equations of the subsection "Multiplication and Pulse Pattern."

Contrast c may be expressed in terms of frequency ratio. The input contrast c_e and the output contrast c_a are defined by the equations

$$c_e = \frac{f_{e2}}{f_{e1}} \quad \text{and} \quad c_a = \frac{r_{a2}}{r_{a1}}.$$

Contrast enhancement may be defined by the expression

$$|V| = \frac{c_a - c_e}{c_e}.$$

Contrast enhancement has been evaluated by using this equation and is plotted in Fig. 10C. If both input frequencies are approximately equal, little contrast enhancement occurs. With increasing difference between input frequencies, contrast enhancement increases.

Summary

Hodgkin and Huxley's mathematical membrane model of the giant axon of *Loligo* has been extended by an excitatory input channel and an inhibitory input channel which permit the alteration of the sodium and potassium conductances by presynaptic pulses. This mathematical neuron model has been investigated by means of digital computers.

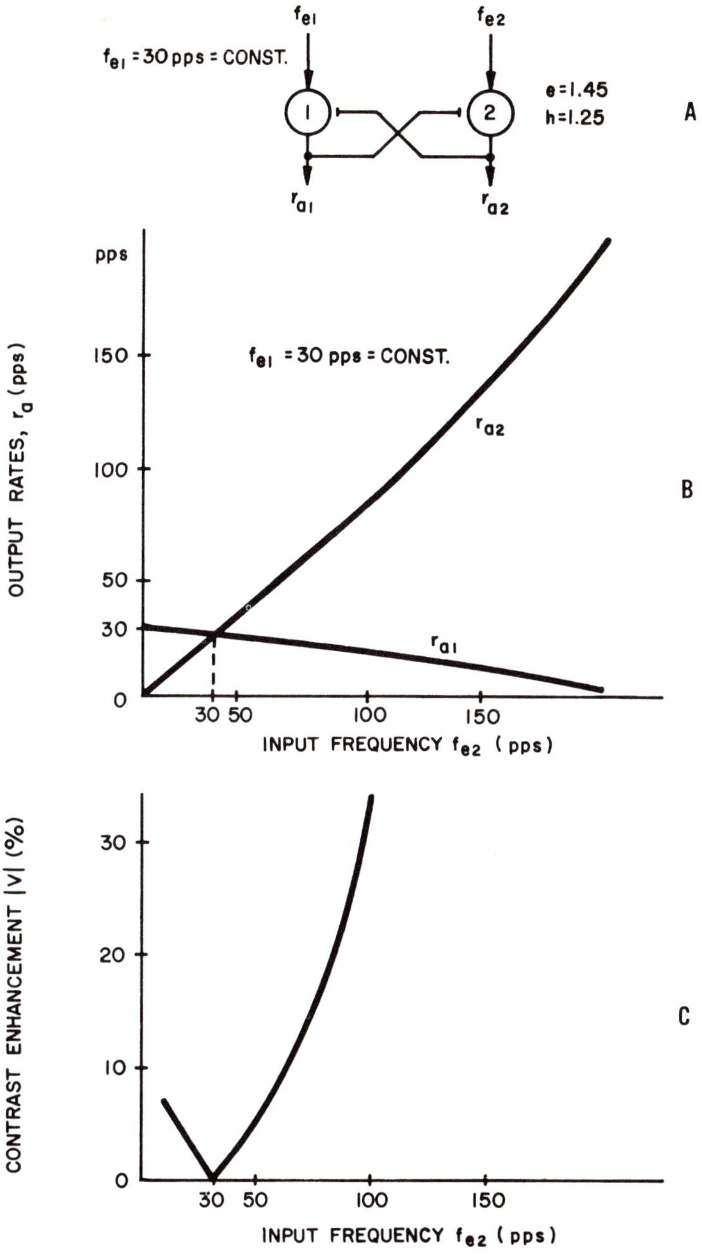

FIG. 10. Lateral inhibition. A. Network of two model cells exhibiting lateral inhibition. Excitatory intensity $e_1 = e_2 = 1.45$; inhibitory intensity $h_{21} = h_{12} = 1.25$. B. Relationship between output rates (r_{a1}, r_{a2}) and input frequency f_{e2}; $f_{e1} = 30$ pps = const. C. Contrast enhancement |V| as a function of input frequency f_{e2}, calculated from B.

The subthreshold activity of the conductance feedback loops of the extended Hodgkin-Huxley system causes a nonlinear characteristic of the excitatory and inhibitory input channels and significant oscillations of postsynaptic potentials similar to the oscillations observed at the giant synapse of *Loligo*.

In motoneurons there is little evidence of nonlinearity and oscillations. This may be explained either by compensating nonlinearities of synaptic input channels or by less subthreshold activity of sodium and potassium feedback loops than with the giant axon of *Loligo*.

An electronic model of a motoneuron was developed which exhibits no subthreshold activity of the sodium and potassium feedback loops. This leads to a linear input characteristic which is quite similar to that of motoneurons. The model was flexibly arranged to permit the testing of various courses of relative refractory recovery.

Testing of the mathematical and the electronic models by periodic pulse trains revealed the great influence of input intensity and refractory recovery on the input-output relations. The input-output relations appearing under testing conditions may be graphically represented by a diagram called the "ν-diagram." With the aid of ν-diagrams, input-output relations of different cells may be compared easily over the complete operation range. The ν-diagrams of neurons would be of great interest.

Patterned output pulse trains were generated by stimulation with two noncoherent subthreshold periodic pulse trains. Patterns depend on input frequencies and input intensities, and provide both amplitude and frequency contrast enhancement. The output pulse rate is proportional to the product of the input frequencies if the input intensities and frequencies do not exceed particular limits.

Contrast enhancement of a simple network consisting of two electronic neuron models exhibiting lateral inhibition has been investigated experimentally by application of two noncoherent periodic pulse trains. Appreciable contrast enhancement has been obtained.

Acknowledgment

The author is very grateful to K. Küpfmüller for valuable discussions and advice; to Deutsche Forschungsgemeinschaft for its support of the work and for making his attendance at this conference possible; to H. Hoehne for digital-computer calculations of the Hodgkin-Huxley system; and particularly to L. D. Harmon for his assistance with the translation. Moreover, the author wishes to express his gratitude to the sponsors of the symposium and to Freunde der Technische Hochschule, Darmstadt, for travel grants.

APPENDIX

The electronic neuron model shown in Fig. 11 (pp. 210–11) consists of an excitatory (T_1, T_2, T_5, T_6) input channel, an inhibitory ($T_3, T_4, T_8, T_9, T_{10}$) input channel, a central membrane circuit (C_6, R_{42}, R_{40}, R_{41}), a threshold device (T_{11}, T_{17}), a sodium feedback path (T_{16}, T_{14}, T_7, T_6, T_{17}, T_5), and a potassium feedback path (T_{16}, T_{13}, T_{12}, T_{10}, T_{17}, T_8), and two parallel output amplifiers (T_{15}, T_{18}). In the resting state all transistors are cut off except T_5 and T_8.

The input stage of the excitatory input channel consists of a common-base amplifier (T_1) which adds the input currents of from 1 to 100 "input fibers." The output voltage, simulating the transmitter substance ($R_{21}C_1$ collector circuit), is amplified by common-collector amplifier T_2, whose emitter current is fed into common-base amplifier T_6, via switch T_5. In the block diagram 1e represents T_1 and T_2. Transistor T_5 is on in the resting state and cut off during the spike-potential period, thus providing an absolute refractory period. Transistor T_6 forms the membrane current i_{Na} by adding the input and the feedback effects. Block 2e represents T_5 and the input of T_6. The collector-base resistance of T_6 is represented by blocks 3e and 5e.

The same explanations hold for the inhibitory input channel. Instead of T_1 there is a two-stage (T_3, T_4) common-base input amplifier providing time constants of 0.5 msec (S_1 open) and 30 msec (S_1 closed). Transistors T_3 and T_4 and common-collector amplifier T_9 are represented by block 1h. Block 2h represents the switch T_8 and the input of T_{10}. The base-collector resistance of the summing transistor T_{10} is represented by blocks 3h and 5h. The membrane potential may be observed at point M (P_3, R_{40}, R_{41}, R_{42}, C_6; block 4 in Fig. 4). The resting value of the membrane potential is -21 V and provides a reverse bias for the threshold transistor T_{11} (block 7). Displacements of membrane potential are limited to -22 V (hyperpolarization) and -10 V (depolarization, spike potential).

Threshold is reached at -19.5 V if all transistors of the positive-feedback path (T_{16}, T_{14}, T_7, represented by block 6e, and T_6) leave cutoff. The positive-feedback path is used to generate the rising phase of the action potential. The falling phase and the after-potentials are generated by the negative feedback path consisting of T_{12}, T_{13} (block 6h). The switching transistors, T_5 and T_8 are operated by T_{17} (block 7) via R_{61}, R_{37} (feedback path IIIe, IIIh).

References

Békésy, G. von, 1960. Neural inhibitory units of the eye and the skin, *J. Opt. Soc. Am.*, **50,** 1060–70.

Bergeijk, W. A. van, 1961. Studies with artificial neurons: II. Analog of the external spiral innervation of the cochlea, *Kybernetik*, **1**(3), 102–7.

Bryant, S. H., 1959. The function of the proximal synapses of the squid stellate ganglion, *J. Gen. Physiol.*, **42,** 609–16.

Bullock, T. H., 1961. The origins of patterned nervous discharge, *Behaviour*, **17,** 48–59.

Cole, K. S., and H. J. Curtis, 1939. Electric impedance of the squid giant axon during activity, *J. Gen. Physiol.*, **22,** 649–70.

Curtis, D. R., J. C. Eccles, and A. Lundberg, 1958. Intracellular recording from cells in Clarke's column, *Acta Physiol. Scand.*, **43,** 303–14.

Derksen, H. E., 1961. A neurone analog simulating the electrical properties of the cat's motoneurone, *Acta Physiol. et Pharmacol.*, **10,** 164–80.

Eccles, J. C., 1953. *The Neurophysiological Basis of Mind: The Principles of Neurophysiology*, Oxford: Clarendon Press.

(*Continued on p. 212*)

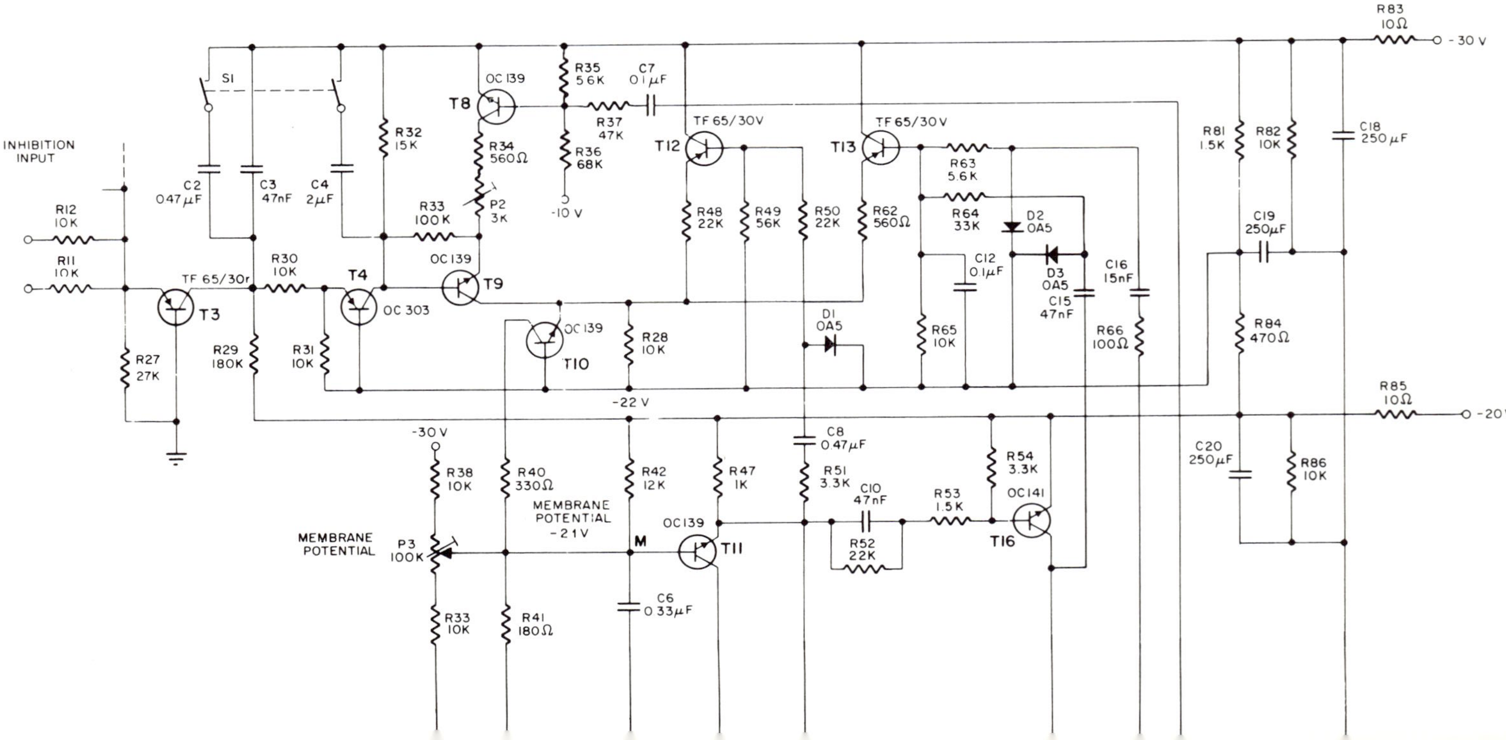
INHIBITION INPUT
R12 10K
R11 10K
TF 65/30r
T3
R27 27K
C2 0.47μF
C3 47nF
C4 2μF
S1
R29 180K
R30 10K
T4
OC 303
R31 10K
R32 15K
R33 100K
T8
OC 139
R34 560Ω
P2 3K
OC 139
T9
R35 5.6K
C7 0.1μF
R37 47K
R36 68K
-10 V
OC 139
T10
R28 10K
-22 V
TF 65/30V
T12
R48 22K
R49 56K
R50 22K
D1 OA5
TF 65/30V
T13
R62 560Ω
R63 5.6K
R64 33K
D2 OA5
D3 OA5
C12 0.1μF
R65 10K
C15 47nF
C16 15nF
R66 100Ω
R81 1.5K
R82 10K
C18 250μF
R83 10Ω
-30 V
C19 250μF
R84 470Ω
R85 10Ω
-20V
C20 250μF
R86 10K
-30 V
R38 10K
MEMBRANE POTENTIAL
P3 100K
R33 10K
R40 330Ω
R41 180Ω
MEMBRANE POTENTIAL -21V
R42 12K
M
C6 0.33μF
OC139
T11
R47 1K
C8 0.47μF
R51 3.3K
C10 47nF
R52 22K
R53 1.5K
R54 3.3K
OC141
T16

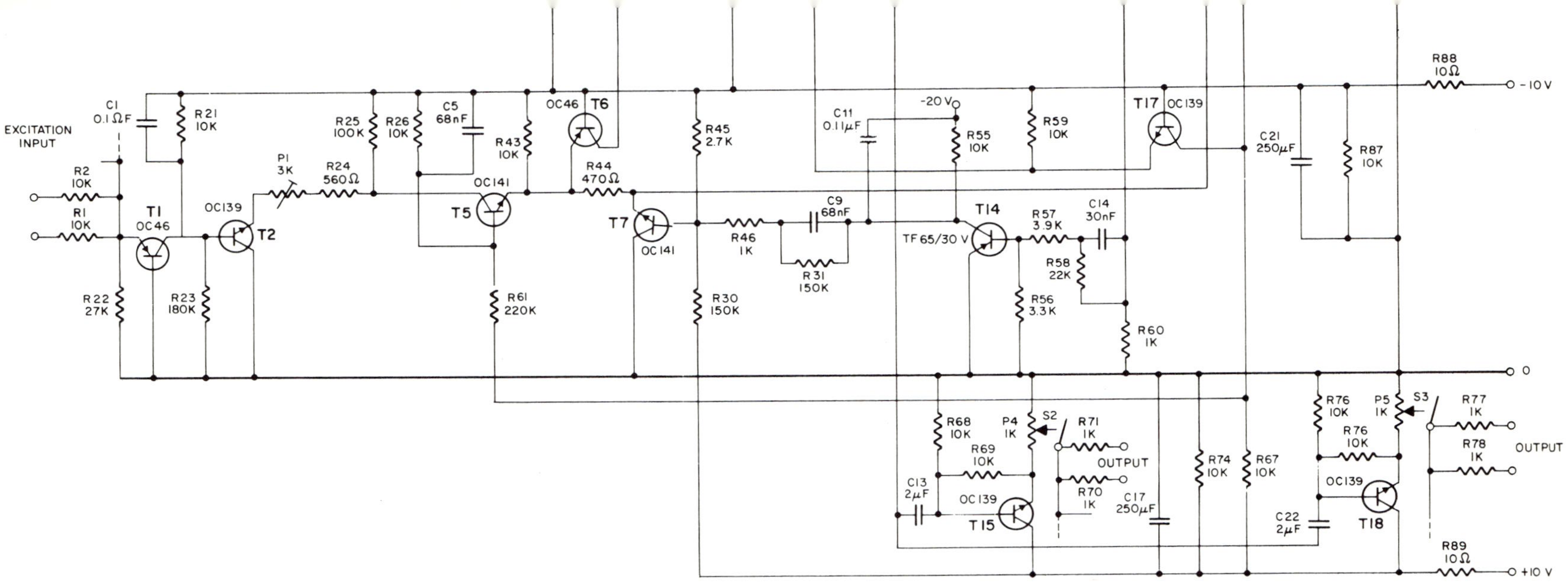

FIG. 11. Circuit diagram of an electronic neuron model of Küpfmüller and Jenik. (The block diagram of this model is shown in Fig. 4.)

Eccles, J. C., 1957. *The Physiology of Nerve Cells*, Baltimore, Md.: Johns Hopkins Press.

Eccles, J. C., 1961. The mechanism of synaptic transmission, *Ergebnisse der Physiologie*, **51,** 299–431.

Fessard, A., 1961. The role of neuronal networks in sensory communications within the brain, in *Sensory Communication*, W. A. Rosenblith, ed., New York: Wiley.

Fuortes, M. G. F., K. Frank and Mary C. Becker, 1957. Steps in the production of motoneuron spikes, *J. Gen. Physiol.*, **40,** 735–52.

Hagiwara, S., and I. Tasaki, 1958. A study on the mechanism of impulse transmission across the giant synapse of the squid, *J. Physiol.*, **143,** 114–37.

Harmon, L. D., 1961. Studies with artificial neurons: I. Properties and functions of an artificial neuron, *Kybernetik*, **1**(3), 89–101.

Harmon, L. D., and R. M. Wolfe, 1959. An electronic model of a nerve cell, *Semiconductor Products*, **2,** 36–40.

Hartline, H. K., and F. Ratliff, 1958. Spatial summation of inhibitory influences in the eye of *Limulus*, and the mutual interaction of receptor units, *J. Gen. Physiol.*, **41,** 1049–66.

Hiltz, F. F., 1962. Analog computer simulation of a neural element, *IRE Trans. Bio-Med. Elect.*, **BME-9,** 12–20.

Hodgkin, A. L., and A. F. Huxley, 1952. A quantitative description of membrane current and its application to conduction and excitation in nerve, *J. Physiol.*, **117,** 500–44.

Jenik, F., 1962a. Electronic neuron models as an aid to neurophysiological research, *Ergebnisse der Biologie*, **25,** 206–45.

Jenik, F., 1962b. Die Überlagerung von Impulsfolgen in Systemen mit einer Schwelle, *Arch. elektr. Übertr.*, **16,** 173–88.

Jones, R. W., C. C. Li, A. U. Meyer, and R. B. Pinter, 1961. Pulse modulation in physiological systems: phenomenological aspects. *IRE Trans. Bio-Med. Elect.*, **BME-8**, 59–67.

Küpfmüller, K., 1959. Informationsverarbeitung durch den Menschen, *Nachrichtentechn. Z.*, **12,** 68–74.

Küpfmüller, K., 1961. Die nachrichtenverarbeitenden Funktionen der Nervenzellen, in *Aufnahme und Verarbeitung von Nachrichten durch Organismen*, Nachrichtentechn. Gesellschaft im VDE, Fachausschuss "Informations- und Systemtheorie," Stuttgart: Hirzel.

Küpfmüller, K., and F. Jenik, 1961. Über die Nachrichtenverarbeitung in der Nervenzelle, *Kybernetik*, **1**(1), 1–6.

Levinson, J., and L. D. Harmon, 1961. Studies with artificial neurons: III. Mechanisms of flicker-fusion, *Kybernetik*, **1**(3), 107–17.

Martin, T. B., 1961. Analog signal processing by neural networks, in *Proceedings of the National Electronics Conference*, vol. 17, Chicago, Ill.: National Electronics Conference, Inc., pp. 317–21.

McGrogan, E. P., 1961. Improved transistor neuron models, in *Proceedings of the National Electronics Conference*, vol. 17, Chicago, Ill.: National Electronics Conference, Inc., pp. 302–10.

Putzrath, F. L., and T. B. Martin, 1961. Speech recognition by neural networks, in *Proceedings of the National Conference*, vol. 17, Chicago, Ill.: National Electronics Conference, Inc., pp. 311–16.

Reichardt, W., and G. MacGinitie, 1961, Zur Theorie der lateralen Inhibition, *Kybernetik*, **1**(4), 155–65.

Schief, R., 1962. Koinzidens-Siebschaltungen. Dissertation, Hannover, Technische Hochschule.

Takeuchi, A., and N. Takeuchi, 1962. Electrical changes in pre- and postsynaptic axons of the giant synapse of *Loligo*, *J. Gen. Physiol.*, **45,** 1181–93.

Taylor, W. K., 1956. Electrical simulation of some nervous system functional activities, in *Information Theory*, Colin Cherry, ed., New York: Academic Press.

Taylor, W. K., 1960. Computers and the nervous system, in *Symposia of the Society for Experimental Biology*, vol XIV, 152–68, New York: Academic Press.

Tischner, H., 1960. Ein Funktionsmodell der akustischen Fasern im Nervus cochlearis, *Zentralblatt f.d. gesamte Neurologie und Psychiatrie*, **158,** 271.

10. Use of Gas-Ion Processes in Modeling Excitable Membrane

H. J. HAMILTON, *General Precision, Inc.*

Models of neurons and neuronal systems may be classified in a number of ways: according to whether they are biologically or technologically inspired, unit-level- or systems-oriented, discrete-state or continuous-variable, formal or nonformal, and so on. Recently, investigators have used mainly the digital computer and electronic circuitry as means for constructing and testing models. Most of this effort appears to be aimed primarily at the formulation and testing of hypotheses concerning particular neural systems; more far-reaching questions about basic theoretical and methodological problems of modeling complex systems are usually side-stepped.

In this paper I shall describe some work dealing with the problem of how to develop effective general-purpose tools and techniques for modeling neural systems. The predominant concern underlying this effort has been with matters affecting the *economy of modeling*. Experience in our Laboratory (Reiss, 1960; Worthy, 1960; Opolak, 1961; Josephson, Reiss, and Worthy, 1961) has caused us to question the long-range utility of purely digital methods in modeling neural systems. The extensive algorithmic computation necessary to obtain reasonably accurate discrete approximations of continuous functions makes this approach economically unsound, despite the advantages afforded by mechanization. We need general-purpose tools which possess not only the operational flexibility and the automatic-control features of digital techniques but also the directness and economy of analog techniques. This suggests that we consider the possibility of mechanizing or automating analogical simulation techniques.

We decided at the outset to search for elementary processes in nature which revealed (or might be made to reveal) a phenomenological equivalence to those intra- and interneuronal processes which we wished to model. This approach promised greater success in achieving the required flexibility and economy than could be expected from an attempt to hybridize conventional digital and analog techniques. We focused our attention primarily on what appeared to be the best-understood facet of neural behavior: all-or-none conduction of nerve impulses along the axon.

This work was partially supported by the Air Force Office of Scientific Research.

A Conceptual Model of the Axon

Before discussing specific analogs we shall construct a primitive conceptual model of the axon. Such a model should satisfy at least six conditions:

1. Energy from an external source is supplied and stored uniformly or quasi-uniformly along a channel.
2. There exist, likewise, along the channel in continuous (or quasi-continuous) homogeneous distribution, kinetic mechanisms which may be activated by (or may excite) similar mechanisms in contiguous areas of the channel but whose behavior is otherwise determined solely by the local properties or state of the energized region.
3. The kinetic mechanisms exhibit a threshold property, responding to a stimulus equal to or greater than the threshold level and quiescent for all other conditions.
4. The activity in any region persists until the potential energy in that region falls below a critical level, after which the region becomes refractory and remains so until the energy is restored and a new activating signal is received. (This is admittedly only one possible assumption concerning the ultimate cause of sodium/potassium inactivation in the axon membrane.)
5. The refractory period is at least as long as the active period.
6. The propagation of excitation proceeds over the length of the channel at a relatively uniform velocity. A schematic representation of such a model is shown in Fig. 1.

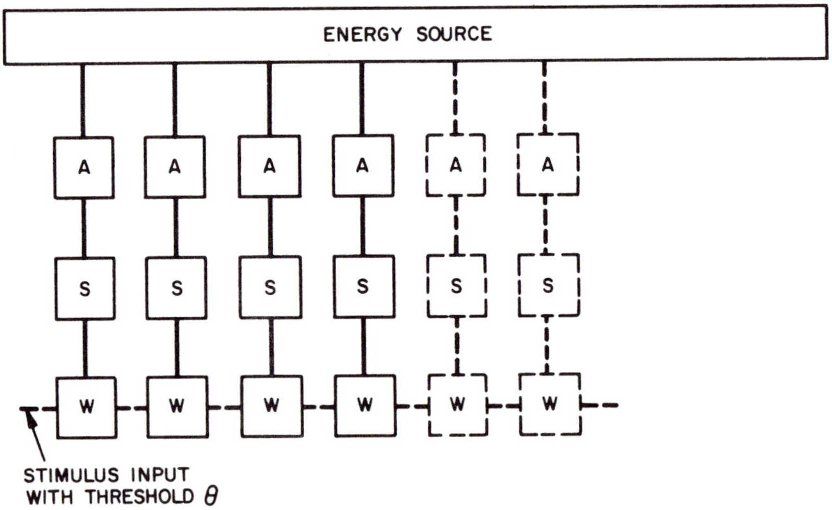

Fig. 1. A primitive conceptual model of the axon. The reservoirs (S) receive energy from an external source through finite conductances (A). An active element (W) may be excited by an appropriate stimulus, and in turn it will excite (dashed lines) contiguous elements while consuming energy from S in the process.

These conditions do not, of course, suffice to describe all properties of axons relative to their function of communicating and processing information in nervous systems. Nevertheless, as a generalized statement of the phenomenon of all-or-none conduction, they are useful in guiding our search for a natural analog.

A Gas-Ion Model of the Axon

If we survey known physical phenomena with the above aim in mind, several interesting possibilities emerge. The famous "iron-wire" model of all-or-none conduction, first investigated by Lillie (1920) and recently studied by Aldrich, McCulloch, Lettvin, and Pitts (1962), serves well to illustrate the concept of a "natural analog." For studies of semiconductor and superconductor systems and of relay models of the axon, see Crane (1960), (1962); Cote (1961); Padovani (1961); and Nagumo, Arimoto, and Yoshizawa (1962).

During the past five years we have been investigating an approach based on the use of ionization mechanisms in low-pressure gaseous systems; it is with such systems that the following discussion is exclusively concerned. My interest here is to present only the basic concepts and some conclusions drawn from our experiments.

Elementary Principles of Two-Element Gas-Discharge Arrays. Contemplation of a conceptual model of the sort sketched in Fig. 1 led us to inquire into the possibility of using gas-discharge mechanisms to model impulse conduction in the axon. To see how this might be possible, let us first briefly review some important properties of low-pressure, cold-cathode gas-discharge systems.

It is well known that electrical conduction in a simple two-electrode gas-discharge device exhibits a threshold property: the conductance is extremely low until the potential between the electrodes reaches a critical value, whereupon it increases abruptly by several orders of magnitude. The phenomenon is the result of an ionization avalanche, which occurs at a critical field intensity. The important features of such discharges, for our immediate purposes, are these: Progressive ionization of the gas molecules produces in a very short time a plasma which generally occupies a large portion of the space between the electrodes, except for a relatively small region near the cathode. The plasma thus formed produces a considerable distortion of the electric field, as manifested by a nonuniform potential distribution across the gap and a marked increase in field intensity near the cathode. This field distortion makes it possible for the discharge, once initiated, to continue with a substantially reduced potential applied to the electrodes, and it figures prominently in the mechanism of non-decremental propagation along an array of electrodes, as we shall see. There are, of course, many factors which determine the character or mode of the discharge. We are concerned here with what is called the *normal glow discharge*. In this mode the intensity of ionization is limited (in general by external circuit parameters), and the discharge cannot proceed into thermally destructive modes. The normal glow is characterized by a relatively steep function relating current to applied potential (see Fig. 2). This behavior is explained in terms of

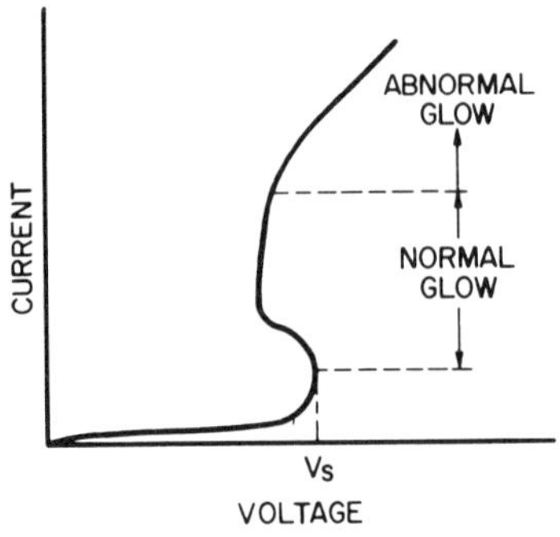

FIG. 2. Low-pressure, cold-cathode discharge characteristics.

geometrical considerations which in substance amount to this: The plasma column widens rapidly with only a small potential increase and projects farther and farther over the surface of the cathode, yielding an increasing emission of electrons from new areas of the cathode. If the external resistance is low, the discharge proceeds through the normal glow mode with extreme rapidity and may cause a catastrophic breakdown. If, however, the external resistance is sufficiently high to prohibit sustained discharge, a relaxation mode prevails. Under these conditions, the ionization processes continue, once initiated, until the energy stored in a capacitor or in the self-capacitance of the electrode is depleted to a level insufficient to maintain the discharge. A subsequent discharge will take place only when the capacitance again charges to threshold potential. There is no need to elaborate on the principles of the relaxation oscillator; it is sufficient to note that in such systems the dynamics of the discharge itself are determined solely by the local physical and geometrical properties of the electrode–gas system.

Consider now a hypothetical linear array of identical relaxation oscillators so interconnected that a discharge occurring at any unit in the array will quickly induce a discharge in its immediate neighbor, or neighbors. Such a system would appear to closely resemble the conceptual model of the axon discussed earlier if we further specified that the relaxation oscillators were each to be operated at a potential just below that necessary to produce spontaneous firing.

To achieve a real, workable system, we must find a simple, effective means for adjacent-element stimulation under the conditions specified. In principle one might approach this in two ways, by electrically interconnecting isolated discharge systems or by direct electron–ion coupling between electrodes sharing a common gaseous environment. The second approach appeared the more interesting from several points of view, and we discovered in our first experiments (Hamilton, 1960) that very simple arrays could indeed conduct impulses along their length without decrement or other cumulative distortion of the signal. It is in fact far easier to demonstrate lateral stimulus mechanisms in such systems than it is to explain them theoretically. A formal analysis would have to take into account:

1. the mean velocity and extent of lateral spread of the plasma,
2. field distortion due to electrode potential change,
3. photoionization of molecules in adjacent segments,
4. photoemission and ion-impact emission at adjacent cathode elements,
5. geometrical factors,
6. time course of the discharge,
7. physical properties of the gas,
8. statistical and formative time-lag factors affecting discharge onset.

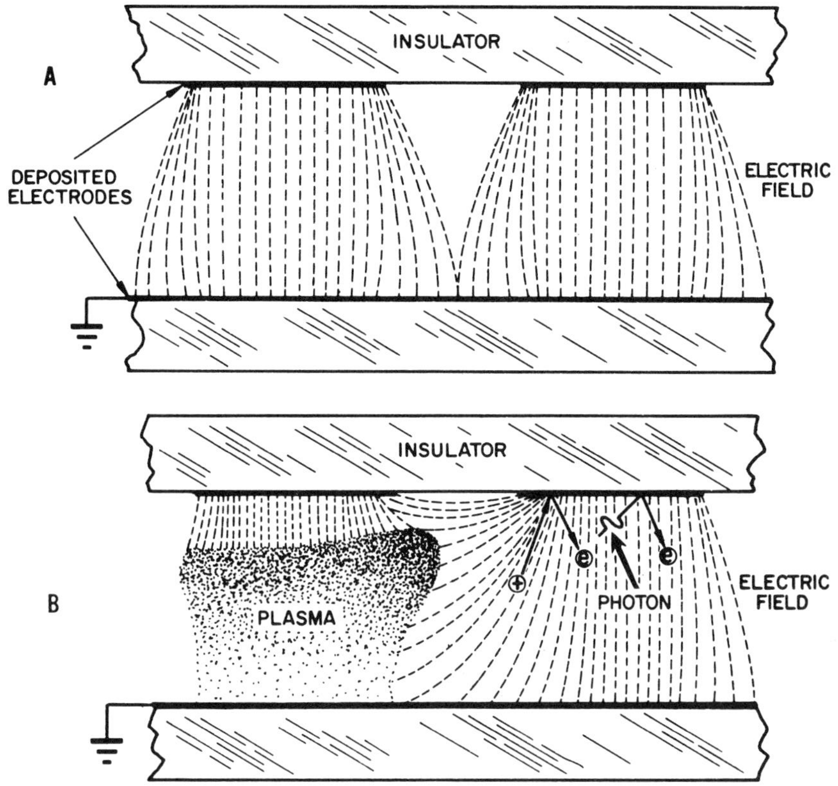

FIG. 3. Factors affecting discharge propagation in two-element arrays. A. Quiescent-state field distribution. B. Discharge-state conditions. A discharge initiated in the gap at the left produces a plasma which spreads to the right because of coulomb forces. The expanding plasma distorts the electric field in the gap and produces photons, some of which yield photoelectrons, ⓔ, at the adjacent cathode. Ions, ⊕, intercepted by the cathode may also yield free electrons.

In Fig. 3 I have attempted to illustrate the contribution of some of these effects in bringing about discharge propagation. In Fig. 3A the upper electrodes (cathodes) are biased relative to the common anode at a level just below the firing potential, and the field in the gap along the array will have a quasi-uniform distribution. In Fig. 3B a discharge has occurred in the gap at the left, producing a dense plasma, which spreads to the right, partly because of coulomb repulsion within the plasma and partly because of the attractive field in the gap at the right. In addition to having pronounced effects on the field distribution, the plasma emits photons which serve both to sustain the original discharge and to produce free electrons, ⓔ, at adjacent cathodes by photoemission. The impact of positive ions, ⊕, also accounts for some of the free electrons in the adjacent gap. The net effect of these phenomena acting in concert is to cause the plasma wavefront to spread to the right (we have ignored here the possibility of bilateral spread). If the parent discharge is sufficiently intense and long-lasting, the lateral diffusion of the space-charge cloud will be relatively great,

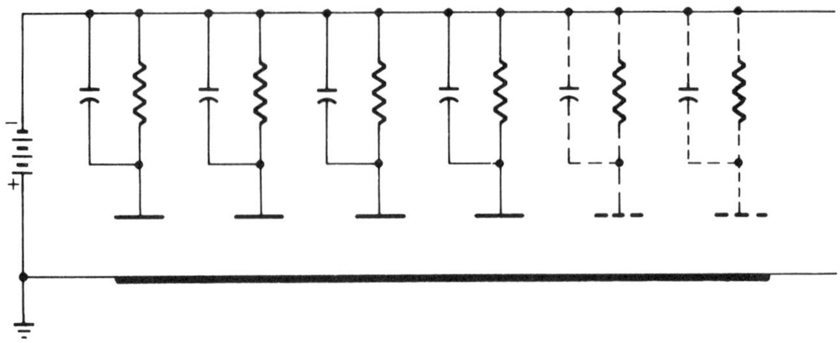

FIG. 4. Schematic circuit of a simple two-element array capable of conducting impulses in an all-or-none manner.

and the resulting field distortion and electron emission processes will be such as to trigger a new avalanche in the adjacent gap. Ultimately, this condition places a minimum requirement on the energy available at each electrode so that impulse propagation is ensured.

Properties of Two-Element Arrays. Figure 4 is a schematic representation of a simple two-element array. The apparent simplicity of such systems, thus diagrammed, belies their actual behavioral complexity.

We have experimented with numerous configurations of two-element gas-discharge arrays. The most successful systems were of the form illustrated in Fig. 5. The system shown in Fig. 5A is perhaps the most obvious arrangement, and it is, in fact, the approach followed in our earliest efforts. The principal drawback is the difficulty of uniformly constructing large systems of this type.

The arrangement shown in Fig. 5B again entails individual assembly of elements, but the deposition of metallic coatings on the inner and outer surfaces of the capillaries is easily accomplished, and such systems offer certain distinct advantages. One such advantage is that the discharge area is readily accessible for the insertion of probes to be used either in controlling or in monitoring the behavior of the discharge. The concentric electrode structure may also serve double duty, providing a convenient capacitor in which to store the energy to support the ionization processes.

Structures similar to that illustrated in Fig. 5C may be fabricated as units without the need to assemble electrodes individually. This, of course, is a distinct advantage in constructing large systems. But the method has other advantages as well; in such configurations, capacitive and resistive elements in each electrode system may be fabricated by mass-production deposition techniques, and flexible control over these parameters may be readily achieved. This control is extremely important in view of our stated long-range objectives.

Despite differences in outward appearances, the behavior of these systems is similar. It should be noted first that very little energy is consumed in the inactive or unexcited state; typically, losses may be of the order of 10^{-9} w/cm or less. For two-element arrays the minimum linear energy density necessary to obtain reliable propagation of the impulse is about 1 erg/impulse/cm. It is interesting to compare this figure with a crude estimate of the energy requirement for the axon based on a consideration of ΔE (the change in membrane potential during an impulse) and the capacitance of the membrane. Assuming an axon several microns in diameter, a ΔV of 0.1 volt, and a capacitance of 1 μF/cm^2, an energy density of the order of 5×10^{-5} erg/impulse/cm is required for the axon. (This figure is about two orders of magnitude above that associated with the heat produced in a fiber due to the passage of a single impulse.)

Another pertinent property of gas-discharge arrays is that they enter a refractory state immediately following conduction of an impulse. This effect places an upper limit on the frequency of impulses which may be transmitted (as in the case of the axon); the stimulus level partly determines that limit. We may also speak of "fatigue" in these systems. Here fatigue is defined as an increase in the duration of the refractory period; it occurs when the reservoir

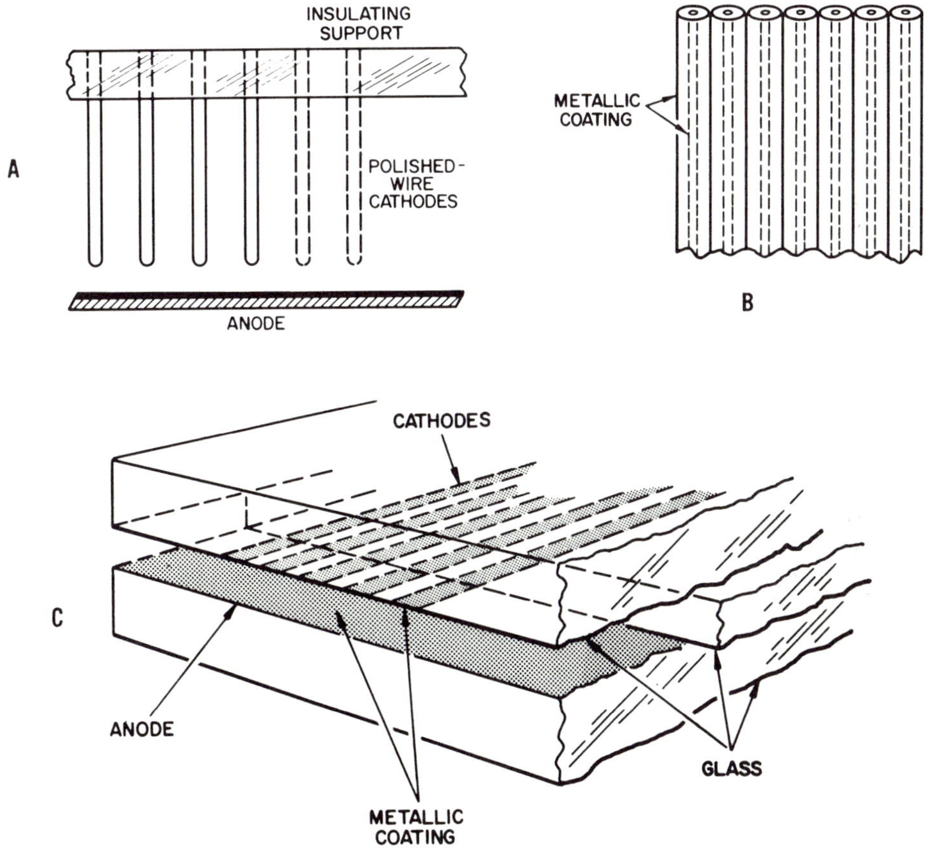

FIG. 5. Configurations of two-element arrays. A. Wire cathodes supported above a common anode. B. Capillary tubing with internal and external metallic coating. Discharge occurs across the end of the capillary, and the coaxial system serves also to store the energy necessary to support the discharge at each electrode. C. Metallic electrodes deposited on glass or other suitable dielectric material. Discharge occurs in the gap separating the cathode elements from the common anode surface.

of energy in the "external source" is limited in extent and rate of supply because the "source" is, in turn, dependent on more primitive sources.

In addition to these relatively obvious features of gas-discharge arrays, other characteristics, less obvious or less well-understood, have been observed. For instance, such systems often show a propensity for firing in bursts or volleys. A volley consisting of from two to ten impulses is not uncommon; on occasion, as many as a hundred impulses have been observed. In general, the pulse height declines after the first impulse, and the pulse interval increases gradually throughout the burst. The relatively long interval between pulses (several milliseconds) would appear to rule out any explanation based on the presence of residual, uncombined electrons and ions in the gap. We have considered the possibility that cathode surface contamination may be responsible for the volley effect. (It is well known that both the onset of discharge and the character of discharge are quite sensitive to monolayer contamination of the cathode.) However, efforts to substantiate this hypothesis have failed.

Integrative Control of Spike Initiation

With the discovery of practical means of achieving all-or-none conduction, we thought it advisable to find out whether or not the techniques thus employed could also be applied to modeling the integrative properties of the entire neuron. In particular we wanted to know whether it was possible for spatial and temporal integration of excitatory and inhibitory impulses from several units to control the activity in another. This led to the experiments described briefly in this section.

Cylindrical elements of the sort pictured in Fig. 5B were used. The first problem was to find a simple method of converting a depolarizing impulse, as obtained at the output of a unit, into a hyperpolarizing stimulus (or an effective equivalent), as required to excite a unit. (It may be recalled here that an increase in potential difference between electrodes in excess of the normal resting potential is required to trigger a discharge. More will be said on this subject later.) It turned out that this could be accomplished easily by using a probe biased at the same potential as the cathodes and positioned as shown in Fig. 6. When the probe is made less negative (driven positive relative to the bias level), the unit fires. On the other hand, if the cathode is made less negative (driven positive relative to the bias level), the unit is inhibited.

Thus positive-going impulses produced at the cathodes of the excitatory generators (units *a*, *b*, *e*, *f*) are integrated and the resultant potential is applied to the control probe of unit *c*. Isolation diodes are used to provide unidirectional signal flow. The effectiveness of each excitatory input may be individually determined by appropriately setting the value of the coupling resistance (R_e). The charging and leakage time constants of the integration networks are determined by the values of R_e, C, and R_ℓ. A similar situation holds for the integration of "inhibitory" impulses, though in Fig. 6 we have shown only a single inhibitory input. Quite obviously, there is nothing very sophisticated in this approach to signal integration. From a practical point of view, the significant feature of this technique is its simplicity in using the output of a unit

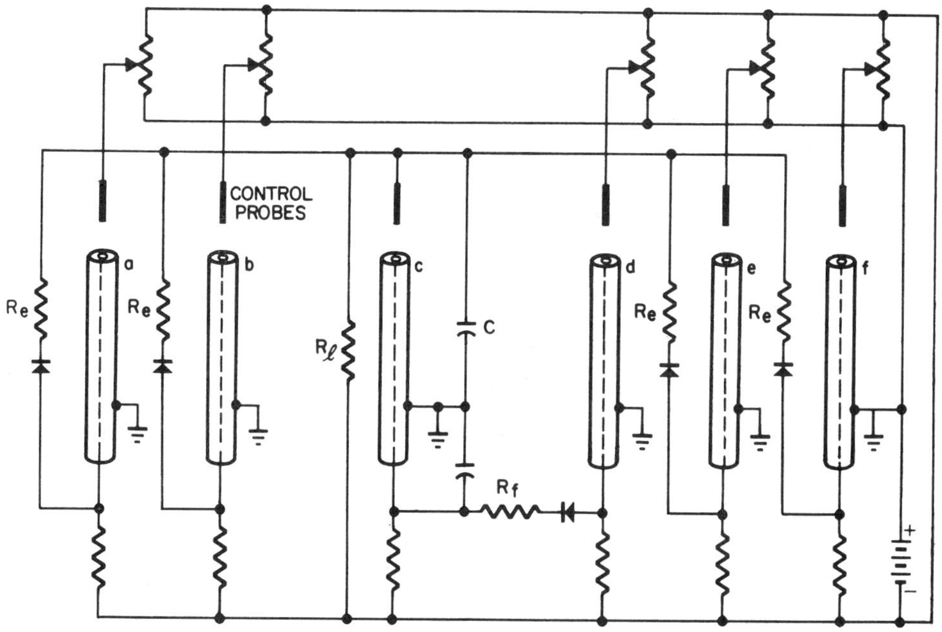

FIG. 6. Technique for separate integration of excitatory and inhibitory signals. Coaxial discharge elements labeled *a*, *b*, *d*, *e*, and *f* are impulse generators. The output of *d* is connected through a diode and resistor R_f to the cathode of *c*. Positive-going impulses are integrated and tend to inhibit the firing of *c*. The outputs of all other generators are summed and applied to the control probe of *c* and tend thereby to have an excitatory influence on *c*.

to achieve either an excitatory or an inhibitory effect. Thus a positive-going impulse from any unit can produce either an excitatory effect or an inhibitory effect on another unit. If the impulse is applied to the probe, the effect is excitatory; if to the cathode, it is inhibitory. The fact that the probe is biased at the same negative potential as the cathode and that depolarizing it tends to cause the unit to fire is what makes this simple, direct approach possible. It should be noted that the probe position relative to the face of the coaxial unit is not critical for the operation of the system. Variation in spacing by 50% or more produced no marked change in the effectiveness of the probe as a control element. It should also be noted that the probe, under normal operating conditions, does not participate in the discharge once it has been initiated. That is, energy is not supplied by the probe to support the discharge; the probe potential is not altered by the discharge. In general, this is fortunate, for it means that the integrated excitatory input at any given time may be capable of long-term influence over the activity of the unit. This is perhaps roughly analogous to the condition frequently observed in neurons, wherein the spike does not invade the soma or the dendritic regions and thereby wipe out or appreciably alter postsynaptic potentials.

Thus far we have made only exploratory tests on systems of the sort illustrated in Fig. 6. Nevertheless, these tests have established that the technique is straightforward and that it provides a highly effective and flexible means of signal integration. It was possible, for example, to vary the excitation threshold smoothly over a very wide range (requiring as few as two or three or as many as several thousand incoming impulses to initiate a discharge or a small burst of impulses). Likewise, the inhibitory mechanism appeared to function well; it was possible to override all but the strongest excitatory stimuli with the single inhibitory input.

Principles of Three-Element Arrays

These investigations brought a variety of problems and questions into focus. Notwithstanding our original intent to seek only a phenomenological model of the axon, we were disturbed that the properties of gas systems should be in some respects remarkably similar to those of axons and in other respects quite different. For example, the resting potential of the cathodes is negative, and a spike appears as a positive-going impulse, as is the case in the axon. However, in our model a hyperpolarizing stimulus is required at a cathode element in order to initiate an impulse, whereas in the axon an action potential is triggered by a depolarizing stimulus. And related to this problem is another: the model displays even worse passive-cable properties than the axon; i.e., subthreshold stimuli are subject to extremely high transmission losses. There appears to be no actual equivalent of electrotonic spread of subthreshold signals. Furthermore, in the axon the relatively high conductance of the axoplasm is generally thought to play an important role in the mechanism of nondecremental conduction, whereas any direct conductivity between cathode segments in the model tends to inhibit all-or-none conduction. There are other, more subtle flaws in the model. Consideration of these differences and a growing concern for the complexities of subthreshold behavior prompted us to re-examine our model in the hope of discovering a more fundamental analogy.

It became apparent after some study that the model might not be in error so much as our interpretation of it. This is best revealed in the answer to the question "Where can we place a probe to observe the equivalent of the transmembrane potential?" Let us begin by looking at the potential distribution in the gap of an idealized two-electrode gas-discharge system. We shall assume that the gas pressure is such that the minimum firing potential (for a gap spacing d_0) is V_s, well above the knee of the Paschen curve, as shown in Fig. 7A. Under these conditions, we may imagine (Fig. 7B) a plane P, at a distance d_p from the cathode, where the minimum firing potential is V_p. With a potential $V_0 < V_s$ applied to the anode the system will remain in a quiescent state. But if it were possible to increase the potential at the plane P to a value slightly above V_p, a discharge would result. The potential at that plane would increase rapidly to a value nearly equal to V_0 until the energy stored in C fell below that necessary to maintain the discharge, whereupon the system would again become quiescent. A grid placed in this plane would serve this purpose. Now if we reference our potentials to the anode (i.e., we make the anode ground), we see that by depolarizing this (negative) grid we cause the unit to fire; firing, in turn, gives rise to a much greater depolarization. (By a suitable choice of reference relative

FIG. 7. A. Paschen characteristic for two-element discharge system. B. Potential distribution for two-element discharge system.

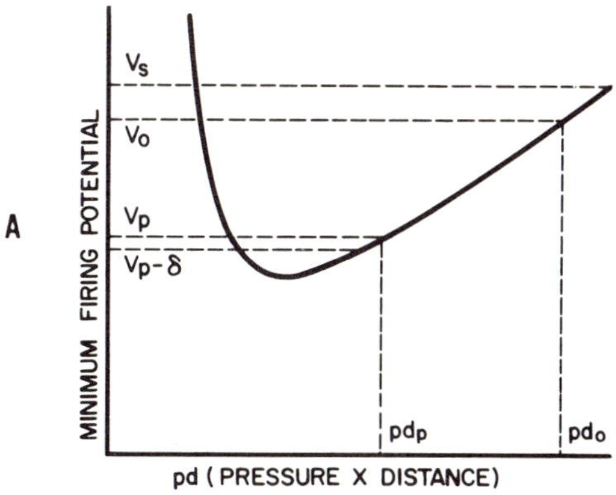

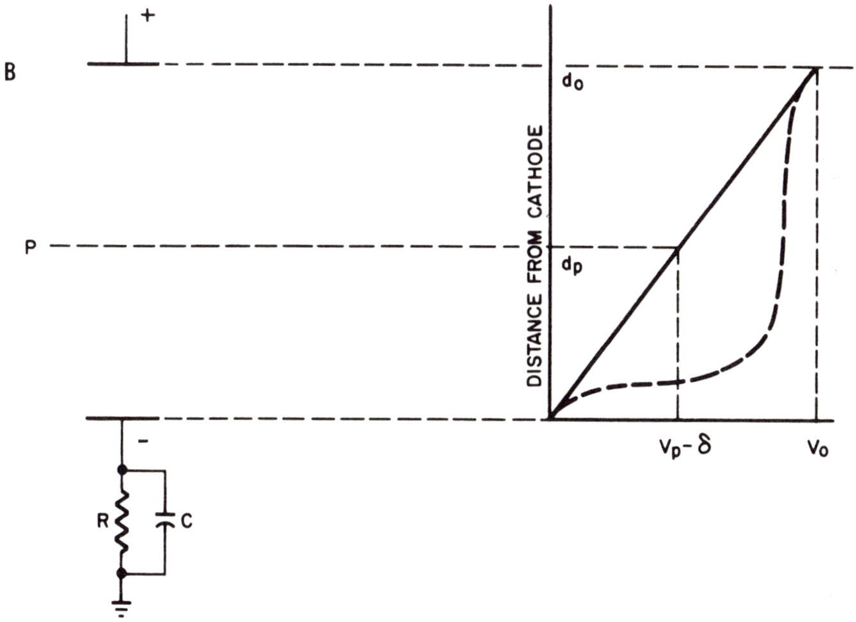

to the resting potential of the grid we could, in fact, produce not merely an additional depolarization of the grid during discharge, but a momentary reversal in polarity. This, of course, in no way changes the operation of the model, but it does provide a closer resemblance to electrophysiological data.) Alternatively, one may view such a system as two juxtaposed discharge mechanisms sharing a common electrode and a common gaseous environment. The triggering of one leads almost instantaneously to the ignition of the other. It is apparent that the grid, or more generally the region in which we put the grid, is the place to observe the equivalent of the transmembrane potential. The grid is not needed to effect propagation of impulses in an array, as we know, but it or some suitable probe is required in order to measure the potential in that region. Also, if we placed a series of probes in this area along the length of the array, we would be able to observe prepotentials resulting from the advancing wavefront

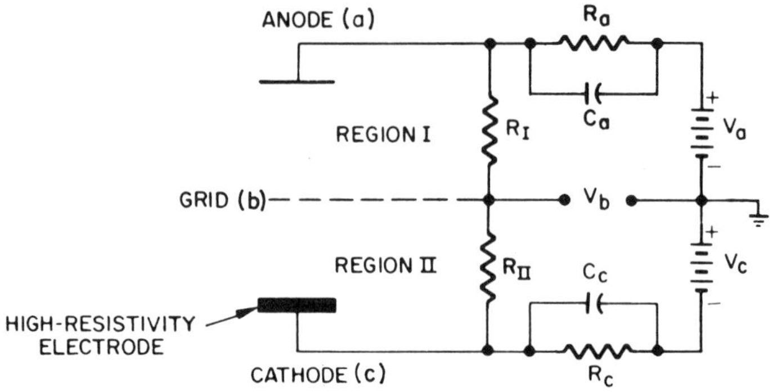

FIG. 8. Schematic of a three-element gas-discharge system. A low-porosity grid (b) divides the space between anode (a) and cathode (c) into two weakly coupled regions (I and II).

or lateral spread of the ion cloud. Thus, in a qualitative way at least, the two-element array provides a more satisfying analog than was at first realized.

But in following this line of reasoning, we have been moving steadily toward more fundamental considerations, from a search for a phenomenological equivalent of all-or-none conduction toward a search for a physical analog of excitable membrane.

In the preceding discussion we invoked a hypothetical grid to facilitate a reinterpretation of the operation of two-element arrays. I should like now to examine the behavior of three-element systems embodying the above general principles. Consider, in particular, the system sketched in Fig. 8, which has the following features: (1) the center electrode is a low-porosity grid which provides weak particle coupling between regions I and II; (2) the cathode is made of a high-resistivity material which effectively limits the electron emission rate from its surface. We assume the resting potentials of the electrodes to be adjusted so that the system is in a subthreshold state as described earlier, and we ask: "How will the system respond to an imposed electrical stimulus on the grid?" Let us consider first a hyperpolarizing stimulus (assuming that the grid potential V_b is negative). When the grid potential is made still more negative, the field in region II is decreased and that in region I increased. But if the field in region I is normally well below that required to trigger an avalanche (as it may be in practice), there is essentially no response by the system even for a relatively large stimulus, and the current in all electrodes remains very small. A sufficiently strong hyperpolarizing stimulus would, of course, result in a discharge in region I and subsequently in region II.

For a depolarizing stimulus the situation may be very different. When the grid is made less negative, the field in region I is decreased and that in region II increased. If, however, the resting field intensity in region II is only just below that necessary to establish an electron avalanche, a slight depolarization of the grid triggers a discharge of limited intensity in region II. Most of the energetic electrons produced in this discharge are collected by the grid, but some small percentage escape into region I, where they may initiate (if their energy and density are high enough) a new and much more intense electron avalanche. A qualitative description of these events is shown in the upper part of Fig. 9. A depolarizing impulse is applied to the grid at time t_0. At t_1 a current-limited discharge is initiated in region II. The delay, $t_1 - t_0$, corresponds to the statistical time lag in the initiation of an avalanche. It is the time required, following

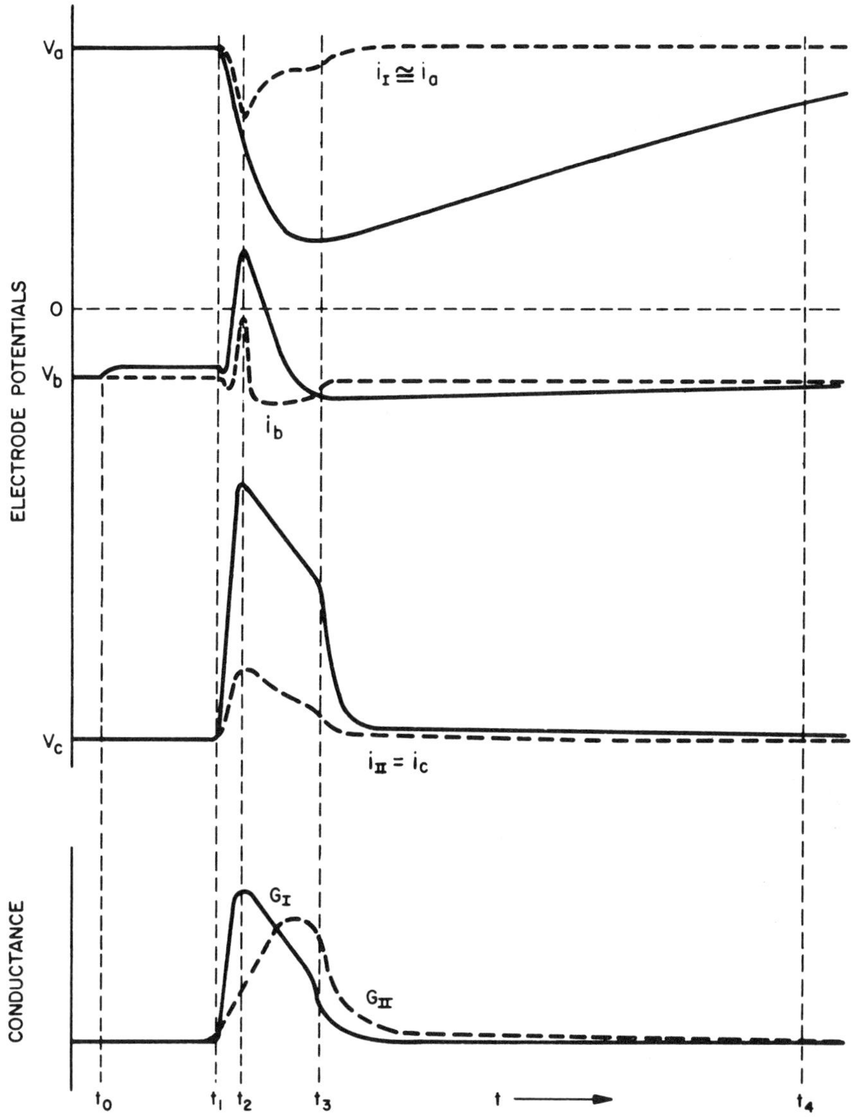

FIG. 9. Time course of electrode potentials and conductance for regions I and II in the three-element system shown in Fig. 8. Electrode potentials (V_a, V_b, V_c) are shown as solid lines. The reference potential is shown as a horizontal dashed line above V_b. Curve i_{I} (uppermost dashed curve) is the electron flux in region I. Curve i_b is the current in the grid circuit. Curve i_{II} is the electron flux in region II. The bottom set of curves depicts conductance in regions I (solid curve) and II (dashed curve). The depolarizing stimulus is applied to the grid at time t_0.

the application of an overpotential, for a free electron to appear near the cathode surface preliminary to the formation of an avalanche. The effective region of avalanche (more accurately, discharge) nucleation between cathode and anode increases as the applied overpotential is increased. Consequently, the probability that a free electron will appear within this region in a given time interval also increases. The net effect in our system is to cause the time delay, $t_1 - t_0$, to decrease as the magnitude of the depolarizing stimulus is increased.

Once the discharge in region II commences, electrons collected by the grid cause a momentary leveling-off or reversal in the grid potential and a brief negative (downward) current in the grid circuit. Very quickly, however, the escaping electrons produce an intense avalanche in region I which causes the grid polarity to reverse rapidly (and likewise the grid current). At time t_2 the discharge intensity in region I peaks, and subsequently declines, owing to the depletion of energy in C_a and the resulting decrease in field strength. At this point the discharge, which has been continuing at a relatively low level in region II, causes the grid potential to repolarize rapidly with the current in the grid circuit again reversing as shown. At time t_3 the field intensity in region II drops below that necessary to support the discharge, and the currents in all electrodes drop to a low value. At the same time the potential V_c at the cathode surface recovers quickly, and the rate of repolarization of the grid potential decreases to a value determined by the time constants of the associated passive elements of the network. For certain adjustment of these parameters, we may expect the electrodes to repolarize in the manner illustrated, i.e., with the grid experiencing a prolonged after-hyperpolarization and with the anode and cathode repolarizing slowly. (We should expect incidentally that the system would be in a relatively refractory state during most of this period because of the subnormal potential between the grid and the cathode. Because the potential at the anode is subnormal during this period or at least part of it, we should also expect a lower peak for any impulse produced.)

In the lower part of Fig. 9 I have sketched the time course of the conductance of regions I and II. The conductance in region I should peak at t_2 when the current is maximum and the potential difference between the grid and the anode battery is minimum. The electron flux in region I continues at a low level after t_2 (even though the current in the grid circuit reverses), and the conductance should therefore continue to diminish from t_2 to t_3. In region II the potential difference between the grid and the cathode battery should drop to its lowest value at time t_3. The electron flux at this time remains comparatively high, and we should therefore expect the conductance in region II to peak at some time just prior to t_3 and to decline rapidly thereafter.

I shall next examine the behavior of the system when the grid potential is suddenly clamped at some arbitrary level. For hyperpolarizations we should again expect little effect, with the grid current remaining very low. When a depolarizing step is applied, however, the electron flux in both regions should increase nonlinearly as the grid is made less negative. For depolarizations equal to or greater than the threshold, a discharge is triggered first in region II and then in region I. The electron flux in I is momentarily much more intense than in II, but as the energy in C_a (Fig. 8) is dissipated the ionization intensity in region I quickly declines to a comparatively low level. In region II, on the other hand, the limited discharge continues as long as the grid depolarization is maintained. Qualitatively, we should expect the conductance time courses in regions I and II to be as shown in Fig. 10. The amplitude of the conductance envelopes would, of course, depend upon the level of the applied clamping voltage.

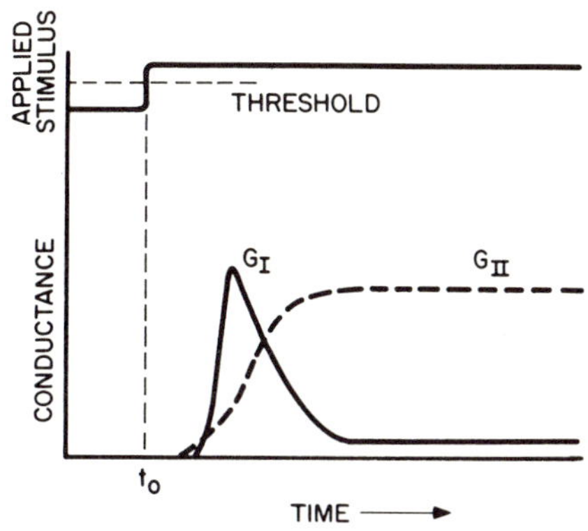

FIG. 10. Expected time course of conductance in regions I and II when the grid is suddenly clamped at a potential above threshold.

Experiments with three-element systems of the sort sketched in Fig. 8 provide general confirmation of the theory just presented. Figure 11 shows the time course of the grid potential of a three-element system using argon at a pressure of 21 mm Hg. A small stimulus lasting 20 μsec is applied through a high-impedance probe to the grid, and after an additional period of about 25 μsec, a discharge is initiated. It is to be noted that the depolarization of the grid, following the stimulus, occurs in three stages. First a small prepotential appears; this prepotential initiates an event which appears to resemble the graded or local potentials found in the neuron. (Figure 12 most clearly illustrates this intermediate effect. These spikes were generated spontaneously rather than in response to an applied stimulus as in Fig. 11. Spontaneous activity results when the normal potential difference between the grid and cathode is adjusted to a value slightly in excess of the threshold potential.) In the third stage the grid potential rises very rapidly to its peak value. Needless to say, we were greatly surprised to find the local-potential effect when we first carried out these experiments, since we had made no effort to extend the analysis of such systems beyond the more obvious first-order effects. We have observed, further, that the spike occasionally fails to materialize, in which case the grid depolarizes only partially, and then repolarizes very slowly. We have, as yet, no explanation for this behavior.

When a spike does occur, the grid repolarizes very quickly and, in fact, overshoots or hyperpolarizes, as seen in Figs. 11–14. The duration of the spike itself is about 2 μsec, whereas the period of the after-hyperpolarization depends upon the time constant of the elements of the network, and is, in general, long in comparison with the spike duration. The over-all time course of the impulse

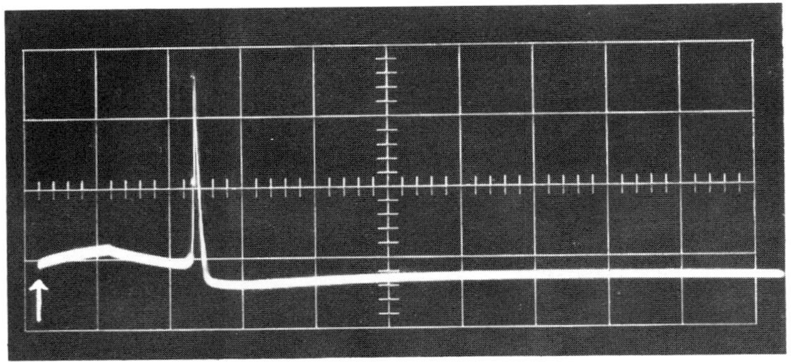

FIG. 11. Spike triggered by 20-μsec depolarizing stimulus applied to the grid (vertical sensitivity 20 V/cm; sweep 20 μsec/cm). Spike retouched.

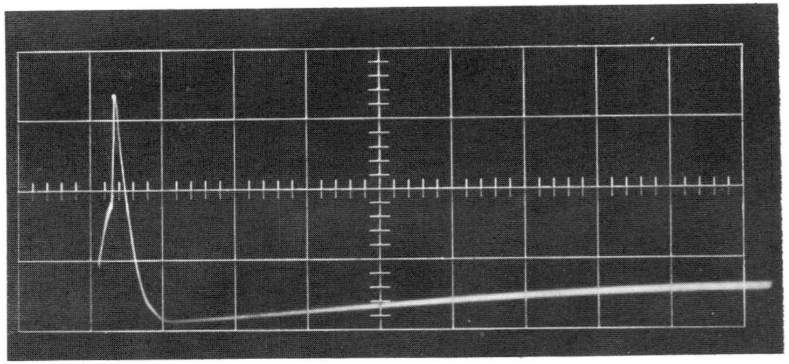

FIG. 12. Spontaneously generated spike (vertical sensitivity 20 V/cm; sweep 5 μsec/cm).

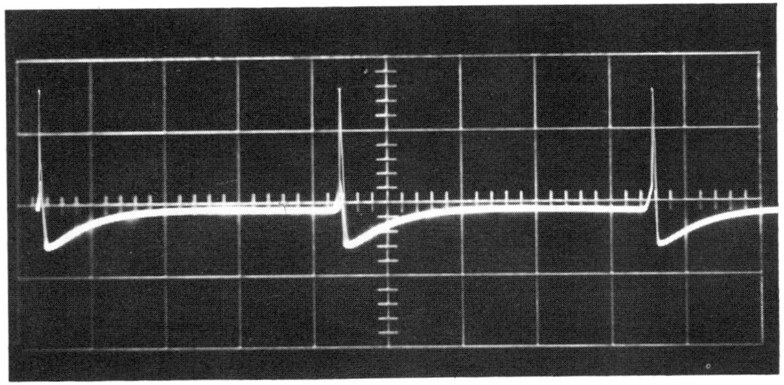

FIG. 13. Series of spontaneously induced spikes (vertical sensitivity 20 V/cm; sweep 50 μsec/cm). Spikes retouched.

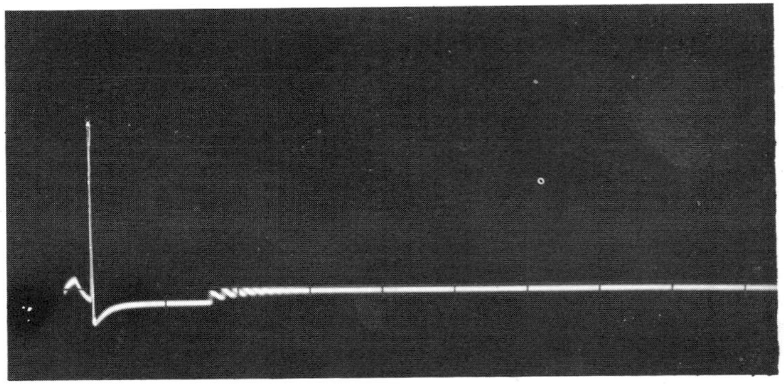

FIG. 14. Stimulus-induced spike showing subthreshold after-effect. Spike retouched.

is best illustrated in Fig. 13, which shows a series of three spontaneously induced spikes. Finally, Fig. 14 shows the peculiar subthreshold after-effects which, under certain conditions, may accompany a stimulus-induced spike. The cause of this strange behavior is not yet known.

Instead of trying to explore the many important and subtle theoretical details which a thorough technical discussion would demand, I shall conclude by suggesting a simple transformation which makes it possible to interpret the behavior of the gas system in terms of the "ionic hypothesis" of the membrane. (The reverse order of reasoning—i.e., interpretation of membrane processes in terms of known gaseous ionization mechanisms—provides a potentially fruitful area of study, but I shall not attempt to pursue this matter any further here.)

Let us begin by identifying the dominant charge carriers (or more properly, particles which determine the electrical current) in each system. The ionic hypothesis assumes that the principal components of the current which flows through the membrane in the active state are due to the movement of Na^+ and K^+ ions. In the gas system, on the other hand, the current is determined almost entirely (approximately 99%) by the movement of electrons, and the contribution of the relatively slow gas ions may be neglected. We note, then, a fundamental distinction: one system is based on *two-carrier*, the other on *one-carrier*, dynamics.

Let us compare next the forces operating on these carriers. There are two driving forces acting upon Na^+ and K^+ ions in the membrane, an electric gradient and a concentration gradient. These forces are superposed in the same space, namely, across the membrane. The electric field tends to accelerate the two carriers in the same direction (inward), whereas under normal conditions the respective concentration gradients tend to accelerate Na^+ ions inward and K^+ ions outward. In contrast to this situation, there is in the gas system only one force, an electric field, tending to accelerate electrons everywhere in the same direction. However, we must now distinguish two contiguous regions of space, with the grid acting as a common boundary. We can most readily compare the dynamics of these two complex systems by identifying those operations which tend to depolarize or hyperpolarize the membrane on the one hand or the grid on the other. Thus, an efflux of K^+ ions from the axon would, other factors ignored, tend to hyperpolarize the grid. Tentatively, then, we may equate electron flux in region II of the gas system with potassium flux outward across the membrane. Similarly, we know that an influx of sodium ions tends to depolarize the membrane of the neuron and that the emission of electrons from the grid (acting as a cathode) into region I of the gas system results in a depolarization of the grid. We may accordingly equate electron flux (actually only that portion emitted from the surface of the grid) in region I of the gas system with sodium current flowing inward across the axon membrane. This view agrees in a general way with what we know about the conductance in regions I and II, as discussed previously.

Thus far we have found a qualitative correspondence between the *two-carrier* and *one-carrier* systems, and we see that transformations from a *one-space* to a *two-space* geometry and from a *two-component* to a *one-component* driving

TABLE 1

TRANSFORMATIONS BETWEEN THE NEURON AND THE GAS SYSTEM

System	Charge Carrier	Type of Force	Active Region	Net Driving Force in Resting State
Neuron	K^+	{Electric field, Ion-concentration gradient}	Membrane	Diffusion (outward)
Gas	Electron	Electric field	II (Fig. 8)	Electric
Neuron	Na^+	{Electric field, Ion-concentration gradient}	Membrane	Electric and diffusion (inward)
Gas	Electron	Electric field	I (Fig. 8)	Electric

force, respectively, are required in order to compare the dynamics of the two systems. The nature of these transformations is summarized in Table 1.

Finally, what can be said about the potentials V_a, V_b, and V_c, which determine (at least in the resting state) the electric gradients in regions I and II? In a rough approximation we can compare V_a to E_{Na^+}, the sodium-concentration battery in the ionic hypothesis. Similarly, V_c corresponds to E_{K^+}, the potassium battery, and, of course, V_b corresponds to E, the transmembrane potential. We should hasten to add here that a more accurate comparison would be obtained by substituting $V_a - V_f$ for V_a (where V_f is the "cathode fall potential") and V_i for V_c (where V_i is the first ionization potential of the gas). These differences are significant but I think not fundamental.

Summary

The research which has been briefly reviewed in this paper aims, ultimately, at the development of powerful general-purpose tools for modeling neural systems. For the present, we have directed our attention to the problem of developing a natural analog of excitable membrane which could serve as a flexible basis for modeling the varied electrical activity of the neuron.

In this paper I have tried to show how gas-ion processes may be used to model the phenomenon of all-or-none conduction observed in the axon of the neuron. I have briefly traced the course of our reasoning, commencing with a discussion of elementary principles of ionization mechanisms in relatively simple two-element conduction arrays, and concluding with a qualitative analysis of the operation of three-element systems. An elementary transformation of particles and forces employed in the ionic hypothesis of the membrane may be used to advantage in describing the behavior of three-element systems of the type discussed.

Acknowledgments

I should like to thank Richard Vinetz for his assistance in recent phases of this research, and William Pollock for his contributions in the design and construction of experimental apparatus. Finally, my thanks to E. R. Lewis for his helpful discussion and criticism.

References

Aldrich, J. A., W. S. McCulloch, J. Y. Lettvin, and W. H. Pitts, 1962. "The Iron Wire Neural Nets," Massachusetts Institute of Technology, Research Laboratory of Electronics Quarterly Progress Report No. 64, Cambridge, Mass.: The Laboratory, pp. 303–5.

Cote, A. J., 1961. A neuristor prototype, *Proc. IRE*, **49**, 1430–31.

Crane, H. D., 1960. "Neuristor Studies," Stanford Electronics Laboratories Technical Report No. 1506-2, Stanford, Calif.: The Laboratories.

Crane, H. D., 1962. Neuristor—A novel device and system concept, *Proc. IRE*, **50**(10), 2048–60.

Hamilton, H., 1960. "Physical Research on Neural Analogies," in Information Systems Group, Laboratory for Automata Research Semi-Annual Report No. 1, Glendale, Calif.: General Precision, Inc., pp. 78–89.

Josephson, R. K., R. F. Reiss, and R. M. Worthy, 1961. A simulation study of a diffuse conducting system based on coelenterate nerve nets, *J. Theoret. Biol.*, **1**, 460–87.

Lillie, R. S., 1920. Recovery of transmissivity in passive iron wires as a model of recovery processes in irritable living organisms, *J. Gen. Physiol.*, **3**, 109–29.

Nagumo, J., S. Arimoto, and S. Yoshizawa, 1962. An active pulse transmission line simulating nerve axon, *Proc. IRE*, **50**(10), 2061–70.

Opolak, A., 1961. "The Cyclops System—A Simulation Study," in Information Systems Group, Laboratory for Automata Research Semi-Annual Report No. 2, Glendale, Calif.: General Precision, Inc., pp. 97–111.

Padovani, F. A., 1961. "Neuristor Device Study," in Solid-State Electronics Research, Stanford University, Quarterly Status Report No. 10, Jan.–Mar. 1961, Stanford, Calif.: Stanford Electronics Laboratories, pp. 18–23.

Reiss, R. F., 1960. The digital simulation of neuro-muscular organisms, *Behavioral Science*, **5**(4), 343–58.

Reiss, R. F., 1961. "The Reciprocal Inhibition Study: An Outline of the Digital Simulation Model," in Information Systems Group, Laboratory for Automata Research Semi-Annual Report No. 3, Glendale, Calif.: General Precision, Inc., pp. 39–53.

Worthy, M., 1960. "Synapse Simulation on a Digital Computer," in Information Systems Group, Laboratory for Automata Research Semi-Annual Report No. 1, Glendale, Calif.: General Precision, Inc., pp. 5–15.

11. Computer Data Processing and Systems Analysis Applied to Research on Visual Perception

G. D. McCANN and D. H. FENDER, *California Institute of Technology*

Serious interest in the nervous system and in related research—originally the realm of the biologist, the psychologist, and the medical practitioner—is now spreading rapidly to physical scientists, mathematicians, and engineers. Important advances in the understanding of the functional cerebral processes and of sensory perception offer encouragement to the belief that the highly complex methods by which an animal senses the world around it, analyzes this information, and then matches its physical behavior to the environment can help systems engineers and computer engineers to develop more sophisticated devices. Studies of certain of the simpler reflex systems in animals and insects have provided new types of feedback control systems; knowledge of the information processing at the neuron-network levels in the frog-eye retina has provided new engineering concepts for character- and pattern-recognition devices, and now serious efforts are being made to develop computers which will have adaptation and learning characteristics mediated by electrical networks similar in performance to some of those found in the nervous system.

Comparisons of the concepts employed in the biological nervous system with engineering principles developed in the fields of computer technology, systems analysis, and communication theory reveal a number of similarities, and it is becoming apparent that the natural sciences and engineering can materially aid biological systems research. Many of the important questions regarding both peripheral sensory perception and central brain organization of higher-order animals can only be solved by sophisticated experimental and analytical research techniques in which the physicist, chemist, mathematician, and systems engineer are able to assist the biologist and psychologist. For example, present histological and microprobing techniques do not permit the determination of the detailed structure of many portions of the nervous system where the neural

The Program described in this paper is supported by the National Institutes of Health, United States Public Health Service.

The first part of this paper was published in a preliminary version in the *IEEE Transactions of the Professional Group on Biomedical Electronics*, copyright 1963, IEEE. Permission to use this material is gratefully acknowledged.

elements are small and densely packed. Mathematical methods of analysis and synthesis developed for engineering systems can, however, help to determine structure from response measurements made on large elements of the neural network activated by carefully chosen stimuli. Finally, perhaps the greatest stimulus to this spreading interest is the evidence which now indicates that knowledge of the development and functional properties of the nervous system can form the basis for more effective educational methods.

The California Institute of Technology, like several other institutions, has initiated a study, pursued jointly by the Biology and Engineering Divisions, to facilitate cooperation between the physical sciences and engineering, and the behavioral sciences. That such combined efforts are needed is demonstrated by the way in which electronic neural modeling has been handicapped by insufficient knowledge about biological data processing. The primary role of applied science and engineering in this branch of the life sciences should be to seek understanding of basic principles in sufficient depth, to develop new concepts. This requires participation in basic research on the sensing, data processing, and control functions of the nervous system.

Some Important Requirements for Research on the Nervous System

Three general experimental techniques used in various combinations by biologists have provided the bulk of our present knowledge. One of these techniques is histology, in which the physical structure of neural networks is studied by sectioning, chemical staining to differentiate individual elements, and photomicrography with optical and electron microscopes. Another is the measurement of electrical activity, either in general areas with gross electrodes or in highly localized regions near or within individual cells or fibers with microelectrodes. The third technique is observation of the reactions or motor responses resulting from carefully chosen stimulation of the sensory organs. Through language the patterns impressed at the conscious level of the brain from sensory stimuli can also be studied in humans. These are the established methods of neurobiology and psychobiology; to them we are considering the addition of further experimental methods which will be capable of providing data usable in computer analysis of conceptual mathematical models.

Although our knowledge of the nervous system is already very great, it is still only a collection of fragments in a mosaic of tremendous complexity; thus some of the basic concepts of energy conversion in the sensory receptor cells have been established; data processing in a few interneurons has been described; information transmission in the nerve fibers and the use of this information to control the muscles have been well worked out. But in the case of visual perception we know only a little about the energy-conversion process in light receptors, and histology and microprobing reveal the structure and electrical activity of only the larger cells and axons of the optic system. A certain amount of information has, however, been obtained from cells and fibers right up to the visual cortex of several high-order vertebrates such as cats and monkeys—enough to indicate that more sophisticated methods of stimulation and response measurement, combined with more generous use of computers for data reduction, can significantly advance our knowledge of visual perception. Most nervous-system research to date has employed rather qualitative methods of stimulating the sensory system, but stimuli properly determined from systems-analysis techniques are required, together with suitable methods of measurement.

Another important consideration is the rapid execution and control of experiments: the biological preparation is frequently long and difficult; the specimen may deteriorate rapidly after preparation. It is therefore important to obtain as much information as possible in a short time. Biological responses are also characteristically "noisy" and subject to fluctuations uncorrelated with the applied stimuli; many controlled repetitions of an experiment may thus be required to eliminate uncorrelated responses, and comprehensive methods of statistical analysis and signal detection may be added as powerful methods of analysis. Finally, the complexity of the nervous system makes it difficult to find mathematical techniques adequate for theoretical modeling. Serious efforts are being made to apply computer programming at the thought-process level (Minsky, 1961), and algebraic (Farley, 1962) and electronic analog (Harmon, 1961) models are being used at the neural-network level. Coordination of these methods with experimental research is essential if description of the nervous system is to pass the colloquial stage.

This paper describes the biological-control and data-processing system developed for visual-perception research in the Biological Systems Laboratory at the California Institute of Technology. The results of our investigation on the optomotor responses of the housefly, *Musca domestica*, are also presented, since they illustrate some of the important requirements for data analysis in nervous-system research and bring out the usefulness of our data-processing method.

The Biological Data-Processing Facility

A data-processing system designed to overcome some of the basic difficulties mentioned above has been developed at the California Institute of Technology. Since a large central computing system providing direct communication with remote stations is also under development, it was decided to integrate the two programs. The biological-control and data-processing system is therefore a satellite of the larger central system. The present control system consists of three main computers and associated hardware; these are an IBM 7090 (later to be converted to a 7094), an IBM 7040, and a Burroughs 220, interconnected as shown in Fig. 1. In this system a 32,000-word-core-memory IBM 7040 computer is used partially as a communication-control device for a number of remote stations, one of which is the biological-experimentation system.

The interim system used in the studies reported here combined an IBM 7090, an IBM 1401, and a Burroughs 220 and is shown in Fig. 2. This diagram shows the biological control and data-processing system, whose functions are divided into three major phases:

1. controlling the application of complex patterns of stimulation to the biological preparation and recording the responses of the system,
2. converting pertinent measurements into digital form and transmitting them from the remote experiment location to the central computing facility,
3. controlling central processing and the display of analyzed results at the remote station.

Figure 3 is a photograph of the control system. In its present form all recordings are made on analog magnetic tape during phase 1 above. The system is also designed to process analog magnetic tapes from other research projects.

Digitized data sent to the central system have two possible destinations under the control of the IBM 1401: They may go either to the IBM 7090 via IBM magnetic tape or to the core or tape memory of the Burroughs 220. Simpler types of data analysis required immediately (such as averaging of successive repetitions of the same experiment and the graphical plotting of several quantities

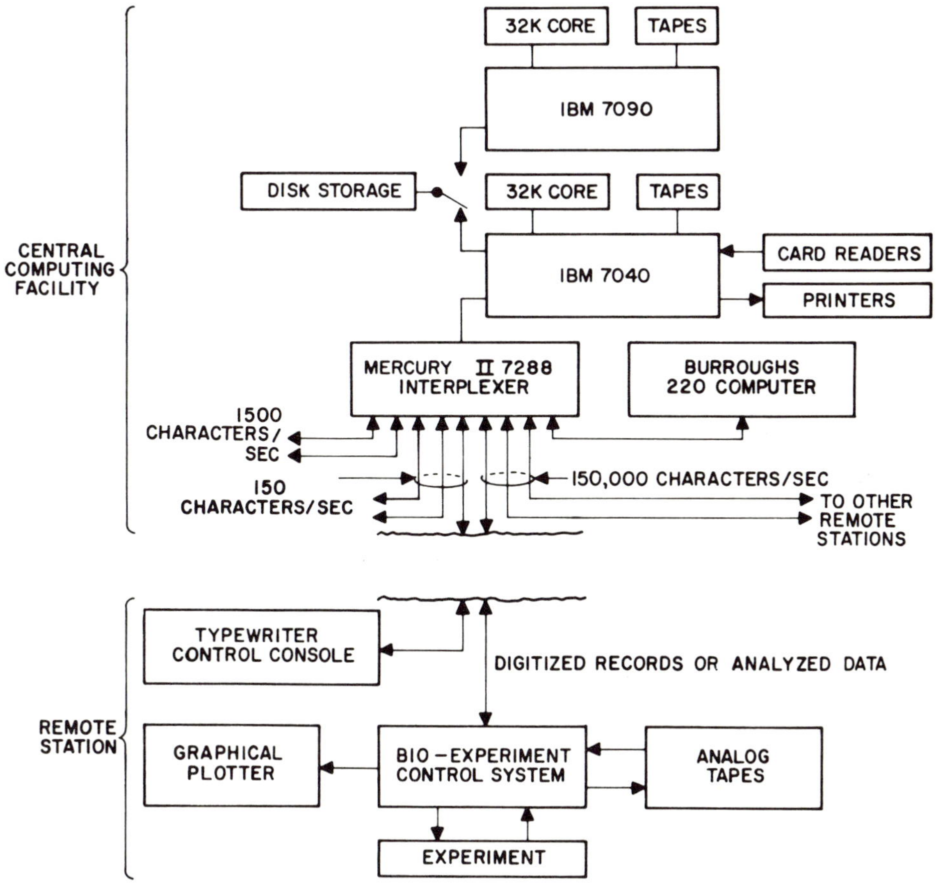

FIG. 1. Schematic of complete computer system for biological experiments.

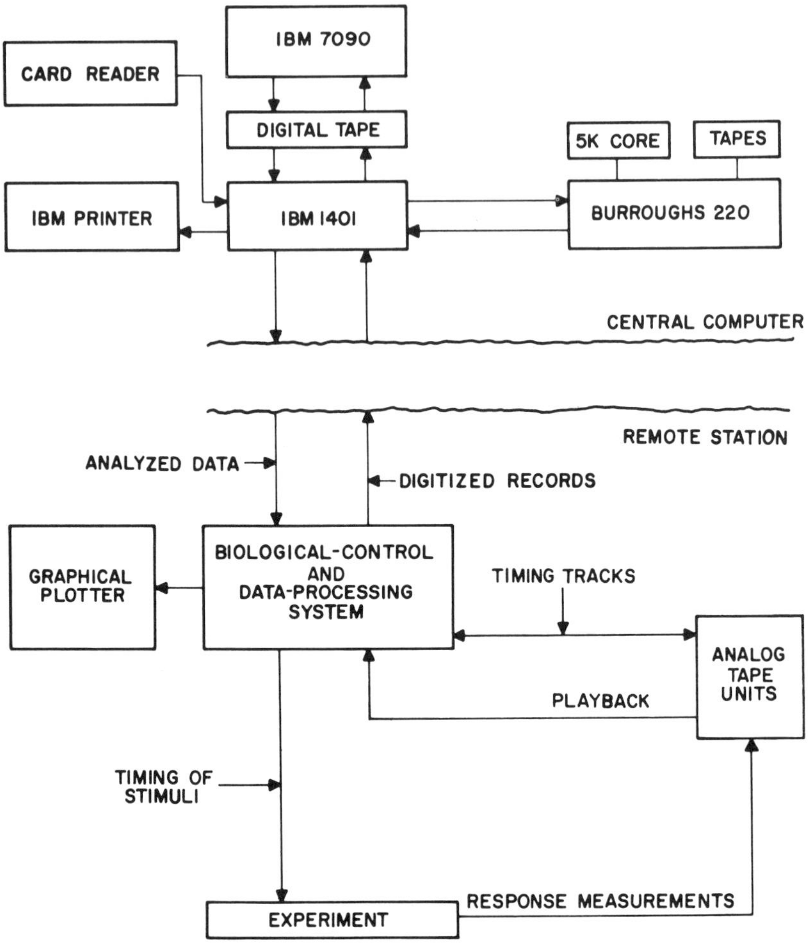

FIG. 2. Schematic of interim computer system for biological experiments.

for comparison) are sent directly to the core memory of the Burroughs 220. Longer analyses such as autocorrelations are carried out by the IBM 7090.

Experiment-Control and Recording Phase. This section of the biological-control and data-processing system is shown in Fig. 4. A time scale is generated either by a 100-kc crystal oscillator, or in the case of repetitive stimulation produced by nonsynchronous-motor-drive devices, by timing pulses derived directly from the stimulus. Such timing pulses can be scaled down to a suitable value by the decade divider and are then sent simultaneously by the clock-pulse generator to one amplitude-modulated timing track of each tape recorder being used in the experiment and to an 18-bit binary-counter register. This counter has outputs to a series of control circuits which can be used as required in the timing and control of the experiment; the desired times for activation of each control circuit are preset by dial controls. A second tape channel may also be

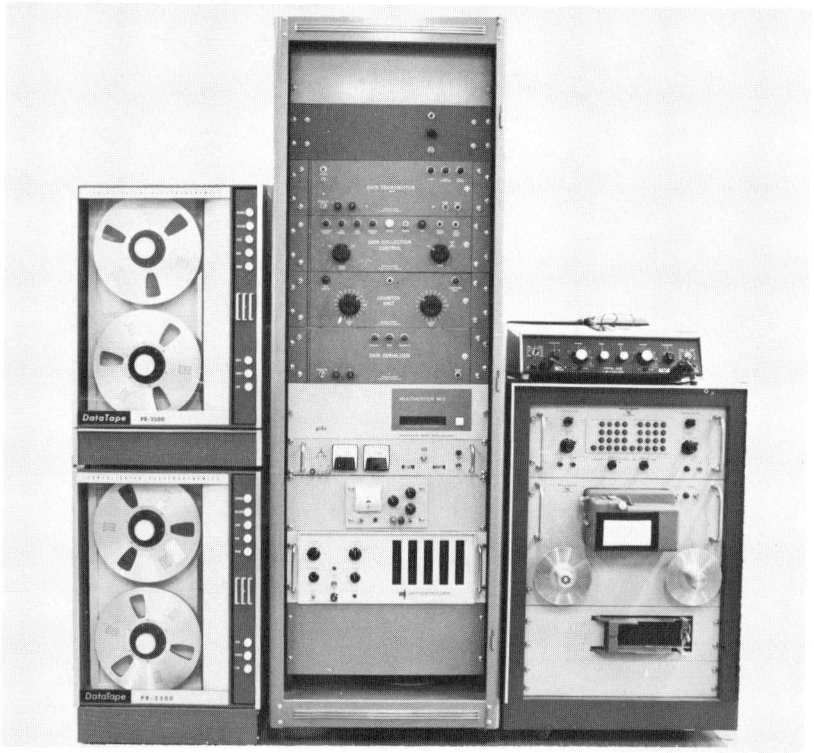

FIG. 3. The remote biological-control and data-processing station.

used for voice descriptions of experiments and for experiment codes that locate each record series on a given tape recorder. This code is used for rapid searching of desired records during playback phases.

The AND gate shown in Fig. 4 provides several modes of experiment control. In one of these the experiment can be started by simply turning on the tape recorders; here it is assumed that the biological experiment has been prepared and all instrumented inputs are functioning. As soon as the tape units are up to speed and completely operative, a control signal appears on the "Tape Ready" line; this signal passes through the AND gate, starting the timing pulses to the timing tracks and counter register. This, in turn, generates the stimulus-control signals. However, if desired, the initiation of the timing pulses can be delayed until signals are received from one or more devices in the experiment. As an example, it might be desirable to synchronize zero time in the experiment with a particular phase of the stimulus, in which case a signal derived from the stimulus would be inserted into the line marked "additional external control."

Digitizing and Transmission Phase. The elements of the system used for digitizing and transmitting data to the central computer facility are shown schematically in Fig. 5. The starting point for signals recorded simultaneously on any section of analog magnetic tape can be rapidly selected by the code track.

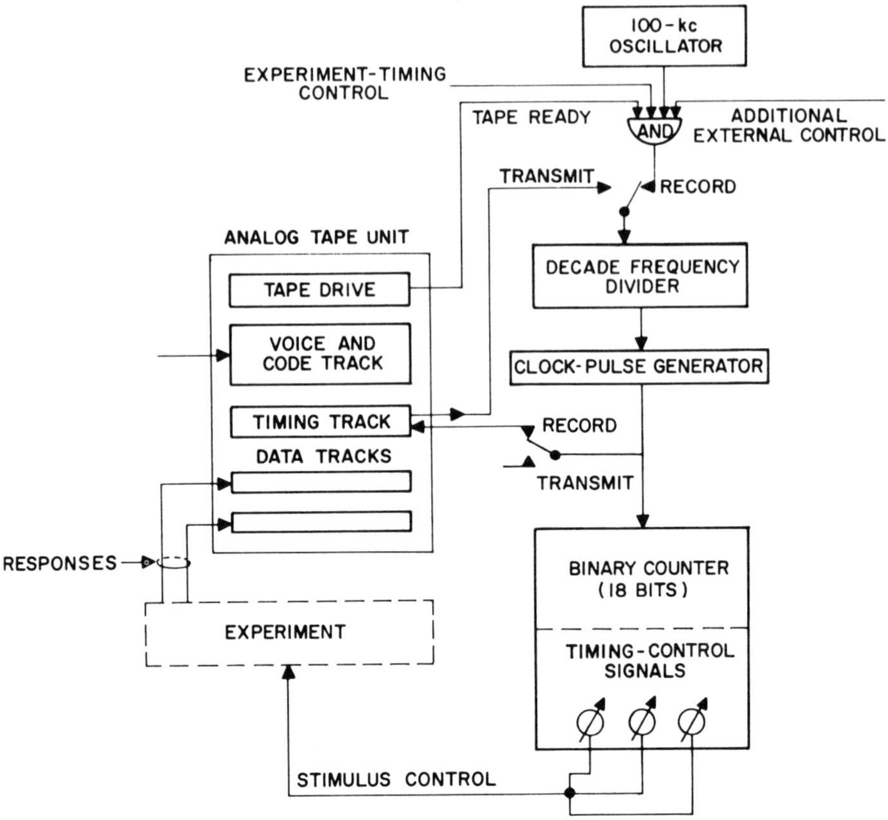

FIG. 4. Schematic of experiment-control and recording phase section of biological-control and data-processing system.

During the digitizing phase the timing track is read from the analog tape and sent into the 18-bit-counter register as shown. From this, the timing control circuit can be used to select automatically the actual start and stop points of the digitizing process in order to control the portion of the record to be digitized.

Two types of digitizing process are used. One of these is straight analog-to-digital sampling at periods determined by a signal from the timing track transmitted through the decade frequency divider and the clock-pulse generators. The A–D converter used in this system can sample at rates of up to 10,000 samples/sec. Its sample time is a period 10 μsec long at the beginning of each sample period.

The other type of data transcription device is called a Time-Of-Event (T.O.E.) recorder. In this mode of digitization, the times of occurrence of a variety of repetitive phenomena can be recorded by means of the 18-bit binary-counter circuit and the corresponding times sent as a series of numbers to the central computer. The range of times over which events can be recorded depends upon the required timing accuracy. For an accuracy of 0.1 msec (the present limit of the device) the range is about 26 sec. For a 1-msec accuracy the range is ten times as great, and so on. In this mode, pulses from the timing track are counted in the 18-bit binary counter: the occurrence of an "event" of the analog type operates the T.O.E. signal generator, which duplicates the reading of the binary counter at that instant in the T.O.E. buffer, and then transmits it in digitized form to the central computer.

The selection of an event whose time of occurrence is to be recorded is made

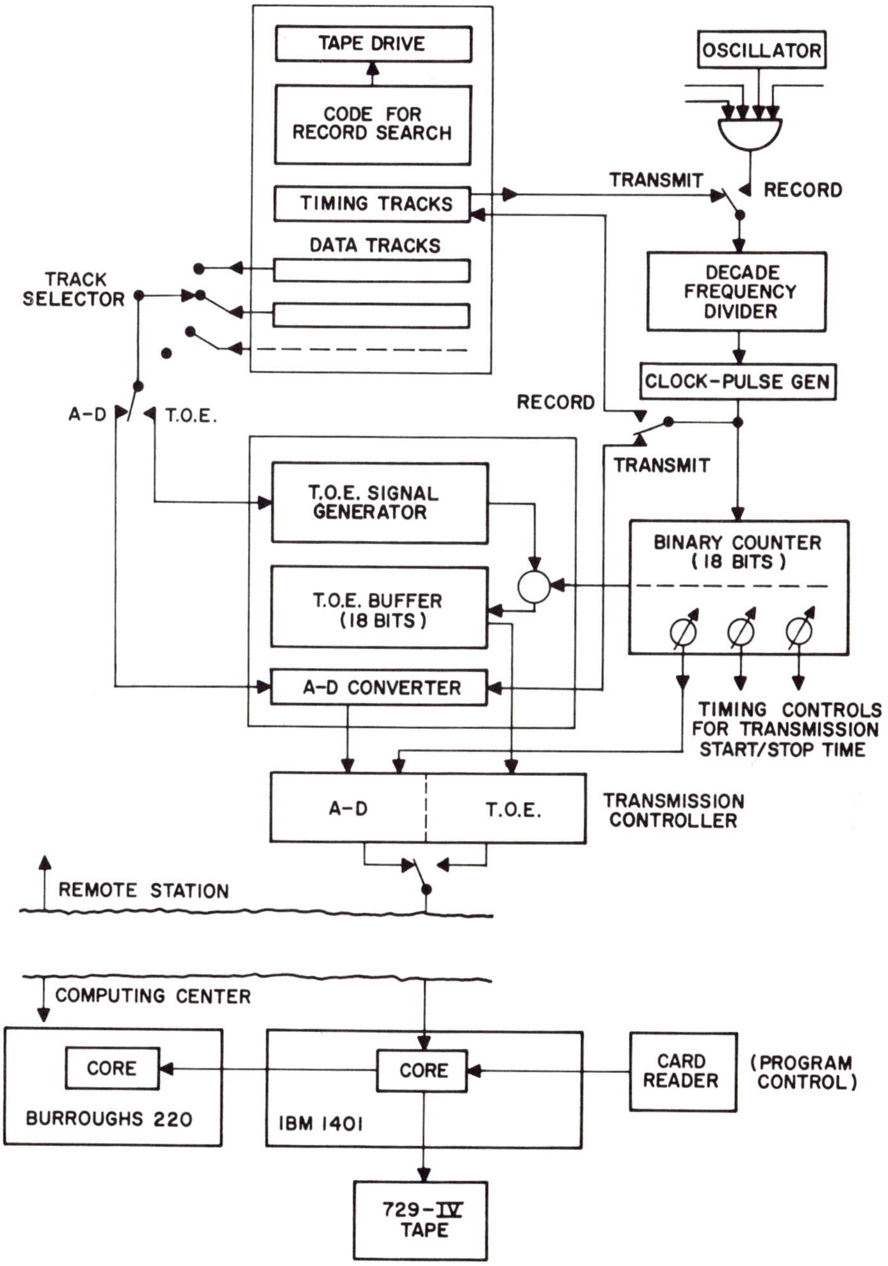

FIG. 5. Digitizing and transmission phase for biological-control and data-processing system.

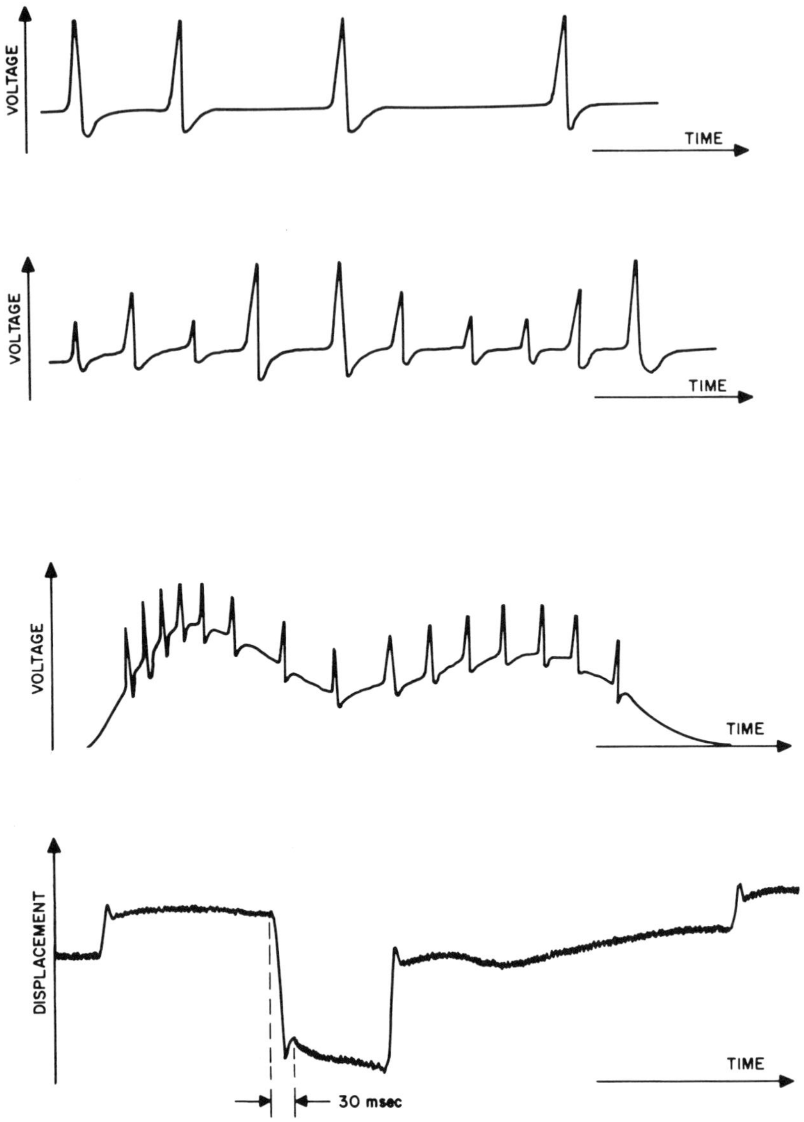

FIG. 6. Types of records from which time-of-event information is extracted.

by a variety of circuits employing both analog and digital logic techniques. These can best be described by some specific examples. Typical records of interest to this program are illustrated in Fig. 6. The first of these is a set of spike discharges obtained from a single fiber or axon. For records of this sort the steady negative potential can either be compensated by a dc bias circuit or, preferably, removed by a high-pass filter. A point on the leading edge of each spike can be selected by a zener-diode preamplifier circuit and fed to a Schmitt trigger circuit which provides a pulse to operate the T.O.E. signal generator.

The second record shown in Fig. 6 is a series of spikes recorded simultaneously from a bundle of fibers. As long as they do not overlap in time, spikes whose peaks fall within a given amplitude range (and, therefore, presumably come from a single axon) can be sorted out by a zener-diode gating circuit operating at two voltage levels. One level can be set as the lower limit and the other as the upper limit for acceptance of pulses. These two levels trip memory circuits feeding an AND logic circuit so that a pulse is identified as such if it reaches the first level and not the second level. An additional short voltage pulse is applied to the AND gate to limit the decision to the front edge of a single pulse and to reject complex tails that might result from two overlapping spikes.

The third record in Fig. 6 was obtained from a light-receptor cell and shows combined spikes and a slow-wave generator potential. The individual spikes can be separated from such a record with a high-pass filter, and the signal is then treated as in the first case described above.

The fourth example of Fig. 6 is typical of records obtained from studies of human eye movements. These movements contain a combination of low-frequency motion (up to about 5 cps) and more sudden step motions (flicks) of either polarity and of varying magnitude. The rise times are less than 30 msec, and differentiation can therefore produce short positive or negative spikes. A bridge circuit converts all of these spikes to the same polarity, and they are then treated as in the first case described above.

Data Transmission Between the Satellite and the Central System. One 6-bit character is sent at a time to the IBM 1401. The 6 bits are sent (with a parity check) through parallel concentric cables between the two buildings housing the research project and the Computing Center. Each sample from the A-D converter is sent in binary form with 11 bits plus sign, thus requiring two characters; the 18 binary bits for a time-of-event require three characters.

The present data-transmission speed to the IBM 1401 is 60,000 characters/sec. This is a limitation of the IBM 1401, and the corresponding value for the IBM 7040 systems can be in excess of 150,000 characters/sec. The average transmission speed through the IBM 1401 is further limited by its core memory, since reception must be halted when the memory is filled until the information has been placed either on IBM tape or in the Burroughs 220 system, or both. When space is allowed for the control program, the IBM 1401 memory is limited to the reception of 7800 characters (3900A-D samples or 2600 times-of-event). The IBM 1401 requires about $\frac{1}{8}$ sec to unload its memory onto tape, and this process reduces the maximum transmission rate to about 30,000 characters/sec. This, however, is faster than the analog tape units can be scanned at present, for the tapes have to be halted and then restarted each time the IBM 1401 core memory fills. The digitizing unit of the biological-control and data-processing system has an additional counter register that keeps track of successive portions of a record if it is too long to send to the IBM 1401 in a single pass. This facilitates rapid buffering between the two units, with a resulting average transmission rate of about 2000 characters/sec. These restrictions on transmission rates are not present in the system shown in Fig. 1, which permits the direct digitizing and transmission of data without initial analog recording. This system has a transmission rate of 150,000 characters/sec. When transcribing from analog magnetic tapes in this system, it is possible to run at suitable tape speeds to continuously transcribe A-D data at 10,000 samples/sec (or 20,000 characters/sec) per channel and then to interplex channels until the limit of 150,000 characters/sec is reached.

Comparison of Analog and Digital Methods of Data Selection. Several specific techniques have been described above for digitizing the time of occurrence of selected features of records. When these are combined with a variety of low-pass, high-pass and band-pass filters, we have a convenient method of minimizing the data sent to the central processor. For instance, a single axon firing at 100 spikes/sec requires 20,000 six-bit characters of information per second when the data are sent to the central computer via the A-D mode, whereas it requires a mere 300 characters of data per second when only the time-of-event information is transmitted. Even a multiple recording of perhaps five axons in a fiber bundle requires only 1500 characters/sec in the time-of-event channel for an average of 100 spikes/sec/fiber instead of the 20,000 required by the A-D mode.

Programs for Data Processing. The usefulness of such a data-processing system as described above is determined by the availability of data-transmission control, data analysis, and data-presentation programs. A very flexible but efficient programming system has been developed to accept the most general sequence of records and to select any combination for analysis after proper scaling factors have been applied. This system has been designed to display information quickly on a graphical plotter so that the experimenter can guide the course of the test. This is an important feature of the system. The visual examination of raw data does not always provide adequate information for such decisions, but the graph-plotting programs assist the experimenter by displaying processed data in a wide variety of formats.

The specific data-analysis routines thus far developed include averaging, standard-deviation calculation, auto- and cross-correlation analysis, Fourier analysis and synthesis, and the compilation of histograms. A series of data-extraction techniques employing sampled-data theory are also employed; these are equivalent to digital filtering and are used for selecting specific frequency components of a record.

Optomotor Responses of the Housefly (*Musca domestica*)

The system described above has been developed in conjunction with the research program on visual perception, which at present consists of three main projects: (1) a study of human visual perception and eye movements, (2) a study of the detailed functional relations between light stimulation of individual ommatidia in the eye of the horseshoe crab *Limulus*, and the neural responses as far up the optic tract as possible, and (3) a comprehensive investigation of the optomotor responses of the housefly, *Musca domestica*. In addition, C. A. G. Wiersma is conducting a separate study of the neurobiology of crustaceans. This work employs analog tape recording and is instrumented so that the tapes may be processed through the computer system. A discussion of the insect optomotor response studies will be used in this paper to emphasize the importance of more precise experimentation guided by systems-analysis concepts, and to illustrate the use of the data-processing system.

The investigation by Reichardt and Hassenstein of the optomotor responses of the weevil *Chlorophanus* has demonstrated the power of systems-analysis techniques in research on the nervous system. The reaction studied in this case is relatively simple, however, and it was considered desirable to attempt to extend the work to a somewhat more complex system. Reichardt (1961) concluded that the information used by the weevil to determine its turning tendency when following revolving cylindrical patterns is essentially a first-order autocorrelation between the outputs of pairs of adjacent ommatidia: the magnitudes of the turning tendencies for most steadily revolving stimuli were constant, and thus one number was the data needed to describe the input-output relation for each stimulus.

The housefly also has a series of reactions that orient it to specific light patterns when in flight. For example, if a fly is fixed at the axis of a vertical-striped rotating cylinder and is able to operate its wings freely and to rotate, it will turn with the cylinder so that the temporal variation of the light stimulus is reduced to a minimum in any region of the visual field; it will also turn in response to fixed spots of light suddenly turned on or off.

Our program consists of the measurement of these over-all system performance characteristics and their integration with histological and microprobing techniques in an effort to obtain detailed knowledge of one aspect of visual perception in this insect. Histology indicates the geometrical arrangement of the ommatidia in the eye: microprobing records the signals generated in these ommatidia and in ganglia cells just below the receptor cells. In addition, it may be possible to record the motor-nerve control signals to the wings. But the over-all response characteristics can also provide information about these stages, particularly if precise stimulation patterns designed to test specific parts of the system can be applied and the flight responses reliably determined. This phase of the project was started about one year ago, before the biological-control and data-processing system was put into operation. The response of the fly to a rotating striped pattern proved to be more complicated than that of the weevil, with a transient component, a steady-state level, oscillating components at wing-beat frequency, responses apparently correlated with the movement of each stripe of the drum, and many more whose origins could not be identified. Representation of this response by a single number, as had been done in the case of *Chlorophanus* by Reichardt, might therefore be totally inadequate. The following sections indicate how the biological data-processing system has been used to identify each component of the responses.

Insect Flight-Torque Measurements. It was decided to study this response characteristic as an "open-loop" experiment in which the insect is held rigidly in position while its flight torque is recorded continuously as a function of time.

A small cardboard tab is glued to the back of the fly, taking care that the glue bridges the head and thorax and so prevents relative motion of the two. The

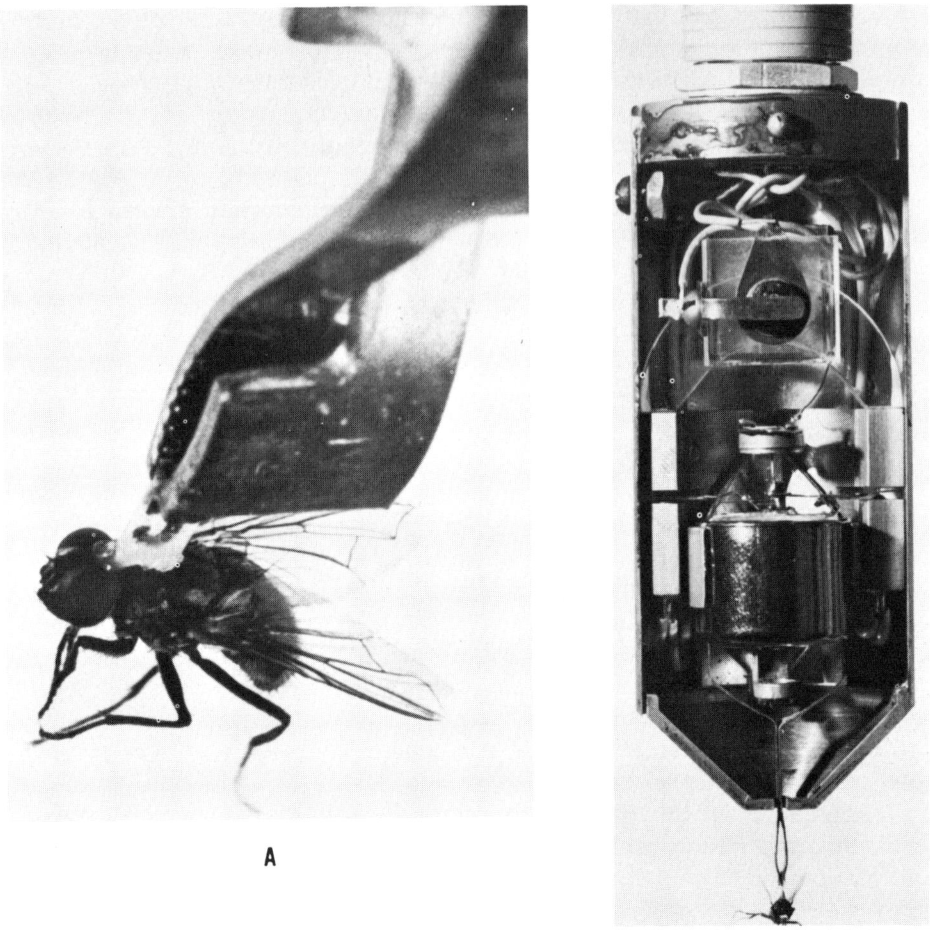

A

FIG. 7. A. Cardboard tab waxed rigidly between head and thorax of the fly. B. Fly mounted in galvanometer torque meter.

B

tab is then fixed to the moving element of a torque meter working on the galvanometer principle. Rotational motion of the meter coil is sensed by a pair of photocells, and the difference signal controls the current through the galvanometer coil so that it remains nearly stationary. The restoring current through the coil is a measure of the deflecting torque (usually about 1 to 10 dyne-cm) which the yaw-motion of the fly applies to the meter. The fly-mounting technique and torque meter are shown in Fig. 7.

Visual examination of successive flight-torque records for repetitions of the same stimulus reveal gross differences in the fine structure of the response; large, apparently erratic fluctuations are present with frequency components from about 7 cps down at least to 0.01 cps. There is another band between 120 and 150 cps, which can be identified with the wing-beat frequency.

Figure 8A shows the high-frequency character of a typical flight torque record. The torque-meter control-circuit response was designed with a sharp

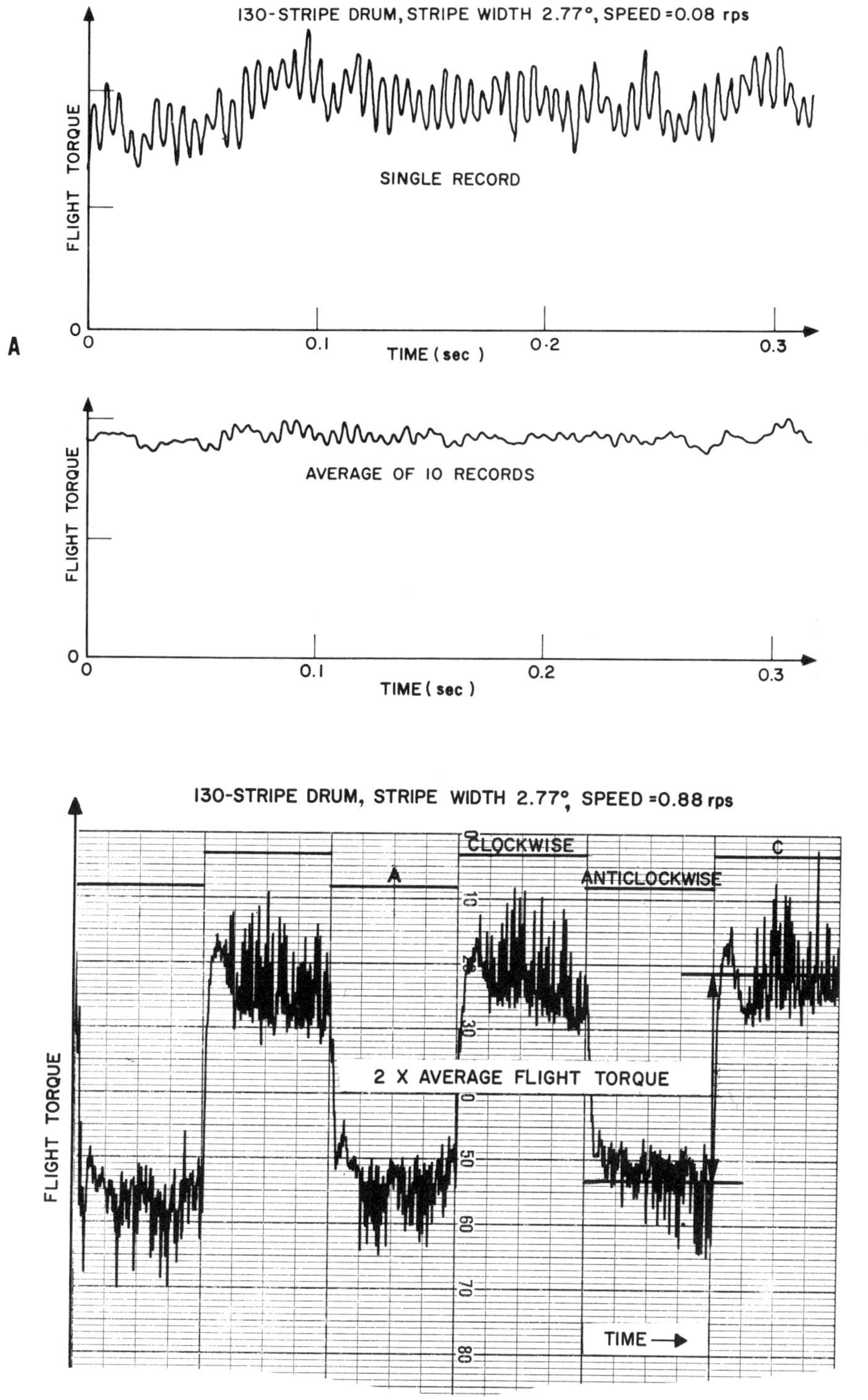

FIG. 8. A. High-frequency flight torque records showing 140-cps wing beats. B. Low-frequency flight-torque records showing response to reversal of rotating drum.

cutoff above 15 cps; thus it only prevents fly movements up to this frequency. The 120-cps to 150-cps component of the flight torque induces very small yaw motions of about 0.06 deg (which have been studied by other techniques), and these, in turn, generate an induced emf in the output circuit enabling us to measure this high-frequency component. A study is in progress of the effect of the resulting high-frequency head motion on the fly's responses to the lower-frequency stimuli. Tentatively, we believe these effects are negligible.

For the remaining studies reported in this paper all frequencies above about 20 cps were removed by analog filtering. A typical record of this type (obtained on a chart recorder) for a black-and-white striped cylinder and a fixed rotational speed is shown in Fig. 8B. Here the drum is rotated first in one and then in the other direction to eliminate slow-torque drift effects. It is apparent that the response is still "noisy"; however, a single number representing the average response to 60 sec of constant stimulus can be derived, as illustrated in Fig. 8B, and these averages can be used to examine one form of input-output relationship.

Responses Characterized by a Single Mean Number. Single characteristic numbers obtained from records as described above can be used to obtain the speed-torque curves shown in Fig. 9. Initial analysis of these records from a variety of cylinder patterns and speeds immediately indicates that the Reichardt model of the optomotor responses in *Chlorophanus* is not adequate in this case. For instance, Fig. 9 shows a response curve for a cylinder with equal-width black and white stripes, each spanning 1.8 deg. Such a moving pattern may produce a negative torque reaction because the angular spacing between ommatidia is similar to the spacing of the stripes on the drum, causing a stroboscopic effect producing apparent motion in the wrong direction. This is shown to be present in Fig. 9 for drum speeds between 0.00003 and 0.01 rps. The

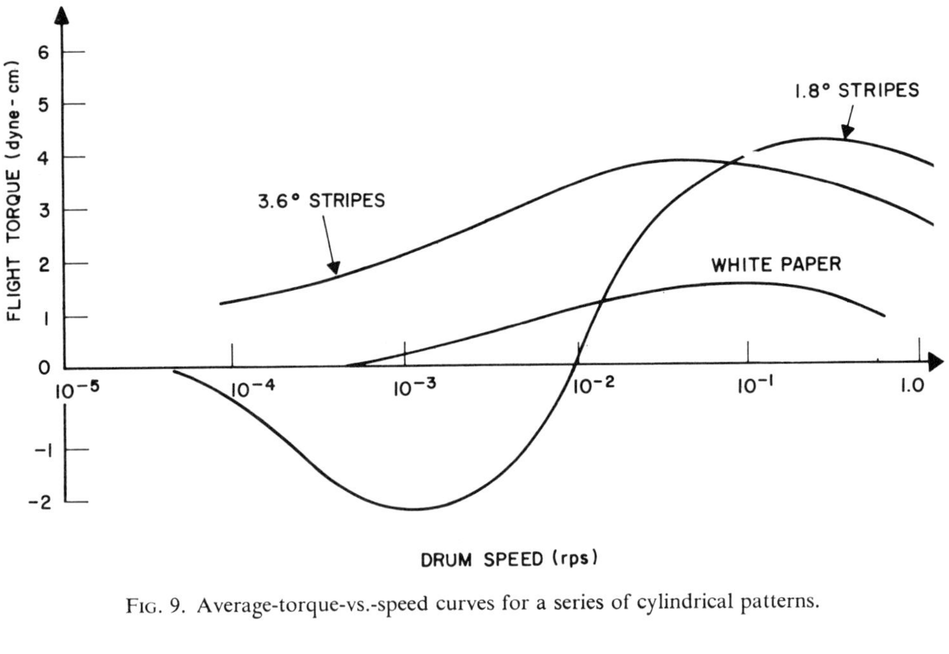

FIG. 9. Average-torque-vs.-speed curves for a series of cylindrical patterns.

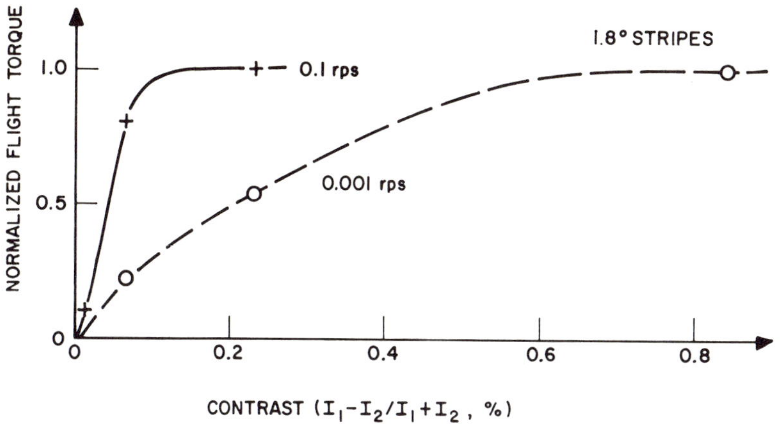

FIG. 10. Saturation effects with variation of pattern contrast (1.8-deg stripes).

Chlorophanus model predicts that the reaction should be negative at all speeds, but in this case we see that for drum speeds above 0.01 rps the torque reaction becomes positive. Reichardt's model assumes equal ommatidial spacing, complete identity of stripes in the pattern, and interaction between neighboring or near-neighboring ommatidia only. A clue to the change required in the model is given by the reaction curve of Fig. 9 for plain white paper, which shows an appreciable positive reaction at about 0.001 rps, the speed at which the response to the 1.8-deg pattern begins to increase rapidly. White paper contains a certain amount of fine-grain low-contrast detail below the resolution limit of the fly's eye and also larger areas of slightly varying luminance, hue, and texture, and it is probably these very-low-contrast patterns of longer wavelength which produce the large reaction components at the higher speed range of the drum.

This effect is further illustrated by the curves of Fig. 10. These show a highly nonlinear and rapidly saturating response to increased contrast of the 1.8-deg striped pattern in the high-drum-speed region but a much more linear effect at the lower speed range. It must thus be assumed that the saturation shown in Fig. 10 for the 0.1-rps drum speed is produced by integration of response over a range of stimuli which are not effective at the higher speeds. This too can be ascribed to the very-low-contrast longer-wavelength "noise" components in the stimulus.

Detailed Examination of Response. The results described above indicate that the system is complex. Although this may mean that experimentation is more difficult, it also ensures that increased information can be obtained about the biological system provided only that suitable analytical methods are used. First, we see that we must generate precisely defined stimuli with a very low noise content; new techniques for doing this are being developed. Then the many components in the response must be analyzed separately, not lumped together by the averaging process used above. The computer data-processing system has been used to separate the large-amplitude noise in the response records from the truly correlated response patterns by the use of two data-analysis techniques. One of these is the direct averaging of the responses obtained by successive applications of the same stimulus; the other is the harmonic analysis of individual records or averaged records in order to display the frequency components of the correlated response and to provide information on the interaction between these components and noise or uncorrelated responses.

Averaging. The experiment-control mode of the biological-control and data-processing system was used to record the response to repeated application of the same stimulus. Responses were measured at 1-msec intervals, and in the analysis the records were synchronized with each other to the same time-interval accuracy. The type of information extracted by averaging many records in this fashion is illustrated by Fig. 11. This shows an individual response and a stage of averaging for a rotating stimulus consisting of a single vertical white stripe, 17 deg wide, on a black cylinder.

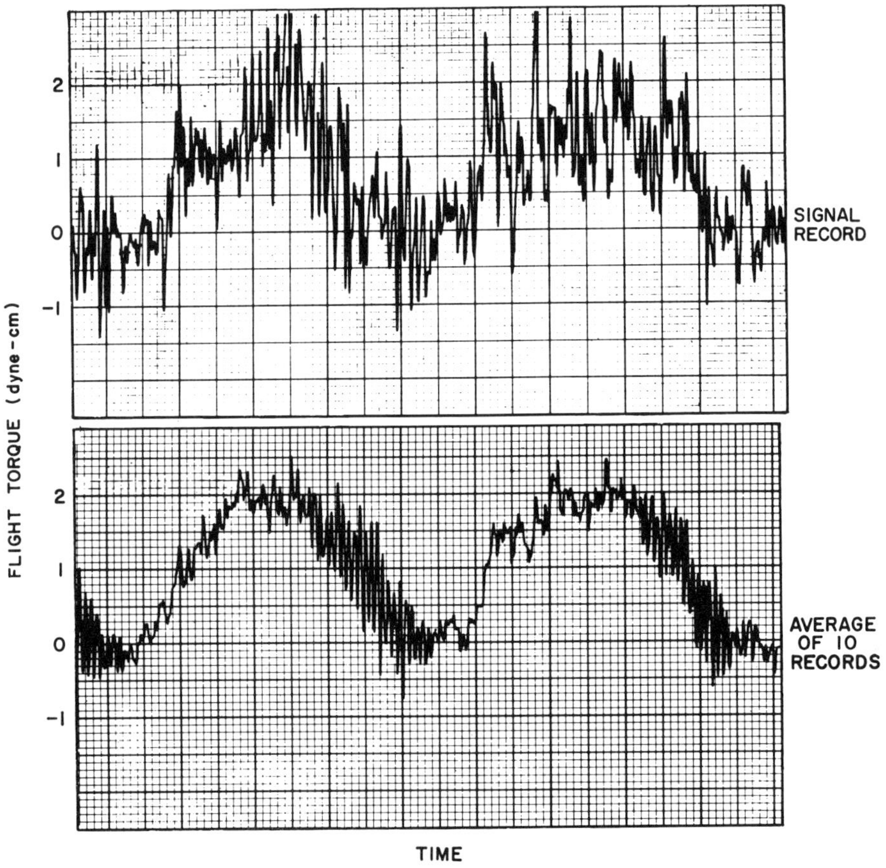

FIG. 11. Response records showing the effectiveness of averaging techniques in the reduction of uncorrelated responses.

It will be noted that in Fig. 11 the averaged response contains both a steady component and a periodic one. The response to the black drum without the white stripe has only a steady component. For a single record this is buried in the noise but extracted by the averaging process. The periodic component of the response obtained after averaging many records contains few of the higher-frequency components present in the individual records. This does not mean, however, that these components are all noise or responses uncorrelated with the stimulus, but merely that they have no fixed phase relationship with the stimulus.

"Hunting" Component of Response. The trace in Fig. 11 representing the average of ten records reveals an interesting "hunting" response in the fly's behavior; this average was actually obtained by selecting records which contained this hunting oscillation to a high degree. The large-amplitude oscillations always occur during the half-cycle of pattern rotation when the stripe is going away from the front of the insect. This component is variable in its amplitude and duration (both for successive applications of the same stimulus and for different stimuli), but its frequency is quite constant and has only ranged from 5 to about 6 cps for some ten flies studied so far. The phase relationship to the stimulus is so variable, however, that the hunting response is considerably reduced by averaging over a large number of samples.

Additional Characteristics of the Averaged Responses. A further interesting feature of the averaged responses for single-stripe stimuli (such as are shown in Fig. 11) is the correlation between the frequency components of the average and the variation of these same components in each individual record. This has been examined in several ways. Figure 12 shows an average plotted directly on two records for comparison. The frequency components of the average curve agree closely with the corresponding components of each record. Note particularly, in Fig. 12, the sudden large-amplitude bursts of the hunting response which end right on the averaged curve. Another indication of the consistency of these averaged responses is obtained by comparing averaged curves generated from different groups of records; very close correlation is obtained for the responses to single stripe stimuli.

Averaged Responses for Uniformly Striped Patterns. We have described in the sections above the response of the fly to a rotating stimulus consisting of a single stripe, and we have shown that this response has a large oscillating component. If the drum consisted of uniformly spaced black and white stripes, we might speculate whether the fly would still show responses to each stripe or whether these responses would be integrated into steady torque. Stimuli have been used in which the basic pattern wavelengths have ranged from 3.6 deg to 45 deg. The amplitude of the periodic component in the averaged records for these stimuli is much smaller than before, and a steady torque reaction is also present, as shown in Fig. 13. This diagram shows that a very small component is still present at the repetition frequency of individual stripes, although the integrating action of the stimulus applied to all portions of the visual field produces a relatively large steady component.

Conclusion. It is the purpose of this paper to describe the philosophy which permits a physicist, mathematician, or engineer to enter the field of biological

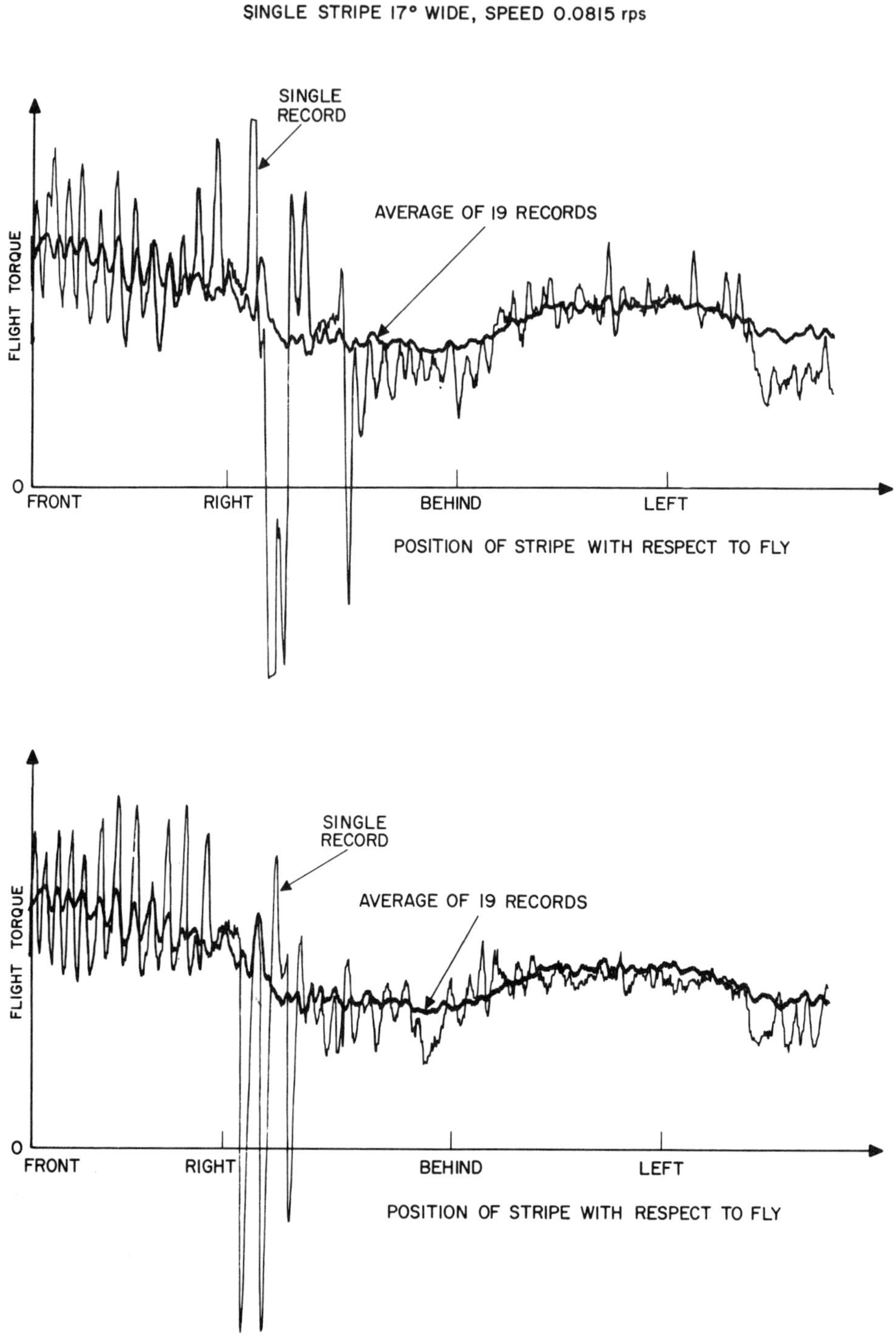

FIG. 12. Comparison of individual records with the averaged response.

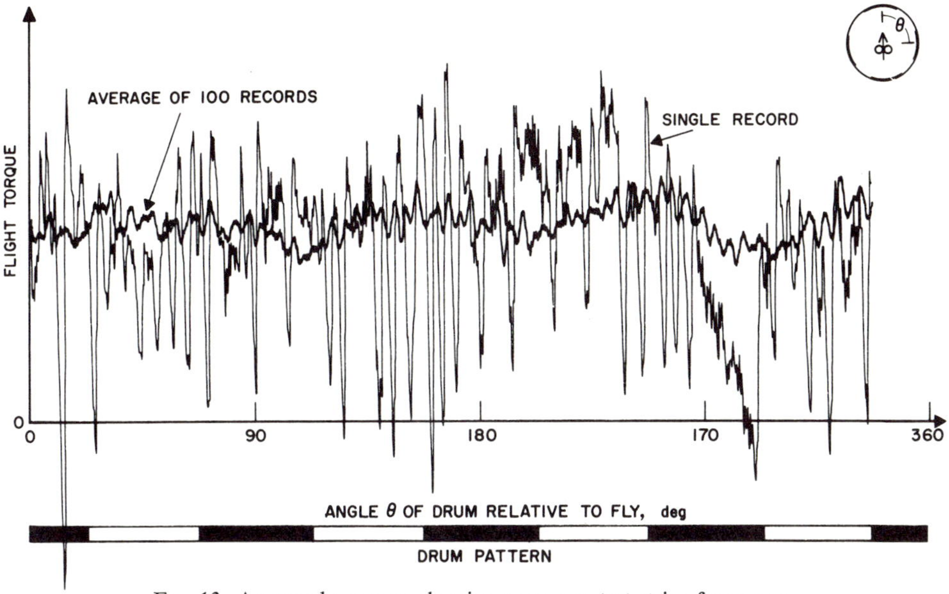

FIG. 13. Averaged response showing component at stripe frequency.

study. Input-output relationships, which can be measured so easily in mechanisms, enable the engineer to say a great deal about what goes on inside the device. Sometimes he can confirm these findings by dismantling the machine or by nondestructive testing; sometimes he cannot, and the input-output relation remains the best available description of the mechanism. The biologist frequently finds himself in the same position, and it becomes logical to attempt to bring over into this field the powerful techniques of systems analysis, information theory, and data processing, which have paid such handsome dividends in the engineering field.

These methods of analysis demand a ready flow of data; we have therefore described a data-collecting and processing system well suited to experimental research in visual perception with particular emphasis on the neural-network-analysis, information-processing, and biological-systems concepts.

As an example of the performance of the system, we might describe a typical experiment on the optomotor responses of the fly. One hundred repetitions of the experiment might take half an hour; digitizing the information (about 200,000 characters) and transmitting to the computer, in the interim system described above, take another half hour; computing the average values and standard deviations and then plotting the results occupies another minute. With the system shown in Fig. 1 the digitizing and transmitting delay is avoided and the result is available about one minute after completing the experiment. Another example might be drawn from our work on human vision. A two-dimensional histogram showing the amount of time that a subject spends looking at any part of a visual display can be plotted on a 100×100 matrix in less than two minutes.

The real test of a system such as this is not the speed with which it performs routine operations but the rapidity with which it advances our knowledge of a particular system. To illustrate this we have described some experiments on optomotor responses in which little more than an averaging process carried out over many repeated applications of the same stimulus has removed the "noise" from the records and has already revealed a number of characteristics of the fly's optomotor responses which had previously remained buried in other uncoordinated reactions.

References

FARLEY, B. G., 1962. Problems in the study of the nervous system, in *Proceedings of the 1962 A.F.I.P.S. Spring Joint Computer Conference*, vol. 21, Palo Alto, Calif.: National Press, p. 147.

HARMON, L. D., 1961. Studies with artificial neurons, I: Properties and functions of an artificial neuron, *Kybernetik*, **1**(2), 89–101.

MINSKY, M., 1961. Steps toward artificial intelligence, *Proc. IRE*, **59**, 8.

REICHARDT, W., 1961. Autocorrelation, a principle for the evaluation of sensory information by the central nervous system, in *Sensory Communication*, W. A. Rosenblith, ed., New York: Wiley, pp. 303–17.

Part II

Theories and Models of Specific Systems

Introduction to Part II

DONALD KENNEDY, *Stanford University*

The second part of this book differs from the first in that it deals with "wet" preparations. Each author of Part II is a biologist committed to asking experimental questions of living organisms. It is the hard fact, however, that only some are also modelers or theorists. Fender, Bliss, Mittelstaedt, and Kunze use cybernetic treatments, and Josephson synthesizes a full computer model, but none of the other contributors proposes a model or discusses a central theory at great length. The first part of this introduction will try to explain why the rest of us are represented in such a book.

Most biologists welcome the attention now being paid to the nervous system by engineer-mathematicians. For one thing, the motivations of the two groups are very similar. Though it is often said that the device-oriented engineer can learn useful new tricks from the study of biological systems, this argument is scarcely persuasive. One suspects that it is the invention of research administrators, who often have to supply the wrong reasons for doing the right thing, and that in reality the biologist and the engineer-mathematician share an interest in nervous systems *purely for their own sake.*

There are several reasons why greater interaction between biologists and engineers should prove profitable. In the first place, the biologist needs help. Neurophysiology is about finished with the taxonomy of action potentials and the ABC's of functional organization; it is moving toward a direct assault on the problem of how unit processes interact to produce behavioral action patterns. As a broad question, this one has many parallels in biology. One of the most distinctive features of living systems is their construction in layers, each one built from units of the one below: molecule to cell, cell to organism, organism to population. Biologists every day face the problem of translation between levels, and must consider the principle of emergence; yet it is surprising how many of us do not deal comfortably with complex interactions between basically similar units. It is with just such transitions that engineers, mathematicians, and physical scientists have always been able to help biologists most. Within the past decade, for example, the use of mathematical models has revolutionized the study of population genetics and population dynamics.

The nervous system is a prime target within the organism for this sort of attack. The practitioners of molecular biology, fresh from spectacular successes at the cellular level, are confident that the problem can be solved by dissection.

According to Joshua Lederberg, for example, "Until now . . . the phenomena of learning (in its neurobiological aspects), neoplasia and senescence could be approached only at the most superficial level. They are about to be transformed in the sense that genetics has been, as epiphenomena of protein and nucleic acid synthesis." But the reductionist approach will not provide all the answers. If current efforts were to give us overnight a macromolecular coding system for memory, we would be all dressed up with no place to go. Somehow, a molecular code must be reached by an input pattern involving many cells; somehow, it produces an equally complex output pattern which is behavior. We must know what these patterns are and how they come to have significance. In the end, that task falls to the student of interactions.

It is for this reason that the coalition between biologist and engineer-mathematician is important. From it, the biologist needs very much more than sophisticated instruction in how to handle data. Though such technical contributions as pulse-interval histogramming and autocorrelation methods have been quickly appreciated and widely used, the biologist needs the *active* participation of people who are versed in the theory of signal processing and who simulate by habit.

It is certainly legitimate to ask whether such "poachers" are not actually ready to take over the entire preserve, converting the former owners into a class of touchingly old-fashioned ranger-naturalists whose main task is the collection of new specimens for the theorists to ponder. There are several reasons for believing that biologists must play a wider role than that. First, their education, though it needs remodeling in some respects, has given them a respect for organic diversity which is crucial to the understanding of living machinery. The biologist has a healthy skepticism of universals; he knows that organisms come up with an embarrassing number of adaptive solutions when challenged with a problem in evolutionary engineering.

Second, the biologically trained neurophysiologist can provide a working knowledge of the constraints which the unit properties of neurons and junctions place upon possible models; in other words, he can keep models in some sort of touch with the systems they are meant to model. He can also apply brakes to other enthusiasms; for example, theoretical analyses of central nervous function frequently forget that behavior is the real business of the nervous system. As Pringle points out in Part I, models can become uselessly esoteric if they ignore the output. One reason for the great success of the cybernetic approach (championed here by Mittelstaedt) is that the analysis *begins* with the output.

Third and most important, the man who works with the organisms defines and establishes the need for models. Modeling is by definition an imitative activity; models cannot be conceived or constructed except from pre-existing data. Ideally, the data gatherer would build his own models; but today most of us need and want help. Perhaps it is this need that best explains why this book is not quite a single unit, yet not quite two books either.

The first contribution in Part II, by Mittelstaedt, employs a real model. In keeping with the approach founded by the late Erich von Holst, to whom the paper is dedicated, Mittelstaedt shows how cleverly conceived behavioral experiments can be used to decide between alternative control mechanisms suggested by a cybernetic model. Specifically, in situations where organisms orient themselves in some quantitative way with respect to some stimulus variable or variables, Mittelstaedt asks what rules govern the central processing of the afferent signal which measures deviation from the "normal" position. Though behavioral experiments can provide important answers to this kind of question, the answers are peculiarly formalistic: they do not tell the curious physiologist what is in the black box, or even where it is; they only sketch limits for its operational properties.

Where Mittelstaedt's main concern is for the output, Cohen's is for the input. The difference between the two papers does not stop there; Cohen's is as cellular as Mittelstaedt's is systemic. Cohen emphasizes the problem of unit-to-system transition by referring to a variety of sensory processing mechanisms—range fractionation, opposing receptors, phasic vs. tonic channels—which depend upon how the central nervous system treats many sensory channels together and integrates the information contained in them. (Many of the illustrations are taken from a new receptor preparation, the crustacean myochordotonal organ, which biologists will welcome with great enthusiasm.) Cohen reminds us that the properties of the input result in active filtering of information, and thus have much to do in defining the final behavioral output.

The next two papers, by Kunze and Fender, deal with the use of cybernetic models in the interpretation of specific behavioral systems. Kunze, in his analysis of the control of optic nystagmus in a crab, defines some unexpected regional limitations in stimulation of the eye with moving patterns to evoke eyestalk movements. He interprets these as indicating nonlinearities in the central processing mechanism. Fender, working with the tracking control system in the human eye, uncouples the normal feedback loops by returning false results. Among other things, this paper—which proceeds in an appealingly direct way to set up and progressively modify a model—demonstrates that central predictive capacity may play a role in some neural control systems.

The brief treatment by Bliss suggests some alterations in the celebrated Reichardt-Hassenstein model of optomotor responses in beetles. The work reported is possibly without precedent, in that an accomplished mathematical system theorist has evaluated a model, made predictions from it about novel situations, tested these by *biological* experiments, and used the results to suggest revisions in the model.

Wilson and Hoyle both concern themselves with the motor side of the same animal, the locust, and both display a preference for combining behavioral analysis with single-unit recording. Yet they almost never view the same output, Wilson looking at flight and Hoyle for the most part at more pedestrian behavior. Both, in a way, emphasize the diffuse nature of output control. Hoyle shows that certain movements are accompanied by synchronous—though subthreshold—frequency changes in motor neurons going to functionally unrelated muscles. This provides a substrate for the evolution of novel behaviors, and in addition suggests a rather diffuse system of central driver neurons; the results may help to explain such behavioral phenomena as the *magnet-effekt* between peripheral rhythms originally described by von Holst. Wilson supports the concept of the organized center for the locust flight system and reports elegant experiments which show that sensory feedback is important for the center in a quantitative rather than a temporally or spatially specific way. He emphasizes, as does Hoyle, that the central connections are probably diffuse ones, rather than rigid reciprocal hookups between final motor nerves; in locusts the same sets of muscles may be used for walking and flying but have opposite agonist/antagonist relationships for the two behaviors.

Vowles, and Kennedy and Mellon, have both dealt with the central nervous system of arthropods. Vowles, combining single-unit recording with the local stimulation techniques developed for insects by Huber, attempts to give a broad picture of regional function in an insect brain. Though tentative in some respects, Vowles' hypothesis of the function of the mushroom body is a good example of physiological identification of "black box" components. Kennedy and Mellon, by contrast, are concerned with finer details of integrative performance at the level of a single kind of central neuron. Hopefully their data can be used to explain certain complex unit responses in sensory systems, such as those recorded by Vowles and others; they also might represent real-world versions of such pulse-coding models as those of Crane and Reiss in Part I. The approach has the defect, however, of being as yet quite uncorrelated with behavioral output.

In the final contribution, Josephson reports the results of a computer simulation designed to help explain the operation of a "simple" nervous system—the coelenterate nerve net—using some of its known biological properties. The simulation has the desirable outcome of specifying some limitations on future biological investigations, and also provides the provocative suggestion that randomness of connectivity may be a property of nervous systems with which the orderly investigator must simply learn to live.

Perhaps it is appropriate that the symposium concludes with a paper which uniquely comes out of a collaboration between "Part I" people and "Part II" people. Some future collection may consist entirely of such efforts; at the very least, it is apparent that active collaboration between biologist and engineer can produce vigorous hybrids at the output.

12. Basic Solutions to a Problem of Angular Orientation

H. MITTELSTAEDT, *Max-Planck-Institut für Verhaltens-physiologie*

In memory of Erich von Holst (d. May 26, 1962).

The Problem

An oriented organism maintains a spatial relation to its environment which is determined by the functioning of the organism itself. The spatial relations which will be treated here are the *angular* relations of the organism's body to a spatially arranged stimulus or field of stimuli. Most frequently, in the oriented state, one of the main body axes points in the direction of the orienting stimulus. In bilateral animals this is usually the longitudinal axis if the animal orients in the dorso-ventral plane, the dorso-ventral axis if it orients in the frontal plane, and either one if it orients in the median plane. Even if external conditions remain unchanged, however, *deviations* from the "basic position," as the normal angular relation will henceforth be called, are maintained for various amounts of time. This angular-orientation capability is of prime importance to any freely mobile animal. It is the basis of an astounding array of orientational performances, from the seemingly simple tendency to head into a light source ("positive phototaxis") to the apparently sophisticated ability of a bee to transmit the compass bearing of a food source by indicating its angle to the sun while allowing for the sun's azimuthal velocity at that particular time and place ("sun-compass").

For the sake of greater precision, the theories to be developed will be restricted to a particular, though ubiquitous, class of orientational capabilities. It is defined by the following criteria:

1. Orientation is maintained against extraneous disturbances by continuous correcting movements.
2. In organisms oriented in the basic position, the strength of this correcting effort (measured as the force exerted by the effectors, the velocity or acceleration of body rotation, or some other relevant variable) is proportional to the sine of the deviation from the basic position.
3. The organism is able to maintain self-determined deviations from the basic position (to be called "commanded deviations") throughout the entire range of 360 deg.

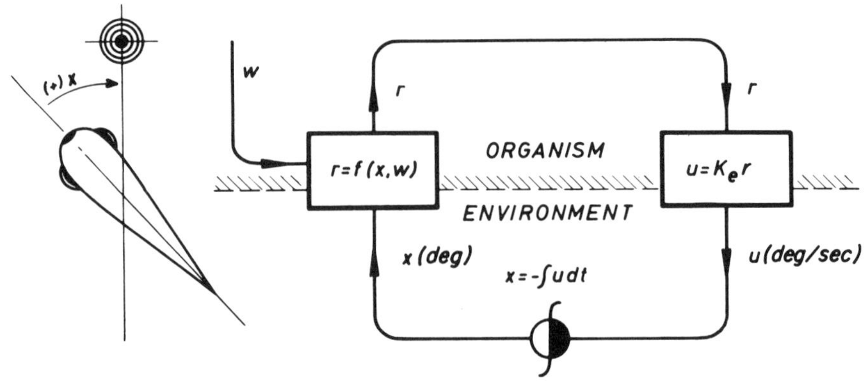

FIG. 1. Information-flow diagram of a control pattern which represents the basic functional properties of the analyzed class of orientation systems. Boxes represent information-processing subsystems, lines represent the processed variables (*not* connecting cables or channels), and arrows on the lines indicate the directions in which changes of the variables cause effects. The dashed and fringed line separates those variables considered to exist within the organism from those outside, and thus indicates where the flow of information enters or leaves the organism. The half-black circle divided by an integral symbol indicates the relation between x and u, namely, reversal of sign and temporal integration.

4. The receptors which collect the necessary information about the angular relation to the orienting stimulus are, in the final steady state of the system at least, in a fixed position relative to the part of the body which carries the effectors.

From these four criteria it follows that the over-all pattern of the information flow which yields the orientational capability is a continuously operating negative-feedback loop whose reference value can be set at any angle (Fig. 1). Its input variable ("control variable") is defined as the deviation x of one of the main axes from the source or field of stimulation in the plane of orientation considered. The "correcting effort," henceforth called "turning tendency" r, is a function of the control variable *and* a central-nervous system variable w, which sets the reference value of the loop in a way yet to be discussed. The turning tendency r eventually determines the output variable. In any concrete case the dimension of the output depends on the physical mechanism by which the effectors operate. For the sake of convenience it will henceforth be assumed to be an angular velocity u. The sign of the deviation x will be called positive (0–180 deg) if the source is to the right of the organism's main axis; the sign of the output velocity u will be called positive if the organism turns to its right (see Fig. 1). Thus the prerequisites from which this account starts can be formulated as follows. We have

$$u = K_e r, \tag{1}$$

where K_e is a velocity coefficient, and

$$x = -\int u \, dt; \tag{2}$$

thus

(3) $$x = -\int K_e r \, dt,$$

so that, at the final steady state of the feedback process, the turning tendency r will be zero. The value of the deviation x at steady state depends on the function that determines the turning tendency r from the deviation x and the reference-setting variable w:

(4) $$r = f(x, w).$$

It is the aim of this account to elucidate this function and the information-flow pattern by which it is realized. It will be noted that this further restricts our topic, because it is only this final steady state and the "structure" of the system which will be dealt with, its dynamics being left to future endeavors.

Solutions

The first systematic attempt to determine this function was made by von Holst, in 1950, for the gravity orientation of fish. He had the ingenious idea of measuring the system by balancing the gravity loop against the loop of the "dorsal light reaction." This reaction tends to turn the dorso-ventral axis of the fish into the center of the total light distribution and thus normally operates in harmony with the gravity loop, which tends to turn the dorso-ventral axis, dorsal fin up, toward the zenith. Von Holst put the two systems into conflict by setting the center of illumination at various angles to the zenith. The animals, instead of choosing between the two sources, assumed a compromise position which was then measured. This compromise position depended not only on the angular deviation α from the zenith ("zenith angle") and the angular deviation β from the center of illumination ("light angle"), but also on the *intensity* of the light and the *strength* of the apparent gravitational force (which was varied by means of a centrifuge). One of the decisive results is shown in Fig. 2. If the light angle β and the light intensity are kept constant, the sine of the zenith angle (α_2) after doubling gravity is one-half of the sine of the zenith angle (α_1) at 1 g, or, in general,

(5) $$F_1 \sin \alpha_1 = F_2 \sin \alpha_2,$$

where F_1 is linearly proportional to gravity in the first experiment (yielding α_1), F_2 in the second one (yielding α_2). Von Holst's main conclusion about the functioning of the fish labyrinth is rather convincing and has been confirmed by direct recordings from the afferent nerve of the utriculus otolith. The afferent output of the gravity receptor [to be called the "afferent variable" (α_a)] is linearly dependent on the shearing component (s in Fig. 2) of the weight of the utriculus otolith, whereas its pressure component (d in Fig. 2) is without effect. By the same method von Holst also elucidated the dorsal light-orientation system; the result is shown in Fig. 3. Under the conventions introduced above, it can be formulated as follows:

(6) $$r = L \sin \beta + F \sin \alpha,$$

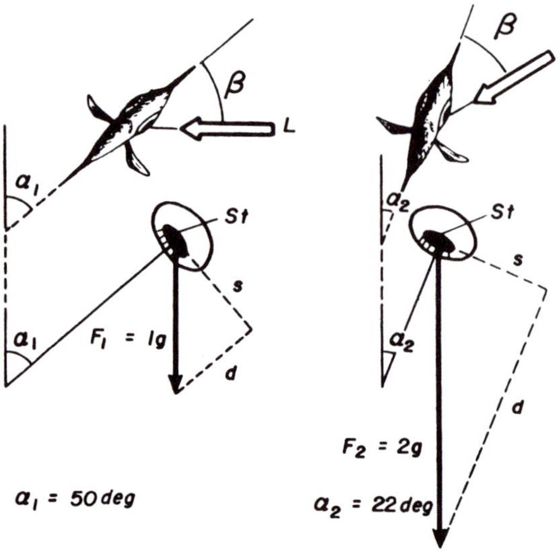

FIG. 2. Experiment of von Holst (1950). The fish swims against a laminar stream. (It is depicted from a head-on viewpoint.) While the light angle (β, relative to the fish) and the light intensity are kept constant, the apparent gravitational force is doubled by means of a centrifuge. The inset shows the position of a utriculus otolith St; s and d are the shearing and pressure vectors, respectively, of the force F exerted by the utriculus otolith (modified after von Holst, 1950).

where L is a coefficient proportional (logarithmically at high intensities) to the light intensity.

In all these experiments von Holst took pains to ensure that the animals tended to assume the basic position if exposed to either light or gravity alone. If the light was below threshold or turned off ($L = 0$), the fish behaved according to the equation

(7) $$r = F \sin \alpha;$$

i.e., they came to rest ($r = 0$) at $\alpha = 0$. If gravity was "turned off" (by surgically removing the utricles), they behaved according to the equation

(8) $$r = L \sin \beta;$$

i.e., they came to rest at $\beta = 0$. Consequently, these equations and the information-flow pattern of Fig. 3 pertain only to fish in which there are no internal actions ("commands") that could set the reference value of the loops differently. That is, $w = 0$ or, on the average at least, w is distributed around zero ($\int w_i = 0$). It is well known, however, that "commanded" deviations from the basic position occur in fish over appreciable amounts of time. In some fish (*Poecilobrycon eques*, *Chilodus punctatus*) the longitudinal axis (and the plane of the utricles) deviates on the average by amounts up to 50 deg from the horizontal plane throughout their lives. The question then arises how "commands" operate on the gravity loop. In 1950 von Holst and the author suggested that the central nervous system generates a message of the same dimension as the afferent variable and *adds* this message, the "command" w , to the afferent variable x_a. The principle of this solution is well known in feedback engineering as the setting of the reference value of a feedback system by additive superposition. The underlying functional organization can be expressed in the symbolic language of control theory as shown in Fig. 4. This solution of our problem

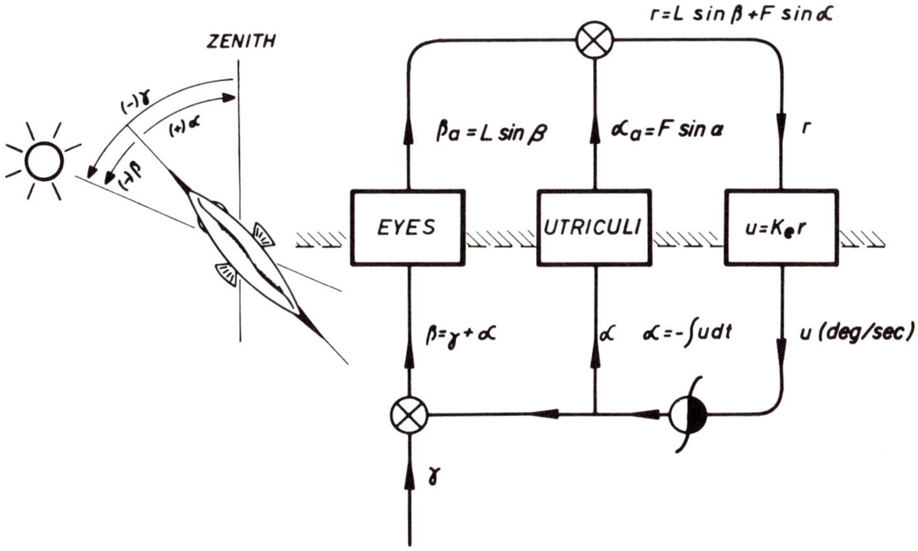

FIG. 3. Control pattern of gravity and light orientation in fish, if commands are absent. Angle α is the angle between dorso-ventral axis and zenith. Angle β is the angle between this axis and the light source. Angle γ is the angle between light source and zenith. All signs are as indicated in parentheses in the inset. (The fish is seen from behind.) Crossed circles indicate that variables are added, or if quadrants are black, subtracted. Note that in the final steady state the eyes of the fish are always in an identical (symmetrical) position relative to the body.

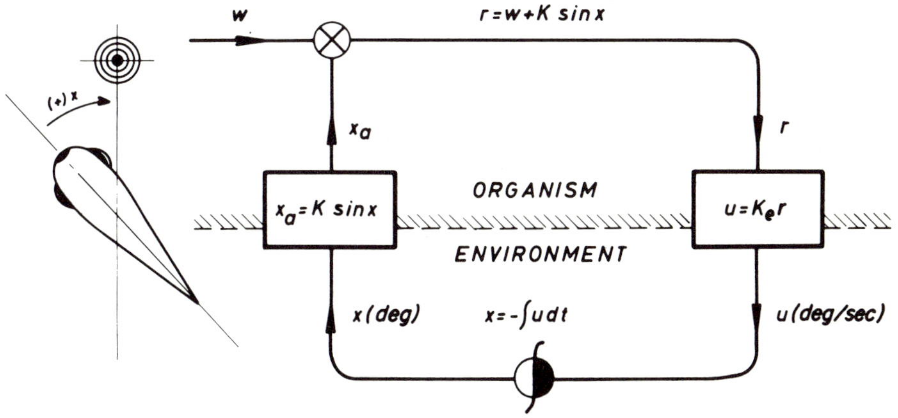

FIG. 4. Control pattern of a theory (von Holst and Mittelstaedt, 1950) of how commanded deviations are produced by the nervous system.

will henceforth be called theory (I). The function $x = f(x, w)$, according to this theory as expressed in Eq. (9), is thus

$$r = w + K \sin x, \tag{9}$$

or for the simultaneous operation of light and gravity orientation,

$$r = w + F \sin \alpha + L \sin \beta. \tag{10}$$

The formula yields the operation of light or gravity alone if we set F or L, respectively, to zero.

The precise formulation of the theory (I) as expressed by Eq. (9) makes rigorous experimental testing possible. One immediate suggestion for a test is to increase gravity in an animal displaying a reliably constant commanded deviation. Schöne (1957) found that crayfish, if released near the surface of a homogeneously lit aquarium, swam toward the bottom at a rather constant deviation of the longitudinal axis below the horizontal. If gravity was increased, the deviation decreased according to $F_2 \sin \alpha_2 = F_1 \sin \alpha_1$, indicating that the deviation was indeed caused by a constant turning tendency additively imposed on the loop, exactly as Eq. (9) predicts for a constant command. However, in the lifelong-deviating fish mentioned above, Braemer (1958) found significant discrepancies between the recorded effect of increased gravity and the one predicted by Eq. (9).

Another remarkable consequence of theory (I) [Eq. (9)], one which can be readily tested, becomes clear upon examination of a plot of the turning tendency r against the deviation x at various commands w, as shown in Fig. 5A. If the command w is zero, any deviation from $x = 0$ and $x = 180$ deg will evoke a turning tendency r which is so directed as to return the animal to $x = 0$ deg by the shortest way possible. This $x = 0$ deg position will therefore be called the "stability position," and the other zero-crossing (here at $x = 180$ deg) will be called the "lability position." If a command is imposed on the loop, the stability position does indeed shift as desired. The lability position, however, is displaced in the opposite direction. Furthermore, the steepness of the slope dr/dx at stability position, which naturally determines the sensitivity of the loop in detecting deviations, declines with increasing values of the command w until, at a commanded deviation of 90 deg, it is zero. Any attempt to command the system beyond $x = 90$ deg must result in permanent rotation. That obtuse angles to the source of stimulation cannot be commanded in this way is not as great a handicap, however, as the decline of sensitivity and the ipsilateral shift of lability position, because it could be overcome by reversing the sign of the feedback, a well-known maneuver in phototactic arthropods (Jander, 1957).

Phototaxis offers a rather elegant test of the expected drastic change of lability position. Most insects tend to maintain a commanded deviation relative to a light source for some period of time ("photomenotaxis"). If the light source is then experimentally displaced, the insect resumes its previous commanded deviation. According to theory (I), however, the "rule of the shortest way" back into that commanded position no longer applies (see Fig.

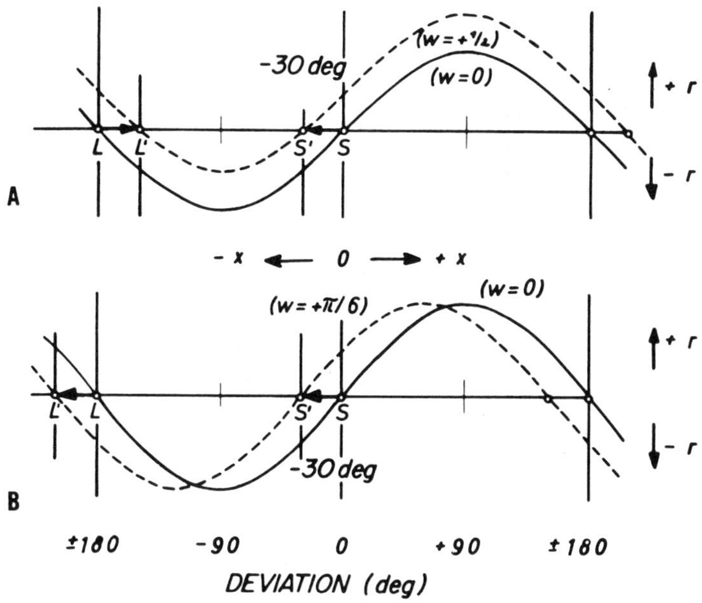

FIG. 5. The turning tendency r (causing right turns if positive) is plotted against the deviation x at $w = 0$ and at $w = +\frac{1}{2}$ in A and at $w = +\pi/6$ in B. The effect of such a command (dashed curve) according to theory (I) is shown in A. The effect according to theories II and III [Eqs. (11) and (13)] in B. Note that the stability position S at which the animal eventually comes to rest shifts in the same way (to S') in both A and B, whereas the lability position ($L \rightarrow L'$) shifts in opposite directions.

5A). In ants Jander (1957) has indeed found shifts of the lability position in the direction expected from theory (I). It is doubtful, though, that their amount, which Jander's method could not measure, is as great as is demanded by (9). In other arthropods, however (von Buddenbrock, 1935, 1952), the rule of the shortest way apparently holds also for commanded deviations.

In all cases, then, where the rule of the shortest way is valid throughout, where the sensitivity of the loop is the same at all commanded deviations, and where obtuse deviations can be reached without reversal of feedback, theory (I) must be rejected. Two alternative solutions have been offered so far (Mittelstaedt, 1961, 1962). According to theory II [Eq. (11)], the afferent variable x is proportional to the angle x of the deviation instead of to its sine, and the necessary sine transformation is performed *after* command w has been added to the afferent variable. The diagram of Fig. 6 shows the functional organization of the second solution. The coefficient relating x_a to x must be constant and will, for convenience, be assumed to be unity. Thus the function of interest, $r = f(x, w)$, according to theory II, is

(11) $$r = K \sin (w + x),$$

or, for the joint operation of gravity and light,

(12) $$r = F \sin (w + \alpha) + L \sin (w + \beta).$$

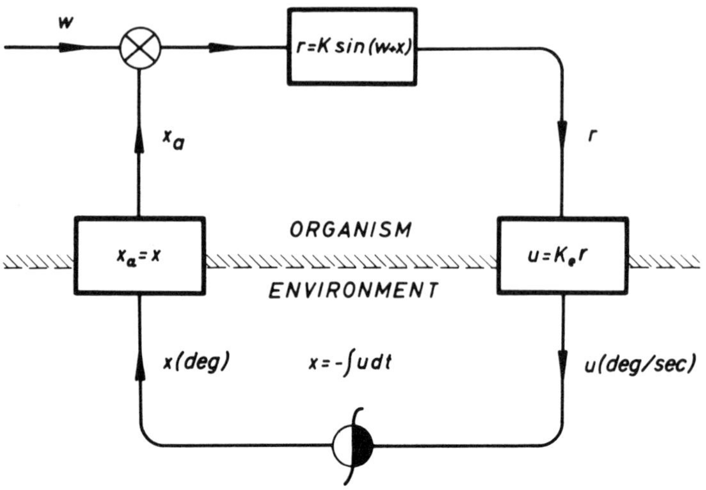

FIG. 6. Control pattern of theory II (Mittelstaedt, 1961, 1962). The command w is added to an afferent variable (x_a), which is proportional to the angle x itself instead of to its sine. The sine transformation subsequently takes place in the central nervous system.

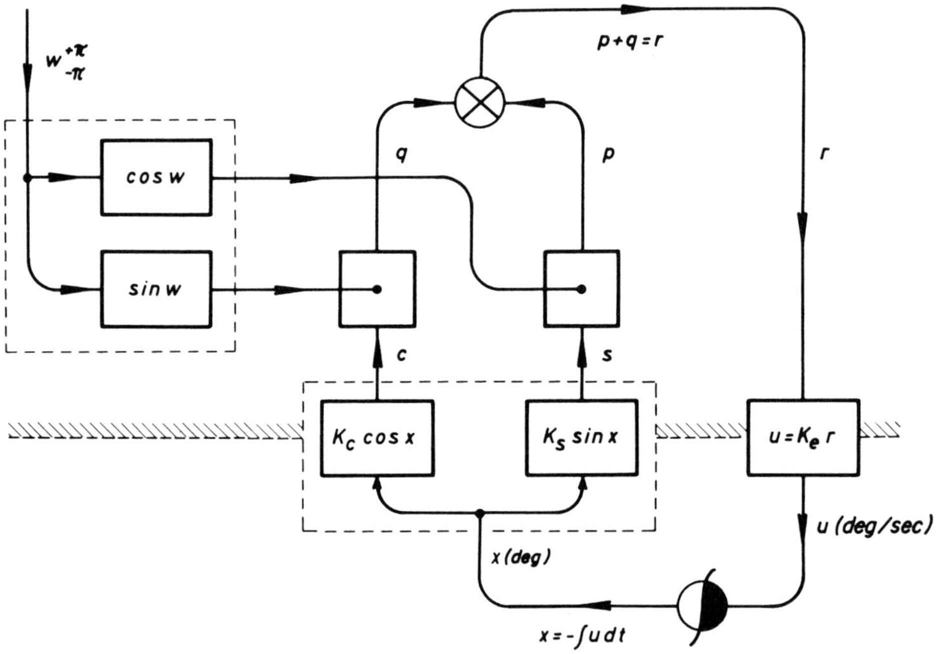

FIG. 7. Control pattern of the bicomponent theory (Mittelstaedt, 1961, 1962). The command acts reciprocally on the sine and the cosine components (s and c) of the deviation x, which are then added. The command is either equivalent to one angular variable w, which is then split into sine and cosine components, or it is represented as a pair of variables which are firmly coupled in an equivalent relation. Square boxes with a central dot indicate multiplication of the input variable connected to the dot (e.g., $q = c \sin \alpha$). See Eq. (13).

Again, the isolated effects of light or gravity are obtained by setting F or L to zero.

As an alternative, I have suggested the "bicomponent theory" (Mittelstaedt, 1961, and Fig. 7). In this theory, (III), the control variable is split into a sine *and* a cosine component and then modulated reciprocally by the command in a fashion equivalent to Sherrington's principle of reciprocal innervation of limb muscles. The command activates the sine component while it inhibits the cosine component and vice versa, and the animal turns until the sum of the two components is zero. If the inhibition of the sine component is total (at w_{max}), this seesaw is completely dominated by the cosine component, which forces the animal into a deviation of 90 deg. If both components are equally strong, the animal comes to rest at 45 deg, and if, at $w = 0$, the inhibition of the cosine component is total, the animal, under the exclusive action of the sine component, obeys the rule elucidated by von Holst. The process of reciprocal modulation works ideally if the effect of the command is equivalent to the split of an angular variable into a sine and a cosine component, which are then multiplied "crosswise" with the sine and cosine components of the control variable and added to yield the turning tendency r. Thus the function of interest $r = f(x, w)$, according to the bicomponent theory, is

$$r = K_s \sin x \cos w + K_c \cos x \sin w, \tag{13}$$

and for joint operation of light and gravity,

$$r = (F_s \sin \alpha + L_s \sin \beta) \cos w + (F_c \cos \alpha + L_c \cos \beta) \sin w. \tag{14}$$

Equations (11) and (13) are equivalent in the sense that they yield identical relations between the command w and the variable x, provided that the coefficients of the sine and cosine components in (13) are equal ($K_s = K_c = K$; $F_s = F_c = F$; $L_s = L_c = L$). In both cases the stability and lability positions move by identical amounts in the same direction (see Fig. 5B) if a command is imposed on the loop, although the underlying information-flow patterns are quite different.

Evidence

As already indicated by the experiments referred to above, it can hardly be expected that the same one of these three basic solutions will be always found realized throughout the animal kingdom. A decision between them in any concrete case can be made on the basis of three groups of criteria: first, the different dependence of the turning tendency r on the deviation x and the command w (compare A and B of Fig. 5), which distinguishes theory I from theories II and III; second, the split into two components, which distinguishes theory II from the others; and third, the effect of variations in stimulus intensity, which distinguishes among all three. For the first group of criteria, the examples given above may suffice, so that this last part of the paper will deal only with the other two.

The split into components can be most readily checked in the gravity system of vertebrates. Here the utriculus otoliths, as shown by von Holst (1950) and by Löwenstein and Roberts (1949), do indeed provide an afferent variable proportional to the sine of the angle between the dorso-ventral axis and the zenith.* Furthermore, two additional pairs of otoliths, in the lagena and the sacculus, are arranged perpendicularly to those in the utricles, so that if they operated under the same principle, they would yield a cosinusoidal afferent variable. Recordings of Löwenstein and Roberts (1949) from the afferent nerve of the ray's lagenae are in good agreement with this hypothesis. A particular merit of the bicomponent theory is that it offers an explanation for the apparent paradox that, electrophysiologically, the lagenae and the utricles both show the typical characteristics of positional receptors, whereas von Holst, in his classical experiments (1950), found almost no influence of the lagenae on the utricle-dominated gravity orientation. According to (13) and (14) this is to be expected if, as in von Holst's experiments, the command w is zero. The small effects of the lagenae found by Schoen and von Holst (1950) can also be understood on the basis of the bicomponent theory, (III). The most convincing piece of evidence, however, has been discovered by W. and H. Braemer (1958) in experiments with lifelong-deviating fish. They removed both utricles and found that the fish, which normally swims nose up at 50 deg (see Fig. 8), deviated permanently by 90 deg. This follows from Eq. (13) under the condition that the utricles are the *sole* providers of the sine component. Since that condition has already been demonstrated by von Holst (1950), the evidence is conclusive.

The third and last group of criteria, the effects of variations in stimulus intensity, offers the most sophisticated set of experiments and interpretations in this area. The basic fact here is that (switches of system and reversals of feedback excluded, of course) no normal bilateral animal has ever been found to change its angle of orientation with a change of stimulus intensity *if it is oriented in the basic position* ($\alpha = 0$, or $\beta = 0$). As can be seen from the equations, this is in agreement with all three theories introduced. Demonstration of intensity effects thus requires deviations from the basic position. Two kinds

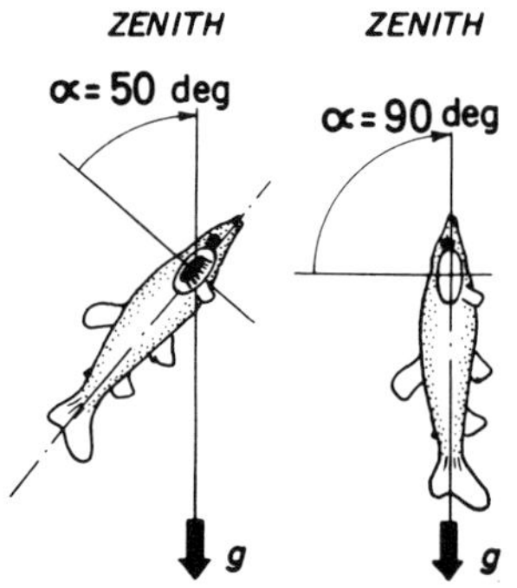

FIG. 8. Normal position of a lifelong-deviating fish before (left) and after (right) extirpation of both utriculus otoliths (modified after Mittelstaedt, 1961).

* In all probability they provide two separate messages for deviations around two perpendicular axes. The interaction of these messages in truly two-dimensional orientation can be easily envisaged from the viewpoint of the bicomponent theory. Results relevant to this problem are still too sparse, however, to warrant a detailed theory at this time.

of deviations will be compared in this respect: those which originate from the conflict between two orientation systems with diverging basic positions while commands are absent, as in von Holst's classical experiments (to be called "provoked deviations"); and those which originate from central-nervous-system actions on the loop ("commanded deviations"). According to their reaction to a change of stimulus intensity under conditions of provoked and commanded deviations, the animals can be classified into three main groups: (a) those in which both provoked *and* commanded deviations change with changing intensity, (b) those in which provoked deviations are intensity-dependent whereas commanded ones are not, and (c) those in which both kinds of deviation are independent of stimulus intensity.

An example for the fourth logically possible combination, the inverse of (b), has not yet been found. As in the terminology suggested by Jander (1963), the orientational capability of the first group (a) will be called "protaxis" and that of the third group (c) "metataxis." For the intermediate behavior of the second group (b), I suggest "mesotaxis" as an appropriate name.

The behavior of group (a) is well in agreement with theory (I) if the coefficients K, L, and F of Eqs. (9) and (10) depend on the stimulus intensity in the way von Holst (1950) and Braemer (1957) found for provoked deviations of fish [linearly (F) or logarithmically (L, at high intensities)]. Protaxis contradicts theory (II), however, because commanded deviations [see Eq. (11)] are necessarily independent of the value of the coefficient K. Protactic behavior is compatible with the bicomponent theory (III) only under the condition that the two coefficients K_s and K_c depend reciprocally on the stimulus intensity: Either K_s increases with increasing intensity while K_c decreases or vice versa: K_c is directly, K_s inversely, proportional to intensity. It remains to be seen whether such an assumption can be justified by experiment. It should be noted, though, that the behavior resulting from the second possibility has indeed been observed (Jander, 1963). Some prophototactic insects, while running at an acute angle ($\beta < 90$ deg) toward a light source, *increase* the light angle if the intensity of the source is increased!

The behavior of group (b) clearly stands in contradiction to theory (I). It is in agreement with theory (II) under the assumption that L and F in Eq. (12) are proportional to the respective stimulus intensities. It is difficult to see from an evolutionary viewpoint, however, why the effort to provide intensity independence of the afferent variable α_a, β_a should have been made, if it is to be spoiled by introducing an intensity dependence *after* the inflow of the command w. This implausible arrangement would even require an extra channel of information. Mesotactic behavior follows from the bicomponent theory (III) if the two coefficients of the bicomponents both depend on the intensity *in the same way* (K_s/K_c = const.). A clear-cut example is the behavior of the crustacean *Mysidium gracile* (Jander, 1962). Its *mean* provoked deviation (dorsal light reaction against gravity) depends on the light intensity, whereas the standard error of the scatter around this mean (which is due, in the undisturbed aquarium, to internal actions and thus attributable to "commands") is constant and independent of the light intensity of a source oriented at 0 deg or 90 deg to the vertical. This

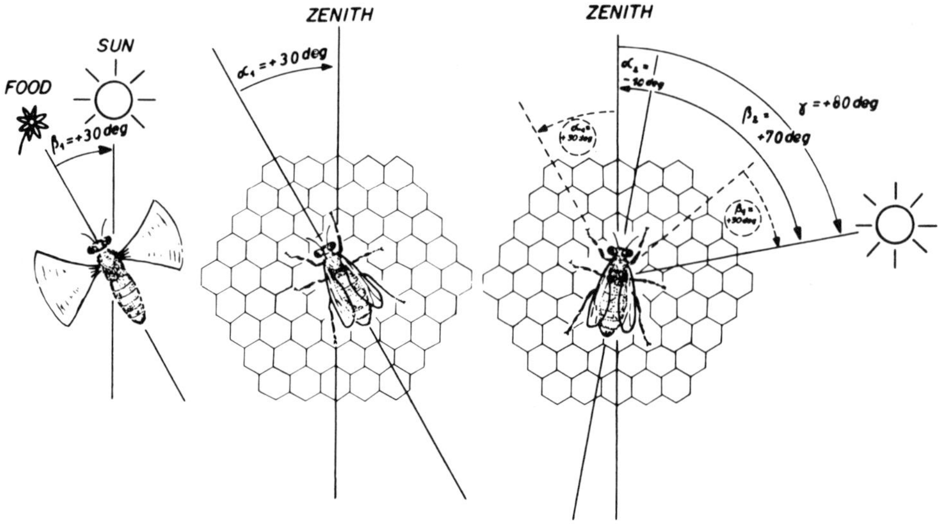

FIG. 9. Experiments of von Frisch (1948, 1962) with bees indicating the deviation angle of a food source from the sun (β_1, left picture) by "dancing" on the vertical honeycomb in a dark or homogenerously lit room, at the same angle to the zenith (α_1, middle picture). In the critical experiment the vertical honeycomb is exposed to the sky, so that both light and gravity act on the bee. Note that now (right picture) the dancing direction (α_2) exactly bisects the angle between the *dashed* lines.

is to be expected from theory III and follows from Eq. (14) under the plausible assumption

$$\frac{F_s}{F_c} = \frac{L_s}{L_c} = 1. \tag{15}$$

The behavior of group (c) is compatible with any of the three theories, but only under the condition that the coefficients are zero if the stimulus intensity is zero, or smaller than a given threshold value, but finite and independent of stimulus intensity above that value. Metatactic animals such as ants and bees show the most advanced orientational capabilities. From them a wealth of qualitative and quantitative data is already available. In our context a result of von Frisch (1948) in the honeybee is of particular importance, which he recently confirmed and elaborated (von Frisch, 1962). As von Frisch had shown earlier, bees are able to indicate the difference (β_1) between the sun's azimuth and that of a food source by running on the vertical honeycomb, in the dark, in figure-8's whose average midline deviates by the same angle (α_1) from the zenith (i.e., $\alpha_1 = \beta_1$). On a horizontal comb under the sun, the blue sky, or an artificial light source, they indicate the azimuth angle (β_1) directly. If we expose them—and this is the point—to light and gravity simultaneously by making them dance *under the sun* on a *vertical* honeycomb, the bees indicate the arithmetic mean of the angle they would have danced with the gravity influence removed (in the vertical plane the sun's elevation is seen as azimuth) and the one they would have danced (Fig. 9) in the dark,* or, mathematically,

* In the 1962 repetition of this experiment von Frisch had to screen off the sun and instead pit the polarization pattern of the blue sky against gravity in order to reproduce this result. With the sun added, the bees "switched off" the gravity component and oriented to the sun only ($\beta_2 = \beta_1$).

(16) $$\alpha_1 = \beta_1 = \alpha_2 + \frac{\gamma}{2} = \beta_2 - \frac{\gamma}{2},$$

where α_1 and β_1 are the deviations at $L = 0$ and $F = 0$, respectively, α_2 and β_2 the deviations if both systems are operating simultaneously, and γ the angle (more exactly, the azimuthal difference at the plane of orientation) between the light source and the vertical. Since

(17) $$\beta_2 = \gamma + \alpha_2,$$

the result can also be formulated thus:

(18) $$\alpha_1 = \beta_1 = \frac{\alpha_2 + \beta_2}{2}.$$

This result is in agreement with theory (II) and follows directly from (11) and (12) if the coefficients L and F are equal and constant above the threshold of stimulation ($L = F$ for $F, L > 0$). It is in agreement with the bicomponent theory and follows from (13) and (14) if all four coefficients are constant and equal ($F_s = F_c = L_s = L_c$ for $F, L > 0$). It stands, however, in irreconcilable contradiction to theory (I).

For the "language" of the bees there is no compelling evidence available at the present time to support a choice between theories (II) and (III). There are some indications, however. Von Frisch (1948, 1961) found, for instance, that β_1 and α_1 are not always exactly equal but deviate by various small amounts up to (rarely) 15 deg. If these "translation errors" (which, fortunately for the bees, are compensated for by reverse translation of the recipients!) are plotted over β_1, a characteristic pattern arises (von Frisch, 1961). In the statistical mean of a large number of measurements, the "errors" turn out to be zero at $\beta_1 \approx 0$ deg, plus and minus 60 deg, plus and minus 90 deg, plus and minus 120 deg, and 180 deg. A pattern of this kind emerges according to the bicomponent theory, (III), if the quotient L_s/L_c differs slightly from F_s/F_c, and if at least one of the input components deviates slightly from the exact sinusoidal or cosinusoidal function in the direction of a sawtooth function.

The strongest evidence in favor of a theory with more than one component stems from the ability of the bee to indicate compass bearings compensated for the sun's azimuthal velocity at such widely differing geographical latitudes as 0° and 49°. It is difficult to see how the basic pattern of theory (II) could be adapted to this performance, whereas theory III offers a natural basis for it. The bicomponent theory of the "sun-compass" (Mittelstaedt, 1962) is in fact in good agreement with the existing relevant findings in arthropods (Lindauer 1957, 1959, 1960; Kalmus, 1956; New, 1962) and fish (Braemer and Schwassmam, 1963).

References

Braemer, W., 1957. Verhaltensphysiologische Untersuchungen am optischen Apparat bei Fischen, *Z. vergl. Physiol.*, **39**, 374–98.

Braemer, W., and H. Braemer, 1958. Zur Gleichgewichtsorientierung schrägstehender Fische, *Z. vergl. Physiol.*, **40**, 529–42.

BRAEMER, W., and H. O. SCHWASSMANN, 1963. Vom Rhythmus der Sonnen orientierung am Äquator (bei Fischen), *Ergebnisse der Biologie*, **26,** 183–201.

BUDDENBROCK, W. VON, 1935. Die Physiologie des Facettenauges, *Biol. Revs. of the Cambridge Phil. Soc.*, **10,** 283–316.

BUDDENBROCK, W. VON, 1952. *Vergleichende Physiologie* (Band I, Sinnesphysiologie). Basle: Verlag Birkhauser.

FRISCH, K. VON, 1948. Gelöste und ungelöste Rätsel der Bienensprache. *Naturwiss.*, **35,** 12–23, 38–43.

FRISCH, K. VON, 1950. Die Sonne als Kompass im Leben der Bienen, *Experientia*, **6,** 210–21.

FRISCH, K. VON, 1961. *Sprache und Orientierung der Bienen.* A. Wander-Gedenkvorlesung (Band 3), Berne, Switzerland: Verlag H. Huber.

FRISCH, K. VON, 1962. Über die durch Licht bedingte "Missweisung" bei den Tänzen im Bienenstock, *Experientia*, **18,** 49–53.

HOLST, E. VON, 1950. Die Arbeitsweise des Statolithenapparates bei Fischen, *Z. vergl. Physiol.*, **32,** 60–120.

HOLST, E. VON, and H. MITTELSTAEDT, 1950. Das Reafferenzprinzip (Wechselwirkungen zwischen Zentralnervensystem und Peripherie), *Naturwiss.*, **37,** 464–76.

JANDER, R., 1957. Die optische Richtungsorientierung der roten Waldameise (*Formica rufa* L.), *Z. vergl. Physiol.*, **40,** 162–238.

JANDER, R., 1962. The swimming plane of the crustacean *Mysidium gracile* (Dana), *Biol. Bull.* **122,** 380–90.

JANDER, R., 1963. Insect orientation, in *Annual Review of Entomology*, vol. 8, Palo Alto, Calif.: Annual Reviews, 95–114.

KALMUS, H., 1956. Sun navigation of *Apis mellifica* L. in the southern hemisphere, *J. Exptl. Biol.*, **33,** 554–65.

LINDAUER, M., 1957. Sonnenorientierung der Bienen unter der Äquatorsonne und zur Nachtzeit. *Naturwiss.*, **44,** 1–6.

LINDAUER, M., 1959. Angeborene und erlernte Komponenten in der Sonnenorientierung der Bienen. Bemerkungen und Versuche zu einer Mitteilung von Kalmus, *Z. vergl. Physiol.*, **42,** 43–62.

LINDAUER, M., 1960. Time-compensated sun orientation in bees, in *Cold Spring Harbor Symposia on Quantitative Biology* (vol. XXV), 371–77.

LOWENSTEIN, O., and T. D. M. ROBERTS, 1949. The equilibrium function of the otolith organs of the thornback ray (*Raja clavata*), *J. Physiol.* **110,** 392–415.

MITTELSTAEDT, H., 1961. Probleme der Kursregelung bei freibeweglichen Tieren, in *Aufnahme und Verarbeitung von Nachrichten durch Organismen*, Stuttgart: Hirzel Verlag, pp. 138–48.

MITTELSTAEDT, H., 1962. Control systems of orientation in insects, in *Annual Review of Entomology*, vol. 7, Palo Alto, Calif.: Annual Reviews, pp. 177–98.

NEW, D. A. T., and J. K. NEW, 1962. The dances of honeybees at small zenith distances of the sun, *J. Exptl. Biol.*, **39,** 271–91.

SCHOEN, L. and E. VON HOLST, 1950. Das Zusammenspiel von Lagena und Utriculus bei der Lageorientierung der Knochenfische, *Z. vergl. Physiol.*, **32,** 552–71.

SCHÖNE, H., 1957. Kurssteuerung mittels der Statocysten (Messungen an Krebsen), *Z. vergl. Physiol.*, **39,** 235–40.

SCHÖNE, H., 1962. Optisch gesteuerte Lageänderungen (Versuche an Dytiscidenlarven zur Vertikalorientierung), *Z. vergl. Physiol.*, **45,** 590–604.

13. The Peripheral Organization of Sensory Systems

MELVIN J. COHEN, *University of Oregon*

The behavior of a successful animal is related to its environment in such a way as to favor survival. A limiting factor in developing this adaptive behavioral linkage may be the manner in which environmental information is made available to the organism. In order to understand behavior it is therefore important to know what aspects of the external world can be detected by an organism and the way in which these environmental features are presented to the central receiving system. In animals with definitive nervous systems, the pertinent facets of the external world are converted into biological signals in afferent paths leading to the central nervous system. A type of coding exists, therefore, whereby certain events in the sensory paths come to represent specific happenings in the external world. The nature and organization of this code is a primary concern of this paper.

There are several stages in the translation of physical events into biological signals. The actual transducing process whereby the adequate physical stimulus is initially converted into a biological signal occurs in either primary sensory endings or secondary sense cells. Studies of generator potentials in a variety of receptors have provided an understanding of the transducing mechanism and the manner in which the generator signal initiates a propagated all-or-none impulse (Edwards and Ottoson, 1958; Eyzaguirre and Kuffler, 1955a, b; Fuortes and Poggio, 1963; Loewenstein, 1958). I would like to focus our attention at the level of the nerve impulse in the first-order sensory neuron. At this point the external world is symbolized in the form of digital all-or-none events which can be easily monitored in many different systems. It is here that we have the greatest amount of first-order sensory data, and studies at this level offer the most definitive information on the initial organization and nature of the coding process.

What properties of a stimulus must be known to the animal before an adaptive response can develop? I make no attempt here to specify all possible properties of sensory systems, but rather only those that appear crucial to me and that can be analyzed by examining data obtained from the nerve impulses in

The work described in this paper has been supported (in part) since 1958 by PHS research grant NB-1624 from the National Institute of Neurological Diseases and Blindness, U.S. Public Health Service.

first-order sensory neurons. These are as follows:

1. *Specificity resolution* (information about the form of the stimulating energy).
2. *Intensity resolution* (related to the immediate past or to an absolute reference value).
3. *High sensitivity* (A large response to a small stimulus increment).
4. *Reliability* (The constant relation of a given biological signal to a specific physical event).

Consider now the way in which these information requirements are actually achieved in a variety of sensory systems. The classical functional concept of the first-order sensory neuron is that it utilizes frequency modulation of digital events (namely, nerve impulses) as the primary means for conveying information (Adrian, 1928). The discharge frequency is related directly to stimulus strength, and the specificity is provided by the transducing element's being highly sensitive to only one form of energy. One of the main purposes of this paper is to shift our attention from this relatively simplified viewpoint of intensity and specificity resolution to the more complex peripheral patterns of organization now apparent in many preparations.

The second major objective is to call attention to the non-sensory role played by afferent systems, i.e., their generalized function in setting levels of excitability within the central nervous system. This aspect of afferent-system physiology has come to play an important part in our concept of central integrative mechanisms (Lindsley, 1961) but is frequently ignored by investigators concerned primarily with specific sensory function. I believe it is quite important to consider afferent systems in their totality, taking into consideration specific sensory as well as non-sensory functions in order to fully understand their role in shaping animal behavior.

Modes of Organization

Figure 1 illustrates the organizational categories we shall be concerned with here. I propose to discuss these by using examples from a variety of preparations.

A number of these organizational concepts can be illustrated by data from the myochordotonal-organ proprioceptive system in the walking legs of the crab *Cancer magister* Dana. The myochordotonal organ, first described by Barth (1934), has only recently been investigated physiologically (Cohen, 1960a, b, 1963; Hwang, 1961; Raj and Cohen, 1962). As seen in Fig. 2, the sensory system consists of a specialized muscle, the accessory flexor, which is coupled by a long tendon to the tendon of the main flexor muscle. The accessory flexor muscle is therefore in parallel with the main flexor muscle. An elastic strand is attached to the proximal head of the accessory flexor, and a group of sensory neurons are embedded in this strand. Action of the joint between the meropodite and carpopodite (M-C joint) causes length and tension changes in the accessory flexor muscle and the attached sensory neurons. These neurons are sensitive to stretch and therefore alter their discharge with movement of

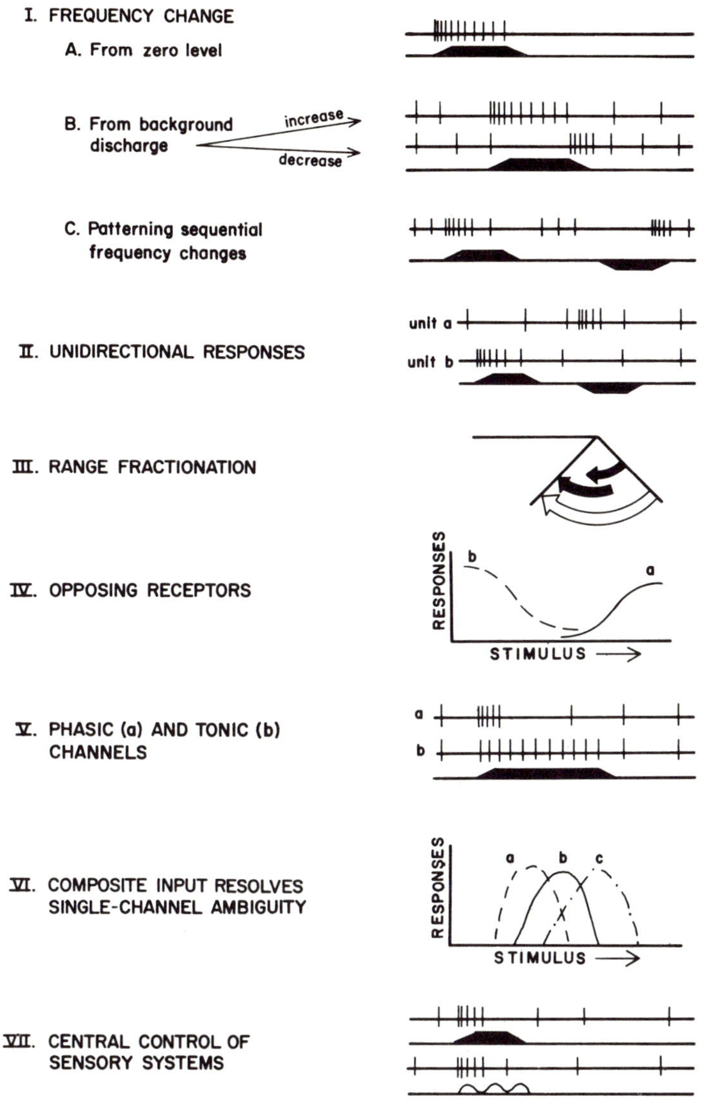

FIG. 1. Types of organization in sensory systems. In I, II, V, and VII, the vertical lines represent action potentials and the heavy bar on the lower line indicates the duration of a physiological stimulus. In IC and II the stimulus bars above and below the baseline indicate stimulus changes in opposite directions from a physiologically neutral level (i.e., flexion or extension). Part VII shows the response of the same unit to a physiological stimulus (black bar) and to electrical stimulation (wavy line) of the efferent nerve to a receptor. In III the full extent of the stimulus range for joint receptors is shown by the open arrow. The response of two units to different fractions of the range is shown by the black arrows. In IV, unit (a) responds to an increase in stimulus, unit (b) to a decrease in stimulus. In VI, curves (a), (b), and (c) are response curves from three different units.

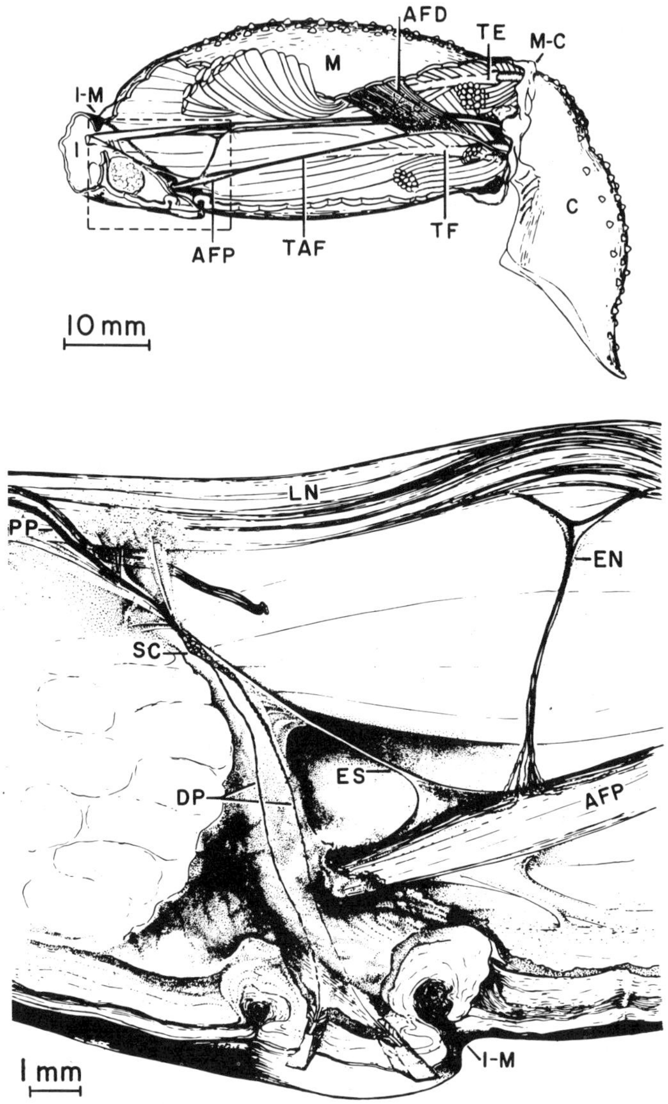

FIG. 2. Anterior view of the left walking leg in the crab *Cancer magister*, showing the myochordotonal-organ proprioceptive system. Dorsal is above, medial to the left. A. View of the ischiopodite (I), meropodite (M) and carpopodite (C). The receptor muscle is the accessory flexor which consists of a proximal head (AFP) and a distal head (AFD) joined together by a long tendon (TAF). The accessory flexor tendon attaches distally to the tendon of the main flexor muscle (TF). B. Expanded view of the boxed area shown in A. An elastic strand (ES) is stretched taut between the proximal accessory flexor head and the leg nerve (LN). Sense-cell bodies (SC) are attached to the strand, and anchor their distal processes (DP) in the cuticle, whereas their proximal processes (PP) join the leg nerve to enter the central nervous system. Action of the mero-carpopodite joint (M-C) indirectly changes the tension on the stretch-sensitive distal processes and alters their discharge pattern. Efferent neurons (EN) connect the leg nerve to the accessory flexor muscle. Joint I-M is the ischio-meropodite joint (modified from Cohen, 1963).

the M-C joint. Recordings have been obtained from the proximal processes of the sensory neurons just before they enter the main leg nerve.

The Multiplicity of Labels in a Single Channel

The response of a receptor activated only by movement of the joint and not by different static joint positions is shown in Fig. 3. The unidirectional nature of the response is seen clearly in the upper record, where a burst of activity is evoked only by extension and not by flexion. Records A through E illustrate the range-fractionation principle, as well as unidirectionality. Unidirectionality is shown by the lack of response to flexion in A and B. In record C, even though extension is taking place, the unit is not yet within its sensitive range, and no response is evoked over the specific 10 deg traversed. However, in D and E the movement occurs within the active range, and full responses are evoked during these extensions. Thus, although the unit is sensitive to extension, it responds only over a particular fraction of the full physiological range. Several units, each with a different response sector, are therefore

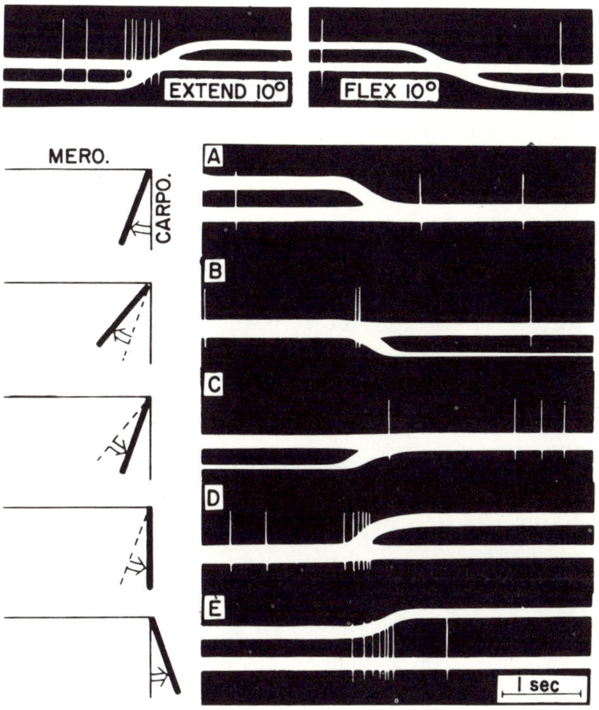

FIG. 3. Records from a myochordotonal-organ extension-movement receptor, illustrating unidirectionality and range fractionation. The top record shows unidirectional response when the joint is extended 10 deg from rest and then flexed back to starting position. Lower beam signals joint movement. Records A–E are portions of a continuous record from the same unit during a sequence of 10-deg movements of the carpopodite (carpo.). The diagrams at the left show the direction and extent of movement (from Cohen 1963).

needed to cover the complete stimulus range. This type of receptor is also sensitive to the velocity of movement and fires with a higher frequency for increasing rate of movement. Complementary units showing the same response properties but active only during flexion and not during extension have also been demonstrated (Cohen, 1963).

Consider now the specific aspects of the stimulus that may be coded in this single channel. The information provided is (1) occurrence of movement, (2) velocity of movement, (3) direction of movement, and (4) the particular fraction of the stimulus range being traversed at any given time. The classical concepts of a qualitative label (movement) and a quantitative measure of intensity (velocity) based on frequency changes are satisfied. In addition, two other stimulus parameters are encoded in the same channel by utilizing unidirectionality and range fractionation.

Why should the composite input from many different units be required to cover the entire physiological range for any given stimulus? What advantage does this method have compared with a system in which a single receptor responds over the entire stimulus range? One can speculate that range fractionation may be of value when the physiological stimulus spans a wide energy spectrum which must be covered with high resolution. Under such circumstances a highly sensitive single unit subjected to the full energy flux might easily be damaged, while a less sensitive unit would have a high threshold and low resolution. Subjecting individual units to only a fraction of the stimulation energy permits the single receptor to respond with high resolution to small stimuus increments within its limited range, without risk of damage. Coverage of the full span thus requires consecutive inputs from many channels. I have suggested (1963) that range fractionation evolved to permit high resolution of a stimulus covering a wide energy range. Sensory systems such as temperature receptors in the tongue (Dodt and Zotterman, 1952), various arthropod-limb proprioceptors (Pringle, 1956; Barber, 1960; Wiersma, 1959), and spatial equilibrium receptors in vertebrates (Lowenstein and Roberts, 1949) and invertebrates (Cohen, 1955, 1960b) utilize range fractionation and seem to fit within this generalization.

Separate Channels for Phasic- and Tonic-Stimulus Parameters

A second receptor type in the myochordotonal organ is the position receptor, which has a different steady-discharge frequency for each specific maintained position of the M-C joint. The response of such a receptor to different tonic joint positions is shown in Fig. 4. There is some slow adaptation as each new position is reached, but the adapted level is significantly different for each of several joint locations. Note that the response during movement consists of a gradual change in discharge frequency. Several sensory systems dealing with spatial equilibrium and proprioception in both vertebrates and invertebrates use separate channels for signaling the phasic and tonic aspects of the stimulus (Barber, 1960; Cohen, 1955, 1960b; Kuffler, 1954; Lowenstein and Sand,

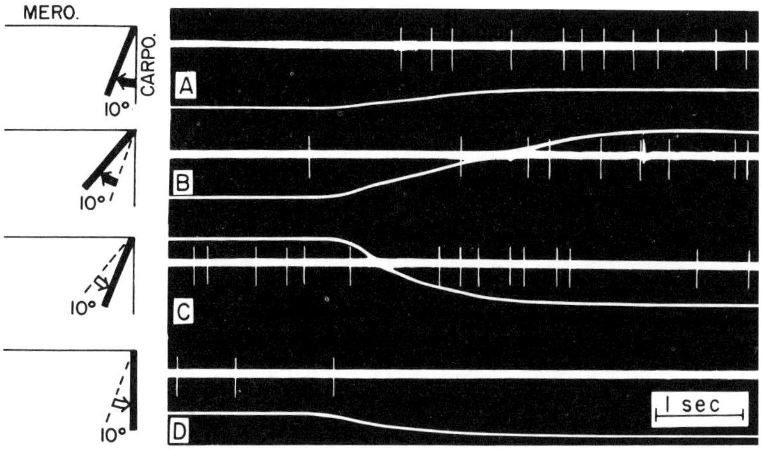

FIG. 4. Response of a myochordotonal-organ position receptor. Lower beam signals M-C joint displacement. Segments from a continuous record with a 20-sec time gap between the end of one record and the start of the next. There is considerable adaptation at each new position, but the adapted frequency is different for each position (from Cohen, 1963).

1940; Pringle, 1955, 1956). Other sensory systems such as those dealing with thermal, chemical, and tactile stimuli seem to have only one receptor type responding to both phasic and tonic aspects of the stimulus (Dodt and Zotterman, 1952; Cohen, Hagiwara, and Zotterman, 1955; Loewenstein, 1958). Why should certain modalities require separate channels for handling phasic and tonic stimulus parameters while others use only one receptor type for both aspects of the stimulus? The answer may lie in the different discharge patterns of the phasic and tonic receptors. As seen in Fig. 3, the response of the phasic receptor is a high-frequency burst. The tonic receptors, on the other hand, show a gradual change in impulse frequency during the phasic aspect of the stimulus (Fig. 4). The high-frequency bursts of the phasic receptors may utilize temporal summation in the central pathway to evoke a behavioral response with relatively short latency. The presence of special phasic channels may be needed where delicate, precisely controlled responses must be evoked with a latency short enough to be relevant to the changing stimulus. This would apply to proprioceptive and spatial-equilibrium systems, in which separate phasic and tonic channels are known to occur.

Opposing Receptors

In some sensory systems the physiological stimulus range is covered by two receptor types. One receptor is active at the low end of the stimulus range, and the other responds at the high end of the range. These receptors show opposing responses; the type covering the higher stimulus level responds with a higher discharge frequency as the stimulus increases, whereas the receptor covering the low end of the stimulus range responds with an increased frequency as the

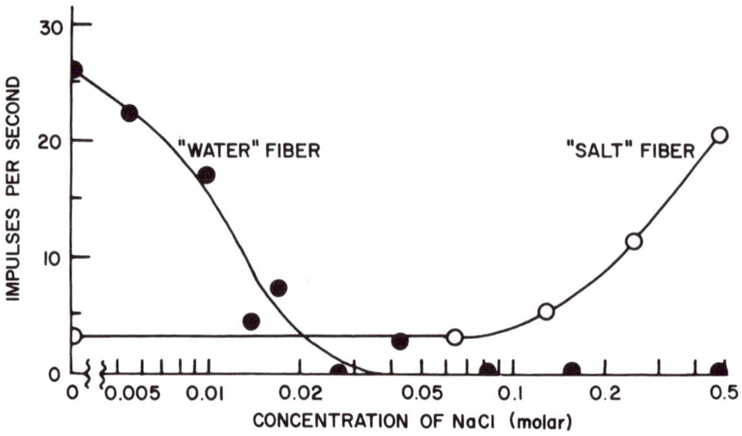

FIG. 5. Response from single fibers in the chorda tympani nerve of the cat to saline solutions flowed over the tongue. The "salt" fiber increases its activity with higher concentrations, whereas the "water" fiber fires at a higher rate as the saline concentration drops. This illustrates the opposing-receptor type of organization (from Cohen, Hagiwara, and Zotterman, 1955).

stimulus level decreases. An example of such an opposing receptor system is seen in Fig. 5, which is taken from chemoreceptor units in the cat's tongue (Cohen, Hagiwara, and Zotterman, 1955). The entire saline range is covered by two different types of units. The so-called "salt" fiber increases its discharge frequency as the saline concentration rises, whereas the "water" fiber fires at a higher rate as the saline concentration decreases. A similar type of organization is seen in thermal receptors of the cat's tongue (Dodt and Zotterman, 1952,) where the "warm" and "cold" fibers show the same kind of opposing-receptor configuration.

In the saline and thermal receptors mentioned above, the stimulus level at which the opposing receptors overlap and show their lowest discharge frequencies corresponds roughly to the portion of the physiological thermal and saline range maintained as optimal by the homeostatic mechanisms of mammals. The internal thermal and saline environment of mammals is regulated very precisely. With proper central connections, such an opposing-receptor system could operate on a null-return principle to fix temperature and salt concentration within the very narrow limits required in mammals. This type of mechanism could function over long periods of time and would be independent of environmental fluctuations because of the opposing nature of the receptor organization. The confirmation of such a hypothesis depends upon the accumulation of a great deal more comparative sensory data in animals with and without homeostatic mechanisms.

Patterning of Impulses in a Single Channel

When assessing sensory responses at the level of the propagated action potential, one commonly refers to average discharge frequency calculated over some arbitrary interval of time. However, in certain sensory systems it appears that relatively long-term sequential fluctuations in frequency (patterned discharge) may be an important part of the information code. An example of this can be seen in the acceleration receptors of the crustacean statocyst which is involved in spatial equilibrium (Cohen, 1955, 1960b). As seen in Fig. 6, the organ consists of a cuticular sac filled with water and housed in the basal segment of each antennule. A crescentic arrangement of chitinous hairs occurs on the floor of the sac, with the distal process of a single sensory neuron terminating in the base of each hair. A sand mass, loosely coupled to the cyst floor, lies within the hair crescent so that the tips of some of the hairs make contact with it. When the animal shifts its position in the gravitational field, about either the long or the transverse axis, the sand mass pushes or pulls on the hairs, thereby changing the force exerted on the attached sensory nerve terminal and altering its discharge pattern. The thread hairs, located along the posterior-medial wall, are seen in Fig. 6 as fine white lines. They float free in the fluid of the cyst and are not in contact with the sand mass. These hairs are phasic receptors responding only to angular acceleration and not to different maintained positions of the animal. The response of a group of thread hairs to the animal's being rolled about its long axis is shown in Fig. 7. The units have a background discharge with the animal in its primary resting orientation. A side-down movement, relative to the recording side, results in a definite sequential pattern of frequency changes. This pattern consists of a burst of impulses at the onset of movement followed by a depression of activity, below the background level, at

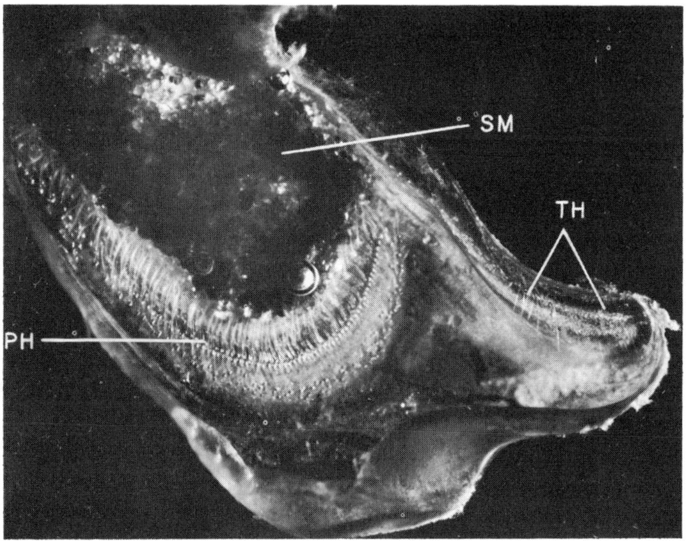

FIG. 6. Dorsal view of the left statocyst in the lobster *Homarus americanus*. Anterior is to the top, medial is to the right. The roof of the cyst is removed and the crescentic arrangement of the position-receptor hairs, PH, is seen on the cyst floor. The sand mass, SM, lies within the hair crescent on the cyst floor. The thread hairs, TH, which are the acceleration receptors, are seen as fine white lines at the right, attached to the posterior-medial wall of the cyst. The longest dimension is approximately 7 mm.

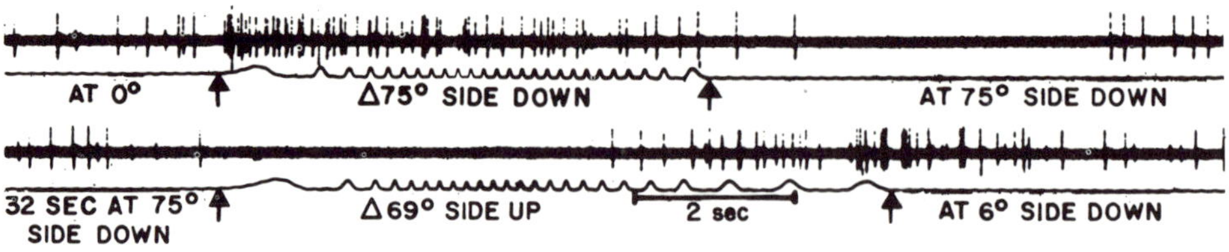

FIG. 7. Response of several thread-hair acceleration receptors in the lobster statocyst to rotation of the animal about its long axis. Lower beam signals displacement; one deflection for every 3 deg. Arrows indicate onset and end of movement. A. side-down rotation; B. the reciprocal side-up rotation. Note the different sequential patterns of frequency changes evoked by the opposite movements (from Cohen, 1955).

the end of movement and then a gradual return toward the background level even with the animal held in the tilted position. The reciprocal movement (a side-up roll returning the animal approximately to its original position) causes a reversed sequence of frequency change in the same units. This involves a depression of activity, below the background level, at the onset of movement followed by a burst of activity at the end and then a gradual return to the resting level. Although this is shown in a multi-unit preparation, single-unit records clearly indicate that such reciprocal discharge patterns do occur in a single sensory unit. Thus the same channel can provide information that the animal is rolling about its long axis, and whether the roll is side-down or side-up. Presumably the sequence "background discharge → burst → depression → background discharge" signals side-down rotation. The reciprocal sequence "background discharge → depression → burst → background discharge" signals rotation in the opposite direction. If the central receiving mechanism can make use of such sequential changes, this is obviously an economical way of utilizing one channel for more than just a simple label and intensity resolution within that label. A precisely similar pattern of coding has been described by Lowenstein and Sand (1940) for the acceleration receptors of the vertebrate semi-circular canals.

The Composite Input Pattern from Many Channels

The information provided by a single sensory neuron may often be imperfect in specifying the form of stimulating energy or in signaling stimulus intensity. These ambiguities within the single unit may be largely eliminated in sensory systems containing many parallel channels if the composite input is considered, rather than the activity in one line alone.

The Sharpening of Stimulus Specificity. In first-order sensory neurons from chemoreceptors in the cat's tongue, a given neuron may respond to several different kinds of gustatory stimuli (Cohen, Hagiwara, and Zotterman, 1955). Figure 8 shows that a single sensory neuron in the cat chorda tympani nerve responds to distilled water, quinine, or dilute acid solutions flowed over the tongue. The three stimuli activating the same sensory neuron can be differen-

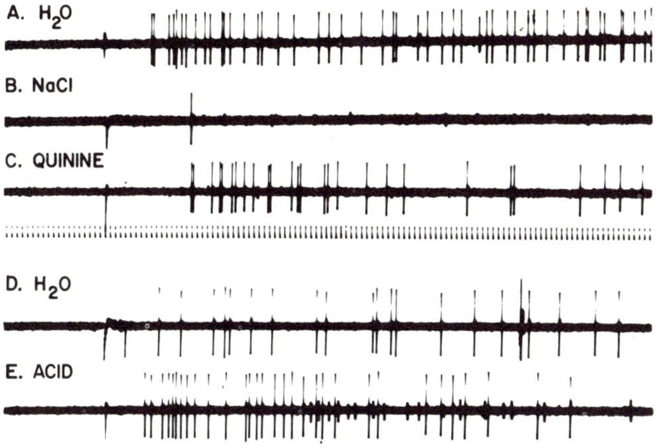

FIG. 8. Responses of single "water" fibers in the cat's chorda tympani nerve to various solutions flowed over the tongue. Records A–C are from the same single fiber. Records D and E are from a different fiber. Note that this fiber type responds to several different gustatory stimuli such as distilled water, quinine (bitter), and acid (sour) solutions. This single sensory channel lacks stimulus specificity. Time marker 50 cps. (Modified from Cohen, Hagiwara, and Zotterman, 1955.)

tiated by the animal, to judge from behavioral observations. The question then arises: How can the animal distinguish between the three different stimulating solutions which activate the same sensory neuron? If one examines all of the receptor types contributing to the gustatory input and plots them in Table 1, it becomes evident that for any given specific stimulus which the animal can differentiate behaviorally there is a unique composite input pattern. The plus indicates an increase in discharge frequency for the stimulus shown and the zero indicates a lack of response. A given fiber type may present ambiguous information by responding to more than one quality of gustatory stimulation. However, the composite input pattern from all fiber types is unique for each separate stimulus. The composite input pattern evoked by salt stimulation differs from that caused by acid stimulation, even though a given fiber type responds to both stimuli. Thus the total input pattern for a given modality may provide information about stimulus specificity that is lacking in any single channel.

TABLE I

GUSTATORY FIBER TYPES IN THE CAT CHORDA TYMPANI NERVE AS DEFINED BY THEIR RESPONSE TO VARIOUS GUSTATORY STIMULI

Stimulus	"Water" Fiber	"Salt" Fiber	"Acid" Fiber	"Quinine" Fiber	Sensation Evoked
H_2O (salt < 0.03 M)	+	0	0	0	→ Water
NaCl (0.05 M)	0	+	0	0	→ Salt
HCl (pH 2.5)	+	+	+	0	→ Sour
Quinine	+	0	0	+	→ Bitter

From Cohen, Hagiwara, and Zotterman (1955).

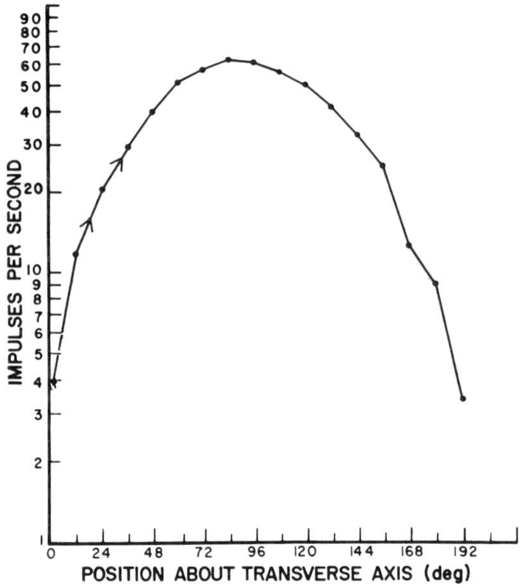

FIG. 9. Response of a single position receptor from the lobster statocyst to a slow continuous rotation about the transverse axis. The bell-shaped response curve indicates that the same discharge frequency can occur at two separate stimulus positions. This represents an ambiguous response to stimulus intensity in single sensory channels.

Resolving Intensity Ambiguity. Another type of single-channel ambiguity that may be resolved by the composite input pattern is found in tonic receptors which show a bell-shaped response curve, such as position receptors in the vertebrate otolith organs (Lowenstein and Roberts, 1950) and the invertebrate statocyst. An example of this is seen in Fig. 9, which is taken from a position receptor in the lobster statocyst (Cohen, 1955). This receptor is from one of the hairs on the floor of the cyst normally in contact with the sand mass (see Fig. 6). The bell-shaped nature of the response curve indicates that the same discharge frequency is reached at two separate stimulus levels. Therefore, if the animal were dependent upon this single unit for detecting position about the transverse axis, it could not distinguish a slightly head-up position (about 24 deg) from a position almost flat on its back (about 160 deg), since the receptor fires at about 20 impulses/sec at either location. There are approximately 400 position receptors in each statocyst (Cohen, 1960). Their curves overlap but reach maxima at different stimulus strengths (see Fig. 1, category VI). Therefore, stimuli at two different levels may have the same frequency in a given receptor, but the composite input pattern from many channels with overlapping response curves is unique for any given stimulus intensity.

The Dual Role of Afferent Systems

The classical picture of a sensory system is that it provides specific information about the physical environment. The organism then uses this specific information as a basis for developing discrete psychic or motor behavioral responses adapted to the environment. This function of providing specific information for the formulation of definite behavioral patterns can be referred to as the sensory function of afferent systems. However, a second role of sense organs as generalized excitors of the central nervous system (*Stimulationsorgane*) has long been postulated (see Wolsky, 1933, for review). As early as 1892, Ewald pointed out the continuous influence of the labyrinth on the neuromuscular system and developed a concept of the "*Tonuslabyrinth.*" He demonstrated that loss in muscular tone and strength after labyrinthectomy is seen not only in muscles specifically involved in labrynthine reflexes but in others as well, such as those concerned with chewing and swallowing (see Lowenstein, 1936, for review). Bethe (1897) reported generalized loss of tone and muscular weakness in Crustacea following removal of the statocysts. These behavioral studies clearly illustrated that in a wide variety of animals, organs having to do with spatial equilibrium exert a generalized tonic influence on body musculature even with the animal in its primary resting orientation.

With the application of electrophysiological techniques to the study of sensory systems, it became apparent even in the pioneering work of Adrian (1926) that most sense organs send a continuous background discharge into the central nervous system. This type of activity is often referred to as "spontaneous" or as "occurring under steady-state stimulus conditions." I would prefer to call it background activity and limit this term to discharges from receptors that are in their physiological neutral or resting state. This usage differentiates it from the slowly adapting response evoked by steady-state stimulation of tonic receptors such as position units in the labyrinth and joints. Receptor responses evoked by an adequate stimulus and signaling a deviation from the physiologically neutral condition are then superimposed upon this background activity. The generalized tonic effect of receptors on motor function, demonstrated by behavioral experiments, can be ascribed to this continuous background barrage impinging on the central nervous system.

The classical case described by Strümpell (1877) of a tendency toward somnolence in a patient with progressive destruction of sensory paths perhaps first pointed out the relationship between psychic function and background sensory activity. The recent extensive studies on sensory deprivation indicate clearly the importance of sensory background activity in the maintenance of normal psychic and motor states (Solomon *et al.*, 1961).

The discovery of a control system in the brain-stem reticular formation, capable of generally influencing central excitability, gave more importance to the background discharge from receptors (French, 1960). A collateral afferent path, paralleling the specific sensory pathways, sends small samples of information into the reticular formation from all major sensory tracts (Starzl, Taylor, and Magoun, 1951). The level-setting function of this master control system is dependent in large part upon receiving continuous stimulation provided by the background activity of receptors. Thus the generalized alteration of central excitability levels associated with sleep, arousal, and attention is related to background activity from the receptors reaching the reticular formation.

We have the concept, therefore, of afferent systems engaged in two kinds of activity: (1) providing specific sensory information about the environment, and (2) setting excitability levels within the central nervous system. This duality is reflected centrally by the sensory information being channeled along two routes: (1) the classical specific sensory tracts and (2) the nonspecific collateral afferent path.

Sensory Function of Afferent Systems

The influence of background discharge on the sensory function of afferent systems has been discussed in several publications, and I shall present only a single example of the type of argument considered (Granit, 1955; Bullock, 1957; Cohen, 1955, 1960b; Fitzhugh, 1958). To distinguish a signal from random background discharge, a central receiver depending upon a single channel would have to wait for several consecutive impulses to arrive at a high frequency, thereby being penalized by a long response time. To reliably detect small stimulus increments and to evoke a response with a short latency, a sensory system with background discharge must be redundant and must utilize several parallel input channels converging on the same central receiver. Thus, from the standpoint of providing specific sensory information, the background discharge is often considered as "noise" and something to be coped with by invoking special properties of the peripheral and central links in the system. However, if this background activity is treated from the standpoint of setting central excitability, then it becomes a pillar upon which this particular function is built. The level-setting function is frequently ignored in the discussion of afferent systems, and I should like now to consider some of the implications of this aspect of receptor activity.

The Central Control of Sense Organs

The relation of central excitatory level to background discharge in first-order sensory neurons leads to the question; What determines the characteristics of the background discharge? In certain sense organs such as those involved with somesthetic or chemoreceptive functions, the background discharge may be a secondary consequence of the receptor's being at or just above its threshold in order to provide maximum sensitivity to small transient stimuli. In these situations there does not seem to be any mechanism for directly altering the background discharge at the receptor or first-order neuron. In other sensory systems, most notably certain proprioceptors in the vertebrates (Granit, 1955) and in invertebrates (Wiersma, Furshpan, and Florey, 1953; Raj and Cohen, 1962), the excitability of the receptor can be altered directly by the central nervous system. This permits a controlled modification of the sense-organ discharge, which may be independent of the adequate environmental stimulus.

An example of the efferent control of sensory discharge is seen in Fig. 10 as taken from the crustacean myochordotonal organ (Raj and Cohen, in preparation). As shown in Fig. 2, the muscular component of this receptor receives two efferent axons from the central nervous system. Stimulation of the thick efferent fiber increases the tension in the receptor muscle just as it increases during extension of the M-C joint. The top record in Fig. 10 shows the response of a receptor to an extension of the M-C joint. Extension causes a passive increase in tension of the receptor muscle and a burst of sensory impulses. In the middle record, the M-C joint is held fixed, and the thick efferent fiber to the receptor muscle is stimulated. This evokes a tension change in the receptor muscle similar to that produced by actual extension of the joint. The increased tension in the receptor muscle, in turn, causes the sensory neuron to respond with a higher discharge rate as if the joint were extended. The bottom record shows a similar response evoked by a higher rate of stimulation to the efferent nerve, with the joint held fixed. This increases the rate of rise in receptor-muscle tension and causes a higher discharge frequency in the sensory neuron.

Thus electrical stimulation of the efferent nerve can mimic the sensory response evoked by moving the joint. Presumably the central nervous system is also capable, therefore, of mimicking the action of the usual physiological stimulus for this sensory system. By long-term alterations of discharge frequency in the efferents to the receptor muscle, the background activity of the slowly adapting sensory neurons can be held at different levels. The ability of the central nervous system to affect both the level-setting and sensory aspects of receptor input raises several important questions.

The Concept of Reality. If the physical world is presented to the central nervous system as a code symbolized by discharge patterns in sensory neurons, then anything other than the adequate stimulus that modifies sense-organ activity could seriously alter the organism's concept of the environment and its response to it. How can an adaptive behavioral pattern exist when the sensory symbol of the external world can be evoked by the central nervous system independent of what is happening in the environment?

The vertebrate muscle spindle is the sensory system about which we know the most with regard to the central control of sense organs (Granit, 1955;

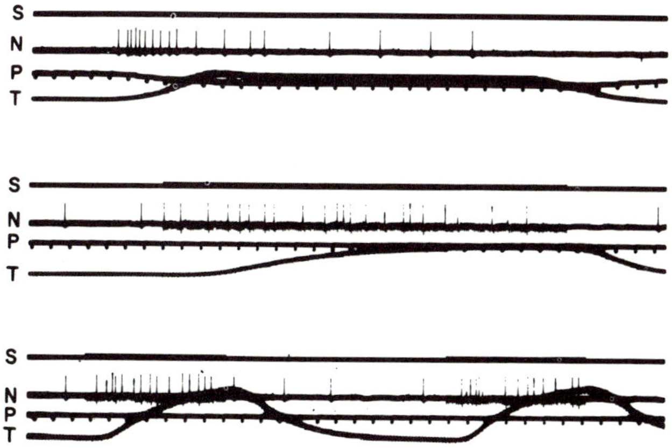

FIG. 10. Response of a single myochordotonal organ proprioceptor to joint movement and then to stimulation of the efferent nerve to the receptor muscle. (See Fig. 2 for anatomical orientation.) Three records are shown, with four events in each record. Event S represents the electrical stimulation of the efferent nerve to the receptor muscle, event N the sensory neuron action potentials, event P the position of the M-C joint (extension signaled by downward deflection); event T represents the tension in the receptor muscle (upward deflection indicates an increase in tension). The time mark on beam "P" is 250 msec (from Raj and Cohen, in preparation).

Eyzaguirre, 1962). The stretch reflex has long been considered the mechanism responsible for postural tone in vertebrates (Sherrington, 1906). Stretching a main limb muscle (extrafusal muscle) excites the afferent neurons of muscle-spindle receptors. These receptors then reflexly act back upon the limb muscle through the alpha motor neurons and cause it to contract. In this way the sensory fibers from the muscle-spindle receptors provide for negative feedback in the stretch-reflex servo loop, which returns the extrafusal muscle to some previously determined length.

Leksell (1945) presented clear evidence that the activity in the muscle-spindle afferents can be modified by stimulating the small gamma motor neurons connected to the specialized muscle of the spindle receptor (intrafusal muscle). This provided the first definitive demonstration of the central control of receptors. The gamma motor neurons control spindle discharge by direct excitation of the receptor-muscle fibers. The spindle afferents, in turn, act back upon the alpha motor neurons controlling the extrafusal muscle. These findings led to the intriguing idea that the primary role of the muscle-spindle receptor may be as part of an effector mechanism used for both the tonic setting of muscle length and the initiation of phasic movements (Eldred, Granit, and Merton, 1953; Merton, 1953). The indirect initiation of extrafusal muscle contraction by exciting the spindle receptor would allow a movement to take off from the postural platform established by the stretch reflex without breaking the servo loop and thus losing its advantages during movement (Merton, 1953). The gamma motor-neuron system to the muscle spindle can be influenced by higher regions of the central nervous system (Granit, 1955). This in turn means that the specific and generalized effects of the muscle-spindle sensory input can be altered by the descending barrage of impulses reaching the special motor neurons of the receptor muscle.

The control of receptor function by the central nervous system, as described above, makes it difficut to imagine how such a modifiable sensory input can provide absolute information about joint and muscle action. The use of a frequency code under these conditions is very difficult to visualize. For a given muscle length or joint position, a spindle afferent may discharge with a certain frequency at one time and a different frequency the next time these same conditions are reached, depending upon the bias placed on the spindle by the efferent nerves to the receptor muscle. One possible resolution of this problem is that the central nervous system monitors its efferent influence on the sense organs and integrates this with the receptor input. Information about absolute stimulus levels such as joint position and rate of movement might then be derived from this integral rather than from the input alone. Such a mechanism has yet to be demonstrated. With regard to proprioception, a more likely solution is that the absolute information is provided from a second group of sense organs which are not under central control. Such receptors exist in the vertebrates as Golgi tendon organs (Kuffler, Hunt, and Quilliam, 1951; Granit, 1955). These sense organs can essentially duplicate the information provided by the passive spindle and may provide the absolute-reference type of sensory information needed for coordinated locomotion. The spindle, on the other hand, may play a primary effector role and participate very little as a classical sensory receptor.

A precisely analogous situation obtains with regard to proprioception in the crab leg. As pointed out above, the myochordotonal organ provides information about the action of the M-C joint. This organ is under the control of the central nervous system via efferent axons to the receptor muscle, just as with the vertebrate spindle. There is, however, another group of sense cells also serving the M-C joint and essentially duplicating the information provided by the passive myochordotonal organ (Wiersma, 1959; Wiersma and Boettiger, 1959). These sense organs may be termed MC_1 and MC_2 and considered analogous to the vertebrate tendon organs in that they are, as far as can be determined, not under the influence of the central nervous system. Thus, at least for the M-C joint, which appears to be the key joint for active locomotion in the decapod Crustacea, there is a precise parallel with the vertebrate proprioceptive mechanism. Two groups of receptors are present; one group can be modified by central efferent neurons, whereas the other is influenced only by the adequate proprioceptive stimulus.

Summary

(1) Attention is drawn to the versatility of afferent system organization at the level of the first-order sensory neuron. The composite input for a given modality is considered to be a complex entity made up of specialized channels and capable of encoding several aspects of the stimulus in addition to intensity. The types of afferent organization are summarized and illustrated by examples from a number of different preparations.

(2) Functional specialization of individual afferent channels allows a single sensory neuron to encode several stimulus parameters. A movement receptor in a crustacean joint can signal the following: whether the joint is moving or being held at some new steady position (separation of phasic and tonic channels); the velocity of joint movement (intensity); the direction of joint action (unidirectional response); and the particular fraction of the total stimulus range activated (range fractionation).

Thermal and saline receptors in mammals show an opposing-receptor type of organization wherein the full physiological range for each stimulus is covered by two kinds of receptors. One responds at the high end of the range and increases its activity as the stimulus increases. The other type covers the low end of the range and increases its activity as the stimulus level drops.

(3) It is suggested that these organizational properties may actually determine the type of behavior developed by an animal. Thus range fractionation may permit high sensitivity to small stimulus increments in a modality where the stimulus covers a wide energy range. A given receptor is subject to only a fraction of the stimulus energy and can be highly sensitive without risk of damage.

The precise homeostatic regulation of the saline and thermal environment in mammals may be related to opposing-receptor organization if the input is processed centrally on a null-return principle. Such a system would tend to maintain these internal environmental factors within very narrow limits in the face of widely fluctuating external conditions.

(4) Single sensory channels may be inherently ambiguous with respect to either intensity or stimulus specificity. This ambiguity is greatly reduced by utilizing the input from many parallel channels. A single fiber carrying taste information from the cat's tongue may respond to several different gustatory stimuli that can be differentiated behaviorally by the animal. The composite input pattern from all the receptor types is unique, however, for each specific gustatory stimulus.

The bell-shaped response curves of many spatial equilibrium receptors means that a single channel may fire at the same frequency at two different spatial orientations. However, if there are many parallel channels with slightly different but overlapping response curves, the composite input is unique for any given point.

(5) The role of afferent systems in setting the general level of central-nervous-system excitability is pointed out. The maintenance of normal psychic and motor states is related to the continuous background discharge found in most receptors. The direct control of receptor activity by the central nervous system in both vertebrate and invertebrate proprioceptive systems is discussed. It is suggested that this control provides a means for regulating the sensory background discharge and thereby indirectly setting central excitability. In the proprioceptors subject to central control, the main role may be as part of an effector system used to reflexly initiate or sustain movement. The detection of joint and muscle action may be relegated to parallel receptors not under central control.

References

Adrian, E. D., 1926. The impulses produced by sensory nerve-endings, Part I, *J. Physiol.*, **61**, 49–72.

Adrian, E. D., 1928. *The Basis of Sensation*, New York: W. W. Norton.

Barber, S. B., 1960. Structure and properties of *Limulus* articular proprioceptors, *J. Exptl. Zool.*, **143**, 283–305.

Barth, G., 1934. Untersuchungen über Myochordotonalorgane bei dekapoden Crustaceen, *Z. Wiss. Zool.*, **145**, 576–624.

Bethe, A., 1897. Das Nervensystem von *Carcinus maenas*, *Arch. für mikroskop. Anat.*, **50**, 460–546.

Bullock, T. H., 1957. Neuronal integrative mechanisms, in *Recent Advances in Invertebrate Physiology*, B. T. Scheer, ed., Eugene, Ore.: Univ. Oregon Pubs., pp. 1–20.

Cohen, M. J., 1955. The function of receptors in the statocyst of the lobster *Homarus americanus*, *J. Physiol.*, **130**, 93-4.

Cohen, M. J., 1960a. A proprioceptive system in the legs of the crab *Cancer magister*, *Anat. Rec.*, **137**, 346.

Cohen, M. J., 1960b. The response patterns of single receptors in the crustacean statocyst, *Proc. Roy. Soc.* (*London*), **B-152**, 30–49.

Cohen, M. J., 1963. The crustacean myochordotonal organ as a proprioceptive system, *J. Comp. Biochem. Physiol.*, **8**, 223–43.

Cohen, M. J., S. Hagiwara, and Y. Zotterman, 1955. The response spectrum of taste fibres in the cat: A single fibre analysis, *Acta Physiol. Scand.*, **33**, 316–32.

Dodt, E., and Y. Zotterman, 1952. Mode of action of warm receptors, *Acta Physiol. Scand.*, **26**, 345–57.

EDWARDS, C., and D. OTTOSON, 1958. The site of impulse initiation in a nerve cell of a crustacean stretch receptor, *J. Physiol.*, **143,** 138–48.

ELDRED, E., R. GRANIT, and P. A. MERTON, 1953. Supraspinal control of the muscle spindles and its significance, *J. Physiol.*, **122,** 498–523.

EWALD, J. R., 1892. Bedeutung des Ohres fur die normalen Muskelkontraktionen, *Zentrbl. Physiol.*, **5,** 4–6.

EYZAGUIRRE, C., 1962. Motor regulation of the vertebrate spindle, in *Muscle Receptors*, D. Barker, ed., Hong Kong: Hong Kong Univ. Press.

EYZAGUIRRE, C., and S. KUFFLER, 1955a. Processes of excitation in the dendrites and in the soma of single isolate sensory nerve cells of the lobster and crayfish, *J. Gen. Physiol.*, **38,** 87–119.

EYZAGUIRRE, C., and S. KUFFLER, 1955b. Further study of soma, dendrite and axon excitation in single neurons, *J. Gen. Physiol.*, **38,** 121–53.

FITZHUGH, R., 1958. A statistical analyzer for optic nerve messages, *J. Gen. Physiol.*, **41** (4), 675–92.

FRENCH, J. D., 1960. The reticular formation, in *Handbook of Physiology*, vol. II, J. Field, H. W. Magoun, and V. E. Hall, eds., Washington D.C.: Am. Physiol. Soc.

FUORTES, M. G. F., and G. F. POGGIO, 1963. Transient responses to sudden illumination in cells of the eye of *Limulus*, *J. Gen. Physiol.*, **46,** 435–52.

GRANIT, R., 1955. *Receptors and Sensory Perception*, New Haven: Yale Univ. Press.

HWANG, J. C. C., 1961. The function of a second sensory cell group in the accessory-flexor proprioceptive system of crab limbs, *Am. Zool.*, **1,** 453.

KUFFLER, S. W., 1954. Mechanisms of activation and motor control of stretch receptors of lobster and crayfish, *J. Neurophysiol.*, **17,** 558–74.

KUFFLER, S. W., C. C. HUNT, and J. P. QUILLIAM, 1951. Function of medullated small-nerve fibers in mammalian ventral roots: Efferent muscle spindle innervation, *J. Neurophysiol.*, **14,** 29–54.

LEKSELL, L., 1945. The action potential and excitatory effects of the small ventral root fibers to skeletal muscle, *Acta Physiol. Scand.*, **10,** 1–84.

LINDSLEY, D. B., 1961. Common factors in sensory deprivation, sensory distortion, and sensory overload, in *Sensory Deprivation*, P. Solomon, P. E. Kubzansky, P. H. Leiderman, J. H. Mendelson, R. Trumbull, and D. Wexler, eds., Cambridge, Mass: Harvard Univ. Press.

LOEWENSTEIN, W. R., 1958. Generator processes of repetitive activity in a pacinian corpuscle, *J. Gen. Physiol.*, **41,** 825–45.

LOWENSTEIN, O., 1936. The equilibrium function of the vertebrate labyrinth, *Biol. Rev.*, **11,** 113–45.

LOWENSTEIN, O., and T. D. M. ROBERTS, 1949. The equilibrium function of the otolith organs of the thornback ray (*Raja clavata*), *J. Physiol.*, **110,** 392–415.

LOWENSTEIN, O., and A. SAND, 1940. The individual and integrated activity of the semicircular canals of the elasmobranch labyrinth, *J. Physiol.*, **99,** 89–101.

MERTON, P. A., 1953. Speculations on the servo-control of movement, in *The Spinal Cord* (CIBA Foundation Symposium), J. L. Malcolm, J. A. B. Gray, and G. E. W. Wolstenholme, eds., Boston: Little, Brown.

PRINGLE, J. W. S., 1955. The function of the lyriform organs of arachnids, *J. Exptl. Biol.*, **32,** 270–78.

PRINGLE, J. W. S., 1956. Proprioception in *Limulus*, *J. Exptl. Biol.*, **33,** 658–67.

RAJ, B. S. D., and M. J. COHEN, 1962. Efferent control of a proprioceptive system in the crab leg, *Am. Zool.*, **2,** 550.

SHERRINGTON, C., 1906. *The Integrative Action of the Nervous System*, New Haven: Yale Univ. Press.

STARZL, T. E., C. W. TAYLOR, and H. W. MAGOUN, 1951. Collateral afferent excitation of reticular formation of brain stem, *J. Neurophysiol.*, **14,** 479–96.

STRÜMPELL, A., 1877. Ein Betrag zur Theorie des Schlafs, *Pflug. Arch. Ges. Physiol.*, **15,** 573–74.

WIERSMA, C. A. G., 1959. Movement receptors in decapod Crustacea, *J. Marine Biol. Assoc. U.K.*, **38,** 143–52.

WIERSMA, C. A. G., and E. G. BOETTIGER, 1959. Unidirectional movement fibres from a proprioceptive organ of the crab *Carcinus maenas*, *J. Exptl. Biol.*, **36,** 102–12.

WIERSMA, C. A. G., E. FURSHPAN, and E. FLOREY, 1953. Physiological and pharmacological observations on muscle receptor organs of the crayfish *Cambarus clarkii* Girard, *J. Exptl. Biol.*, **30,** 136–50.

WOLSKY, A., 1933. Stimulationsorgane, *Biol. Rev.*, **8,** 370–417.

14. Eye-Stalk Reactions of the Ghost Crab *Ocypode*

PETER KUNZE, *Yale University*

Many Crustacea have excellent visual perception of movement. Surprisingly, only a few data are available in this field, although optomotor responses have been widely used for determination of visual thresholds. The influence of the pattern of visual input on the nystagmus of *Carcinus* was investigated by von Buddenbrock and Friedrich (1933). They reported an increase in the number of animals showing nystagmus with decreasing length of period of striped patterns (the smallest period used was 30 deg). More complete data on the same problem exist for *Lysmata* and *Leander* (Hassenstein, 1954), in which eye-stalk responses were measured. The reaction rose with increasing length of period from zero to a maximum (at a period of about 26 deg) and declined again.

The size of the moving field seems to play an important role for Brachyurans. For movement around the vertical axis of the animal, the moving field must be of considerable width (100–180 deg) to evoke the typical optomotor reaction (*Carcinus*, von Buddenbrock and Friedrich, 1933; *Goniopsis*, Waterman, 1961; *Uca*, Kunze, 1963). The height of the moving field may be small, but it must be presented within a range of 5 deg below and 20 deg above eye level (*Uca*, Kunze, 1963). No data are available on the influence of movement velocity on optomotor reactions in crustacea.

In the experiments reported in this paper, the influence of the above-mentioned parameters (velocity and period of stripe patterns, size and position of moving field) was investigated in one genus. *Ocypode* was chosen for its notably sharp perception of movement and because, being a littoral crab, it did not require the experimental complications of work in sea water.

Material and Methods

The experiments were carried out in Hawaii during July and August 1961. The experimental animals belong to the two species of the genus *Ocypode* found on the island of Oahu, *O. laevis* and *O. ceratophthalma*. Small animals, with a carapace width of 12 to 15 mm, were selected. Their eyes were small enough and close enough together in relation to the radius of the rotating drum

---

Author's present address: Zoologisches Institut der Universität Freiburg.

This work was supported by NSF grant 9690. The principal investigator was T. H. Waterman, Dept. of Zoology, Yale University. Work also reported as Hawaii Marine Laboratory Contribution No. 198.

to be taken as a single point in first-approximation calculations of visual angles. Because these animals were immature, their species could not be distinguished with certainty, and the results therefore apply to a mixture of the two species. Any possible differences must lie within the observed variations.

For the experiment the animal's carapace was fixed dorsally with a heated mixture of beeswax and resin to an adjustable lead rod which suspended the animal in a normal position (the anterior edge of the carapace slightly raised) in the center of a vertical drum. The legs of the animal did not touch ground. The claw on the same side as the eye that was fastened to the measuring apparatus was cut off to prevent interference with the measurement.

The drum (24.5 cm in diameter), which had vertical stripes on its inner surface, could be rotated in either direction. A synchronous motor and fixed gears allowed 13 discrete speeds, increasing in 2 × steps from .087 to 360 deg/sec. The lower and upper edges of the drum were screened so that the vertical field in which movement was presented subtended an angle of 27°10′ above and 27°10′ below the level of the crab's eyes. Behind the animal a bar supporting the lower screen covered an area of about 15 deg horizontally. The device was designed to accept additional screens for further reduction of the visual field.

The reaction of one eye stalk of the experimental animal was studied with an apparatus originally designed to measure turning responses of flying insects to optomotor stimuli (Kunze, 1961). It consisted of two electric coils joined together, one above the other, and suspended by thin metal torsion bands which allowed them to rotate together around a vertical axis. The lower coil was placed in an alternating magnetic field so that angular displacement of the coil would induce in it an alternating current proportional to the sine of its angular displacement. (At small angles the current was directly proportional to the angular displacement.) This current in the lower coil was amplified and then passed through a phase-dependent rectifier that used the current producing the alternating field in the lower coil as a reference. In this way displacements to the right and to the left were made to produce direct currents of opposite signs. These were fed back to the upper coil, which was situated in a permanent magnetic field. The resulting torque was adjusted to compensate for the original turning force on the lower coil. Hence the compensating current, which was measured with a recording pen, indicated the direction as well as the strength of any force that produced a torque on the torsion bands.

A clamp extending downward from the lower coil held a small wooden splint eccentrically, pointing downward. The tip of the ghost crab's eye was glued to that splint with a drop of beeswax and resin. The vertical axis for horizontal eye-stalk movements coincided approximately with the turning axis of the coils.

A heavy stand supported the measuring apparatus and a micromanipulator which was used to position the crab for connection of its eye stalk to the measuring apparatus. By means of a screw the measuring apparatus and the animal could be lowered simultaneously into the center of the drum.

Illumination was provided by three frosted 25-W bulbs 20 cm above the upper edge of the drum. The light intensity in the center of the upper end of the drum was 120 ft-c. The experiments were carried out in an otherwise unlighted room.

Results

General Properties of the Eye-Stalk Response. In *Ocypode* as in other stalk-eyed crustacea, the eye stalks respond to horizontally moving patterns by nystagmus movements. At the moment the pattern starts to rotate, the eye stalk exerts a torque in the direction of pattern movement (Fig. 1). This force increases at a high rate in the beginning, and then more slowly. Suddenly this process is interrupted by a strong torque in the opposite direction. Such nystagmus movements are repeated as long as the visually perceived movement lasts. In *Ocypode* it may go on for more than 30 min.

In many experiments the rate of nystagmus after 15–20 sec was lower than the initial rate. At high nystagmus rates (1 nystagmus per sec and more) nystagmus may stop suddenly for several seconds after about 1 min. However, it is resumed for short periods at the original rate. (The interruptions may be caused by fatigue.)

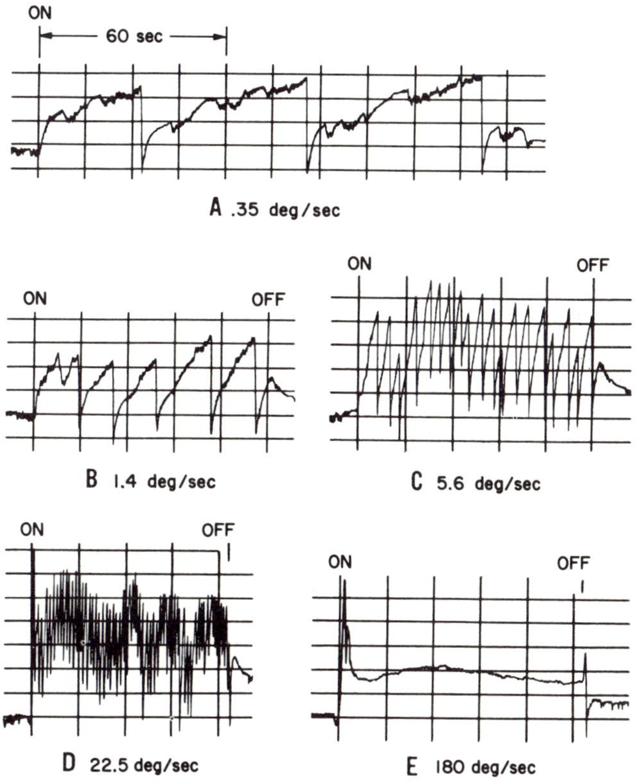

FIG. 1. Records of optomotor responses of an *Ocypode* eye stalk to a striped pattern moving from right to left at the angular velocities indicated below each record. All recordings were made on the same animal. The low-amplitude, high-frequency changes particularly obvious in the top record must be attributed at least partly to mechanical pickup of small leg movements.

The different speeds of the moving pattern elicited responses which differed characteristically (Fig. 1). The rate at which the eye stalk turned in the direction of movement was higher with higher speed, and the fast (return) phase of the nystagmus started sooner. The average duration of one complete nystagmus of the reactions in Fig. 1 is given in the following table.

Speed of pattern (deg/sec)	.35	1.4	5.6	22.5
Average duration of one nystagmus (sec)	54.4	14.2	3.8	1.1

The amplitude of the nystagmus remains approximately constant up to pattern speeds of 5.6 deg/sec. The slight decrease in amplitude in the 22.5-deg/sec record is due to the limiting frequency response of the pen recorder.

At a stripe speed of 180 deg/sec, nystagmic movements occurred immediately after the start of pattern movement, but there was no subsequent nystagmus during the next 75 sec. However, when the motor turning the stripe pattern was switched off, nystagmic movements were again recorded. Thus in this experiment nystagmus occurred only when the drum was gaining or losing speed and hence turning at some intermediate rate which induces strong eye-stalk responses. This does not prove that the animal did not *perceive* movement at all during the period of constant pattern speed, but only that the movement stimulus did not evoke nystagmus. Thus the reactions of the eye stalk to the

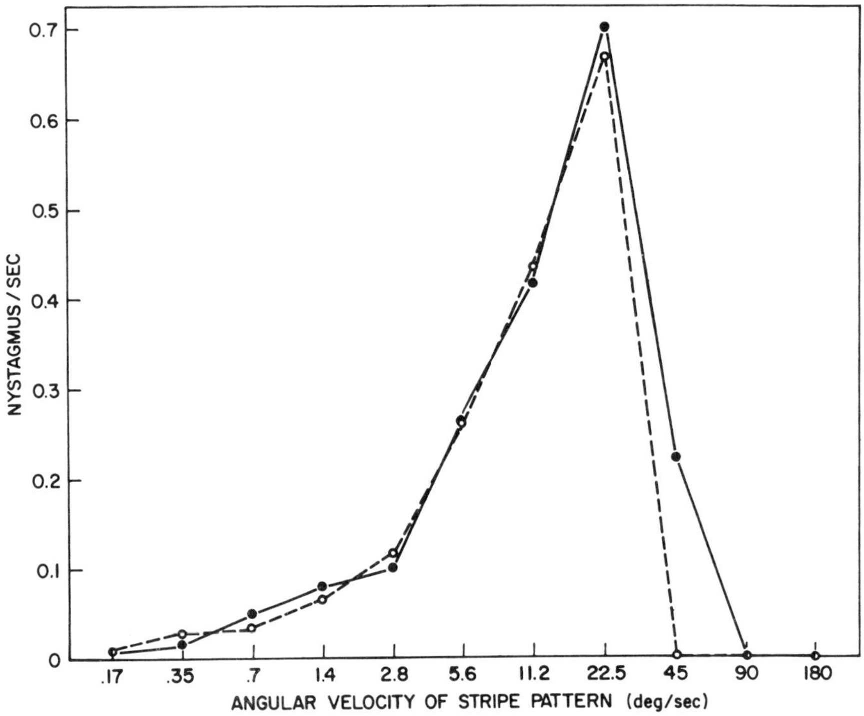

FIG. 2. Reaction to different direction of visual movement. Solid line: stripe pattern rotating to the right. Dashed line: stripe pattern rotating to the left.

rotating pattern proved to be graded with respect to the speed of the pattern, the time between nystagmic beats being a measure of the reaction. For convenience this will be expressed as the average number of nystagmi per sec (rate or frequency of nystagmus) during one experiment.

In order to test whether the direction of movement influences the reaction, a pattern was shown moving first in one direction and then, after a pause of about 2 min, at the same speed in the opposite direction. In both cases the measurement was done on the same eye stalk of the experimental animal. Figure 2 is a plot of typical results. The animal shows essentially the same average rate of nystagmus at a given speed for both directions of movement. Naturally, for movement to the left the following phase of the eye stalk was directed to the left and the return phase to the right, and the reverse was true for movement to the right.

Influence of Velocity and Period of the Moving Pattern on Nystagmus. Eight different patterns were tested. Each consisted of vertical black and white stripes of equal width. The patterns differed in their period λ, which is defined as the angle subtended by one black and one white stripe. During one set of experiments a crab was presented with a particular pattern at different speeds, in a random sequence. The experiments were separated by pauses of 1–3 min. Eye-stalk movements during the first 15 sec after the start of drum rotation were not evaluated, since they might have been influenced by the initial acceleration of the drum.

The results of this series of experiments are shown in Fig. 3, in which the average rate of nystagmus was plotted against pattern velocity for the eight patterns tested.

In general the curves are continuous, with one maximum. In all curves the reaction increases gradually with velocity. On double-logarithmic coordinates the increasing portion of the curves is almost linear, with approximate average gradients of (A) .76, (B) .86, (C) .88, (D) .82, (E) .78, (F) .83, (G) .73, and (H) .5.

The curves are characterized by a point at which the reaction steeply declines. Doubling of the pattern velocity at these points decreased the average reaction in A to 11%, in B to 12%, in C to 12%, in D to 15%, in E to 6%, in F to 26%, in G to 53%, and in H to 40%.

In most of the curves, the steep decline begins at the point of maximum reaction. The curves of F seem to constitute an exception to this rule. There each curve has a plateau spanning three points. In the curves of A–E, G, and H the "true" maxima can be assumed to be between the highest points, but the plateau in the curves of F cannot be caused by the spacing of the discrete velocities of the movement stimulus. Since this group of curves is the highest, what we are observing may be a cutoff effect due to saturation in the data-processing parts of the C.N.S., in signal transfer from the C.N.S. to the effectors, or in the effectors themselves.

Usually when the rotation velocity was greater than that producing maximum responses, eye-stalk nystagmus was less regular. Often a torque in the direction of pattern movement was observed in the absence of the return phase of the nystagmus (compare, for instance, Fig. 1E).

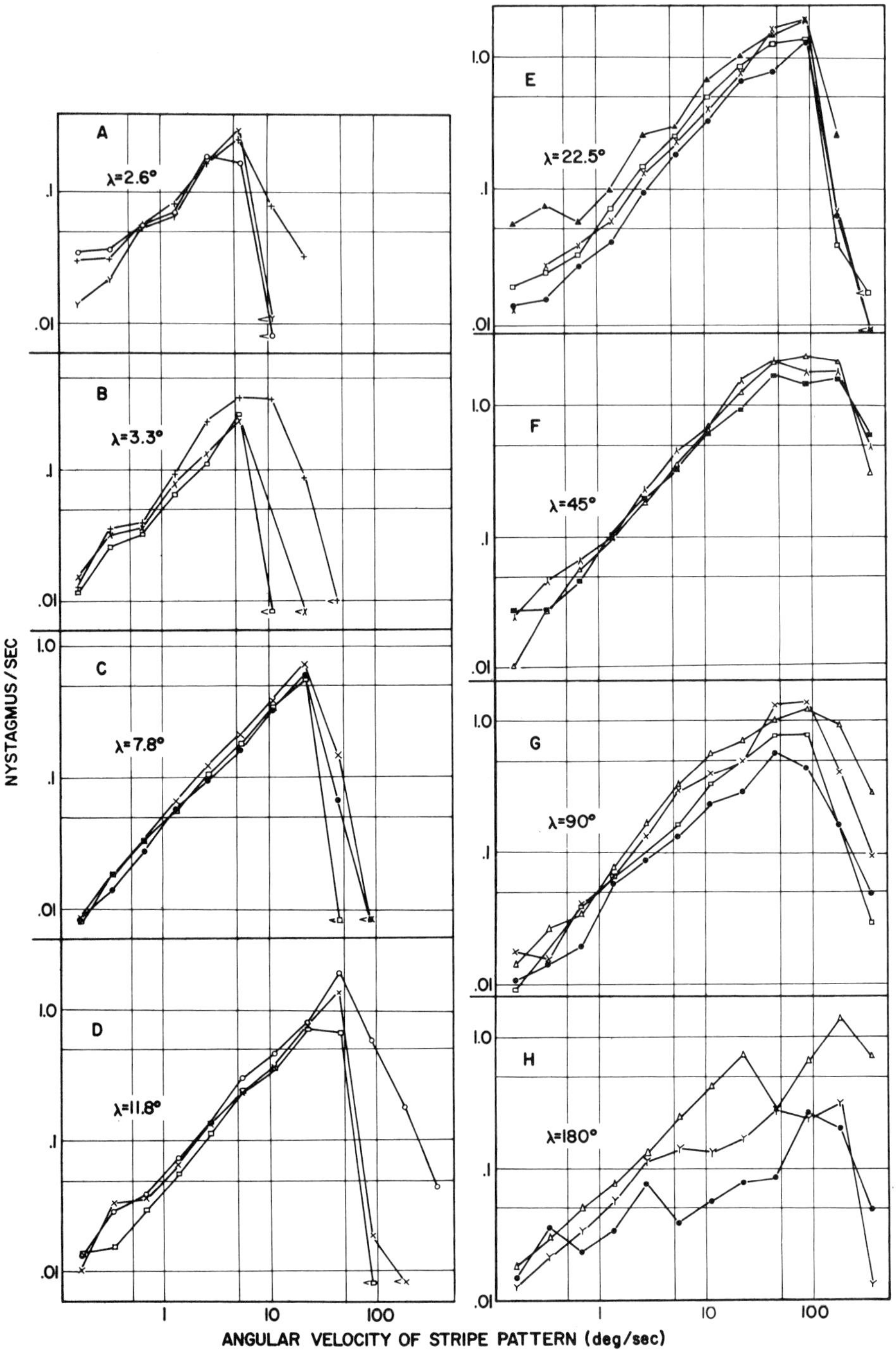

FIG. 3. Reactions to stripe patterns (black and white stripes of equal width) with varying period λ at different velocities of the pattern movement. Each curve represents one crab. The sign $<$ to the left of a data point shows that no return nystagmus phase occurred during a period of movement stimulus which is equivalent to the reciprocal of the corresponding value on the ordinate (always more than 1 min).

The different groups of curves have their maxima at different velocities. The trend is for maxima to move to higher velocities with longer periods λ. The exceptions are the curves in G and H of Fig. 3, whose maxima are, on the average, at the same velocities as the curves in E and F, respectively.

The influence of velocity and period on eye-stalk reaction may be summarized as follows: For a given period the amplitude of the reaction (rate of nystagmus) is determined by the velocity and has a single peak. Changing the period alters the amplitude of the response peak and the velocity at which it occurs.

Influence of Size and Position of the Moving Field on Nystagmus. In order to change the height of the moving field, screens of black cardboard were inserted in the previously mentioned device to screen out the upper and lower edges of the rotating drum. The distance between screen and drum axis was 7.9 cm. In all of these experiments the stripe pattern with $\lambda = 7.8$ deg was used at the constant speed of 11.25 deg/sec. The strength of reaction that occurred with this pattern and speed is well within the range of the crab's normal responses and sufficiently far from regions of threshold and possible saturation.

For tests of the *vertical* visual field of the crab the screen between crab and rotating drum was extended step by step from bottom to top. There was almost no change of reaction as long as the upper edge of the screen did not reach the level of the crab's eye. However, when the screen was raised above that level, the average nystagmus frequency decreased rapidly (and simultaneously the nystagmus sequence became more and more irregular). The reaction approached zero when the height of the moving field was reduced to the upper quarter.

With screens extending downward from the top, the reaction started to diminish after the upper quarter of the field was screened off. The nystagmus reaction reached zero when the upper half of the field was covered.

A somewhat different technique allowed a more accurate measurement of the quantitative change of the reaction. The screen which covered part of the rotating drum remained stationary, and the animal was moved up and down in small, controlled steps. For practical reasons, the lower edge of the crab's eyes was chosen as a reference for the position of the eye. Figure 4 shows a typical result of this experiment.

A screen covering 180 deg (or more) of the *horizontal* extent of the moving field almost completely eliminates the nystagmus response. (When the crab was free to move about in a small glass dish in the center of the rotating drum, the optomotor walking response was also eliminated.) Increasing the moving field to 210 deg restored the nystagmus reaction to the frequency observed without screens.

Each of the crab's eyes has a horizontal visual field of an estimated 320 deg. Therefore the question arose whether for a full reaction each of the eyes must be presented with a moving field of at least 210 deg, or whether different moving fields, each seen by only one eye, interact in their stimulating value. For this experiment the inner half of each eye of the crab was coated with opaque black paint. A screen covered the moving field of the animal laterally from medio-frontal to medio-caudal (180 deg). In this situation, movement of the pattern produced only an insignificant reaction (Fig. 5). Then an opening in the center

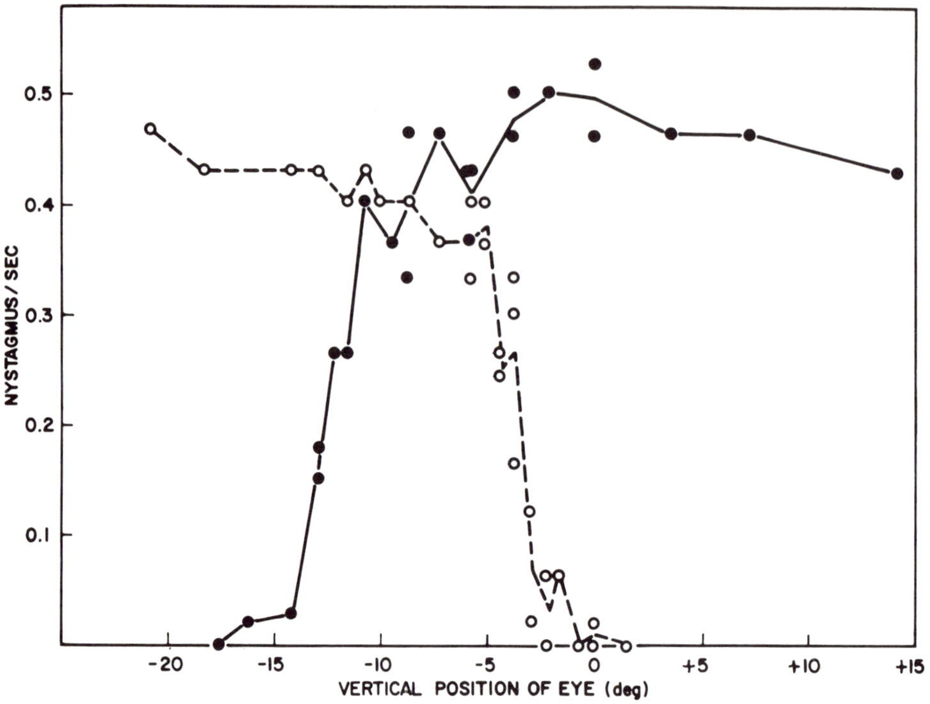

FIG. 4. Change of reaction with different vertical position of a crab relative to the moving field. The abscissa represents the position of the lower edge of the eye in degrees above (+) or below (−) the edge of the screen. Solid line: screen covers lower area of the crab's visual field. Dashed line: screen covers upper area of the crab's visual field.

of the screen was horizontally extended step by step. The reaction arose very quickly, reaching a plateau with a 30-deg opening, which gives a total moving field of 210 deg (180 deg seen by one eye, 30 deg seen by the other). Thus the widths of the moving field seen by the two eyes were additive in their effect on the reaction.

The same experimental setup was employed to test the contributions of different areas of the eye to the reaction. The 180-deg screen laterally surrounded the eye which was attached to the measuring apparatus (and therefore immobile). It contained an opening 14 deg wide. Movement of the pattern was presented at different horizontal positions of this opening within a total range of 100 deg. A 14-deg opening lies in a steeply increasing region of the reaction curves in Fig. 5, so that any change in reaction as a function of the position of the opening should be easily detected. Figure 6 shows typical observed reactions.

The reactions remained fairly constant within the tested range. As in the experiments of Fig. 5, the direction of pattern movement had no influence on the nystagmus frequency.

Discussion

The velocity range in which *Ocypode* reacted to moving periodic patterns corresponds to that known for other arthropods, namely, insects (Mittelstaedt,

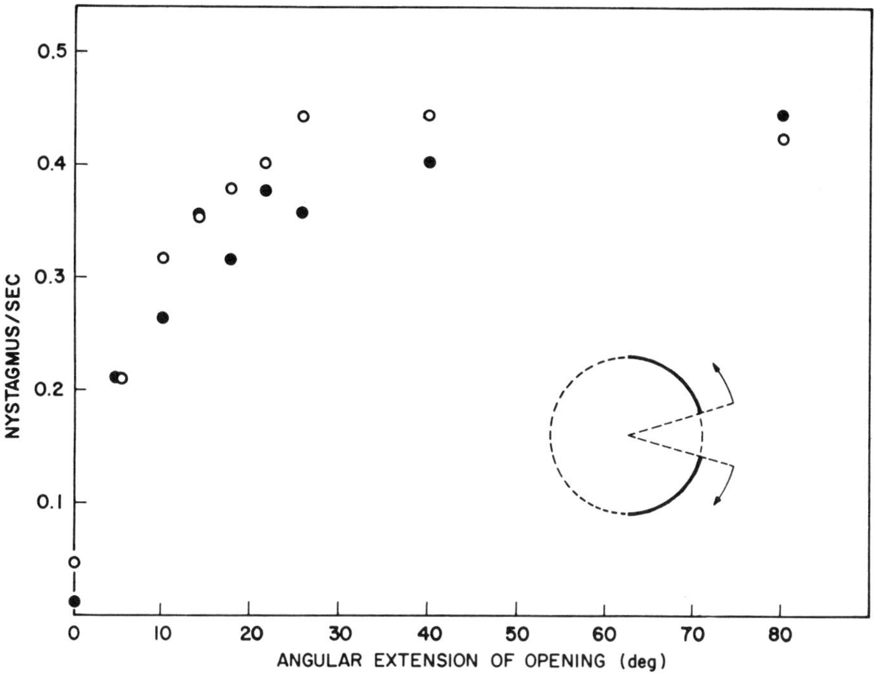

FIG. 5. Reactions on different widths of moving fields. The inner half of each of the crab's eyes was blinded. A screen covered 180 deg of the horizontal visual field. The abscissa represents the angular dimension of the opening in the center of the 180-deg screen.
○ Pattern movement to the left. ● Pattern movement to the right.

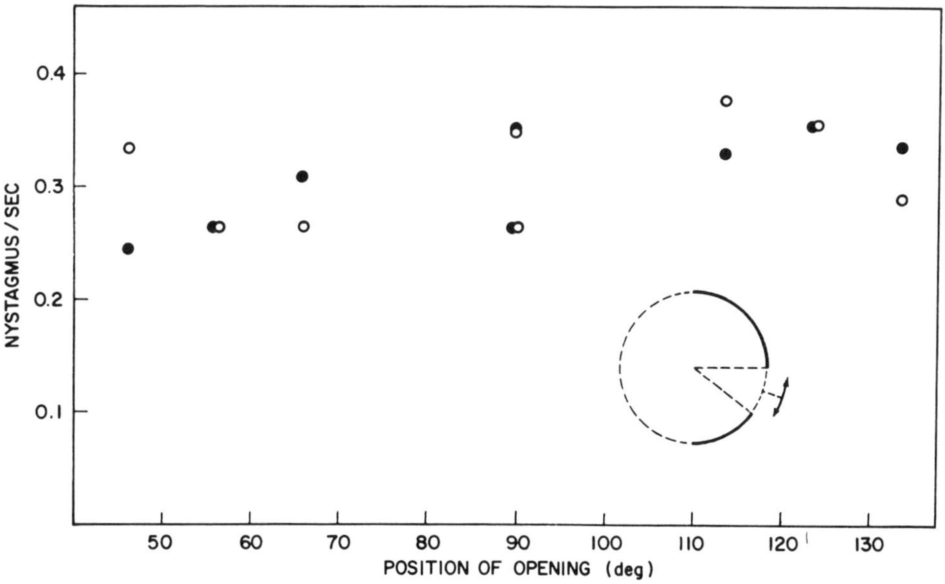

FIG. 6. Reactions for different horizontal positions of a moving field. The animal is the one used in Fig. 5. Both eyes were blinded on the inner side. The screen surrounding one eye contained a lateral opening of 14 deg. The abscissa represents the horizontal position of the center of the 14-deg opening.
○ Pattern movement to the left. ● Pattern movement to the right.

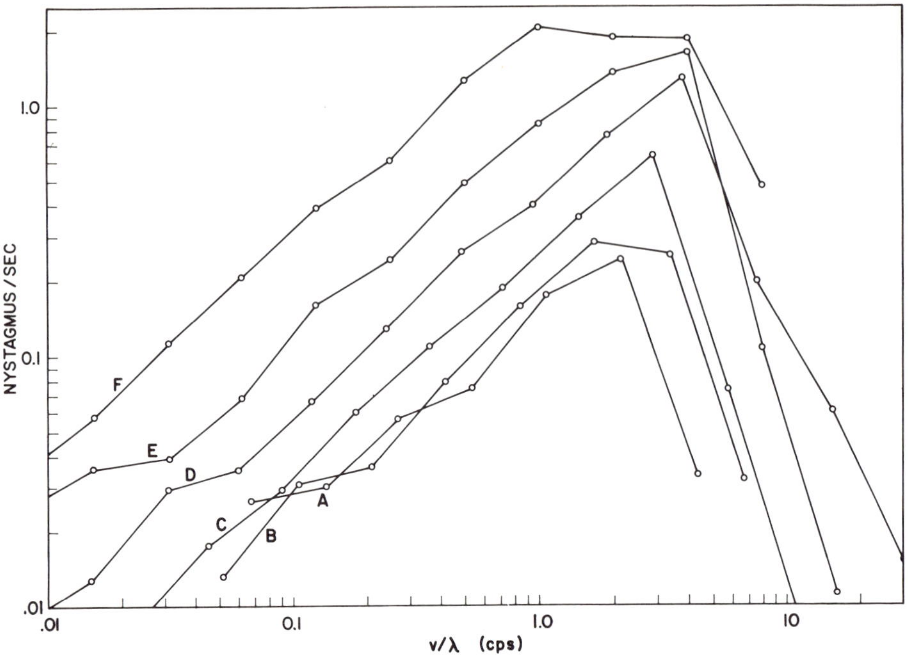

FIG. 7. The average reactions in A–F of Fig. 3 plotted as functions of v/λ (v = angular velocity, λ = period of the stripe pattern).

1951; Hassenstein, 1958; Kunze, 1961). The reactions to the lowest velocity (.08 deg/sec) corresponded so precisely with the stimulus that reactions to still lower velocities can be expected with certainty; nor did the limit of the optomotor reaction observed seem to be reached at the highest velocities used.

A close similarity seems to exist between insects and *Ocypode* in the way in which the period of the pattern influences velocity perception. In all investigations on insects (Gaffron, 1934; Hertz, 1934; Hassenstein, 1959; Kunze, 1961), movement of regular patterns of black and white stripes evoked maximal reactions at the same ratio of velocity to period of the pattern.

The reaction curves A–F of Fig. 3 (G and H seem to be exceptions and will be discussed later) reach their maxima, with growing λ, at increasingly higher angular velocities. In Fig. 7 the average reactions of Fig. 3A–F were plotted as a function of the ratio of the angular velocity to the period of the stripe pattern. The curves start to decline steeply at almost the same values of v/λ. (The differences occur within the range 2.2–4, i.e., within the range of experimental variation of velocity.)

One may ask which link in the chain of functions between stimulus and reaction cuts off the reaction at this particular value of v/λ. Apparently the muscles of the eye stalk cannot be responsible, since the nystagmi reach different maximal frequencies with different values of λ. (Evidence of a reaction-limiting property of the effectors would have at first appeared in curve F of

Fig. 7.) For the same reason, the parts that convert the continuous-movement stimulus into the periodically changing nystagmic movement cannot be the limiting ones. Limiting must take place either in the parts of the C.N.S. that carry out movement perception or in the sense organs themselves. Since v/λ is the light flicker frequency for each ommatidium, the second alternative seems the more probable. The issue could be decided by direct electrophysiological tests (e.g., measurement of ERG as an indication of flicker fusion frequency). A similar problem was solved in this way by Kirschfeld (1961) in *Chlorophanus*.

The properties of the reaction curves shown in Fig. 7 are generally in accord with the relations established by Varjú (1959) for the stimulation of compound eyes by moving patterns of equidistant dark and light stripes; they have the same position on the v/λ axis. Each curve follows approximately the same function (on double-logarithmic coordinates they are almost parallel). Their amplitudes differ as a function of λ by a multiplicative factor (on double-logarithmic coordinates the curves are shifted in the direction of the ordinate). A quantitative comparison with the investigations of Varjú would require knowledge of the functional connections of ommatidia for movement perception (Hassenstein, 1954). These are not known for *Ocypode*.

The reactions G and H of Fig. 3 show some differences from all other reactions. They reach their point of steep decline, on the average, at 1 cps, compared to about 3 cps for all other reaction curves. They increase and decrease, in both cases, with the lowest gradients. Therefore these differences seem to be orderly rather than accidental. Because of the black and white stripes, the ommatidia were stimulated with sequences of square-wave light pulses. It is conceivable that harmonics in the sequence of light changes influenced the reaction to patterns with long fundamentals (λ) but that they had less influence, or perhaps no influence, on the reaction to patterns with short fundamentals. It could also be supposed that the particular spatial relations in the patterns with large λ influenced the reaction in a special way. The experiments in which the width of the moving field was changed proved the importance of the spatial arrangement for movement perception.

The observed changes of reaction with changes in width of the moving field must have been caused by functional nonlinearities in the system for movement perception. The experiments shown in Fig. 6 prove that different elements of the eye contribute in the same way to the reaction. Hence possible differences in the density of ommatidia or differences in the properties of ommatidia cannot be responsible for the sudden decline of the reaction as the width of the moving field is reduced below 210 deg. The mode of functioning of parts of the C.N.S. concerned with movement perception must be the cause.

However, the change of reaction with different *vertical* positions of the moving field (Fig. 4) can be related to the morphology of the eye of *Ocypode*. The eye has an ellipsoid shape (Fig. 8). When the eye stalks are erect, the long axis of the eye is almost vertical. Therefore most of the ommatidial axes are almost horizontal, only a small number of them looking down or up. This is also made apparent by the form and position of the black pseudopupil in Fig. 8; it is formed by ommatidia with almost parallel axes looking in the same

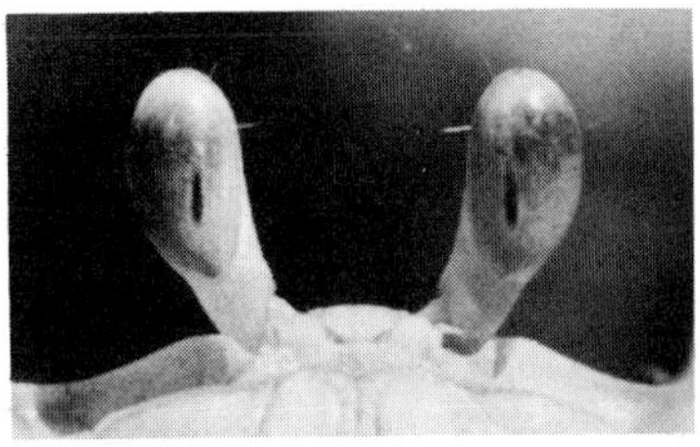

FIG. 8. Eyes of an *Ocypode* (frontal view, four times actual size).

direction. Strong changes of the reaction should be expected if screens are extended vertically in this region where nearly parallel ommatidial axes have the highest density. Similar relations were observed in *Uca* (Kunze, 1963).

Summary

Eye-stalk reactions of ghost crabs (*Ocypode*) to rotating patterns of equal-width vertical black and white stripes were measured mechanically. Continuous movement of the pattern elicited nystagmus, i.e., movements in the direction of pattern movement and quick return movements in the opposite direction (Fig. 1). The nystagmus is of high regularity and can be maintained for very long periods (more than 30 min). Stimulus movements in both directions elicit reactions of equal strength (Fig. 2).

The reaction (frequency of nystagmus) is graded. It changes continuously with the angular velocity of pattern movement and the period of the pattern (Fig. 3). Maximum reactions are obtained at similar ratios of velocity to period of the stripe patterns (Fig. 7). The optimum flicker frequency is about 3/sec.

Two stripe patterns with long periods (90 deg and 180 deg) are exceptions to this rule. With those patterns, maximal reactions were obtained at flicker frequencies of 1/sec.

The strength of maximum reactions increases with the period λ of the patterns up to $\lambda = 45$ deg, and decreases with greater λ (Fig. 3).

The reaction is almost absent if more than half the width of the stripe pattern is covered by a screen. A free visual field of at least 210 deg is necessary to evoke the maximum response for a particular pattern velocity (Fig. 5). The reaction of the animal remains constant at different horizontal positions of moving fields (Fig. 6). Hence the strong effect that the width of the moving field has on the animal's reaction seems to be caused not by morphological properties of the eye, but by nonlinear data processing in the C.N.S.

Moving fields elicit reactions only if presented in an area between zero and 17 deg above eye level (Fig. 4). This is in accord with the astigmatic form of the crab's eye (Fig. 8).

ACKNOWLEDGMENTS

I wish to thank T. H. Waterman for originally stimulating this work and for valuable help. Thanks are also due the staff of the Hawaii Marine Laboratory.

References

BUDDENBROCK, W. VON, and H. FRIEDRICH, 1933. Neue Beobachtungen über die kompensatorischen Augenbewegungen und den Farbensinn der Taschenkrabben (*Carcinus maenas*), *Z. vergl. Physiol.*, **19,** 747–61.

GAFFRON, M., 1934. Untersuchungen über das Bewegungssehen bei Libellenlarven, Fliegen und Fischen, *Z. vergl. Physiol.*, **20,** 299–337.

HASSENSTEIN, B., 1954. Über die Sehschärfe von Superpositionsaugen (Versuche an *Lysmata seticaudata* und *Leander serratus*), *Pubbl. staz. zool. Napoli*, **25,** 1–8.

HASSENSTEIN, B., 1958. Die Stärke von optokinetischen Reaktionen auf verschiedene Mustergeschwindigkeiten, *Z. Naturforsch.*, **13B,** 1–6.

HASSENSTEIN, B., 1959. Optokinetische Wirksamkeit bewegter periodischer Muster. (Nach Messungen am Rüsselkäfer *Chlorophanus viridis*.), *Z. Naturforsch.*, **14B,** 659–74.

HERTZ, M., 1934. Zur Physiologie des Formen- und Bewegungssehens, *Z. vergl. Physiol.*, **21,** 579–615.

KIRSCHFELD, K., 1961. Quantitative Beziehungen zwischen Lichtreiz und monophasischem Elektroretinogramm bei Rüsselkäfern, *Z. vergl. Physiol.*, **44,** 371–413.

KUNZE, P., 1961. Untersuchung des Bewegungssehens fixiert fliegender Bienen, *Z. vergl. Physiol.*, **44,** 656–84.

KUNZE, P., 1963. Der Einfluss der Grösse bewegter Felder auf den optokinetischen Augenstielnystagmus der Winkerkrabbe, *Ergebnisse der Biologie*, **26,** 55–62.

MITTELSTAEDT, H., 1951. Zur Analyse physiologischer Regelungssysteme, *Verh. Dtsch. Zool. Ges. in Wilhelmshaven*, pp. 150–57.

VARJÚ, D., 1959. Optomotorische Reaktionen auf die Bewegung periodischer Helligkeitsmuster, *Z. Naturforsch.*, **14B,** 724–35.

WATERMAN, T. H., 1961. Light sensitivity and vision, in *The Physiology of Crustacea* (vol. II), T. H. Waterman, ed., New York: Academic Press.

15. The Eye-Movement Control System: Evolution of a Model

DEREK H. FENDER, *California Institute of Technology*

The study of eye movements, leading to their qualitative and quantitative description, now stretches back over more than a century. It has long been apparent to every worker in this field that the eyeball and the direction of gaze are positioned by a very sensitive control, yet it is only recently that an attempt has been made to analyze the structure of the system by using the concepts of feedback-control theory (Fender and Nye, 1961; Stark, Vossius, and Young, 1962; Sünderhauf, 1960; and Vossius, 1960).

The control mechanisms of the eye enable the visual system to achieve maximum acuity by positioning the image of an external object on the central fovea. This can be done if the system employs a signal proportional to the displacement of the desired portion of the image from the fovea and controls the movement of the eyeball in such a way as to reduce this discrepancy to zero. In addition, the eye-control system is able to distinguish between retinal-image movement produced by voluntary rotation of the eye or body and image movement resulting from the displacement of objects in the external world. This indicates that there must be at least one other informationally independent feedback pathway.

Thus the control system of the eye involves an interconnected network of functions extending up to central levels and enabling us to perceive a stable world within which the assessment of position and velocity may be made. In principle the *minimum* organization necessary to achieve the observed elements of behavior can be deduced solely from a study of the feedback sources. In this context the problem of visual stability depends for its solution first on knowledge of the origin and nature of the feedback signals and then on knowledge of how the information derived from these signals is processed.

Experimental Procedures

The usual techniques for studying the behavior of servo systems involve the measurement of the response of the system to certain standardized inputs

This research is supported by a grant from the National Institutes of Health, United States Public Health Service.

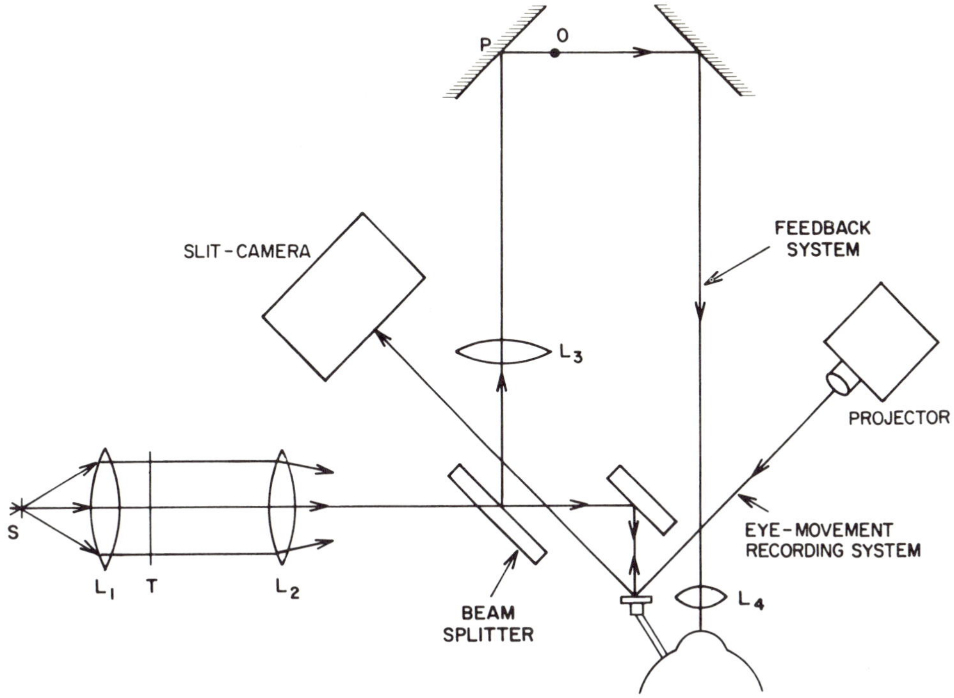

FIG. 1. Apparatus for varying the magnitude of retinal-image feedback and for recording eye movements.

when the various feedback loops are opened or their transmission characteristics modified. In the eye-movement control system, the loop most readily available for experimental study is the feedback generated by retinal-image motion; three methods, widely differing in their characteristics, have been used for this.

Fender and Nye (1961) used the equipment shown in Fig. 1. In normal vision, when the eye rotates through an angle θ, the image of a stationary point in the external world is displaced on the retina through a distance $d = R\theta$, where R is the distance of the rear nodal point of the dioptric system from the retina. The apparatus illustrated in the figure has the effect of multiplying the retinal displacement by a factor F, so that the modified displacement is given by

$$d' = FR\theta. \tag{1}$$

The magnitude of F can be adjusted to any positive or negative value or to zero.

The source S illuminates the target T with a beam of collimated light; the source is then refocussed by the lens L_2 onto a mirror attached to a contact lens worn by the subject. When the eye and the contact-lens mirror rotate through an angle θ, the reflected beam of light rotates through an angle 2θ. The image of T is formed at O, and for a rotation θ it is displaced through a distance $2\theta f_3$, where f_3 is the focal length of lens L_3. If the image at O is viewed through a lens L_4 of focal length f_4, then the image appears to rotate through an angle $2\theta f_3/f_4$. The angular displacement of the image with respect to the eyeball is therefore

$$\theta' = \theta[(2f_3/f_4) - 1]. \tag{2}$$

The resultant linear displacement of the image of T on the retina is

$$d' = R\theta' = R\theta[(2f_3/f_4) - 1]. \tag{3}$$

From Eq. (1) we have $d' = FR\theta$. Therefore

$$F = [(2f_3/f_4) - 1]. \tag{4}$$

If the ratio of the focal lengths (f_3/f_4) is made equal to $\frac{1}{2}$ (i.e., the system forms a telescope of magnification $\frac{1}{2}$), then $F = 0$, and from Eq. (2) θ' is also zero. Thus the image seen by the subject rotates through the same angle as the eye itself, the image of T is stabilized with respect to the retina, and any retinal-image feedback loop is open-circuited. By adjustment of the ratio f_3/f_4, values of F in the range $\infty > F > -1$ can be achieved with a roof prism at the point P (Fig. 1). Alternatively, a range of $-\infty < F < -1$ can be achieved if a plane mirror is substituted for the roof prism.

The response of the eye to periodic displacements of the target under various values of F is recorded by means of another beam of light separated from the telescopic system as shown. The sensitivity of the feedback and recording system is about ± 15 sec of arc with a bandwidth of 0–400 cps. The field of the equipment is ± 5 deg; it operates equally well for motion of the visual axis in any direction, but the necessity for wearing a tightly fitting contact lens severely limits the number of subjects available for the experiment.

Rashbass and Westheimer (1961) have used an apparatus in which the subject views a target consisting of a 3-mm white spot on the face of a large cathode-ray tube 2 m from the eye. The spot may be deflected horizontally with applied voltage signals.

The horizontal movements of the eye are recorded by measuring the distance between the temporal limbus and the image of a light source formed by specular reflection in the anterior surface of the cornea. Seen from the front, a point in the limbus moves through the same lateral distance as the corneal reflection of a distant light source when the eye makes side-to-side translational movement. When the eye makes a rotational movement around a vertical axis, however, these two landmarks move through unequal lateral distances. The separation of a point on the limbus from the corneal reflection is therefore a function of rotational eye position only; it is unaffected by translational shifts of the eye.

A short vertical line produced by a high-frequency vertical scan on a subsidiary cathode-ray tube is imaged by means of a lens on the anterior surface of the eye. The line scans the eye horizontally, and the light reflected from the anterior surface of the eye is measured by two photomultiplier cells. During a single scan the output of one cell undergoes a change when the image crosses from the sclera to the cornea. In the other cell an output pulse is generated the instant a specular reflection of the image from the cornea is formed at this particular cell. In the summed output of the two cells the time interval of the

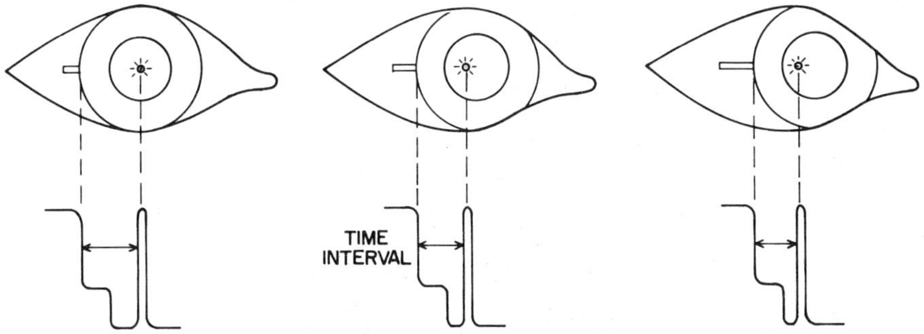

FIG. 2. Use of reflections from limbus and cornea to measure eyeball rotation (redrawn from Rashbass and Westheimer, 1961).

two changes is a measure of the rotational eye position (Fig. 2). The time interval is isolated electronically, and a signal proportional to the angular position of the eye is available. This signal is adjusted so that the output is equal to the voltage required to drive the spot on the stimulus cathode-ray tube through a distance corresponding to the movement of the eye, thus removing retinal signals generated by image motion. The noise level is less than 5 min of arc, the frequency response of the recording system extends at least to 30 cps, and the field of the equipment is 8 deg. However, the method works only for side-to-side motion of the visual axis.

Stark, Vossius, and Young (1962) used a system in which they measured the eye position by detecting the difference between diffuse light reflected by the sclera and that reflected by the iris. Part A of Fig. 3 shows the arrangement of

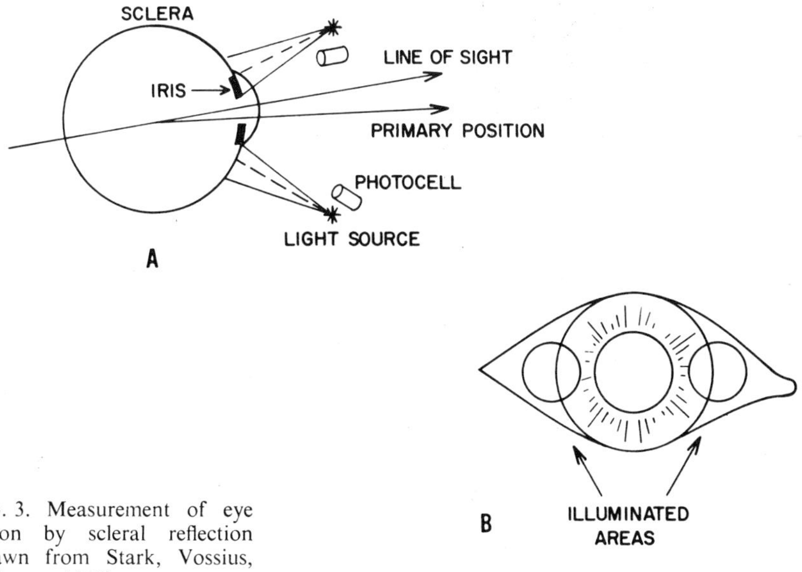

FIG. 3. Measurement of eye position by scleral reflection (redrawn from Stark, Vossius, and Young, 1962).

light sources and photocells mounted on a pair of goggles worn by the human subject; B shows how light is reflected from either side of the eye. It is clear that the areas of sclera illuminated by each lamp, and hence the amounts of reflected light received by each photocell, are functions of the angle of sight.

A mirror galvanometer, positioned directly below the subject's head, is used to project the target. If a signal is fed from the eye-movement recorder to the galvanometer input with the correct gain, the target can be made to move through the same angle as the eyeball, thus opening the retinal feedback loop (but for horizontal displacements only).

Feedback Loops in the Eye-Movement Control System

Major investigations of the response of the eye to targets moving with sine-wave oscillatory motion were carried out by Fender and Nye (1961) and by Stark, Vossius, and Young (1962). Although the two experimental techniques are greatly dissimilar, the results obtained are in substantial agreement. The motion of the eye when following such a target in normal vision is shown in Fig. 4; clearly the response of the eye differs from that of a mechanical system, since it contains some smooth following components, which are not of the same frequency as the input, and some flicks, which may be likened to step functions and have no simple relation to the input. In the discussion which follows, the component of the output which has the same frequency as the input is considered to be the response of the system. Other components of the output are considered in later sections.

Another variation from the normal analytical techniques of feedback control systems should be noted here. In examining a control mechanism, it

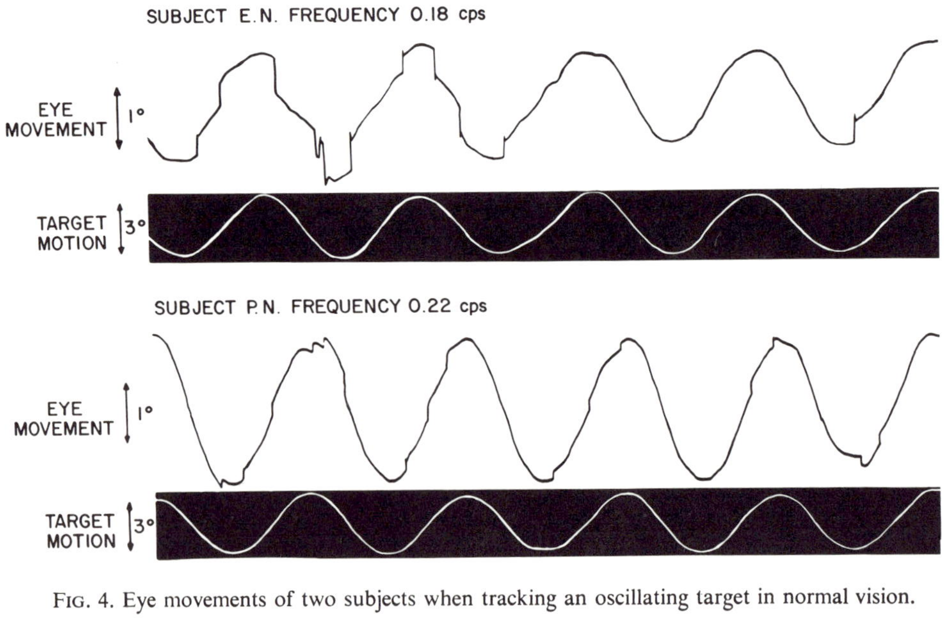

FIG. 4. Eye movements of two subjects when tracking an oscillating target in normal vision.

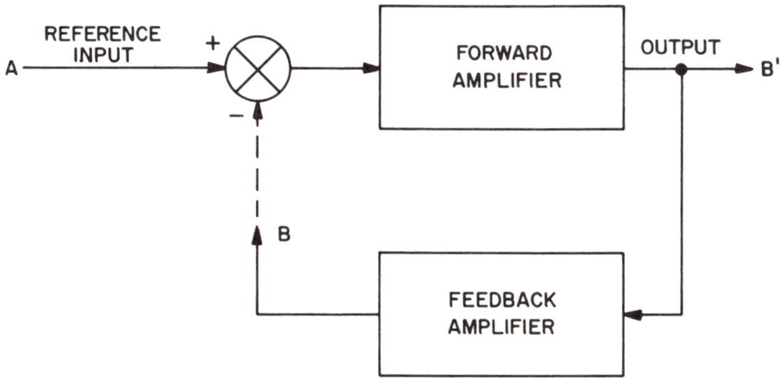

FIG. 5. Simple feedback circuit.

is customary to break the feedback loop at the position of the dotted line shown in Fig. 5; the system is then tested by applying an input at A and measuring the response at B. The ratio of these gives the so-called open-loop transfer function of the system; all methods of analysis by Bode plot [Chap. 3]* or Nyquist diagram [Chap. 5] assume that the measurement has been made in this way. But in the human eye-movement control system we are as yet unable to make a measurement at point B, for this is located somewhere in the cortex; instead the eye movement is measured at point B′ whether the feedback loop is closed or open. If the transfer functions are expressed in terms of the complex variable s [Chap. 2] and written symbolically as the functions $G(s)$ for the forward amplifier and $H(s)$ for the feedback amplifier, then the system has the following transfer functions.

(i) Normal open-loop measured from A and B′:

(5) $$T_1 = G(s)H(s).$$

(ii) Open-loop as used in this paper, measured from A to B′:

(6) $$T_2 = G(s).$$

(iii) Closed-loop measured from A to B′ (in this measurement, normal theory and the usage of this paper coincide):

(7) $$T_3 = \frac{G(s)}{1 + G(s)H(s)} \qquad \text{[Chap. 4].}$$

It will be seen, therefore, that measurements (i) and (ii) will coincide only if $H(s)$ is unity at all frequencies. However, if T_2 and T_3 can be measured, then $H(s)$ can be calculated by drawing a comparison between Eqs. (6) and (7).

The gains of the system in the open-loop ($F = 0$) and closed-loop ($F = 1$) conditions are shown in the Bode plot of Fig. 6; if the values from these curves are inserted in Eqs. (6) and (7), it can be established that the magnitude of $H(s)$ is very nearly unity throughout the frequency range at which following

* For the convenience of the reader whose contact with the literature of feedback control systems is limited, references on this subject are given in square brackets thus [Chap. 3] to a standard text: Bower and Schultheiss, 1958. Many other textbooks also treat the same points.

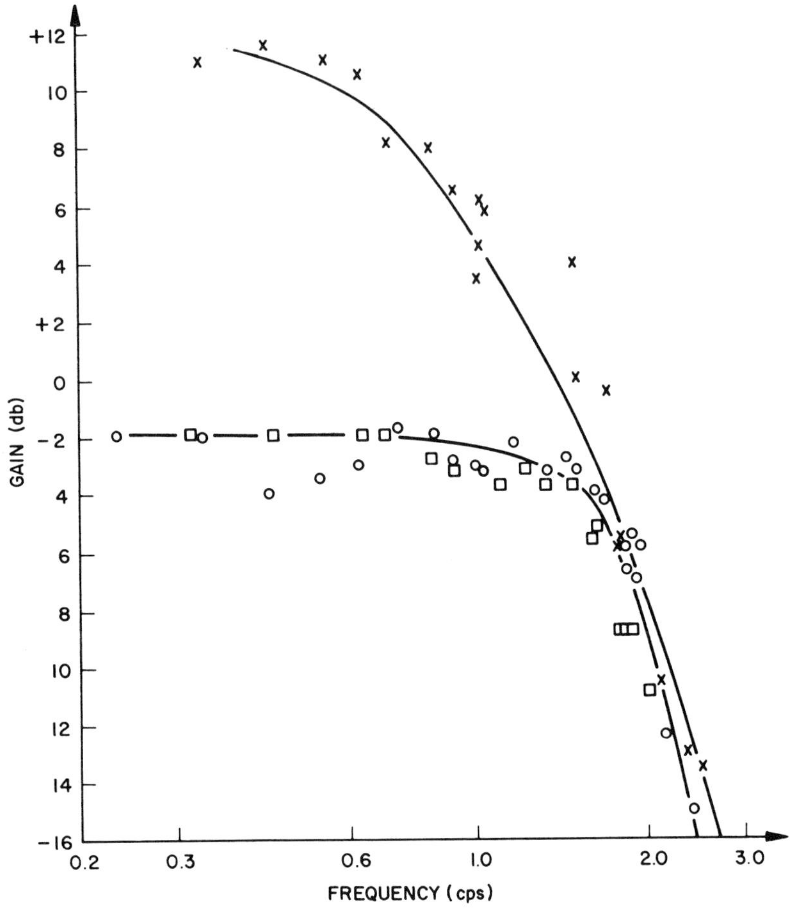

FIG. 6. Bode plot showing variation of gain with frequency and feedback (horizontal motion). $\times$, $F = 0$. $\bigcirc$, $F = 1$. $\square$, predicted values for $F = 1$.

eye motions can be elicited. The phase shift in the feedback loop is also very nearly zero degrees; this can perhaps best be demonstrated from the Nyquist plot of the open-loop response $T_2 = G(s)$ shown in Fig. 7, for if $H(s)$ is unity, then the closed-loop transfer function T_3 is given by

$$T_3 = \frac{G(s)}{1 + G(s)}. \tag{8}$$

At any frequency the function $G(s)$ is represented in the diagram by the vector **A** and $1 + G(s)$ by the vector **B**; the ratio of these two lengths can be derived as in Eq. (8) and used to predict the gain of the closed-loop system [Chap. 7]. These values are marked on the Bode plot of Fig. 6, and the correspondence with the measured values is good, even at the higher frequencies, where phase shift in the feedback loop is most likely to be troublesome.

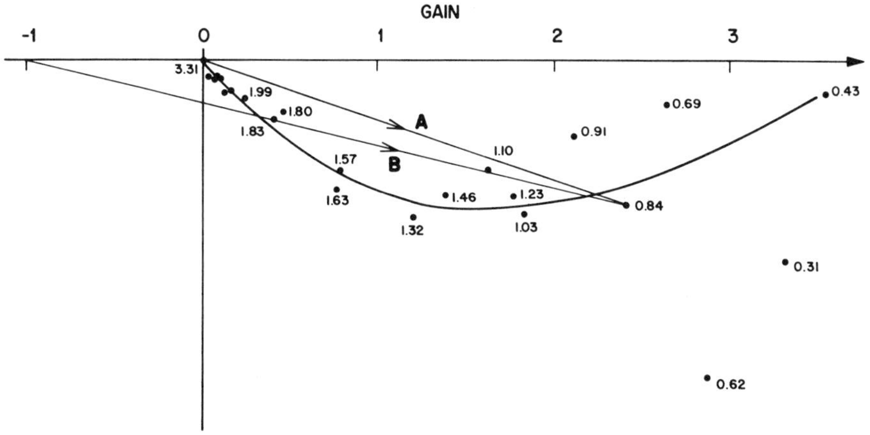

FIG. 7. Nyquist diagram showing the open-loop phase lag of the eye-movement control system. The frequency in cps of each observation is noted at the data point.

At this stage the model of the eye-movement control system can be presented as in Fig. 8, with just one box representing the forward transmission of the system combined with unity negative feedback derived from retinal-image displacement. The "forward amplifier" represents all the energy changes which take place during the transmission of a signal from the cortex through the motor neurons and their synaptic junctions to the eye muscles, and also the dynamics of the eye muscles and of the eyeball itself. However, the open-loop transfer function T_2 as measured in these experiments describes the performance of the forward amplifier; hence we can derive a certain amount of information about $G(s)$ from the Bode and Nyquist plots of the open-loop transfer function. If we approximate the Bode plot with straight-line asymptotic sections whose slopes are multiples of 6 db per octave as in Fig. 9, then each such asymptote indicates an information transfer in the control system through a simple filter section whose electrical analog is shown in Fig. 10 [Chap. 3]. These filters act as integrators and tend to smooth out signals changing more rapidly than the time constants of the circuits. The time constants are given by $\frac{1}{2}\pi f_c$, where f_c is the frequency of intersection of successive portions of asymptotes. From Fig. 9 we see that the steepest asymptote has a slope of 24 db per

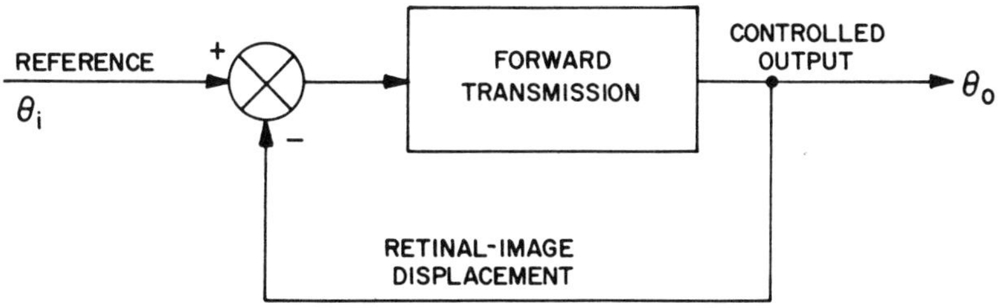

FIG. 8. Eye-movement control system with unity retinal-image feedback.

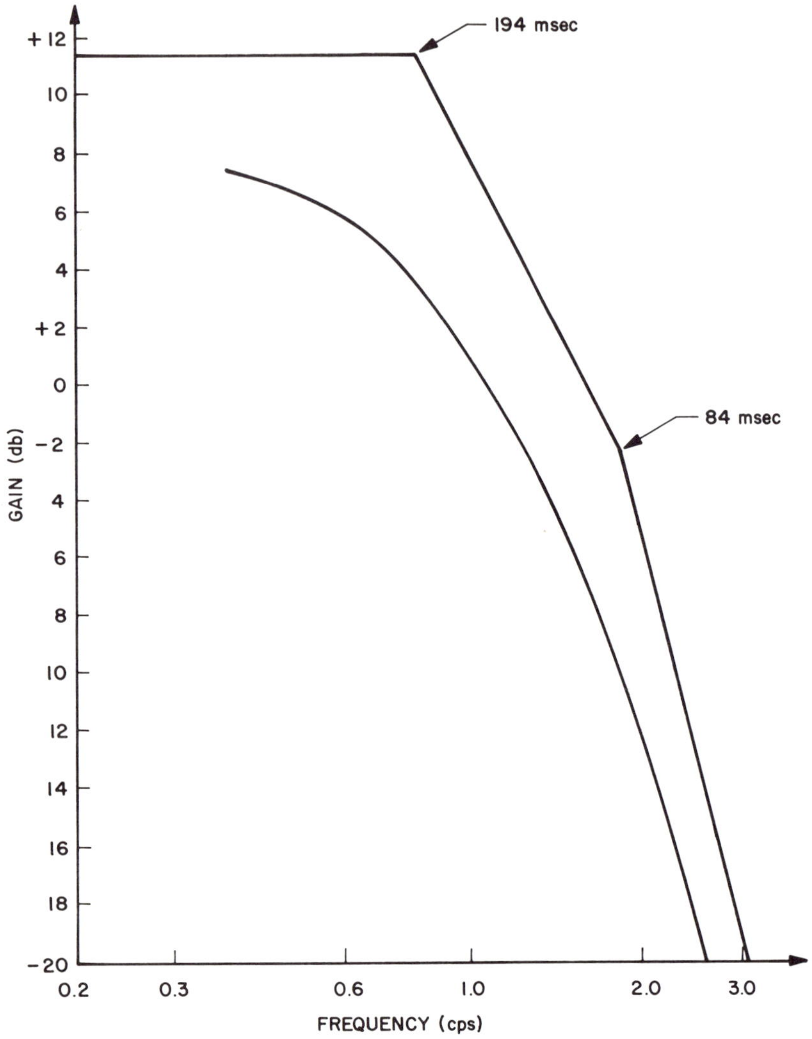

FIG. 9. Asymptotic approximation to Bode diagram.

octave, indicating four information-transfer points, and that these occur in pairs having time constants of 194 msec and 84 msec, respectively.

This treatment is not restricted to those portions of the oculomotor tract where the signal is of an electrical nature; the use of these analog filters merely indicates that in any particular stage of the transmission system the information content of the signal, whatever it might be, is modified in the same way as the information content of an electrical signal passing through the network shown in Fig. 10. Speculation on the physiological significance of the information-transfer stages and their time constants is deferred to the next section.

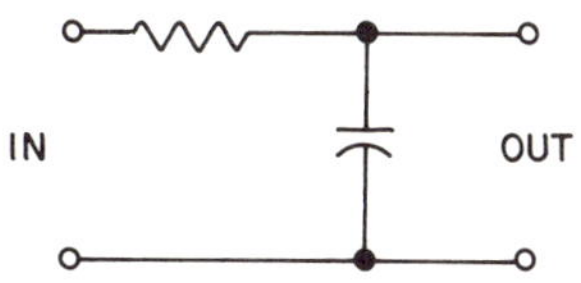

FIG. 10. Simple filter element.

In carrying out the above analysis we have used only the gain characteristics from the Bode plot; unfortunately, if we turn our attention to the phase characteristics as shown in the Nyquist diagram (Fig. 7), we reach a different conclusion. Each of the filter elements discussed above can introduce a phase lag of 90 deg, hence four would give a phase shift of up to 360 deg, but we see that the Nyquist diagram only enters the lower right-hand quadrant, indicating a maximum phase lag of 90 deg and thus only a single filter element (in engineering practice an element of this sort may consist of a linear element having first-derivative negative feedback around it). We have thus reached an impasse: To satisfy the observed gain characteristics of the system we must postulate at least four filter stages in the forward transmission path, but the phase characteristics allow no more than one. A possible solution (and here we become speculative) is to employ a system with a number of information transfer stages having a positive-feedback, or regenerative, loop around them; in this way the large phase lags of the multistage system can be avoided. Before including feedback loops of this type in the model, however, we would be wise to consider the possible sources of feedback in the optomotor system, to identify those which have been tested by experiment, and to evaluate the physiological evidence for others of a more tentative nature.

Feedback Sources in the Oculomotor Tract

Retinal-Image Motion. The displacement of the image from some rather ill-defined region in the fovea acts as an error signal generating unity negative positional feedback for tracking tasks. This has been clearly demonstrated by Fender and Nye (1961) and by Stark, Vossius, and Young (1962), whose work has been reviewed in the preceding sections. The ability to extract such an error signal appears to exist in a very limited region of the retina; subjects find it almost impossible to track a moving target while maintaining fixation a few degrees away from it. One subject did, however, learn to track a target reasonably well at very low frequencies while "fixating" it with his blind spot.

It has recently been shown that retinal-image motion also generates velocity signals. If a stabilized target is suddenly moved to one side of the visual axis, a series of flicks is generated (Fig. 11); these flicks are an attempt to restore central fixation but are of no avail, since the image is stabilized. If the target is now returned to the visual axis at constant velocity, a smooth eye movement is immediately initiated *in the same direction* as the target motion, although this is in the opposite direction to target displacement. A somewhat similar effect has been noted by Rashbass (1961), who concludes that the smooth component of eye movements in tracking tasks is generated entirely by velocity signals from the retina.

At this point, however, it should be noted that apparently both the retina and cortex have at least two parts to play in the eye-movement control system, and it is important to distinguish between them. The retina is used

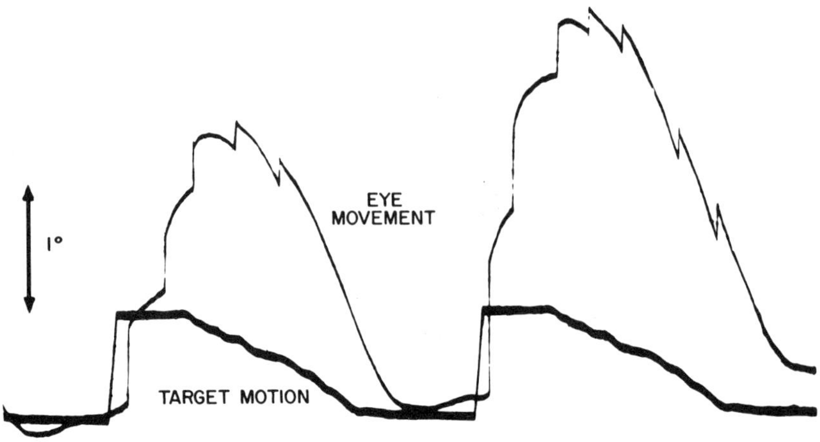

FIG. 11. Eye movements generated by a displaced target moving with constant velocity; retinal-image feedback loop open (record taken by P. W. Nye).

primarily to display the visual world to the cortex. This display is in itself a function of the direction of gaze, θ_0, but if the eyeball is set into any pattern of activity, information derived from the retina is used to control the motion of the eyeball and hence the value of θ_0. In a similar fashion the cortex acts as an analyzer of the visual world projected onto it, but once a course of action is selected, the cortex generates the commands which produce the desired motion of the eyeball. It is desirable that both of these functions of the retina and cortex appear in the model of the eye-movement control system. In this context it is apparent that a moving target might generate a velocity signal in the cortex (acting as an analyzer) or even in the retina, this signal being transmitted to the cortex as one abstraction of the visual world. The cortex initiates appropriate eyeball motions but uses only retinal-image position as a feedback signal to control these motions; there is no evidence in the work reported earlier that retinal-image velocity is used in the feedback loop. The displacement and velocity signals appear to be separate; their effects are superimposable, and it is presumably the velocity signal which generates optokinetic nystagmus.

Motion of the Eyeball. When the retinal-image feedback loop is broken, the motion of the eyeball is still stable in a restricted sense; i.e., when the eyeball is called on to track a sinusoidally moving target, its motion contains a component at the same frequency as the input and of finite amplitude, but as shown in Fig. 12, a drift at a much lower frequency is also present. The oscillatory motion shows rather more than unity gain and appears to be stabilized by feedback signals of (presumably) either position or velocity; it is tempting to equate these signals with those recorded by Cooper, Daniel, and Whitteridge (1955) from the muscle spindles of the extraocular muscles. The nerve-impulse frequency (defined as the reciprocal of the inter-spike interval) generated by these muscle spindles is proportional to the sum of the instantaneous angular displacement of the globe and the first derivative of this displacement.

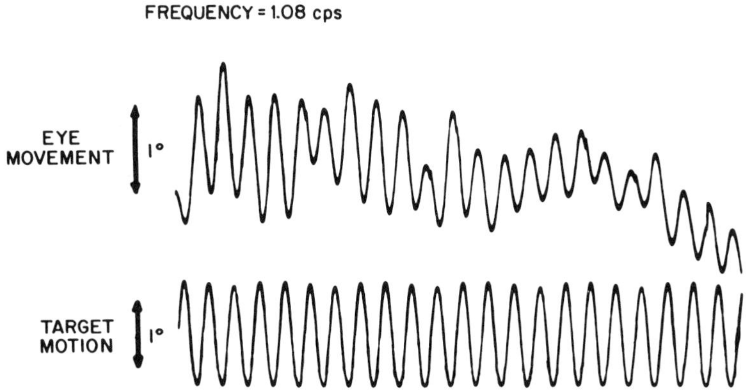

FIG. 12. Eye tracking movements with no retinal feedback.

There is experimental evidence that the positional signal is not used, for if a subject views two similar but separately generated stabilized images, one with each eye, it is found that for a short period the two visual axes move in conjunction. However, this motion quickly breaks down, and the visual axes move independently, sometimes getting as far apart as 30 deg in the horizontal direction and 15 deg in the vertical. There is, of course, no binocular retinal-image disparity to act as a cue in this case, and it appears that any positional signal which might arise from the extraocular muscles is quite ineffective in maintaining the parallelism of the visual axes. It is not possible to reach any *direct* conclusion concerning the derivative signal in this experiment, nor has the histological work of Cooper, Daniel, and Whitteridge (1955) identified the neural pathways taken by these signals. At best, then, we can say that if such derivative information is needed by the eye-movement control system, a source for the signals exists, but the actual use of the signal is not known.

Feedback Within the C.N.S. It was suggested above that the model of the eye-movement control system could be brought more into line with the observed behavior of the eye if we postulated a regenerative loop. This proposal was made in another context by von Holst (1951), who pointed out that an efference copy (equivalent to positive feedback) would enable an organism to differentiate between self-motion and world-motion. The characteristics of such a loop have have not been tested experimentally, but animal experiments such as those conducted on the positional reflexes of fish when in a centrifuge or with otoliths removed, or those on a pigeon in conditions of enhanced gravitational acceleration, do not conflict with this postulate. The output for the efference copy must occur at the input to the extraocular muscles, and it is transmitted to the cortex to be compared with the signal arising from retinal image motion. Thus we must break the forward-transmission channel of Fig. 8 into two blocks, as shown in Fig. 13. Block $G_1(s)$ is the transfer function of all cortical activity of the eye-movement control, and $G_2(s)$ represents the dynamics of the eye muscles and the eyeball.

A system with positive feedback is in general oscillatory and can only be

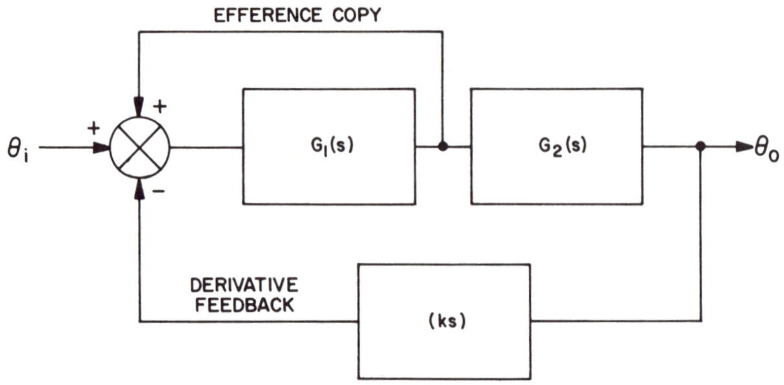

FIG. 13. Block diagram of forward-transmission channel.

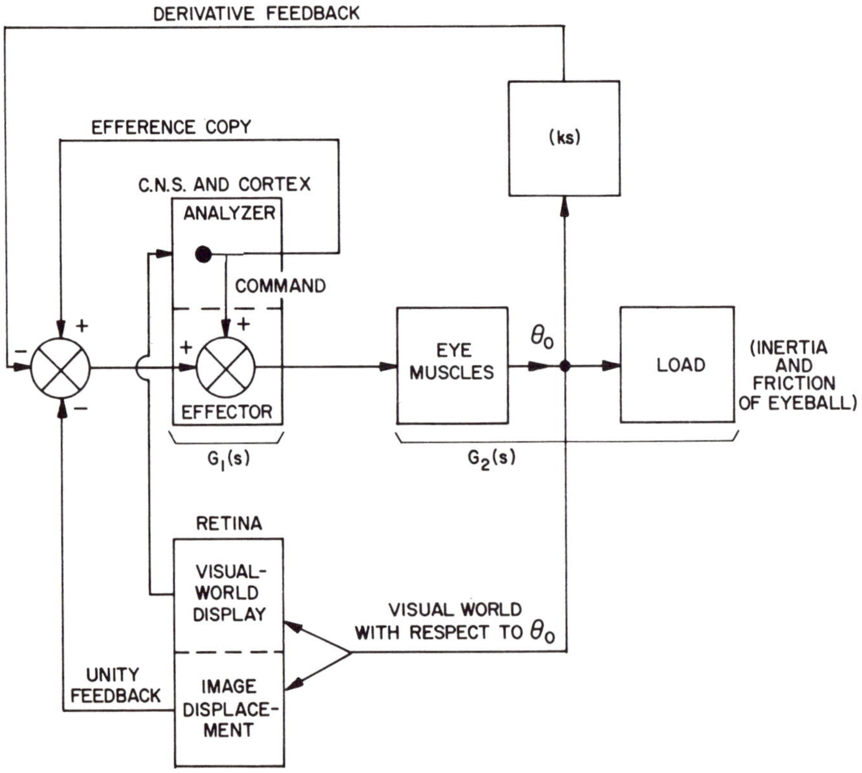

FIG. 14. Block diagram of eye-movement control system.

stabilized by derivative feedback if this loop encloses the unstable element; hence velocity information, represented by the box labeled (ks) [Chap. 2], is taken from the output of the eye muscles and returned as negative feedback to the cortex. The transfer function of this system is given by

$$\frac{\theta_0}{\theta_i}(s) = \frac{G_1(s)G_2(s)}{1 - G_1(s) + ksG_1(s)G_2(s)}\,. \tag{9}$$

If we evaluate this equation, using four filter elements distributed between $G_1(s)$ and $G_2(s)$ and inserting the time constants derived above, we can obtain phase lags of about -100 deg at 2 cps, whereas the measured value is about -40 deg. If phase lags as small as this are to be obtained, then the gain k of the derivative element must be as high as 15, which appears to be quite unreasonable in a biological system.

If the ideas discussed above are incorporated in the model of the eye-movement control system, then a system such as that shown in Fig. 14 results; but although this can be adjusted to give a better fit with observed results than earlier models, it is obviously much more speculative. This, however, is the function of a model; it is used to introduce hypotheses by analogy from another discipline and then to test these speculations by experiment. In this case, the derivative feedback might be attacked next; if this arises from the extraocular muscles, then the loop might possibly be opened by surgical procedures or by the selective action of drugs. Tracking experiments with the retinal feedback loop opened would then elicit the nature of the derivative term. Experiments such as this can reasonably be performed only on animals, but some small progress has been made in this direction.

Predictive Control of Eye Movements

The real eye performs tracking tasks with much less phase lag than any linear mechanical system, despite the additions to the model that we have made so far. A control process can have these characteristics only if it is capable of analyzing the input and predicting the probable future position of the target. Does the eye-movement control behave as a predictive system? This question has been treated succinctly by Stark, Vossius, and Young (1962), who asked their subjects to track an input target oscillation consisting of four to nine simultaneous sinusoids which had no low-harmonic relationship to each other; this produces a complex target motion which is unpredictable. The results of the experiment are shown in the closed-loop gain and phase plots of Fig. 15; the response to unpredictable inputs can be adequately described by the model shown in Fig. 14 if we manipulate its transfer function, using plausible values for the network parameters, but we have already seen that this does not correctly represent the response to predictable inputs. We conclude, therefore, that we should add a predictive element to the model.

We might, however, consider the method by which this prediction is mediated; the eye has two distinct forms of motion, and uses both smooth tracking and saccadic eye movements when following a moving target. Rashbass (1961) showed that these movements are caused by separate systems,

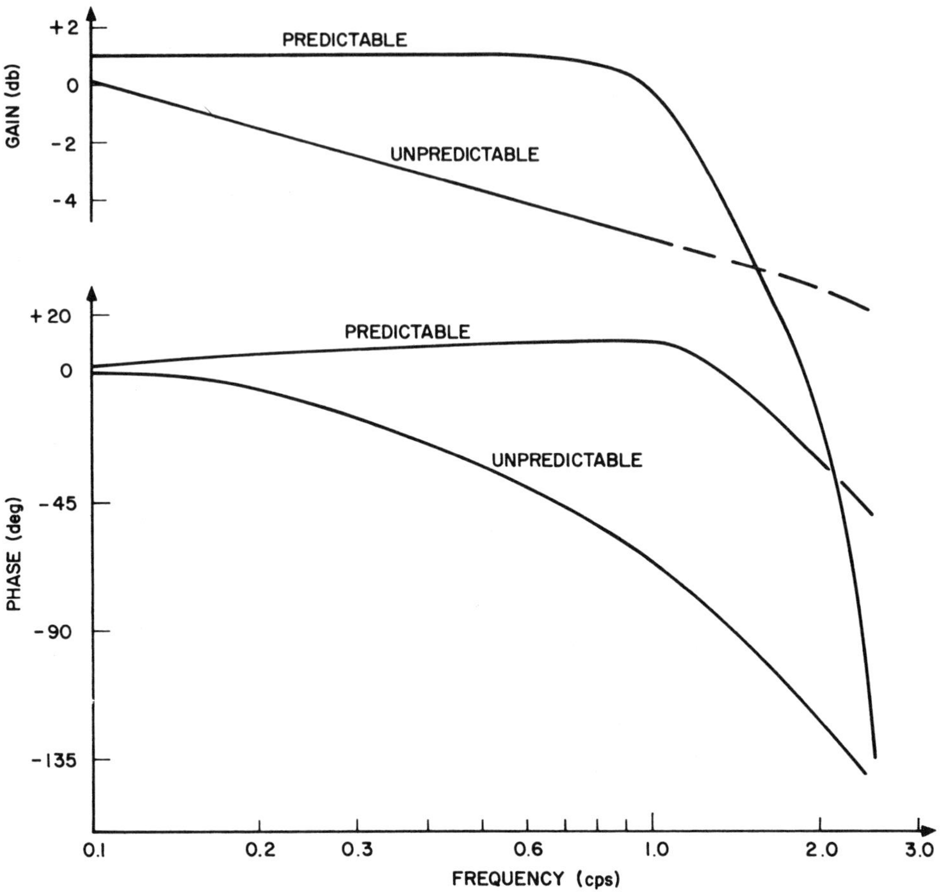

FIG. 15. Response of the eye to predictable and to unpredictable inputs (redrawn from Stark, Vossius, and Young, 1962).

since smooth tracking can be suppressed selectively by the action of hypnotic drugs. It has been suggested by many authors that the purpose of the flick mechanism is to correct errors of fixation during a tracking task. Does the predictive element use the saccades to maintain small phase lags?

It is easy to test the saccades alone, for if the target is moved from side to side in square-wave fashion, the following eye movements consist almost entirely of flicks. If the square waves occur in random time sequence, then the reaction time of each flick is about 200 msec, as shown in the upper histogram of Fig. 16, but if the square waves have a repetition rate of 0.45 cps, then the reaction time tends to zero, as shown in the lower histogram of the same diagram; prediction is thus operative in the flick mechanism.

In general it is not possible to find a stimulus which generates smooth following only. The nature of the following component has therefore been examined in this way: the response to a sinusoidal predictable input consists

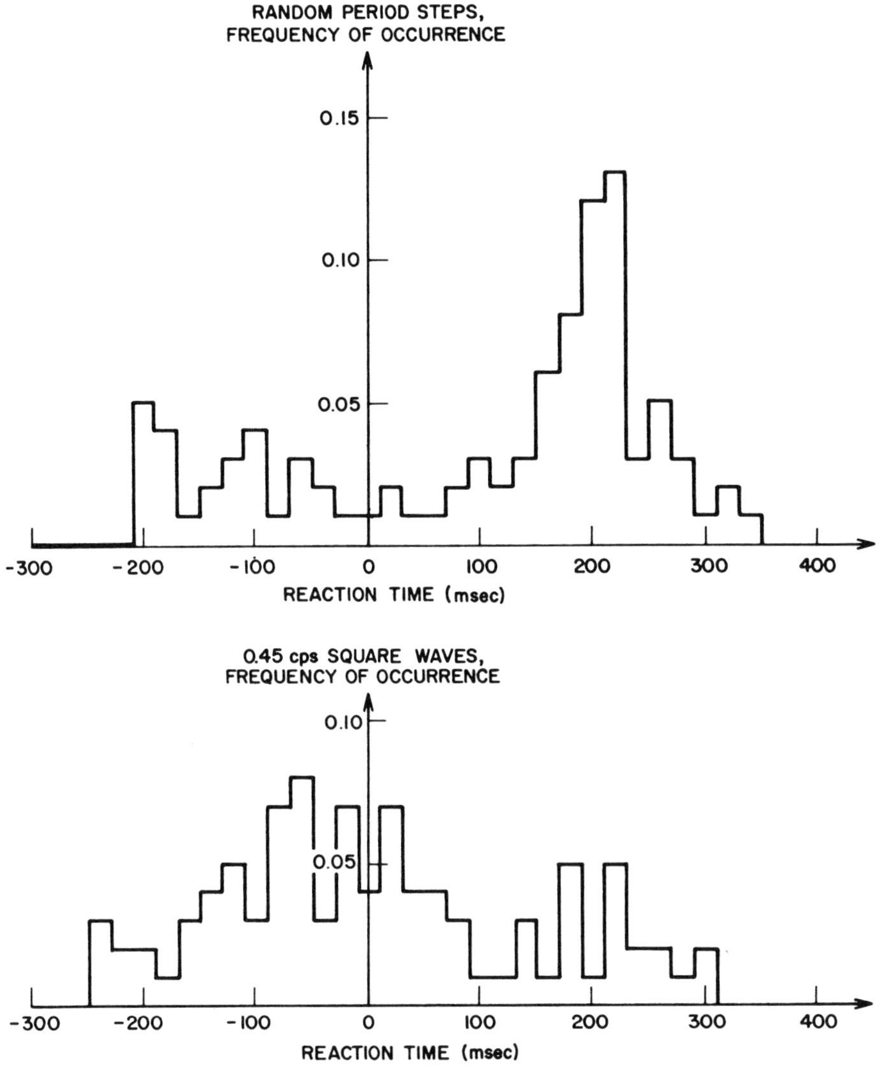

FIG. 16. Histogram showing predictive nature of saccades.

of both smooth components and saccades; the magnitude of each saccade is identified by a suitable computer program, and the value is subtracted from the angular deflection of the eyeball for all subsequent times through the experiment. The result is the same as if all sections of record between saccades were joined smoothly end to end. The results of analyzing a record such as this for amplitude and phase of output are shown in Fig. 17, from which it will be seen that the gain is slightly reduced but the phase is not affected. The smooth-following component is thus also predictive, and the flicks are used to correct errors of amplitude, not of phase. The cooperation of the higher levels of the cortex in predictive eye-movement control can be demonstrated in a number of

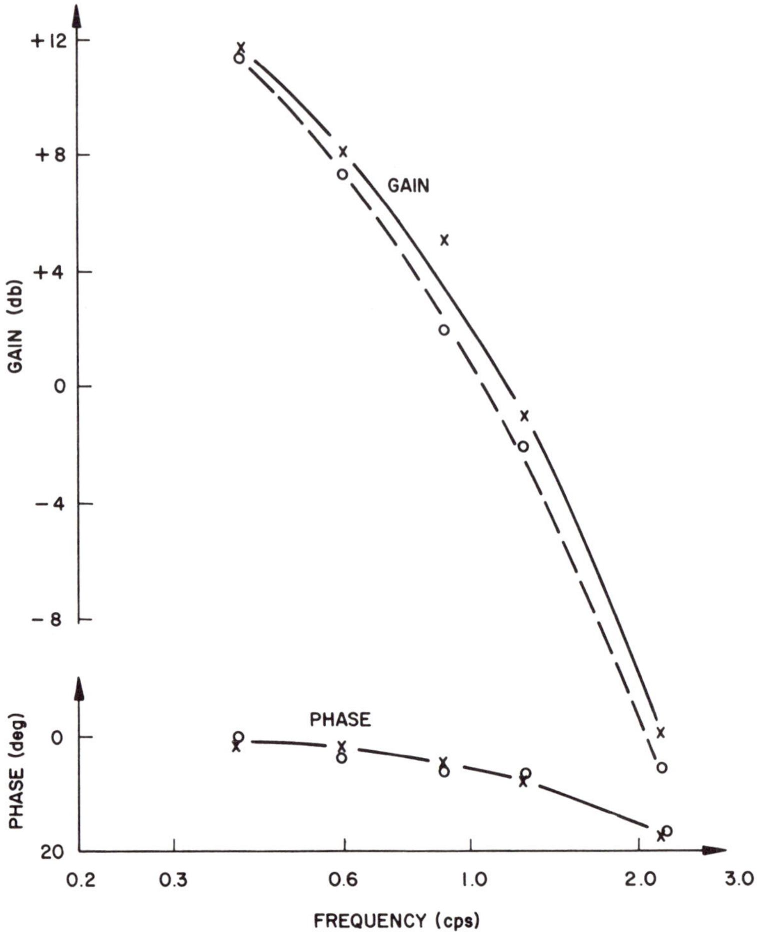

FIG. 17. Bode diagram showing effect of smooth-tracking component only.

ways; for example, if the target brightness is reduced until it is just below the subject's threshold, tracking motion of the eye still exists but prediction is no longer present. This experiment indicates that only if the signal reaches the levels of the cortex responsible for brightness perception can its help be invoked when it becomes necessary to adapt the eye-movement control system to a particular form of input. Finally, then, we may include these ideas in the block diagram of the system, as shown in Fig. 18. The predictive processes are all lumped into the analytic areas of the cortex, whereas the retina is shown as feeding information into several levels of the cortex.

It will be noted that the cortex receives its information via the retina. This raises another interesting point. It has been shown (Ditchburn, 1955; Lettvin, 1960) that perception is suppressed during flicks. In fact, this suppression may even extend from 40 msec before the flick is initiated until 140 msec later (Latour, 1962). This suppression may occur in the retina or in the cortex; if it occurs in the retina, then information reaches the cortex in a discontinuous

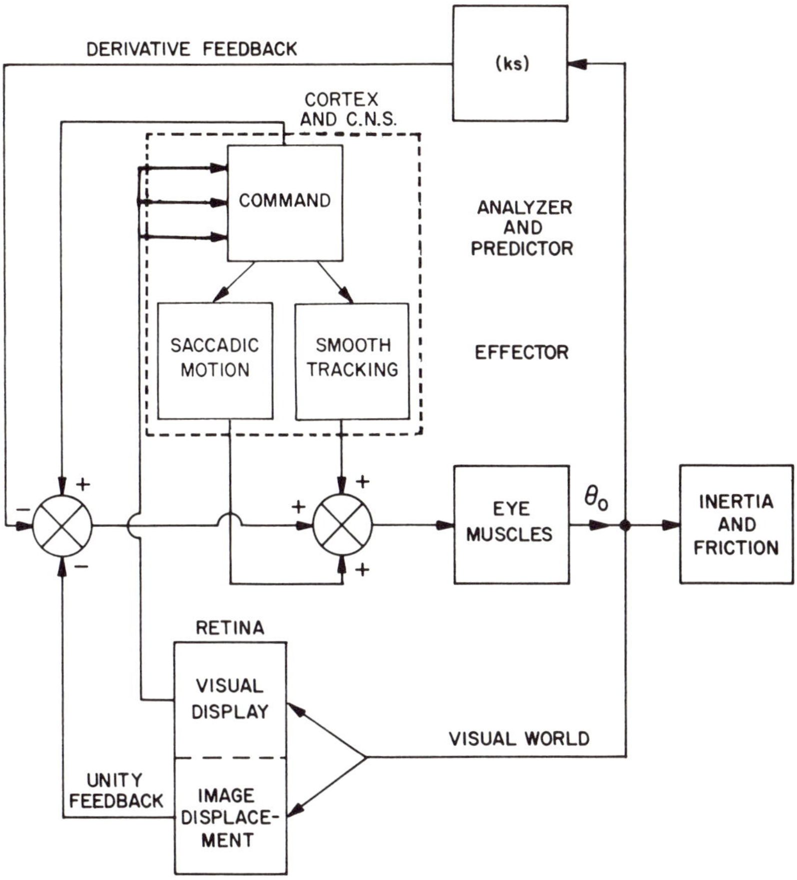

FIG. 18. A possible model of the eye-movement control system.

manner, and it would be more appropriate to treat the system as a sampled data controller. An analysis of this type has been carried out by Young,* apparently with some success.

Conclusion

In this paper an attempt has been made to describe the stages in the evolution of a model describing the eye-movement control system, delineating clearly at every step the postulates which have been exhaustively tested by experiment and found to be true, those which are physiologically plausible, and those which are merely speculative. The value of any model lies in its ability to direct the next step in the theoretical or experimental attack on the real system; perhaps the maximum benefit of the model described above lies in the clear demonstration it offers that the visual system is still a fruitful field of research.

* L. R. Young, personal communication, 1963.

References

Bower, J. L., and P. M. Schultheiss, 1958. *Introduction to the Design of Servomechanisms*, New York: Wiley.

Cooper, S., P. M. Daniel, and D. Whitteridge, 1955. Muscle spindles and other sensory endings in the extrinsic eye muscles; the physiology and anatomy of these receptors and of their connection with the brain-stem, *Brain*, **78**, 564–83.

Ditchburn, R. W., 1955. Eye-movements in relation to retinal action, *Optica Acta*, **1**, 171.

Fender, D. H., and P. W. Nye, 1961. An investigation of the mechanisms of eye movement control, *Kybernetik*, **1**, 81–8.

Holst, E. von, 1951. Zentralnervensystem und Peripherie in ihrem gegenseitigen Verhältnis, *Klin. Wschr.*, **29**, 97–105.

Latour, P. L., 1962. Visual thresholds during eye movements, *Vision Res.*, **2**, 261–62.

Lettvin, J. Y., 1960. M.I.T. Quarterly Progress Report, Research Laboratories of Electronics, July 1960.

Rashbass, C., 1961. The relationship between saccadic and smooth tracking eye movements, *J. Physiol.*, **159**, 326–38.

Rashbass, C., and G. Westheimer, 1961. Disjunctive eye movements, *J. Physiol.*, **159**, 339–60.

Stark, L., G. Vossius, and L. R. Young, 1962. Predictive control of eye tracking movements *Trans. IRE Professional Group on Human Factors in Electronics*, **HFE-3**, 52–56.

Sunderhauf, A., 1960. Untersuchungen über die Regelung der Augenbewegungen, *Klin. Mbl. Augenheilk.*, **136**, 837.

Vossius, G., 1960. Das System der Augenbewegung (I), *Z. Biol.*, **112**, 27–57.

16. Visual-Information Processing in the Beetle *Lixus*

JAMES C. BLISS, *Stanford Research Institute*

Since this material has been published elsewhere (Bliss, 1962), only a summary will be presented here.

Hassenstein (1959), Reichardt (1961, 1962), and Varjú (1959), have studied visual motion perception of the beetle *Chlorophanus viridis* by analysis of the optomotor response. In the basic form of their experiments, the beetle is placed at the center of a rotating striped drum. He is rigidly held by his thorax and a Y-maze globe is given him to hold, as shown in Fig. 1. The strips of the

FIG. 1. Beetle and Y-maze globe at the center of a striped cylinder.

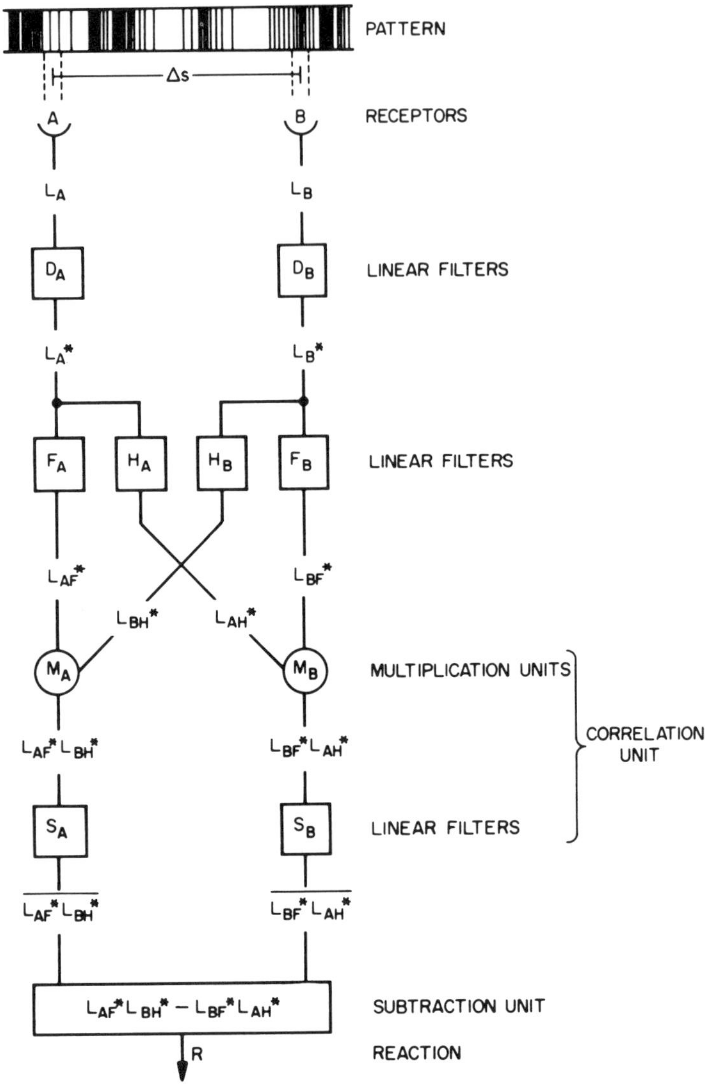

FIG. 2. Reichardt's minimal mathematical model for a two-ommatidia system.

Y-maze globe meet at 120-deg angles, so that as the beetle "walks" he repeatedly comes to divisions in his path which are not in his field of view. The tendency to turn right or left depends upon movement in the visual field. A quantitative measure of the response can be arbitrarily defined as $(W - A)/(W + A)$, where W is the number of choices in the same direction as the movement in the visual field and A is the number in the opposite direction. This measure, called R, can be related to the performance of a postulated mechanism for the perception of movement.

From their behavioral experiments these investigators have been able to formulate the system model for motion perception shown in Fig. 2. The stimuli received by the ommatidia in this model are linearly transformed, and

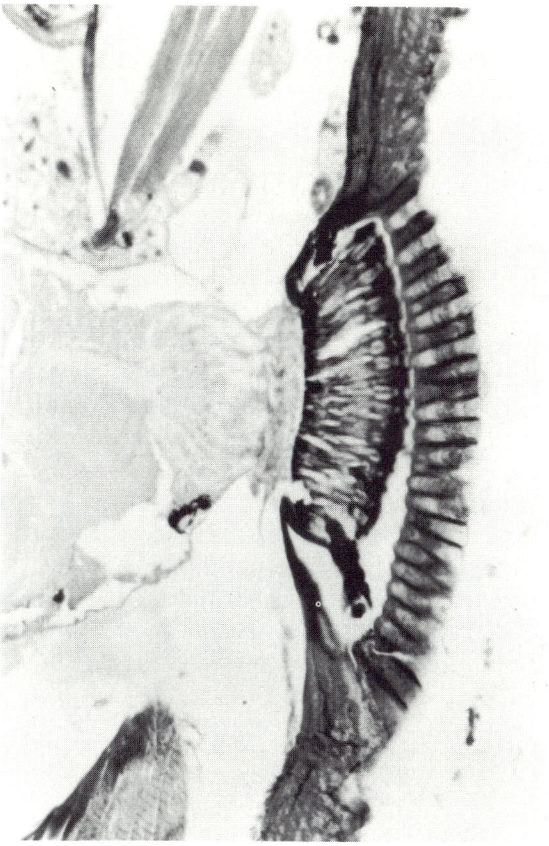

FIG. 3. Ten-micron-thick section of the head of a beetle, *Lixus*.

their interaction in the central nervous system is in accordance with the principle of first-order correlation. This model was able to predict the experimental finding that all visual patterns differing only in the phase relations of their Fourier components produce the same optomotor response as a function of pattern speed.

Figure 3 shows a 10-micron section of the eye of a beetle, *Lixus*. From such anatomical evidence, the angle between centerlines of adjacent ommatidia is estimated to be approximately 4 deg.

Consider Reichardt's mathematical model for a two-ommatidia system, shown in Fig. 2. If the F and H filters are simple lags and the D filter has a transfer function given by $W_D(\omega) = b_D\sqrt{j\omega}$, then for a sinusoidal input pattern the output of the model can be shown to be

$$R(\Delta t) = \frac{B\left(\frac{2\pi v}{\lambda}\right)^2 (a_H - a_F)}{\left[\left(\frac{2\pi v}{\lambda}\right)^2 + a_H^2\right]\left[\left(\frac{2\pi v}{\lambda}\right)^2 + a_F\right]} \sin \frac{2\pi \Delta s}{\lambda},$$

where B = const.,

v = velocity of the sinusoidal pattern,

λ = wavelength of the sinusoidal pattern,

a_H = reciprocal of the time constant of the H filter,

a_F = reciprocal of the time constant of the F filter,

Δs = distance between adjacent ommatidia.

This expression suggests a set of experiments to estimate Δs, a_H, and a_F. If the reaction to sinusoidal patterns of various wavelengths is measured for various pattern velocities, then each reaction curve should be bell-shaped as a function of velocity. The width of the bell is determined by a_H and a_F. Whether the "bell" is positive, negative, or zero is determined by the factor $\sin 2\pi\Delta s/\lambda$.

In one set of experiments the beetle *Lixus* was placed in the center of a striped cylinder. The direction of rotation, speed of rotation, and stripe spacing were varied. There were at least 100 decisions at each speed with each stripe spacing. Figure 4 shows the results. Note that the 4-deg wavelength always gave a positive reaction, the 6-deg wavelength always gave a negative reaction, the 8-deg and 12-deg wavelengths gave reactions close to zero, and the 18-deg wavelength always gave a positive reaction. These results vary with changes in wavelength in accordance with Reichardt's model for *Chlorophanus* if an ommatidial angle of about 4 deg is assumed. However, the bell-shaped reaction as a function of pattern speed predicted by Reichardt's model was not obtained.

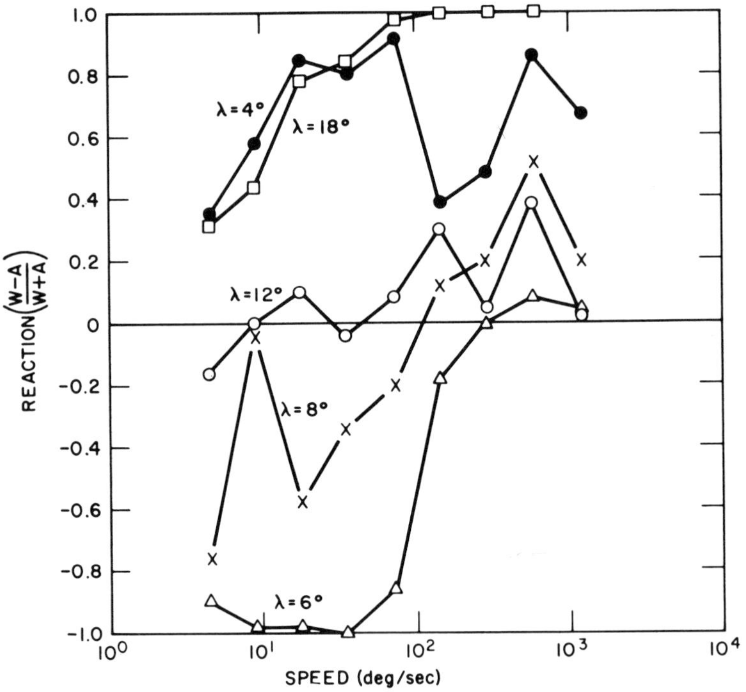

FIG. 4. Reaction vs. cylinder speed for various stripe wavelengths.

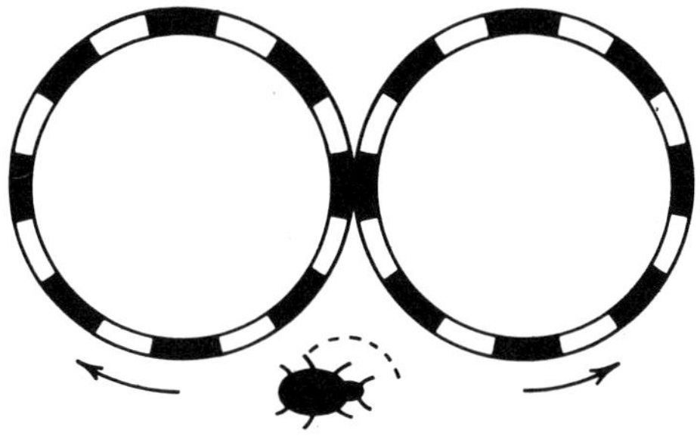

FIG. 5. Experimental setup for the double-cylinder experiment.

Results with the experimental setup shown in Fig. 5 were in agreement with the assumption that the partial reactions of ommatidial pairs are summed. Also, an analysis of sequential decisions showed that the probability of a decision in a given direction being followed by a decision in the same direction was greater than the zero-order probability in that direction.

Electrical recordings from the eye of the beetle *Lixus* were made of the electroretinogram and of spike potentials. The ERG study demonstrated the nonlinearity of this response and suggested a 13-msec pure delay followed by a simple lag network with corner frequency at 4 cps as an approximation for the response to small changes in light intensity. Spike potentials were obtained from a single cell that fired with an increase in light intensity.

The problem of piecing the results of both the behavioral experiments and the electrical-recording experiments into a model that might better describe which functions occur, and the order in which they occur, in the two-ommatidia system is formidable. Reiss has suggested the system shown in Fig. 6 as a bandpass filter which passes frequencies that are harmonics of $1/\Delta T$. Jenik (1962) also discussed such a system. Use of this idea as a component of the model in which the delay is external to the animal is consistent with the experimental results we have obtained.

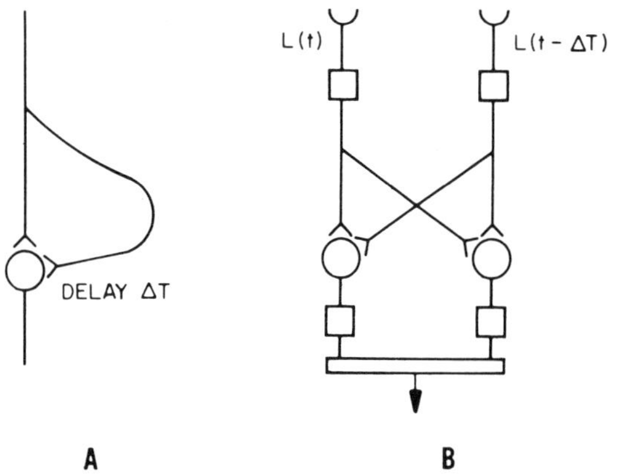

FIG. 6. If the output of the cell in A requires coincidence of the two inputs, then the model passes harmonics of $1/\Delta T$. In B the delay is external to the animal, and other processing is added.

References

BLISS, J. C., 1962. Visual information processing in the beetle *Lixus*, in *Optical Processing of Information*, D. K. Pollock, C. J. Koesler, and J. T. Tippett, eds., Baltimore: Spartan. 1962.

HASSENSTEIN, B., 1959. Optokinetische Wirksamkeit bewegter periodischer Muster (Nach Messungen am Rüsselkäfer *Chlorophanus viridis*), *Z. Naturforsch.*, **14B,** 659–74.

JENIK, F., 1962. Electronic neuron models as an aid to neurophysiological research, *Ergebnisse der Biologie*, **25,** 206–45.

REICHARDT, W., 1961. Autocorrelation, a principle for the evaluation of sensory information by the central nervous system, in *Sensory Communication*, W. A. Rosenblith, ed., New York: Wiley, pp. 303-17.

REICHARDT, W., 1962. Nervous integration in the facet eye, *Biophys. J.*, **2**(2), Part 2, suppl., 121–43.

VARJÚ, D., 1959. Optomotorische Reaktionen auf die Bewegung periodischer Helligkeitsmuster, *Z. Naturforsch.*, **14B,** 724–35.

17. The Origin of the Flight-Motor Command in Grasshoppers

DONALD MELVIN WILSON, *University of California, Berkeley*

Many kinds of behavior, especially rhythmic ones, may be expected to have in their mechanism a cyclic organization such as is depicted in Fig. 1. Starting of the cycle could result either from external input or from spontaneity, usually at the site of the central mechanism. Once the cycle is operating, it is not possible to designate a prime cause, since the chain is closed. A possible feature of such an organization is that the rhythmic period could be controlled by the loop conduction time. The alternative to this scheme is open-chain control in which either only a single cycle follows each external stimulating event or there is some rhythmical center within the chain.

The above constitutes a brief statement of the controversy as to whether animal-locomotion control is central or peripheral. Does cyclical behavior depend upon proprioceptive loops for maintenance and timing or is the central nervous system alone capable of producing the rhythm? The subject has been reviewed by Bullock (1961). He suggests that there may be both autogenic central rhythms and chain-reflex rhythms in different cases and in combinations such that a sensory feedback might modulate a centrally paced pattern. Several cases of central rhythms are now known (see, for example, Hughes and Wiersma, 1960; Horridge, 1961a). Purely reflexive mechanisms are claimed but difficult to prove (Gray, 1950). The present paper will be devoted to a description and discussion of the third possibility, an autogenic central rhythm with reflex modulation.

The locust flight-control system has proved relatively easy to analyze at the level which is of interest here. On the other hand, it is sufficiently complex that it does not seem to be only a special and overly simple example. In addition to the advantages of arthropods in general for nervous-coordination studies, we have in the case of locust flight a detailed description of the behavior itself (Weis-Fogh, 1956a, b) and the possibility of correlating nervous events

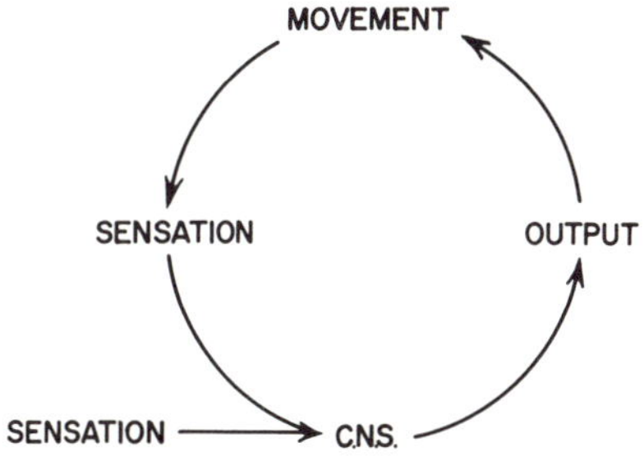

FIG. 1. Typical cyclic organization of behavior.

precisely with limb movements. The factual story reviewed here is presented in detail in the several papers by Wilson (1961, 1962), Wilson and Weis-Fogh (1962), Wilson and Gettrup (1963), and Gettrup and Wilson (1963). These papers will not be referred to again; most uncredited statements either are original or appear in that series.

The Motor Output Pattern

Grasshoppers, including locusts, beat the two pairs of wings at the same frequency in an approximately sinusoidal fashion. Each pair is synchronous, but the forewings lag the hindwings by about 10% of the period. In the desert locust *Schistocerca gregaria*, the frequency averages about 17 beats/sec, corresponding to a period of about 60 msec. The muscles which provide the wing-stroke power are of the ordinary fast, or twitch, type. They are *not* myogenic.* Therefore each wingstroke is the result of specifically phased muscle contractions which are themselves the consequence of muscle action potentials, and the latter reflect the ganglionic output after a conduction delay. The individual muscle twitch lasts only about as long as half the wing cycle (Neville and Weis-Fogh, 1963). The muscles are innervated by about 70 motoneurons. Since each neuron innervates anatomically discrete bundles of muscle, the neuron and associated muscle may be called a motor unit. The whole set of units may be divided into four approximately homologous subsets, one for each wing. Time of the action potential relative to wing position has been measured for many of the individual units, and the remainder have been timed in small groups. The technique is described below. The aerodynamic apparatus was described originally by Weis-Fogh (1956a).

Animals are "flown" suspended in the wind stream from an open-jet wind tunnel. They are mounted on a pendulum which swings fore or aft depending on whether the locust's air speed is greater than or less than the wind velocity. The pendulum contacts microswitches which regulate the compressor of the wind tunnel through suitable intermediate mechanisms. The result is that the locust stands still but chooses his own wind speed. The latter can be metered. The pendulum is suspended from a balanced beam. Changes in aerodynamic lift are measured as small deflections of the beam. A yoke around the animal rotates relative to the pendulum, and body pitch angle is either free or controlled by electromagnets. Rolling and yawing are not possible in the apparatus. It is probable that there are large rolling and yawing forces only when there are observable asymmetries in the leg and abdomen positions or in wing movements, and our measurements are not made under those conditions.

One to several intramuscular (but extracellular) electrodes are implanted. These pick up potentials of several millivolts from the units of the impaled muscle, ten or more times as much as from adjacent muscles. One action potential triggers an oscilloscope sweep, and the unblanking pulse, after an interpolated variable delay, sets off a stroboscopic flash unit, which illuminates

* See Pringle, "Input and Output Elements of Behavioral Models," *this volume*, p. 31.

the animal for photography. A simultaneous photograph of the oscilloscope trace is also taken. One set of observations then includes lift, air speed, body pitch angle, wingstroke amplitude and frequency, instantaneous wing position, and timing of one to several motor units relative to wing position.

The individual muscle units either are quiescent or are activated once or twice (rarely more than twice) per wingbeat cycle. The activation occurs at a precise phase of the cycle. When doubly activated, the second muscle action potential is anti-facilitated; that is, the electrical response is smaller in magnitude but somewhat prolonged. The double activations are usually less than 10 msec apart and produce a single, smooth, twitch-like contraction. The power of the contraction is continuously variable between the limits of about one and two times the value for a singly activated twitch, the variation being a function of the interval between the two closely spaced activations. The paired excitations therefore provide a means for continuous graduation of response from single muscle units during very brief activity cycles. Neville and Weis-Fogh (1963) have shown that at the body temperature of flying locusts the doubly activated contraction lasts for approximately the duration of a unidirectional wingstroke.

Over-all flight power can be graded by the total number of active muscle units, the number of activations per cycle, and—within rather narrow limits—changes in the wingbeat frequency.

The pattern of motor output responsible for the main features of the wing movements is summarized in Fig. 2. The measured action potentials occur at the muscle membrane, the ganglionic output appearing about 2 msec earlier. The muscle contraction begins 1 to 2 msec after the muscle action potential occurs (Neville and Weis-Fogh, 1963). A simple account of the pattern would be characterized as a temporal sequence of firing of four groups of motoneurons, those controlling the hindwing elevators, forewing elevators, hindwing depressors, forewing depressors, hindwing elevators, and so on. Within groups the sequences are not all fixed, and some temporal shifts may be involved in control measures.

In general the muscles act as expected: The elevators begin contracting at the bottom of the wingstroke and the depressors at the top. Some of the depressors, the dorsal longitudinal muscles, usually contract relatively early (that is, during the upstroke). They brake the upward wing movement, absorbing some of the energy supplied by the elevator muscles.

The moving wings can be considered an oscillating mechanical system driven by muscles. Inertia, air loading, muscular force, and elasticity of skeleton and muscle: all play a role in determining velocity and therefore frequency. An optimally efficient frequency for converting muscle work into aerodynamic work exists, and this is probably near the average frequency measured in flight. However, great efficiency in this respect may not always be advantageous to the organism, and in fact during maximum power output (high wind speed and lift) the frequency may be 20 to 30% higher than normal. In this circumstance not only do more units fire more often but the muscles

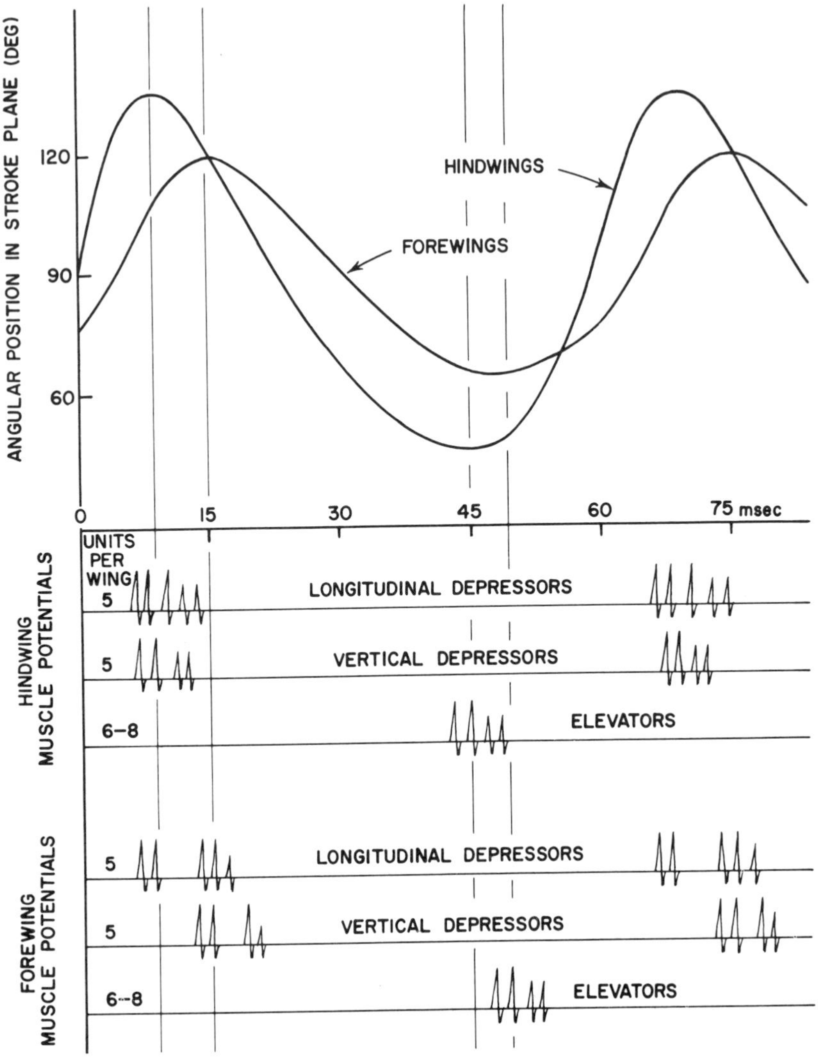

FIG. 2. Angular position of the two wing pairs during $1\frac{1}{2}$ wingbeat cycles and a composite of action potential records from the flight muscles. Each line represents time of firing of several motor units (number at left), many of which are nearly synchronous. The smaller spikes indicate second activations of units already fired once in that cycle.

also fire earlier in the mechanical wing cycle. There is muscular braking at both top and bottom of the wingstroke, and the effect is less efficient production of more aerodynamic power. The wingbeat amplitude is smaller at high frequencies because of the braking activity. The system behaves like a resonant one; in order to increase frequency above the normal one the input is phased earlier and more energy is required to maintain even the same output.

A Mechanism of Lift Control. Although extrinsic to the main thesis of this paper, it seems worth while to point out how this basic pattern can be varied in at least one control maneuver. During flight the locust wing produces some lift during all parts of the wing cycle but far more during the middle third of the downstroke than at other times (Jensen, 1956). Lift of an airfoil is a function partly of angle of attack, and if the wing mechanics are fixed relative to the locust's body, then changing the body pitch angle should change angle of attack and lift. Surprisingly, changes in body pitch amounting to 15 to 20 deg often result in no change in lift (Weis-Fogh, 1956b). The change in body pitch is compensated by changes in the degree of twisting of the wings during the stroke. The twisting is controlled by downstroke muscles which have their attachments in front of (the basalar muscles) and behind (the subalar muscles) the main point of wing attachment to the body. The basalar muscle pulls the wing downward with greater force on the leading edge, causing a pronation and depression. The subalar muscle also produces downstroke power but antagonizes the basalar muscle with respect to twisting. The combined effect of the two muscles determines the angle of attack during the downstroke. Both pairs of wings have this possibility for muscular control of angle of attack during downstroke but not during upstroke (when lift and drag are minimal anyway).

Recordings from the wing-twist controlling muscles during constant-lift performance showed that in the forewings at most three motor units effected a compensation of wing twist for body pitch changes of 15 to 20 deg. At low body angles (animal long axis nearly horizontal) the subalar muscle had both units active, often twice per wing stroke, whereas at high body angles this muscle was largely inactive, leaving the basalar muscle unopposed in producing a strongly pronated downstroke. In the hindwings no compensating changes in muscle activity could be measured, and it was later found that the hindwing twist does not vary.

Maintenance by the forewings alone of constant lift during changes in body pitch angle has an unsuspected consequence. It means that the distribution of lift between the two wing pairs varies with body angle and in such a way that whenever the body angle is pushed away from the natural one, a restoring torque is automatically developed. This is experimentally demonstrable. In the apparatus the animal may be left free to fly in a normal attitude or may be pushed around the transverse axis by electromagnets. When the magnet current is turned off, a restoring movement begins immediately. Thus these aspects of lift control and pitch control are accomplished by a single muscular mechanism. The control of forewing twist seems to require the presence of some sense organs, campaniform sensilla, on the main wing veins. These sensilla respond to upward, forward forces (not movements) on the wings, in the same direction as the maximum aerodynamic pull. Destroying the sensilla destroys the wing-twist control reaction but does not affect other known features of flight performance.

The Sensory Input. It has been established that the grasshopper flight-control system is a closed-loop or feedback system. At the grossest level the feedback may be considered a positive reinforcing stimulus. Flight probably begins ordinarily with a jump. This not only results in a release of inhibition from the leg proprioceptors but also causes an air stream over the head hairs, which excites the flight system. Once flight is established, the forward

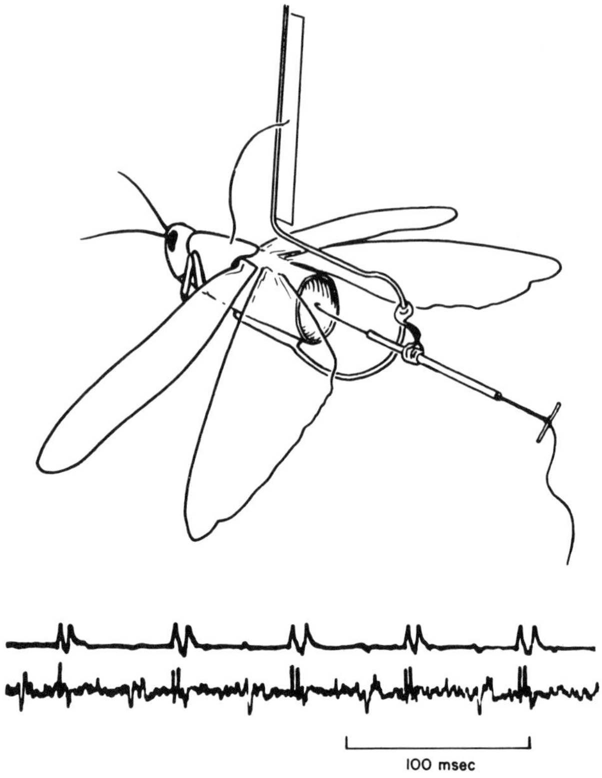

FIG. 3. Drawing of locust mounted for one type of sensory recording. Upper trace: action potential from a downstroke muscle. Lower trace: impulses from the stretch receptor and some muscle potentials.

movement of the animal maintains this wind stimulus, which maintains flight. Acting similarly are inputs from receptors located about the wings, including the wing-hinge proprioceptors to be discussed later. These wing sense organs respond during flight and help to maintain flight once it is started.

The flight-maintaining stimulus of wind on the head may be considered a steady input or at least one in which changes are not phased to the wing movements. This is certainly true in some of the experimental preparations. The internal, or proprioceptive, inputs are not steady, however, and as expected are phased to the wing movements. The discharges of sensory nerves from the several sets of wing proprioceptors have been monitored in the following manner.

It is possible to remove the locust's digestive system and various other organs and use the space remaining for the mounting of electrodes under the sensory nerves, as shown in Fig. 3. After the wounds are sealed, recordings may be made during hours of apparently normal flight. The main wing nerve contains many sensory units. The largest innervate proprioceptive hairs on the tegula and discharge during the last half of the downstroke. In the same nerve are the fibers from the campaniform sensilla mentioned above. These also discharge mainly during the downstroke. Their discharge is probably a measure of lift. In mature animals, as long as wind continues to stimulate the head hairs, both of these sources of wing-phased input may be surgically removed without affecting the gross aspects of the motor-output pattern already described. Neither the main sequencing nor the over-all frequency is changed.

Another wing sense organ is apparently peculiar to the grasshopper-like insects. It consists of some little-known vibration units and an important single-unit stretch receptor. The stretch receptor located at the base of each wing responds to upward movement of the wing with a relatively high frequency of discharge and can signal wing position by a lower-frequency adapted discharge. It is a phasic-tonic receptor. During flight it fires once to several times during the last part of the upstroke.

Destruction of all four stretch receptors always results in a decrease in wingbeat frequency to about one-half of the normal value. Destruction of two or three results in an intermediate frequency. Regardless of which of the four are removed, the effect is equally distributed over the whole system. The individual stretch receptor is not involved in a local, one-wing reflex; its influence is both bilateral and intersegmental. The number of impulses from the stretch receptor during each upstroke depends upon the amplitude and frequency, being greatest during the large, low-frequency excursions of the wings which characterize weak flight. Normally, it discharges only once or twice per cycle. The phasing of the first impulse also depends on frequency. It is relatively early at low frequencies and later at high frequencies, but in all cases it falls within a span of a few milliseconds in late upstroke.

The Inherent Central Pattern

With so much phasic input from the wings available, it was somewhat surprising to find that the only conspicuous effect on the gross motor pattern when all of the wing sense organs are destroyed is a decrease in over-all frequency. A simple experiment is illustrated in Fig. 4. Record A shows the activity in a two-unit muscle at the beginning of a normal flight. Records B and C show the reflex effects due to the prevention of normal wing movements. Record D (the lower trace is from the same muscle) shows the lower-frequency but otherwise normal response after removing the wings and associated sense organs. Lack of dependence on wing sense organs suggested that phasic input was not necessary at all to the production of the motor pattern. A number of experiments substantiated the suggestion, the strongest being the following.

Animals were placed so that a wind stimulus could be presented to the head hairs. Behind the head all tissue was removed except a ventral strip of cuticle supporting the thoracic central nervous system and the tracheae supplying air to the nervous system. The tracheae were perfused with moist air. All traces of flight musculature were removed. Recordings from the stumps of the severed motor nerves still showed a patterned discharge resembling that which occurs

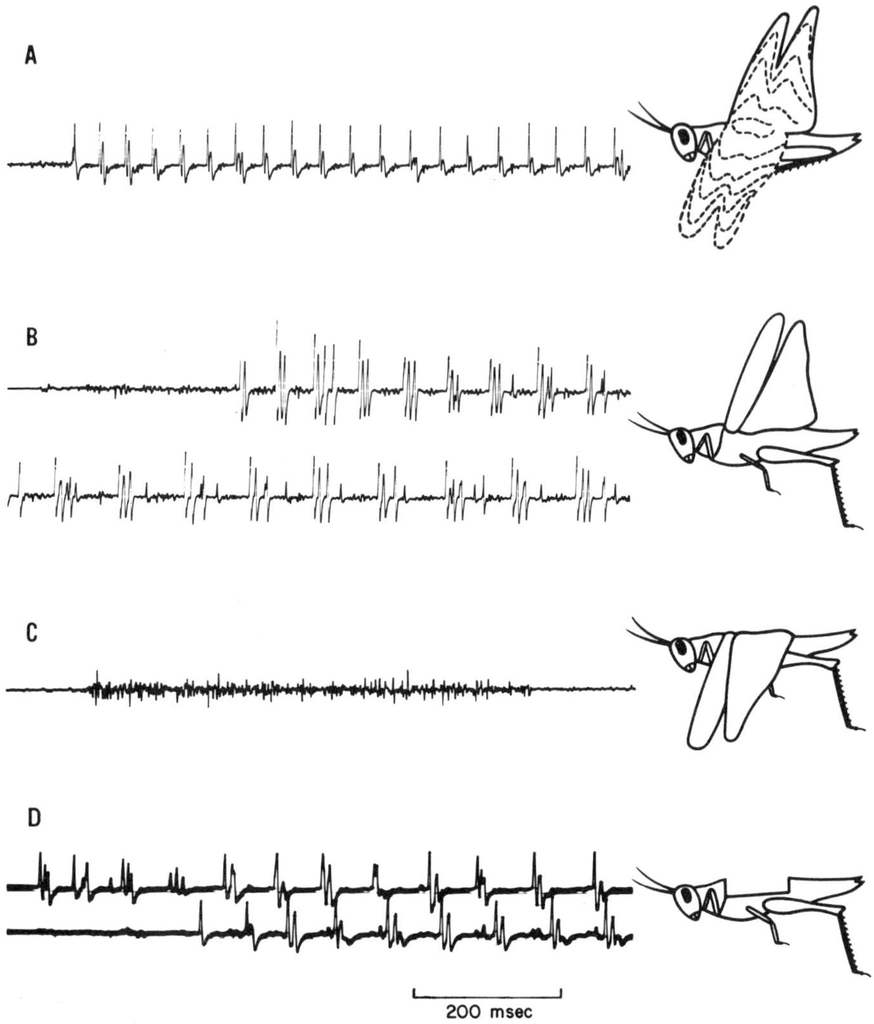

FIG. 4. A. Record from the metathoracic subalar muscle of *Melanoplus* during normal performance. B. Same with the wings waxed in the upward position. C. Same, but waxed downward. (Only elevator muscle potentials are seen). D. Record from a tergocoxal (elevator) muscle (upper trace) and the subalar muscle (lower trace) of metathorax after cutting away the wings and wing-hinge sense organs.

during flight, but at a lower frequency. The pattern did not occur in the absence of stimulation. It required wind on the head hairs or electrical stimulation of the nerve cord either anterior or posterior to the thoracic ganglia. When electrical stimulation was used, diverse patterns of input had no special effects.

Frequency Control. With either a steady input (wind on the head hairs) or a patterned one not phased to the output, the locust thoracic ganglia produce a cyclically patterned output similar to that which occurs during normal flight except that the frequency of repetition is lower. The thoracic stretch receptors are necessary to the maintenance of normal frequency. Any other single source of input may be destroyed without substantial effects on frequency (except in immature animals). With animals flying in nonturbulent air even the head hairs may be put out of action, for under these conditions the wing sense organs are sufficient for good flight. But although the stretch receptors play a dominant role in setting the wingbeat frequency, they are not solely responsible. Just as the four stretch receptors contribute in a spatially non-specific way to the excitation of all parts of the flight system, other inputs also influence frequency, although to a lesser extent. In the highly dissected preparations, output frequency is somewhat dependent upon the velocity of wind stimulating the head hairs or the intensity and frequency of electrical stimulation of the nerve cord. In immature animals which are already able to fly, the hair sensilla on the tegulae contribute significantly to maintenance of normal frequency. Light and sounds also influence frequency. The purpose of these necessarily rather vague comments is to indicate the possibility that output frequency is dependent upon the degree of a general state of excitation of a flight-control system which sums several inputs of unequal importance. Horridge (1961b) has made a more general argument supporting the notion that inputs may have rather widespread non-specific effects.

Coupling of the Neural and Mechanical Cycles. Although a central oscillatory patterns exists, it operates by itself always at a lower-than-normal frequency. It seems possible, therefore, that input phased to the moving wings might not only increase frequency by adding tonically to the general excitation of flight, but also trigger the central-nervous cycle at a specific time relative to the wing position. For example, the stretch-receptor input might set off, earlier than it would otherwise occur, the contraction of one of the muscle groups. The very great precision with which the nervous output pattern was correlated with the wing movements suggested a timing feedback; the stretch receptors were suspected of that role.

A first hypothesis was that the stretch receptors performed in an ordinary stretch reflex; that is, they discharged as the wing was elevated, and their effect was to excite neurons leading to the contraction of depressor muscles. Simultaneous recordings from the sense organs and muscles showed, however, that output preceded input in any one cycle, and therefore that a cycle-to-cycle operation of this kind of reflex was not possible. See Fig. 5.

The converse hypothesis, that the stretch-receptor input effect is summed over longer periods than one cycle and does not have a phasic effect, is thus far supported by all experiments. Shifting the normal input phase by locking the two wings of each side together so that the normal 8-msec difference between segments does not occur had no effect on the timing of output from the two segments. The 8-msec segmental delay in the output persisted. Stimulating the sensory nerves electrically in de-afferented preparations caused frequency increase regardless of stimulus timing.

Figure 5 shows the result of stimulating two of the four stretch receptor

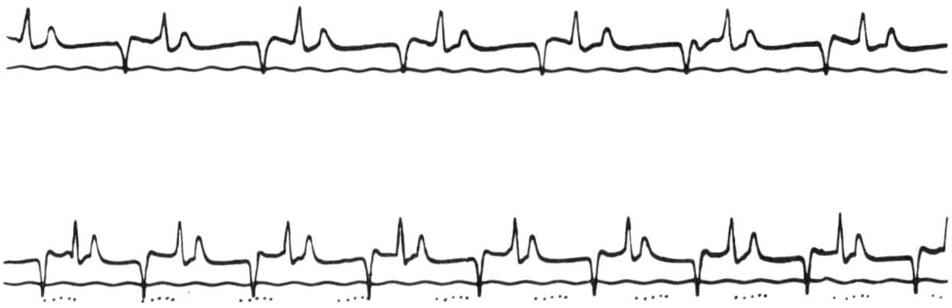

FIG. 5. Effect of stimulating two of the stretch-receptor nerves of a preparation responding to wind on the head. The upward and downward spikes are from antagonistic muscles recorded push-pull. The stimuli are in brief high-frequency trains which undergo all phase relationships with the motor output. The lower traces have a 50-cps time marker and a stimulus marker.

nerves by a pattern in which input and output undergo all phase relationships, but the output frequency is increased by about 30% throughout. The increase takes many cycles to develop, and waning is equally slow when stimulation is stopped.

A study bearing on this problem is that by Gettrup (1963) of the discharge characteristics of the stretch receptors during sinusoidal mechanical stimulation. He found that the phase instant of the first impulse during the stretch was a constant independent of amplitude *at any one frequency*. Large-amplitude stimulation caused more impulses, and the interval between the first two impulses was dependent upon the velocity of stretch. Changes in mechanical-stimulus frequency were accompanied by shifts in the timing of the first discharge. With increasing frequency the receptor fired later and later. This finding, which is indicated in the behavioral recordings as well, should be compared to the muscle activity. As the frequency of wing-beat increases, the stretch-receptor firing comes later in the cycle, but the muscle activation occurs earlier, a situation not easily explained if the stretch receptor triggers the muscle.

What then does keep the wings and central nervous system properly phased? It could be a timing feedback from some other sense organs, but none of these seems to be necessary for frequency control. The best answer seems to be that no timing feedback is involved at all and that it is the central nervous system and muscles which time the wing movements and not vice versa. Such a timing mechanism fits the notion that the wings are a mechanical oscillating system whose phase (and frequency) can be reset by repeated pushes at a new phase (or frequency). It is also consistent with the animal's needs, for a feedback triggering signal would tend to produce a fixed-frequency machine, but though the range of wingbeat frequency is not great, frequency is a significant variable parameter related to gradation of total aerodynamic power.

Observations on the Central Organization

As yet the question "how does the pattern originate?" is unanswered. The early suggestion that it is due to cycling through proprioceptive loops has not been supported. This negation has at least resulted in knowledge of the location of pattern generation, the thoracic ganglia. We can now review some evidence which must be considered in an attempt to model the C.N.S. flight-control mechanism. The data are not yet sufficient for model construction, however.

Another Pattern. Some of the same muscles which move the wings are also attached to the legs and are sometimes used in locomotion involving the latter. In some cases at least, the same motor units, and therefore presumably the same central neurons, are used in either flying or walking. By itself that is not remarkable, since many animals use the same muscles for diverse activities, but in this case the same muscles move different limbs and the intermuscular relationships of synergism and antagonism are different, depending upon activity. The muscles which are synergists for flight are antagonists for walking, and so on. Figure 6 shows the relationship for the main muscle groups of one wing. The point of this in regard to the flight-control system is that any connections between the motoneurons must be labile. Regardless of the great

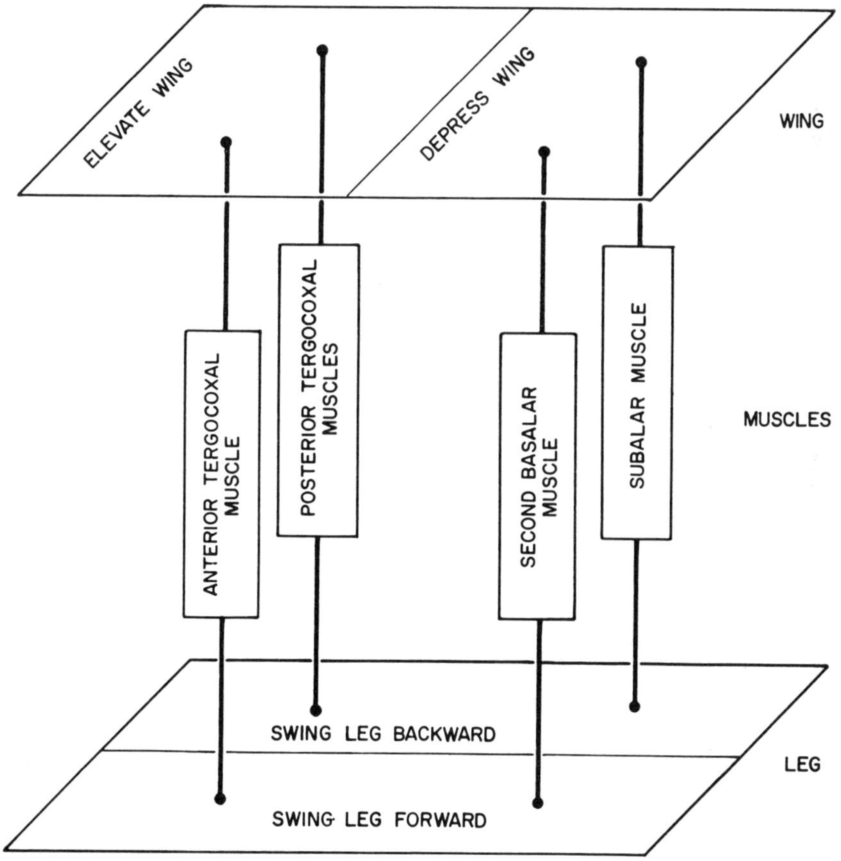

FIG. 6. Diagram of the synergic-antagonistic relationships between the several muscles used in both flight and walking.

precision and constancy of the motor output, the flight system cannot be a set of fixed relay connections at the level of the motoneurons.

A Hierarchy of Centers? If fixed interconnections between the motoneurons do not provide the basis of the flight mechanism, then it is possible that they are driven from "above" through specific connections from special centers controlling flight or walking, and that the motor centers behave as final common paths only. There is, however, no unique flight-control center. Surgical separation of the wing-controlling ganglia leaves two independent systems, either of which may be stimulated to drive its own wing pair in the flight pattern. Unilateral stimulation of one nerve cord may activate the wing muscles of only one side; hence bilateral sites of pattern generation are indicated. Furthermore, during weak stimulation single flight muscles (or small groups) may be activated repetitively at a frequency different from the stimulating one, and one must postulate oscillatory centers for each motor unit. It seems more valuable at present to think of the flight-control system as consisting of a set of coupled oscillators rather than a hierarchy of controlling centers. In this respect the organization of the flight-control system may be comparable to that of the lobster cardiac ganglion (see Maynard, 1955; Bullock, 1957). The flight-control system differs clearly in being composed of reciprocating sets of antagonistic members.

Some Flight Motoneuron Properties. Transganglionic stimulation and recording gives some indication of the nature of the synapses between the sensory and motor neurons but has not yet provided a means for detecting interneurons. For single large-input shocks to the nerve cord the number of output units responding increases with stimulus intensity, but for the most part repeated stimulation is necessary to produce any response at all. During long stimulus trains the frequency of response of any one motor unit depends upon either frequency or intensity of input shocks, where intensity is a combined measure of duration and magnitude. These observations suggest that both temporal and spatial summation occur.

For long trains of input stimuli a motor unit may always discharge in a certain phase with respect to input but not necessarily at the same frequency. There may be a frequency division, with the output unit firing after each second or third input pulse. This division may be due to relative refractoriness of the motoneuron. Measurements of the relative refractory period of the axon show it to be long enough to account for the division phenomenon. The neuron could have the same properties at the spike-initiating site. An important feature of the flight-motor output pattern which may be explained by the refractoriness of the motoneuron is the temporal separation of the double excitations of the flight muscles during one wingbeat. The two output pulses are usually 6 to 8 msec apart. The double excitation produces a single contraction when the interval is that small, the magnitude of the force depending upon the interval. At much smaller intervals the second impulse adds only a small increment of force, and the force goes through a maximum as the interval is increased. At flight temperature the optimum interval is about 6 to 8 msec (Neville and Weis-Fogh, 1963). If the flight motoneurons are ordinary neurons, and they seem to be, then we can expect that their frequency of firing will increase with increased excitation. But increased frequency of the pair of impulses beyond the optimum for power production would have the apparently pointless effect that an increase in excitation would produce less power output. However, a refractory mechanism could effectively limit pulse pairs to intervals as long as, or longer than, the optimum one.

Experiments in which the ganglionic input fibers were stimulated electrically also demonstrate autorhythmicity of the motoneurons (or some otherwise-undemonstrated pacing interneuron). During a long stimulus train the output frequency increases gradually to a steady level, which also declines slowly when the stimulus ceases. There are many cycles of ever-decreasing-frequency after-discharge.

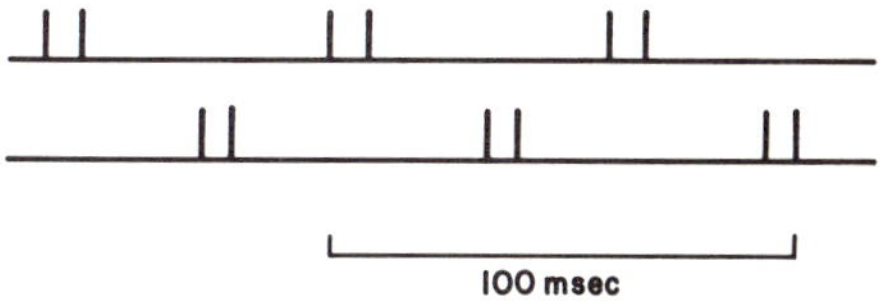

FIG. 7. Temporal firing sequence for a single pair of antagonistic motor units during flight.

Although refractoriness can explain the firing interval for one unit during one cycle, it cannot account for the over-all behavior of that unit. The units may fire in widely separated pairs nearly at the same time as their synergists but reciprocating with the antagonists (see Fig. 7). So far as is known this pattern of repeated pulse pairs occurs only when there is reciprocation between antagonists, but that is not a very demanding specification, for the phenomenon could still be due to either reciprocal inhibition, reciprocal delayed excitation, or one-way interactions. It will be interesting to learn what conditions of interconnection are necessary in the electronic-analog models of neurons in order to produce the above output from two or more units which are each capable of self-paced oscillation when stimulated by a high-frequency or steady input.

Discussion

The loop described in Fig. 1 operates in flying grasshoppers, but the loop is not a necessary part of the mechanism for the generation of the basic flight command. The feedback is necessary in some kinds of control maneuvers and for the maintenance of normal wingbeat frequency and power. The feedback seems to carry more information than is used. The stretch receptor, for example, could inform the C.N.S. about wingbeat frequency, stroke amplitude, upstroke velocity, and the phase setting. In the maintenance of normal frequency, however, the stretch-receptor input is summed over many cycles and seems to have a tonic excitatory effect on the whole flight-control system, rather than a specific timing effect. From this it is apparent that it is necessary to determine what input does, not only what information it carries. It could be generally true that input information is degraded in processes which are not essentially losses due to noise. A point may be raised in connection with the most appropriate direction for analysis of systems of this kind. It is true that one may enter the loop at any point and go in either direction. In traveling the circle it is preferable that no very large jumps be made, for example, directly from sensory input to motor pattern. In the present example, some knowledge of input and output patterns in intact animals led to the formerly rather plausible hypothesis that the system could be adequately defined as a set of temporally linked reflexes. For reasons of economy it is preferable to work through the loop backwards. A single sense organ can provide several kinds of information. Each of these possible uses for input must be followed up when analysis proceeds in the forward direction, but some are dead-end routes. In the other direction, all paths lead somewhere.

The locust's flight-control system seems to be input-phase-independent in regard to the stretch reflex. That is to say, although there is a wing-phased temporal pattern in the input, that phasing is irrelevant. It must be generally true that output temporal patterns *are* significant, and in the case of locust flight this is true; the output is perfectly correlated with the behavior. But in insects with myogenic flight muscles the mechanical cycle could be phase-independent even of the motor output, and apparently it is (Pringle, 1957; Wilson and Wyman*). Some insects with myogenic flight muscles appear to have a phasically unpatterned motor output, whereas in others a clear temporal patterns exists, but like the input pattern it may be unessential.

The only flight activity in insects which *must* be cyclic is the production of power by the muscles. Because of the nature of ordinary sense organs the cyclic movements are likely to be reflected by the input even when unnecessary. Control measures do not need to be cyclic at all. Adjustments of angle of attack, for instance, could be made by smooth, slow contractions of accessory muscles. If there is cycle-by-cycle production of power, but control is tonically imposed ahead of the point of generation of the main oscillatory pattern, then it is possible that in the myogenic fliers the whole nervous-system involvement is primarily a tonic one. There may be no important central pattern in the phasic sense, and in a very unusual meaning the pattern may be peripherally determined. In locusts, with their more ordinary muscles, the main oscillations must be established in the C.N.S., but it is still possible that control is tonic, the controlling input to specific motoneurons being effective over many wingbeats.

Indications of tonic control are occasionally manifested even in the motor output of locusts. For example, in one locust during a very weak flight at very low body pitch angle the main pronating muscle (the single-unit first basalar muscle) contracted only on alternate wing strokes. The lift of the wing must also have alternated between high and low values, but because the momentum of the whole moving animal exceeds considerably the force added during each wing stroke, the alternating variation could be smoothed. Thus large-time-constant temporal summations may occur in at least two stages of the locust's flight-control cycle. The phasic sensory input which regulates wingbeat frequency is temporally integrated, and there is summing and smoothing of

---

* Unpublished data.

the effect of muscular work over many cycles. It is possible that the superimposition of control maneuvers upon a basic phasic pattern may be accomplished primarily by tonic spatial differences within the nervous system, with some parameter-like average frequency in the various lines to different motoneurons (or muscles, in the myogenic case) being the controlling parameter rather than changes in phasing of the individual units.

References

BULLOCK, T. H., 1957. Neuronal integrative mechanisms, in *Recent Advances in Invertebrate Physiology*. B. T. Scheer, ed., Eugene, Ore.: Univ. Oregon Pubs., pp. 1–20.

BULLOCK, T. H., 1961. The origins of patterned nervous discharge, *Behaviour*, **17**, 48–59.

GETTRUP, E., 1963. Phasic stimulation of a thoracic stretch receptor in locusts, *J. Exptl. Biol.*, **40**, 323–33.

GETTRUP, E., and D. M. WILSON, 1963. The lift control reaction of flying locusts, (in press).

GRAY, J., 1950. The role of peripheral sense organs during locomotion in the vertebrates, in *Physiological Mechanisms in Animal Behavior* (Symposia of the Society for Experimental Biology, No. 4), New York: Academic.

HORRIDGE, G. A., 1961a. The centrally determined sequence of impulses initiated from a ganglion of the clam *Mya*, *J. Physiol.*, **155**, 320–36.

HORRIDGE, G. A., 1961b. The organization of the primitive nervous system as suggested by examples of inhibition and the structure of neuropile, in *Nervous Inhibition*, E. Florey, ed., London: Pergamon.

HUGHES, G. M., and C. A. G. WIERSMA, 1960. The coordination of swimmeret movements in the crayfish *Procambarus clarkii* (Girard), *J. Exptl. Biol.*, **37**, 657–70.

JENSEN, M., 1956. Biology and physics of locust flight: III. The aerodynamics of locust flight, *Phil. Trans. Roy. Soc. (London)*, **B-239**, 511–52.

MAYNARD, D. M., 1955. Activity in a crustacean ganglion: II. Pattern and interaction in burst formation, *Biol. Bull.*, **109**, 420–36.

NEVILLE, A. C., and T. WEIS-FOGH, 1963. The effect of temperature on locust flight muscle. *J. Exptl. Biol.*, **40**, 111–21.

PRINGLE, J. W. S., 1957. *Insect Flight*, Cambridge, Eng.: Cambridge Univ. Press.

WEIS-FOGH, T., 1956a. Biology and physics of locust flight: II. Flight performance of the desert locust (*Schistocerca gregaria*), *Phil. Trans. Roy. Soc. (London)*, **B-239**, 459–510.

WEIS-FOGH, T., 1956b. Biology and physics of locust flight: IV. Notes on sensory mechanisms in locust flight, *Phil. Trans. Roy. Soc. (London)*, **B-239**, 553–84.

WILSON, D. M., 1961. The central nervous control of flight in a locust, *J. Exptl. Biol.*, **38**, 471–90.

WILSON, D. M., 1962. Bifunctional muscles in the thorax of grasshoppers, *J. Exptl. Biol.*, **39**, 669–77.

WILSON, D. M., and T. WEIS-FOGH, 1962. Patterned activity of co-ordinated motor units, studied in flying locusts, *J. Exptl. Biol.*, **39**, 643–67.

WILSON, D. M., and E. GETTRUP, 1963. A stretch reflex controlling wingbeat frequency in grasshoppers, *J. Exptl. Biol.*, **40**, 171–85.

18. Exploration of Neuronal Mechanisms Underlying Behavior in Insects

GRAHAM HOYLE, *University of Oregon*

A group of investigators recently emerged with a declared aim to explore and describe the neuronal events underlying those complex behavioral acts of animals termed "instinctive." This interest has clearly resulted from the analysis of instinctive behavior made particularly by Tinbergen (1951), which suggested the existence of only a limited number of kinds of instinctive acts based on relatively simple nervous machinery. Tinbergen himself made a strong plea for neurophysiologists to study ethology and then investigate the neuronal mechanisms involved.

Investigators who have been attracted to this new and exciting field are proceeding simultaneously at several levels (sensory, central, integrative, and neuromuscular) without any prescribed effort at correlation. There is no harm in this approach, but it will not produce, except fortuitously, the complete answers we hope for. Furthermore, each investigator uses either the most convenient preparation or his favorite organism, so that it may not be possible to combine the results in an integrated appraisal.

Somehow, investigators must agree to concentrate on one organism, just as geneticists agreed when they converged on *Drosophila*, if we are to achieve success in behavioral analysis at the neuronal level. These studies must be combined, ultimately, with genetic studies, and experimental developmental morphology. It is very probable that the most suitable organism for all this will be an arthropod. The total number of neurons in the arthropod central nervous system is not large. Wiersma (1957), in a direct count, found 100,000 neurons in the nervous system of a crayfish (excluding the optic lobe). Although no figures are available yet, I estimate the number of neurons in an average insect to be about the same. This limited number means that important functional units contain relatively small numbers of neurons, numbers small enough to permit their simulation in detail by physical models.

What are the relative merits of the various major arthropod groups? I think we may reasonably discount the minor classes, millipedes, centipedes, and the like, and choose among an arachnid, a crustacean, and an insect. Arachnids include spiders, some species of which are large and available in reasonable

The experimental work reported in this paper was done mainly in the Department of Zoology, University of Glasgow, when the author was Jaffé Research Fellow of the Royal Society. Recent work and theoretical developments have been carried out in the University of Oregon and were supported by grant G21451 from the National Science Foundation.

quantities, and also scorpions, which might be very suitable. But we know almost nothing about their basic physiology as yet and cannot assess their suitability for the kind of analysis we have in mind. Crustaceans offer certain advantages: some decapods are large, the nerve fibers can easily be dissected so that recordings can be made from individual fibers, and quite a lot is already known concerning their physiology (especially that of the crayfish). Crustaceans are mainly aquatic, a condition which confers some advantages but also some disadvantages. Experimentally, their behavior patterns have scarcely been touched, and aquatic invertebrates in general appear to have few interesting behavior patterns, compared with their terrestrial relatives. A great disadvantage with crustaceans is that it has not yet been possible to breed and rear them in the laboratory.

We are left with the insects. Insects have diverse behavior patterns which have been closely compared with those of vertebrates (Tinbergen, 1951; Thorpe, 1956); they can learn, and some species can retain learned information for long periods. They are readily available in large numbers and can be bred and reared under controlled conditions in the laboratory. The principal problem in using insects for these investigations is their small size. This at present precludes splitting of nerve trunks and selection of individual nerve fibers to the extent to which this is possible in Crustacea. On balance, however, insects seem to be the most promising experimental animals. We can probably come closer to a complete understanding of the behavior of an insect, in terms of underlying neuronal activity, than of any other class of animal.

Among insects the most experimentally amenable at the present time are cockroaches, stick insects, crickets, and grasshoppers (locusts)—all of the order Orthoptera. Of these the best known physiologically are the cockroach *Periplaneta americana* and the locusts *Locusta migratoria* and *Schistocerca gregaria*. All can be reared in the laboratory in large numbers. Cockroach behavior is coming to be understood better, and has quite complex aspects; for example, they can be trained to run very simple mazes, although they have poor retention. The cockroach *Periplaneta americana* has the great advantage of already being available throughout the world. Recently the cricket has been subjected to extensive studies, and it has the advantage of offering varied, complex, interesting behavior (see *Evolution*, **16,** 1962). Some species of cricket are available the year around.

Regrettably, no single species of animal is so obviously advantageous that it will become an inevitable choice, and in consequence anarchy seems likely to prevail. But just as long as we have to piece together the sensory physiology of a horseshoe crab, the integrative physiology of a sea-hare, the neuromuscular physiology of a locust, and the behavior of crickets, we are going to have a chimera indeed, and will remain far from achieving our goal.

Problems in the Study of Neuronal Mechanisms

Theoretical Considerations

Great progress was made by the recognition of the simple reflex as a major component in the animal's behavior, but the narrow approach resulting from

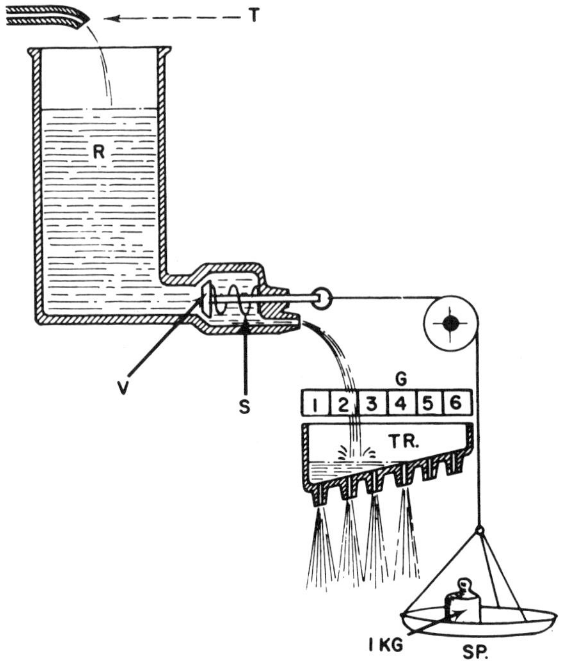

A

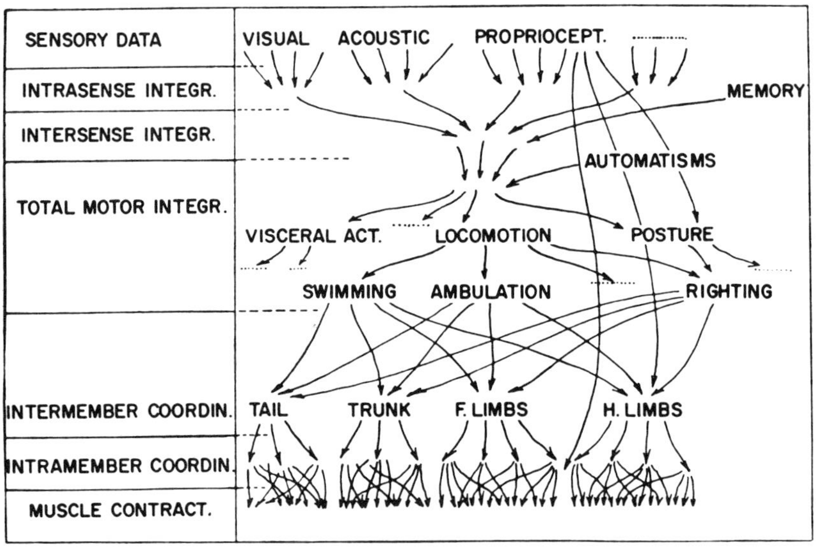

B

Fig. 1. (Parts A and B)

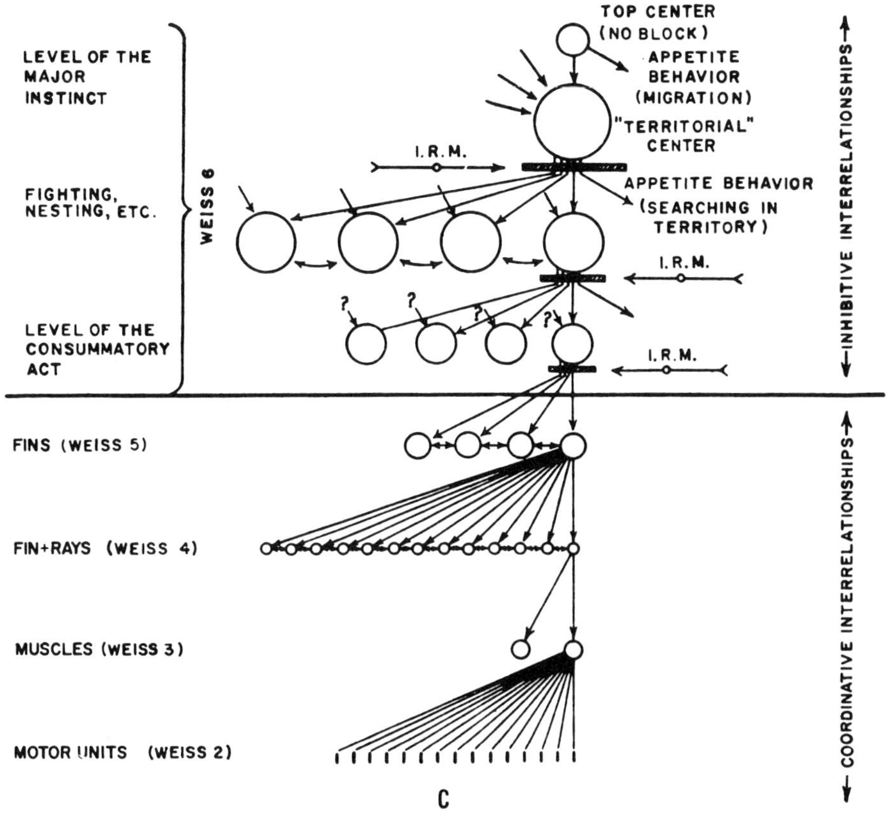

FIG. 1. Models based on ideas of complex instinctive behavior. A. Hydraulic model of Lorenz (1950). "Nervous energy" is accumulated in the central nervous reservoir R. When "released" by an appropriate stimulus the energy flows into channels 1–6, causing successive actions. B. Hierarchical mode (Weiss, 1950). A system of hierarchically organized centers controls behavior. The appropriate stimuli excite the first center, and it in turn excites more than one lower center. These in turn, excite still lower ones, and so on until finally many muscles contract. The key to the functioning of this model would be the detailed neuronal connectivity within and between centers. C. Another concept of the hierarchical mode (Tinbergen, 1950) especially devised to explain behavior in fish.

an overenthusiastic acceptance of the reflex, and the resulting chain-reflex school, has thankfully been replaced by the growing awareness of truly endogenous, centrally determined activity. The work of Lorenz (1950) and Tinbergen (1951) resulting in a new appreciation of central-nervous "centers" has been criticized by Kennedy (1954), Lehrman (1953), and others essentially on the grounds that it has merely disclosed rather more elaborate forms of reflex or given new terminology to old phenomena (see Fig. 1). However most experimental ethologists are satisfied that theirs is a valuable new approach. The final assessment will only be made when an intimate knowledge of the relevant neuronal machinery has been achieved.

The repeated regular elicitation of elaborate but stereotyped behavior patterns by simple electrical stimulation applied via implanted leads to appropriate regions of the mammalian or bird brain stem (Hess, 1943; von Holst, 1963) has further strengthened the notion of central "centers." Similar results have also been obtained by Huber (1960, 1962), who found that various kinds of courtship song or running behavior in crickets could be elicited by local electrical stimulation in the brain, and by Vowles (1961), whose work on bees is reported in this volume.

However, the details of the ways in which these electrical stimuli call forth responses remain obscure, and they are not very amenable to experimental analysis. We may ask the question: What is a "center"? The nearest associated pool of motoneurons is always some distance from the site where stimulation is successful, so it is probably the tracts which are excited and these in turn call forth the responses; the specific region in the vertebrate hypothalamus where an electrode elicits a specific response is not itself the "center" for motor control. Many modern accounts have assumed a hierarchical neural organization based upon a series of subcenters increasing in number while descending from the C.N.S. to the final motoneurons. The intermediate centers would, as a result of their structural interconnections, synaptic delays, spontaneous patterns, and specific properties, give highly patterned outputs calling forth appropriate sequential contractions of muscles.

One of the major tasks of the neuro-ethologists is to determine whether such "centers" exist, and if so to locate them, to elucidate their structure, and to understand their mode of operation. I believe that at the present time this is the highest target at which the neuro-ethologist should aim.

Mode of Attack

Most of the current work, on the motor side, concerns levels of action at the reflex level. Furthermore, it is not concerned with unit or neuronal analysis. For example, the cybernetic approach adopted by Mittelstaedt (1957, 1962, and *this volume*) in his studies on the strike action of the praying mantis, is concerned with a systems analysis of control during a reflex action. Studies of this type, although they are essential parts of our understanding of the nervous system, are probably not likely to succeed in closing the gap between ethology and neurophysiology. The reason for the inevitable failure of the current cybernetic approach is its total disregard, at least at the present time, of specific neuronal events. Treating the nervous system simply as a series of black boxes is interesting, and it may help in our understanding of function, but the outcome of this or any other approach should be to suggest anatomical and physiological investigations which will in turn lead to an interpretation in strictly neurological terms. The proponents of the German school often profess (one wonders how seriously) no interest in linking their ideas to neurophysiology. Their problems are complex enough, they say, without the need to consider unit analysis. My point is that they must investigate problems which are also amenable to neurophysiological analysis if they are to achieve the objectives stated here. Much closer to a true neurophysiological analysis of behavior are the elegant experiments of Wilson (1961, 1962, and this volume) on the neural control of flight in locusts. Wilson unquestionably has the right approach, but the neural control of the specialized mechanism of flight cannot be extended to become a model for other kinds of behavior. A comparable analysis of a more general behavior pattern involving ordinary locomotory muscles would be very desirable.

Of considerable relevance is the work of Roeder, Tozian, and Weiant (1960) on the copulatory movements made by the decapitated male mantis. Coordinated movements of legs, abdomen, and phallomeres of the decapitated male mantis continue for days and lead to effective copulation if a female mantis is encountered. Efferent nerve activity in the form of rhythmic bursts of a kind which might lead to the copulatory movements are present in motor nerves even after complete de-afferentation of the nervous system. A complex interaction of inhibitory and spontaneously firing units was held to be responsible. The term "copulatory center" may truly be applied to the controlling elements in this case (however diffuse the actual neuronal connections may be), for the same muscles are used in a variety of ways, and the control of the coordinated copulatory activity must be at a "higher level." A careful elucidation of the details (if it can be achieved) would be of inestimable value, a real breakthrough in the field. The most significant recent research in the area is the fine work of Huber (1962) on the central control of sound production in crickets.

Much work has also been done on the sensory side. So far this has been linked with the physiology of the resulting motor responses only in a few instances, for example, in studies on the escape response of the cockroach (Roeder *et al.*, 1960) and the feeding response of blowflies (Dethier, 1962).

In entering this field the investigator must be aware of the philosophical problems involved. It has been suggested that even the lower animals have an element of mind/body dualism, and although currently an unpopular notion it is nevertheless evident that collections of neurons have many unique properties. One of the especially exciting aspects of the field is that a sufficiently extensive knowledge of the functioning of the nervous system of any animal should of itself indicate whether or not all the behavior of that animal can be interpreted mechanistically. It is already evident that much of the behavior of animals is determined by simple rules, and hence we may pose questions regarding the ways in which the underlying machinery functions, propose explanatory hypotheses, and devise experiments to test the hypotheses.

Most experimental psychologists and ethologists currently subscribe to a stimulus–response school of thought. Starting with the premise that muscular contraction is behavior, we may trace back the events causing it. The muscle contracts in response to a motor-nerve impulse; this must be caused by a synaptic excitation, and so on. Recently, reference has been made in the literature to "endogenous activity" as if it existed independently of a stimulus. The movements of respiration may be initiated in a "respiration center," but somewhere in this center cyclically fluctuating physicochemical events must occur. These provide the stimulus, though whether such an endogenous fluctuation may be equated with stimulus in the psychologist's sense is debatable. The copulatory movements of the mantis occur when an inhibition is removed. This is a trigger and must in turn have a simple cause. Even the meanderings of the Grey-Walter mechanical tortoise *Machina speculatrix* are initiated by light striking a photocell.

Kennedy's point (1954) was that probably all behavior can be reduced to stimulus–response terms. He objected, in particular, to the Lorenzian concept of "nervous energy" as analogous to a fluid which could be contained, accumulated, overflow, be released, and so on. However, these concepts, which have no physiological basis, made thinking about behavior much easier and greatly influenced ethology.

The concern of the neurophysiologist must be with the details of the various stimulus-response relationships, vacuum and displacement activities, drives, and the like, and with the replacement of hydraulic models by sound physiological principles. All acts of behavior require subtle sequential control of the timing of many muscular contractions. How is the timing brought about? In the simplest reflex acts the anatomical and physiological features of the relevant neurons and their connections provide an adequate explanation of events. But consider, for example, the reaction of the swimming sea anemone *Stomphia coccinea* when touched by a certain species of starfish, *Dermasterias imbricata.* It goes through the following highly complex sequence of actions: (i) takes in water; (ii) closes its mouth; (iii) contracts its circular muscles and elongates; (iv) expands the disc of tentacles; (v) disrupts the adhesive attaching it (very firmly) to the substratum (this may be done by secreting mucus); (vi) relaxes muscles in the center of the basal disc (so that the base balloons out owing to pressure in the body cavity); (vii) initiates a series of contractions which occur alternately on opposite sides of the body, but with a slow rotation so that each set of longitudinal muscles contracts in turn. These events take from one to several minutes to complete. Anemones prevented from executing earlier movements or swimming nevertheless continue relevant later stages when released. When the animal is bisected suddenly in the longitudinal axis after the reaction has commenced, the 2 halves apparently show contractions appropriate to the phase of activity which would follow had the animal remained whole.

It is possible to cause certain events in the chain of response of *Stomphia* to occur as a result of other forms of stimulation; for example, electrical excitation can cause contraction of the muscles responsible for swimming. But this is rarely followed by more swimming, as it would be if chain reflexes caused the sequence. If, as sometimes happens, a brief electrical excitation (occasionally even a single shock) calls forth a swimming response, the chain of reactions usually starts at the very beginning, with water uptake like that caused by touching *Dermasterias.* Then there is no question of a persistent stimulus (chemical from the starfish) having caused the response. We may also note that in many instinctive acts the organism cannot recommence an interrupted sequence at a logical point but must start again at the very beginning and complete the whole cycle (Fabre, 1912).

In such cases one is led to postulate a store of information in the nervous system by means of which directly, or by reference to which indirectly, the complex behavior is controlled. This store must be run through from beginning to end in a prescribed, stereotyped sequence.

Possible Mechanisms Underlying Complex Instinctive Behavior

As information on behavior patterns grows, it becomes evident that the over-all control of the consecutive firing of motoneurons in an appropriate pattern, and with the right quantitative relations to be effective, may be due only in part to the connectivity in reflex pathways and in associated synaptic delays and the like. Reference must also be made to a store of information in the nervous system. The store of information may be regarded as analogous to a record on magnetic tape. For the complex control of web-building by a spider, this store of information is both permanent and heritable. Hence, in addition to unravelling the nature of the "center," the neuro-ethologist must raise the question of the means by which long-term information required for behavioral control is stored in the nervous system, and also how this information is used in controlling behavior. The information referred to is concerned entirely with motor control. Tape-recorded information may also be required as a source of reference for comparing actual sensory data with previously received sensory data or inherited sensory susceptibility. This type of record has been discussed by Sokolov (1960).

It will be desirable, before proceeding, to clarify the meaning attached to the use of the magnetic-tape model in relation to the nervous system. A great deal of "information" can be stored in neurons as a result of their connectivity, and appropriate successive inputs of the kind which would be provided by changing afferent input in conformity with the chain-reflex model can call forth this information in a temporal sequence. However, this kind of system is a set of closed loops requiring feedback to the information store. The information cannot, in principle, be extracted if the loops are broken. The essence of the tape model is that the nervous system contains stores of information adequate for behavior control and not requiring any form of feedback to the information store for their expression. The magnetized region of real magnetic tape is considered to be a good analogy of the kind of information store envisaged. Such information might be contained in individual cells or groups of cells, perhaps in their RNA, which would go through cycles of activity when appropriately excited. The excitation process is analogous to the switching-on of the tape. The equivalent of the tape-drive mechanism could be essentially located in the cells themselves, because their secretory activities require a real-time scale to unfold. In this case it might be due to controlled secretion of a transmitter agent. There must then be a tape-head, or read-out, mechanism. This could consist of nerve cells which are excited or inhibited by the secretions of the storage cells.

A highly subtle point is that for taped control of the kind suggested the motor-command information could be stored and utilized in either of two quite different kinds of sequence. The simplest, and perhaps most logical, would be a tape containing the profiles of motor commands, "tape-recorded" motor commands which would excite central neurons or inhibit inhibitory neurons in a sensible sequence. This is an open-loop system. An alternative, but quite feasible, system would be the utilization of a tape containing only a reference record of the proprioceptive sensation associated with the correct performance of the act—a "tape-recorded" afferent pattern. This involves a closed-loop control mechanism. Each of these possibilities will be considered.

(i) "*Tape-recorded*" *Motor Command.* It is envisaged that this type of control would involve the storage in the higher part of the nervous system of a number of command "tapes" containing the relevant information. Whenever a given stimulus situation occurred, and in the absence of excessive inhibition, the appropriate tape would be automatically selected (perhaps by reference to sensory tape) and played back. The tape must, like real magnetic tape, play back on a simple real-time scale. The commands it issues would be in the form of excitations, or inhibitions of inhibitions, in efferent centers and channels. The command tape might influence movement directly, or it might drive subsidiary centers at lower levels in the nervous system.

The tapes could play into certain common parts of the nervous system directly linked to motoneurons via simple switching. This would eliminate the great mass of cross-connections required by the connectivity model (see Fig. 1).

(ii) "*Tape-recorded*" *Afferent Patterns.* In this alternative means of control, the higher nervous system is postulated to contain a series of stores of information containing the equivalent of recordings of total afferent patterns, with appropriate time scales. It is not the totality of first-order proprioceptive afference which would be stored, but only the equivalent of the final-order pattern derived after convergence in the central nervous system of all the proprioceptive input.

Nervous systems receive a large number of sensory inputs, most channels of which, especially the proprioceptive, are duplicated many times. This redundancy is indicative of the supreme importance to the organism of feedback. The proprioceptive afference is the animal's guide to what it is doing and how well. It would be surprising if the sensory input did not play a significant role in controlling behavior, so it will be worth while to examine methods by which afference could play a major role in the control of instinctive behavior.

It is proposed that the animal compares the actual converged ingoing information with a "tape recording" in the nervous system, selected according to the immediate environmental situation. This could be based upon previous performances of the act or on an inherited "recording." The "afferent" tape would need to work in conjunction with a more general "command" signal to serve as the driving force. The general command would force some unspecific motor output to occur and at the same time start the playback of the afferent tape. The actual converged afferent information would be compared with the "tape record" and the difference used to modify the motor output until the right movement occurred. A crude analogy would be the use of a very rapid servo system to play Bach fugues on the organ from phonograph records of the fugues. A scanner, driven from the phonograph turntable, would try with enormous rapidity several notes until the right one was found, cutting off each wrong note instantly but letting the right one play its course by reference to the sound output from the phonograph.

One has greater difficulty accepting the second alternative (stored afference) than the first (motor profile). Yet a model of this kind of system is not difficult to envisage. The neuro-motor system in humans is highly ineffective when afferent nerve fibers go out of action, however well an action could be performed previously. This is a basis for the "chain-reflex" hypothesis of complex behavior, but it may also provide support for the notion of afferent tapes. Humans can certainly evolve mental images of the sensations associated with their own motor activities, which must require memory traces of such afference. The human mind can store a large number of afferent tapes: tapes of piano concertos, tapes of plays, tapes of political speeches. The argument here is that these stores exist independently of any feedback loops. Of course neither system need be used exclusively; combinations of both afferent and efferent tapes may be utilized, or one sort may be used for controlling some actions, the other others.

The details of the mode of construction and operation of neuronal "tape recorders" and the neuronal equivalent of recording "tape" are very challenging problems. For the present, the task of the neuro-ethologist must be to determine whether or not such things exist, by carefully studying those animals where there is some possibility of determining the details in neuronal terms.

A New Experimental Approach

A Window into the Nervous System

The rest of this paper will be devoted to experiments which have been devised to examine neuronal mechanisms underlying behavior in insects. The first part outlines the approach and the techniques used, and subsequently experimental data and hypotheses arising out of them are presented.

Complex behavior can generally be obtained only in the intact animal. Complex behavior elicited by direct means, like responses to electric stimuli applied to the brain of the cricket by Huber (1960, 1962) or copulatory movement elicited in the male mantis by decapitation, are available only for a small number of specialized behavioral acts. The question arises how one may obtain quantitative information about the nervous patterns during a variety of complex behavior.

Implanted leads recording from the nerves would provide useful information, but would be difficult to achieve technically. The discharges of the motoneurons themselves can, however, be recorded with great precision simply by taking electromyograms. Insect muscles receive few motor axons, and many functionally important ones have only one or two. Where there are two axons, one is usually a "fast" axon causing large electrical responses and strong twitch contractions; the other is a "slow" axon causing smaller, summating electrical responses and frequency-graded contraction. In a large, powerful muscle a single "slow" axon may be used for controlling its contraction in all ordinary locomotory activity. An electromyogram of such a muscle consists of a single train of pulses and as such forms a simple record which can be interpreted and analyzed. These electromyograms can be recorded from the intact animal by the use of implanted electrodes. The large insects can carry with them fine, trailing wires attached to the electrodes—or forming the electrodes—and thus be permitted a considerable measure of freedom to maneuver and "behave." Recordings of electromyograms offer an important window into the central nervous system, an approach analogous to that used by the pioneer of modern neurophysiology, Charles Sherrington, in 1906.

I started taking electromyograms from freely moving intact insects in 1955. Since then my aim has always been to watch the animal while at the same time observing the electrical activity of locomotory muscles transcribed into oscillograph movement.

It is possible to record from at least two muscles at the same time and so, by recording first from one pair, then another, to build up a composite record of the sequential patterns of electrical activity occurring in the various muscles involved in a given act of behavior. We may then examine these patterns to determine whether or not they are repeated exactly, or nearly so, in the intact organism when the act recurs. This result must be expected if the equivalent of motor-command neuronal tape recordings is used. By contrast, if neuronal tape recordings of afferent patterns are in control, a very variable set of electromyograms might be expected because commands would be determined continuously with reference to a variable feature, the afferent input.

Technical Data

Fine (0.003-in.) lacquered pure copper wire has been used for both recording electrodes and leads, but this is likely to be replaced by plastic-coated silver wire. Pairs of small holes slightly smaller than the thickness of the wires are drilled in the cuticle at selected points over the muscle to be studied. The leads are then pushed into the holes until the tips just protrude through the cuticle but do not interfere with the functioning of the muscle. Small beads of Tackiwax are then placed near the insertions of the wires, and an electrically heated stylus brought near enough to melt the wax so that it forms a neat seal, fused with the cuticular wax. Before attempting this, it is important to soak and dry up any haemolymph which has oozed out. These operations may be performed on the unanaesthetized insect if it is carefully pinned down with the legs extended.

The leads, lightly coiled and about 2 ft long, are plugged into sockets in the roof of a double-walled Lucite cage which has dc heating wires and thermostatic control. The internal dimensions of the cage are 20 $\times$ 20 $\times$ 20 in. (Fig. 2).

When complete control over the environment is required, the cage itself is placed in an electrically shielded soundproof room mounted on shear mounts, which are in turn mounted in a heavy concrete slab supported on sand. The slab supports a second soundproof room surrounding the first. A small

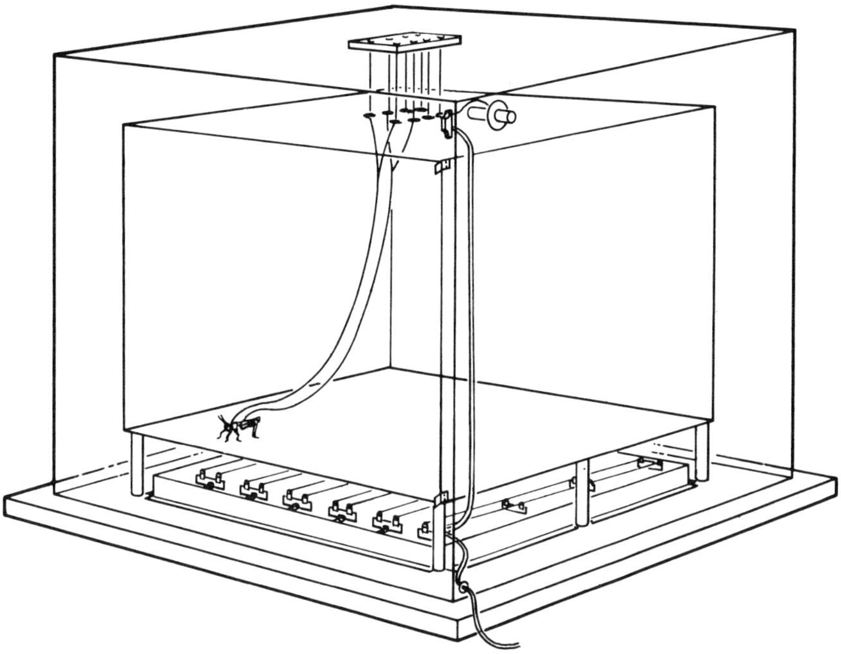

FIG. 2. Experimental cage. The box is double-walled and made entirely from Lucite.

quadruple-glass window permits observation of the movements of the animal from outside.

A large locust, grasshopper, or cockroach can carry two pairs of leads without showing undue impairment of movement. Under favorable circumstances the leads have stayed in place for as long as a week.

Outside the cage the leads enter a recording equipment bank which includes the following items: push-pull preamplifiers, pen oscillograph (as principal display channel) with parallel tape recorder, cathode-ray oscillograph, and audio monitors. The outputs also connect with pulse-shaping units leading into digital recorders which are connected to a common timing circuit and print-out counter (Fig. 3).

The pulses are displayed directly and are also fed to an integrator or digital–analog converter, whose output is displayed on the pen recorder. Every tenth pulse from the counter is also displayed. With these varied displays a rapid quantitative assessment of activity in any pair of muscles may be made. Muscles not related in locomotion but perhaps associated in specific behavioral responses can be investigated together.

Prospects for the Method

This method may be used for assessing the ways in which antagonistic pairs function, the modes of correlation in synergists, and the uses made by the

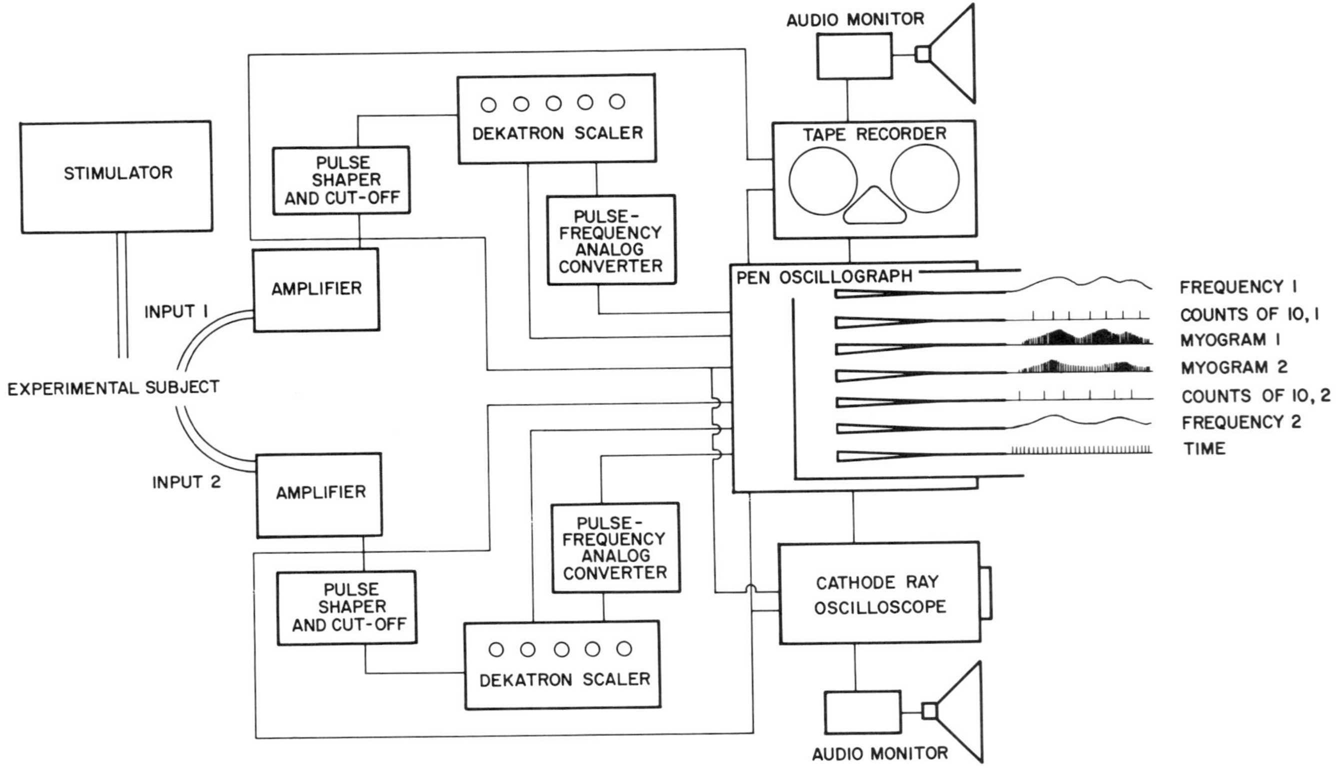

FIG. 3. Outline of instrumentation.

animal of "slow" and "fast" axons. A use that promises to be particularly valuable is in genetic studies of behavioral mechanisms. The technique of implanted leads leaves us with an intact animal which can be used in additional investigations and in breeding. Special features which may be detected in individuals can be made the subject of genetic selection experiments.

In experiments on habituation, conditioning, and other simple forms of learning, a rapid, simple, quantitative method is provided. Furthermore, the animal does not have to produce overt behavior for responses to be detected. The muscles of the apparently non-responding whole animal may be receiving motor-impulse discharges at a frequency too low to cause movements, but the rate of discharge may nevertheless be distinctly affected by stimuli. These subliminal changes are readily detected by this method.

Preliminary Accounts of Results

The following results are presented to indicate the kind of data which have been obtained and some of the features which have emerged. They are not intended to be a definitive statement of results, all of which remain to be published. Detailed findings will be the subjects of separate papers. The study has thus far been concerned mainly with the relationships of antagonistic muscles in achieving movement. The final unit in the behavioral sequence is the individual muscle, and its functioning cannot be detached from that of its synergists and antagonists. Hence a first step in any analysis which starts with efferent output is an understanding of the relationships of the final units.

Peripheral Machinery

Uses of "Slow" and "Fast" Axons. A vertebrate uses recruitment of motor units, as well as increases in the frequency of activation, to achieve greater force and velocity of movement. Arthropods are primarily frequency-operated, for a muscle may comprise a single motor unit, although some muscles have from two to four units. Most have a dual-axon supply to the muscle or to the motor units. In the most powerful muscles the functions of the two axons are clearly separate. In the jumping leg of the locust the fast axon is concerned with jumping and defensive kicks exclusively, all other movements being achieved by the use of the single slow axon. In weaker muscles, fast axons are used primarily to reinforce contractions caused by slow-axon discharges; firing in groups of a few impulses at a time, these fast impulses commonly coincide with peak discharge rates in the slow axons (Fig. 4).

Relations of Antagonists. This has been studied extensively in the extensor and flexor tibiae, especially in *Locusta migratoria*, *Schistocerca gregaria*, and *Romalea microptera*. Simple stimulation experiments, in which reflexes are elicited by touching various parts of the body, while the mechanical responses of antagonistic muscles are simultaneously recorded, show that in about 80% of reflex responses there is a reciprocal inhibitory relation between antagonists even though there is probably no proprioceptive feedback in such preparations.

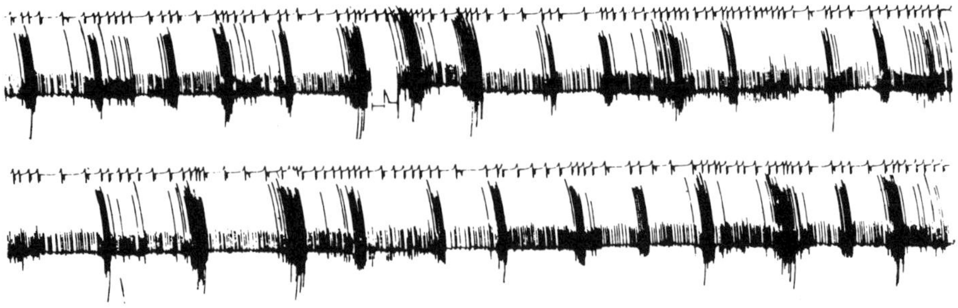

FIG. 4. "Slow" and "fast" axon actions causing contractions associated with continuous locomotion in a weak muscle: extensor tibiae of mesothoracic leg. The small potentials of uniform size are responses of muscle fibers to slow axon discharge. The large potentials occur because the discharge of a fast axon innervates the same muscle fibers but causes greater contraction in them. Each burst of fast potentials corresponds to a step. No two steps are associated with identical patterns; both slow- and fast-axon discharges are very variable. Counts of 10 appear on the lower traces.

That reciprocal relations do not hold in the other 20% of simple responses, means that the inhibitory relationship can be disrupted.

Stimulation of the motor nerve supplying one muscle does not prevent a reflex from occurring at the same time in the antagonist. Since the motor nerve will discharge antidromically, it should inhibit the antagonist if reciprocal connection is made at the level of the last pathway. We must infer that inhibitory cross-connection occurs at a level above the final motor neuron.

General excitatory actions achieved by massive sudden changes, such as switching a light on, blowing over the animal, or tapping the experimental table, commonly affect both members of an antagonistic pair in a similar manner rather than reciprocally. This phenomenon is never better seen than during defecation, when a coincident discharge occurs in antagonists—the animal goes stiff. There is nevertheless some value to considering antagonistic pairs as a unit even though co-contraction is part of their response pattern.

During very delicate actions, antagonists commonly give co-contractions, as they do also in humans (Seyffarth, 1940; Basmajian, 1962). By contrast, casual locomotion is often achieved by a simple reciprocal action in antagonists in each joint (Fig. 5). Indeed, the whole complex of locomotory muscles may be cross-organized on a reciprocal basis. The basic locomotory unit is the tripod of fore and hind legs on one side and center leg on the opposite side, and the two tripods have the same reciprocity found in an extensor–flexor pair. However, this is a very loose arrangement, for patterns of locomotion other than the alternating tripod are utilized, and as pointed out earlier, co-contractions of natural antagonists are also common.

When really fast locomotion occurs in grasshoppers, the simple reciprocal relation breaks down. Then one muscle, commonly the flexor, receives an uninterrupted discharge and is thrown into a state of mild contraction. At the same time the mean discharge rate of the opposed muscle also rises, but fluctuates widely at a high frequency (Fig. 6). The strong contractions evoked overcome the resistance of the opposing muscle, causing movements. Since

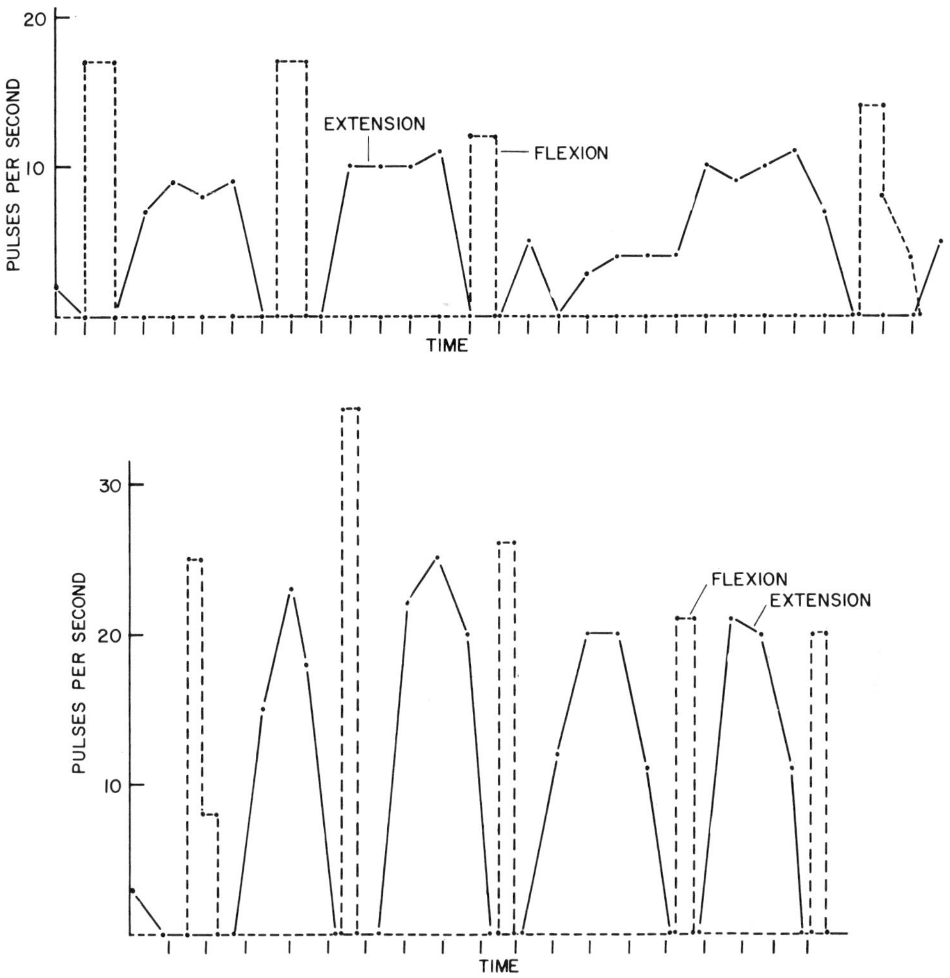

FIG. 5. Electrical activity in extensor (solid line) and flexor (broken line) tibiae muscles during very slow locomotion in two species of insect: upper, *Romalea microptera;* lower, *Schistocerca gregaria.* Complete, simple reciprocal inhibition was evident in both cases. Time intervals, sec.

the opposing muscle is in a state of partial contraction continually, rapid return of the displaced limb occurs when the prime mover starts to relax. It is possible to show by direct experiment on leg preparations that the maximum "stepping" rate of the hind leg of a locust can be increased from under two steps per second to more than three steps per second by this type of control.

Either muscle in the antagonistic pair may be used as a "driver," and changes from the use of one muscle to the use of its opponent have been observed in the intact animal during locomotion (Fig. 7). These observations illustrate the flexibility of the nervous system, even in controlling a simple pattern of movement, and must be taken into account when we consider models simulating central-nervous-system action.

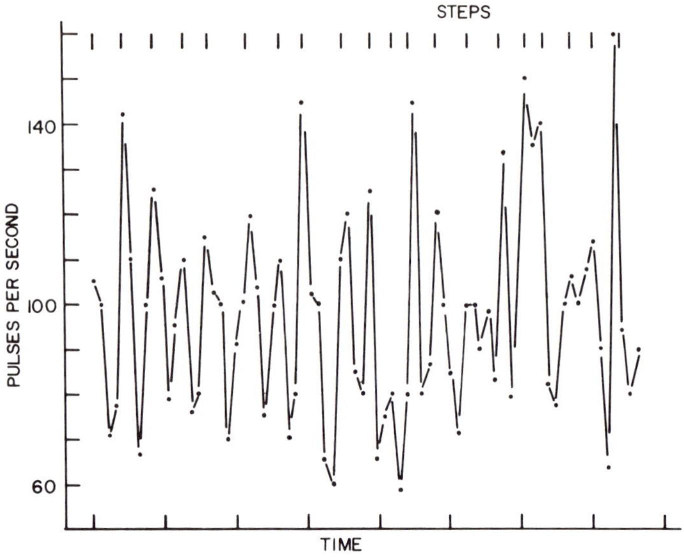

FIG. 6. Discharge rate in "slow" axon during very fast walking in fifth-instar *Schistocerca gregaria*. Frequency does not fall below 60 pulses per sec. Time intervals, sec.

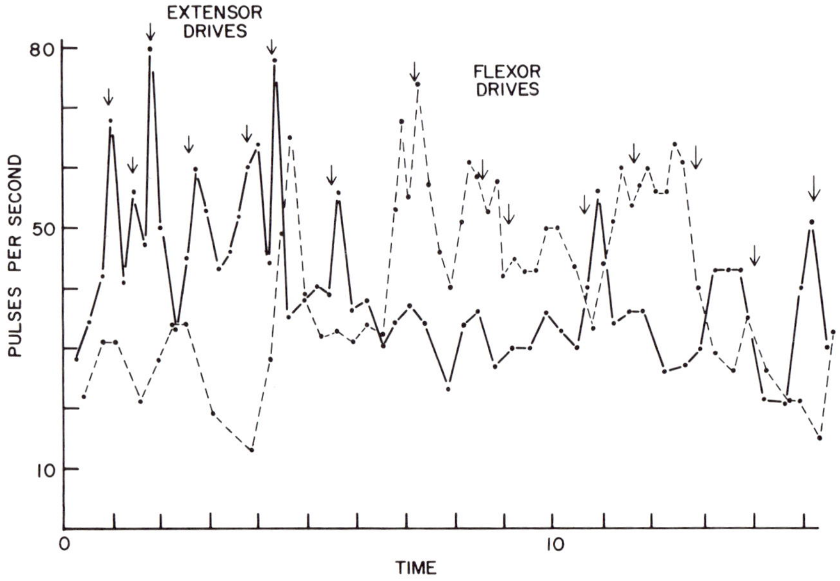

FIG. 7. Frequencies of firing in extensor and flexor tibiae of hind leg of *Schistocerca gregaria* during a continuous "march" with a slight fall in pace. Steps are indicated by arrows. Continuous line: extensor discharge; broken line: flexor discharge (probably due to a single unit). The change in pace was due to a considerable fall in the mean rate of firing of the slow axon to the powerful extensor muscle, combined with a rise in the rate of firing of the flexor unit. Fluctuations in flexor rate then appeared to be responsible for movement, rather than fluctuations in extensor rate. Time intervals, sec.

Central Machinery

Spontaneous Activity. Every muscle examined has shown periods of total electrical silence; these are rare in some, common in others. Major fluctuations in discharge rate, either rises or falls, can be evoked only by external stimulation, especially the addition or withdrawal of food, fluctuations in temperature, sounds, and other physical disturbances. But even when the animal has fed recently and is kept at optimal temperature in dim light in a highly soundproofed stable-air environment, motor-output fluctuations occur in an apparently random manner. These fluctuations often show a close parallel in different muscles (Fig. 8). Occasionally the peak discharges to certain muscles exceed the minimum rate needed for sufficient power development to exceed the natural inertia. Then a movement results, perhaps no more than a slight fidget, but at times a step, or even a little spontaneous walk.

It seems as though no particular set of muscles is the first to move; sometimes a flexor, sometimes an extensor, or a levator, a depressor, a pronator, or a remotor, is the first to shift a limb. All this points to a number of "spontaneous" units in the nervous system, together with inhibitory units, all of them interacting and fluctuating so that the net effect is a series of very irregular outputs.

Only the respiratory rhythm is regular, and it is interesting to find that single impulses sometimes appear in any one of the locomotory muscles in synchrony with respiratory movements. Sometimes the background "random" firing changes to little bursts of impulses, at a frequency much too low to lead to movements, which "sync in" with the respiratory rhythm for long periods. Such activity could be caused by proprioception initiated by shifts in posture associated with respiratory movements or by internal electric-field effects in the neuropile. It seems to point to a rather loose, overlapping, multiple interconnectivity in the nervous system or to the presence of very general interneurons, each making contact with many motoneurons. This finding is further emphasized by simple stimulation experiments. Natural stimulation in any mode applied to any part of the body containing receptors for that mode may cause a significant increase in the slow-axon discharge to any muscle from which recordings have been made.

Prolonged Responses. In these experiments locusts have been found to respond characteristically to temperature changes. If they are at room temperature (20–24°C) and the temperature is slowly raised, they soon become restive and show a period of exploratory behavior which may last several minutes. During this period they "seek out" warm spots and settle there if any are to be found. They show no further locomotory activity if the temperature continues to rise, even to quite high levels (40°C).

They will now, after remaining a short time at the higher temperature (optimally, 37–40°C), show a marked response to a fall in temperature. The fall may be as slight as 1.5°C. This causes them to leave their roosting spot and locomote vigorously. Probably a fall in body temperature is the cause, because a draft of cool air may have a similar effect. A continuing fall in

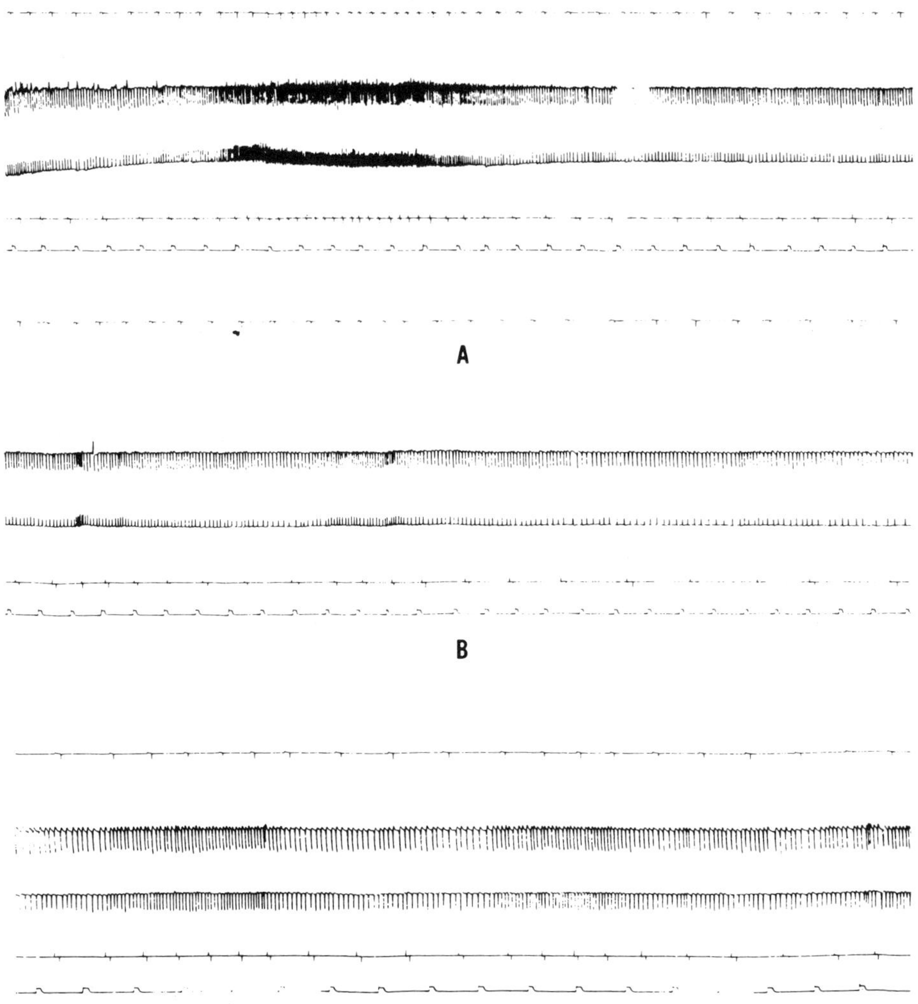

A

B

C

FIG. 8. Continuous recordings from left and right metathoracic extensor tibiae of *Schistocerca gregaria*. Records A and B are of one animal, C of another animal. The discharges are due to the single slow axon and were at too low a frequency to cause movement, except at the beginning of A. Note how closely parallel are rises and falls in discharge rate. Counts of 10 appear on the outer channels. Time tics (bottom line, each part) are seconds.

temperature is not necessary to maintain the response. The movements they then make are different in appearance from the casual ones of pottering and exploration. The insects look "determined" and "purposeful" and the posture is high and rather rigid, in place of the belly-dragging and abdomen-trailing of the potterer, which is especially evident in an animal which has recently fed. This alert appearance is typical of the attitude in marching behavior shown by migrating bands (Ellis, 1951). It lasts for one to several minutes with only a single animal in the cage, or continues for up to 10 to 12 hr if there are several

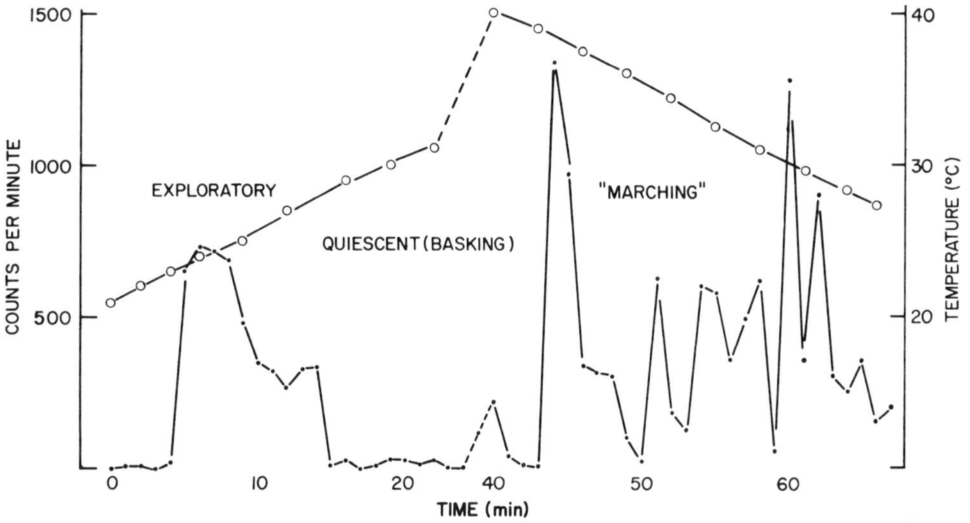

A

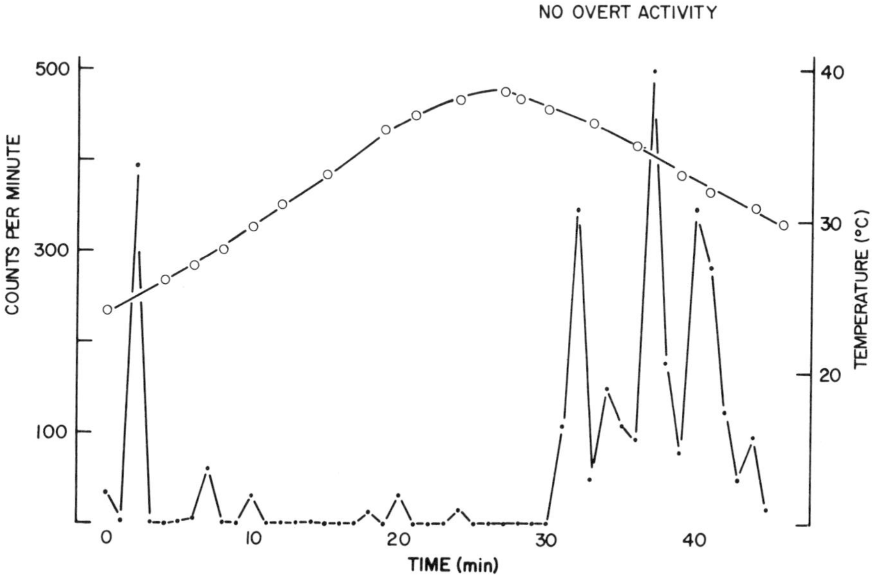

B

FIG. 9. Records of two experiments, each on a different animal, showing the response of fifth-instar nymphs of *Locusta migratoria* to temperature changes. The counts are of impulses due to a slow axon's supplying metathoracic extensor tibiae: Open circles, temperature in cage; dots, counts per minute. A mean count in excess of 500 per minute is associated with overt behavior, in this case exploration or "marching." In record A, on a hungry insect, very active behavior was seen. In B the animal had recently fed and did not move during the whole experiment, but roughly the same pattern of change in discharge occurred at subliminal levels.

others present at the same time. During this time prolonged, continuous, vigorous "marching" locomotion is seen (Fig. 9).

In this kind of activity, the behavioral response far outlasts the stimulus, and there is no possibility that a unique chain of reflex command will keep it going, for the sensory feedback must be similar to that occurring in ordinary walking, which is very sporadic. We must envisage a prolonged, general "driver" discharge commanding locomotion, initiated by the special circumstances—in this case no more than a brief fall in temperature following a prolonged slow rise. The marching can be interrupted by the presentation of food. If feeding is prolonged, the tendency to march is completely abolished; but after a short period of feeding, prolonged marching is immediately resumed (Fig. 10).

Another, though less marked, instance of a prolonged response is found in the reaction of grasshoppers to light. The coxal promotor muscles of *Romalea microptera* show continuous background slow-axon activity at a low level when kept in the light for some hours, but when the light is shut off an immediate brisk discharge occurs, sometimes leading to movement. Spontaneous bursts are frequent up to 2 min after the light is turned off. The main effect, however, is seen in the background rate of firing, which increases markedly. The increase may persist for several minutes. The responses may be changed by keeping the animals in the dark for some hours; then "light-on" leads to a prolonged discharge. These differences are not caused by changes in the sense organs themselves. Internal "switching" of the character of the responses could be brought about by neuro-humoral action. The reaction is affected by temperature; in the early stages of subjection to higher temperatures (104°F) the animals are very active in the light. This activity is inhibited by cutting off

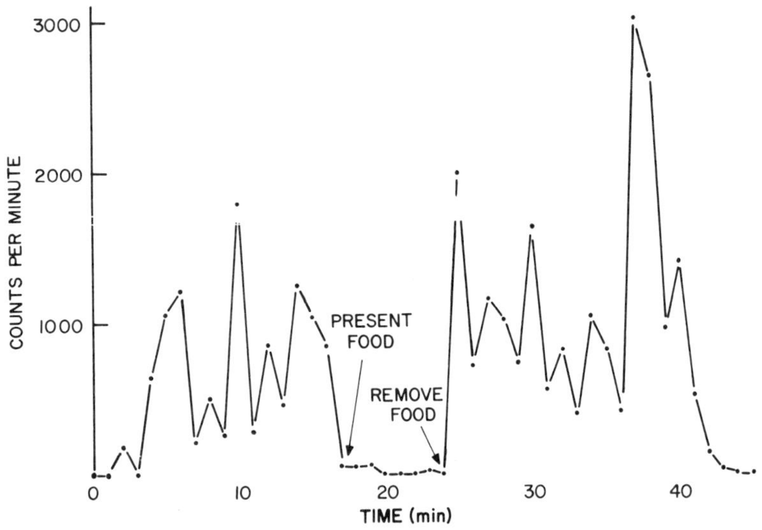

FIG. 10. Interruption of an induced "marching" response of a fifth-instar locust hopper by a brief period of feeding.

the light. The same kinds of reaction are observed whether or not the discharges reach the frequencies required to lead to overt activity.

Subliminal Responses. It was observed during the experiments with rising, followed by falling, temperature that a rise in motor discharge occurred in response to a rise in temperature even though the animal did not, in some cases, show any overt activity. With continued rising temperature the background activity gradually fell, often to zero. After a period at a high temperature, a sudden fall in temperature was always followed by a considerable rise in background activity, leading sometimes, though not always, to movement. It will be apparent that the sequence of changes in rate of firing is similar to that associated with overt behavioral responses referred to above (Fig. 9B). Frequencies in excess of 20 impulses/sec are needed to develop the power needed for locomotion. A mean background count of 5 impulses/sec is not uncommon in the resting animal, and a rise to a mean of 18–22 indicates a highly significant response, although it may not result in any movements. Herein lies a valuable attribute of the technique: the possibility of exploring the animal's responses to situations even when it fails to show overt activity. Although overt activity does not occur, there may never the less be a true "subliminal" behavioral response; the state of the muscles is changed, owing to an increased tone, but the force developed is too weak to cause movement. The existence of complex subliminal activity of this kind provides very strong evidence for endogenous "tapes" controlling motor activity.

Repetition of Complex Acts. Figures 11 and 12 are the records obtained

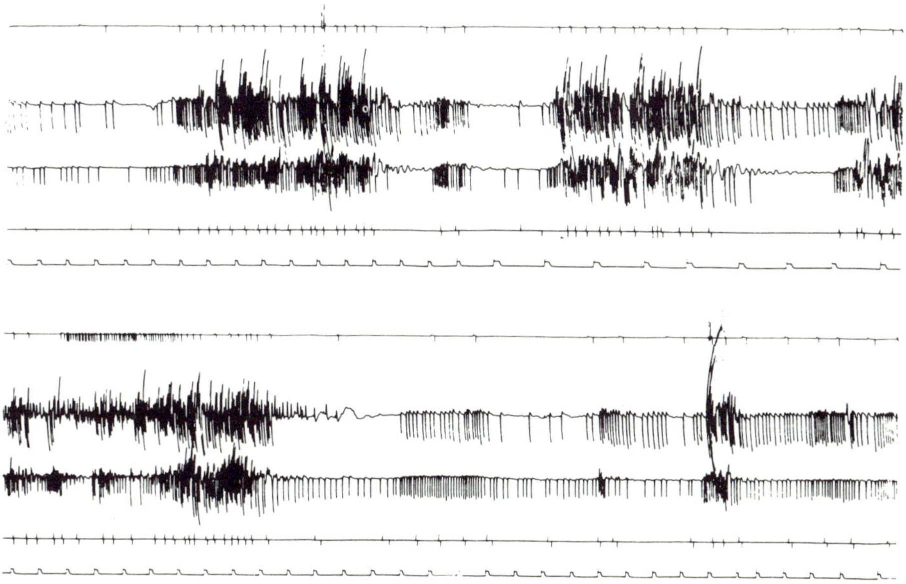

FIG. 11. Continuous record from (upper trace) flexor and extensor (lower trace) tibiae of hind leg of *Schistocerca gregaria* during three "identical" movements. Note the changing of units used by the flexor, the fast-axon bursts used by the flexor and the close parallel between gross rises and falls in frequency in the flexor and the extensor. Counts of 10 are marked on the outer channels. Time: 1 count per second (bottom line).

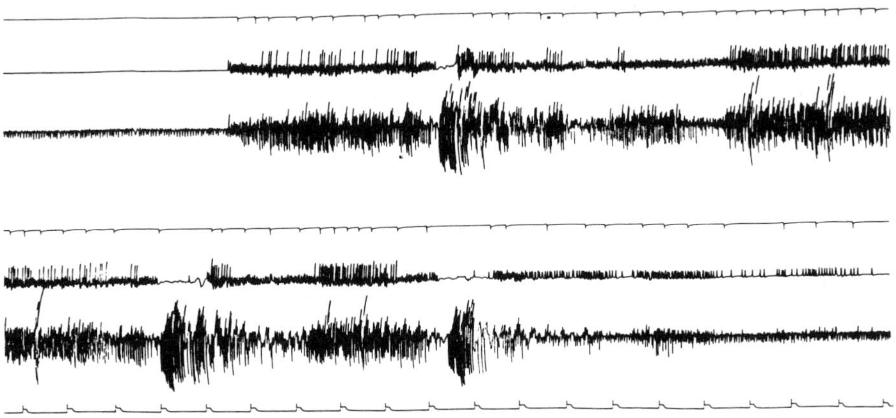

FIG. 12. Records obtained from mesothoracic leg of the second *Schistoserca* during three similar complex maneuvers. Upper trace: extensor tibiae, showing simple, single slow- and fast-axon pattern. Lower trace: flexor tibiae, complex record due to use of two motor units. Record starts with single slow axon active, then a second slow axon and bursts of a fast axon are added. Note the complete inhibition of extensor activity during certain fast flexion movements. The onset of these inhibitions precedes the flexor discharge. Note also the close parallel between gross rises and falls in frequency in the antagonists.

for two different locusts, from an antagonistic pair of locomotory muscles during repetition three times of a motor action (climbing from the underside to the top of a twig which had been rotated). In these cases the brief stimulus was followed by a complex reaction taking about 5 sec to complete. There is a notable similarity among the gross patterns of impulses underlying the movements. In particular, the brief, high-frequency burst of impulses which occurs in the flexor upon achieving the desired position is evident, and this is associated with a marked inhibition of the extensor discharge, which must have been determined by a central, predetermined action, for the inhibition occurs in advance of the powerful flexion. Such actions contain elements which suggest that a stereotyped set of commands has been applied sequentially, following the stimulus.

Postulates Arising from Results

The General Driver Neuron

The observations briefly outlined above have led me to postulate a number of features of neural organization and function in insects, which will serve as a basis for future investigations. The simultaneous appearance of impulses in many muscles, including antagonists, spontaneously or in response to almost any stimulus, can most readily be explained by postulating common interneurons connected to them all either directly or, more probably, indirectly. Such neurons are excitatory, and they may act over long periods. When they fire at a sufficiently high frequency they force the animal into some kind of locomotion or activity. Such a neuron will be termed a *general driver* neuron. The final motor neurons illustrated in Fig. 13 are those which are actually identified in insect limb flexors and extensors (Hoyle, 1955, 1957) and whose discharges in the intact animal have been studied in the present work. The synapses from the internuncial neurons that are excited by general driver neurons are drawn large if they are easily excited, a condition which applies to slow axons, or small if they have high thresholds for excitation, a condition which applies to fast axons. No direct interconnections between these neurons have been observed as a result of studying the effects of antidromic impulses. Reciprocal inhibitory cross-connections are therefore ascribed to the level of the last internuncial neurons. Our experience so far suggests that these are especially strong from the fast extensor channel across to the flexor channels, and from some fast flexor channels across to the slow extensor.

The records from flexor muscles are complex and difficult to interpret, because two or more motor units are commonly firing simultaneously but not synchronously. However, it seems that the same units are involved in most

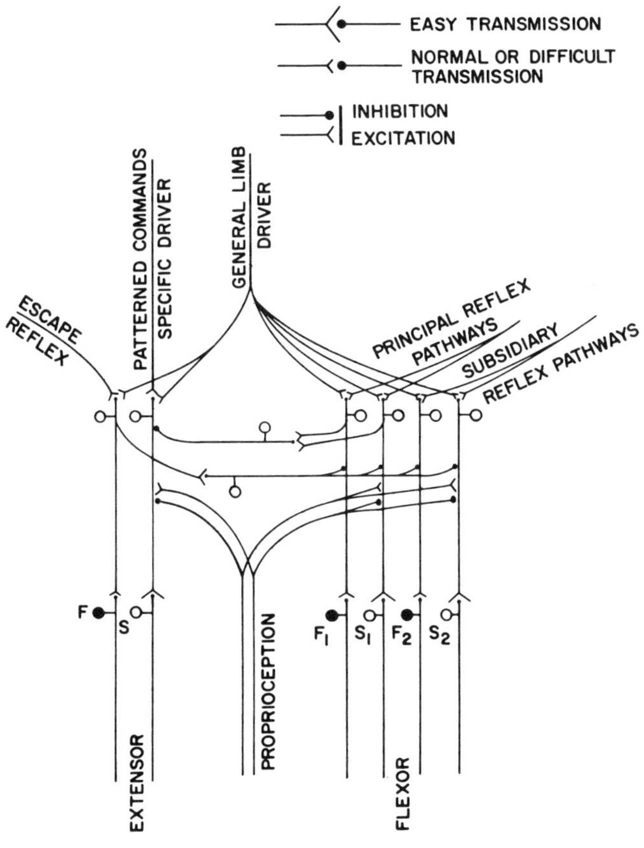

FIG. 13. Scheme for final control of motoneurons, as seen in the metathoracic extensor and flexor tibiae of locusts and grasshoppers. The symbol F denotes a fast axon giving a brisk twitch, and S a slow axon giving contractions requiring facilitation and summation. The extensor receives only one of each, the flexor at least two, supplying different motor units. Slow and fast axons innervate the same muscle fibers. Proprioception and resulting excitatory or inhibitory influences are stylized, but the sites of action are probably correct.

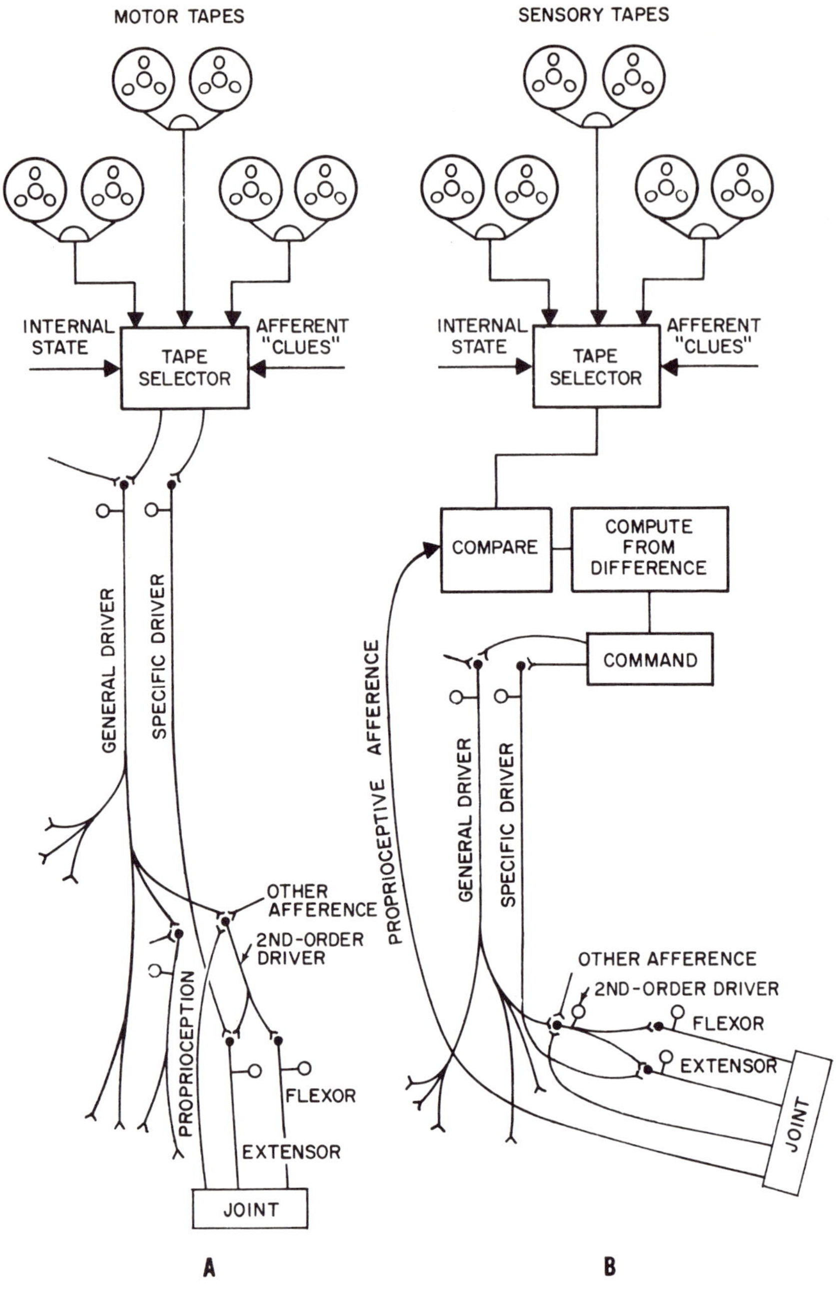

FIG. 14. General and specific driver neurons and sites of introduction of hypothetical "taped" control. A. Control by "motor tape." A *general driver* neuron is envisaged as an efferent interneuron which gives off branches on both sides of the animal. The branches activate second-order general driver neurons concerned with exciting a few muscles on one side only. A "tape selector" is triggered by specific internal and external cues to select a motor tape, which excites the general and specific neurons. The system is open-loop except at the level of the second-order command neurons. B. Control by "afferent tape." The selected tape plays back proprioceptive afferent information relevant to a given motor act. This is compared with actual proprioceptive afference. The difference results in the computation of a command which is passed on to the driver neurons. The system contains two closed loops, a local one at the level of the last interneurons, and a higher one at the level of the comparison center.

ordinary locomotion. Other units are used primarily in unusual postures. For instance, they fire when the insect walks upside down along a twig.

Somehow the details of the model must take account of the fact that at both low and high levels of general excitation, reciprocal relations do not occur. Lack of reciprocity at low levels may be achieved quite simply by requiring an adequate facilitation in the inhibitory pathways before they become effective. Lack of reciprocity at high levels can be achieved by allowing rapid fatigue of a Wedensky-type inhibitory phenomenon in the inhibitory channels, or simple overriding of inhibition by very strong excitatory synaptic action.

The activity of a general driver may be overridden by strong reflex channels. If a marching grasshopper is unduly disturbed, it jumps. If it is attacked by another, it kicks it in the face. Each of these reactions calls for different types of coordinated interruption of the general driver control; the former leads to inhibition of an extensor activity, followed in turn by flexion of the hind tibiae, inhibition of the hind flexor tibiae, and finally a brief and fast extensor-axon excitation leading to a leap. The marching is instantly resumed on landing. The defensive kick calls for beautifully coordinated movements in many leg muscles, and careful timing of the extensor thrust on one side only. It is not a simple reflex, because any attempt to elicit it artificially produces only a jump. The nearest approach to it which can be elicited reflexly occurs in response to persistent tickling of a side of the thorax. This leads to a lifting of the hind leg until the tarsus is located at the source of irritation, at which moment a single kick occurs. The location is extremely accurate, even in the decapitated insect.

The distinguishing feature of a general driver neuron is that it simultaneously activates functionally opposed muscles, such as extensors and flexors, or groups of muscles which must contract in sequence in order to produce useful movements. There may be two levels of action of such drivers, one at the level of a single limb or even a single joint, the other at a higher level, concerned with all the limbs or all the wing muscles.

The Specific Driver Neuron

Activation of general driver neurons can probably account for the initiation and control (from "motor tapes") of many simple actions, especially food-seeking, mate-seeking, exploratory behavior, marching (of locust hoppers), and migration. It would seem, however, that for more subtle complex behavior, subject largely to endogenous control, such as sexual displays, a more direct form of excitation of the motoneurons would be required. "Tapes" of motor command would need channels which did not carry activity from local reflex circuits. Since there is already some evidence (from the present work) that such channels exist, interneurons involved will be given a designation and termed *specific driver* neurons (Fig. 14).

Suppression of both general driver activity and some reflex activity must be envisaged as necessary for the adequate performance of movements by impulses in specific driver neurons; otherwise, the detailed information contained in the tape would be lost among the converging signals. Suppression of general activity is readily observed in the animal; it occurs, for instance, in the male locust during copulation, though the female may walk around the cage and eat during the act. The general activity is suppressed in the female during egg-laying.

A control system using "sensory tape" and a fast, difference-operated servo would probably send its computed commands to the muscles via specific driver neurons.

Evolution of Behavior Mechanisms

Once the existence of well-defined responses too weak to give rise to overt movement was recognized, it was possible to search for this kind of response following a variety of stimuli, though it is usually difficult to distinguish a potential behaviorally meaningful effect. Owing to the apparent widespread interconnectivity, e.g., of "general driver" neurons, such responses have been found (Fig. 15). In these responses—say, for the wing elevators to a scraping sound—there is the germ of new behavior. In a given specimen, a small lowering of thresholds in the relevant part of the nervous system, or a parallel rise in responsiveness of the muscles which open the wings (either of which could be genetically determined), would cause that animal to raise its wings in response to the sound. The movement thus evoked might, in turn, serve by chance as a stimulus to a neighbor of the opposite sex, and a courtship pattern would be established. If this led to mating it might confer a selective advantage and become an established behavior pattern. In this way one can readily understand how animal behavior patterns evolved; it is necessary to postulate a

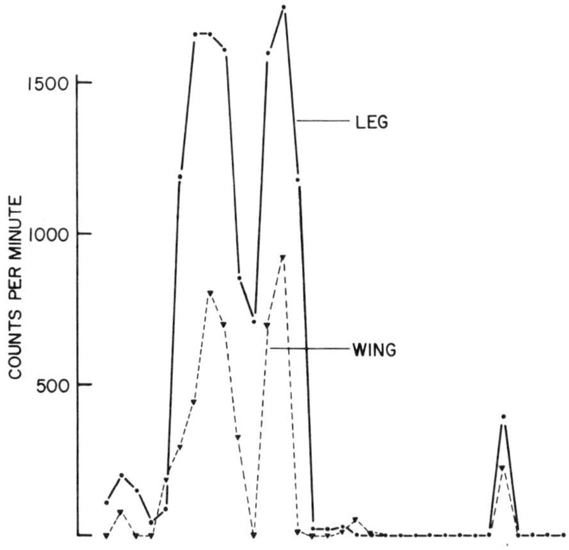

FIG. 15. Parallel increase in rate of firing in muscles not closely related functionally during temperature-induced marching in *Locusta*. Solid line: Extensor tibiae of hind leg; broken line: wing elevator (also a coxal-muscle tergal pronator), mesothoracic leg.

great deal of interconnectedness—for example, via general driver interneurons—and also variation in synaptic transmission properties in the nervous system, but these are exactly what we find.

Conclusions

The nervous systems of the insects studied may contain elements which are the equivalent of neuronal "tape" records of motor instructions. These are, however, not common or conspicuous in action. This may be because there is no mechanism for switching off the activity of local reflex pathways or of general driver neurons when such a "tape" starts to play back. Such "tapes" probably act mainly by causing the excitation of *general excitor interneurons* one or two synapses removed from the final motoneurons, and affecting several neurons, including those of antagonistic pairs, at the same time. At this level, the tape command mixes algebraically with other excitatory and inhibitory neuronal influences. In addition to general excitor neurons, some control sequences probably occur via instructions given more directly to individual muscles through the mediation of *specific excitor interneurons.* Whether the activity of motor-command tapes can occur in the complete absence of sensory information needs to be explored.

It also appears probable, on the basis of data obtained thus far, though more difficult to conceive in regard to mechanism, that the animal's nervous system contains the equivalent of neuronal tape recordings of sequences of sensory patterns. The stored proprioceptive afference would provide a standard of reference with which the immediate proprioceptive afference may be compared, the difference forming the basis for computing the motor command required to "correct" the movement. In such a mechanism the motor output would vary unpredictably and erratically, though still leading to sensible patterns of movement. The mechanisms proposed may be regarded as details of a stimulus–response sequence of events underlying behavior, but with strong, independent endogenous control.

The principal way in which the schemes proposed here differ from earlier hierarchical ones is in requiring only a relatively simple cross-connectivity in the motor nervous system. Neural functioning need not depend on each "higher nervous center's" sharing its connections to lower centers with a multitude of others. Instead, the several information stores (neuronal "tapes") could all feed into a few common effector pathways. However, in spite of the optimism with which even these beginnings lead us to continue our exploration, I feel impelled to end on a note of caution.

We start with a belief that the living organism is an automatic mechanism whose actions are determined entirely by cause and effect in the nervous system and conclude that a complete description of behavior must in principle be possible. We may trace this approach primarily to the philosophy of Descartes. There is another school of experimental psychology, based upon the "cognitive" faculty, which permits control from the mind. Assemblages of neurons possess properties which are more than the sum of the properties of the parts, including, in man, such conceptually difficult phenomena as consciousness, will, prediction, and inference. We are not accustomed to endowing the lower organisms we study with any of these higher functions. Yet we have no reason to deny their nervous systems the capability of these functions, at least in some measure, since we are sure they exist in our own nervous system, and there are no known significant physiological differences between the units of which the two kinds of brain are composed.

Should this be the case, and should an insect have an awareness of itself in space-time, albeit in a very crude way, with a capacity to "will" itself into some action, that will introduce an element of indeterminacy into any investigations we make on the whole organism and therefore make it impossible to give a complete explanation of its behavior. Nevertheless, we have undoubtedly a wonderful technical armament for making studies on the nervous system, and great strides can surely be made in applying them, provided our conceptual framework can be expanded to take care of the data we can so readily accumulate.

Summary

1. A plea is entered for the serious study of neuronal mechanism underlying the instinctive behavioral acts of animals.

2. This can best be achieved by the concerted study by several independent investigators who agree to converge their attention on one suitably amenable experimental animal.

3. Current models of central-nervous-system mechanisms underlying behavior are considered briefly, and a new one is added.

4. The latter supposes the nervous system to contain the equivalent of magnetic storage tapes of behavioral information, together with appropriate playback and read-out mechanisms. Two different sorts of tape may be envisaged ,"motor-command tapes" and "afferent tapes," each utilizing quite different kinds of associated neuronal machinery. "Motor tapes" operate in open-loop systems. "Afferent tapes" must function through a closed-loop arrangement, but the loop does not include the tape region itself.

5. A means is described for experimentally investigating some of the hypotheses associated with the models by making physiological studies of a novel type on certain species of insect.

6. The results of work using these techniques are briefly described. They are summarized as follows:

a. Parallel activity appears simultaneously in several different muscles having a variety of functional relationships to each other. This suggests the existence of generalized interneurons making contact with many motoneurons.

b. Reciprocal relationships are found in the responses of antagonistic muscles, but the inhibition is not mediated at the level of the motoneuron itself. The synapses responsible for the reciprocal relationship are themselves readily inhibited.

c. Activity appears in the motor system below the level for liminal expression of movement. Such activity may be similar in pattern to that which would cause an overt response if it occurred at a higher general level of frequency.

d. Activity appears in muscles unrelated to a given set when the latter are active in causing a meaningful movement, reinforcing the generalized interneuron concept.

e. This "irrelevant" activity provides for the evolution of new behavior.

f. There are several different functional ways of pitting different muscle groups against or with one another to achieve a given, or closely related, type of movement. Many different ways are observed to be used if the same movements are elicited repeatedly, or occur naturally, in the intact animal.

7. The results suggest that "motor tapes" are used in certain actions of locusts.

8. Other results, especially those summarized in 6f, are best interpreted on the basis of a "sensory tape" type of control.

References

Basmajian, J. V., 1962. *Muscles Alive*, Baltimore: Williams & Wilkins.

Dethier, V. G., 1962. Neurological aspects of insect behavior, in *Roots of Behavior*, E. L. Bliss, ed., New York: Harper.

Ellis, P. E., 1951. The marching behavior of hoppers of the African migratory locust in the laboratory, *Anti-Locust Bull.*, **7**, 1–48.

Fabre, J. H., 1912. *Social Life in the Insect World*, London: Fisher Unwin.

Hess, W. R., 1943. Das Zwischenhirn als Koordinationsorgan, *Helv. Physiol. Acta*, **1**, 549–65.

Holst, Erich von, and Ursula von St. Paul, 1963. On the functional organisation of drives, *Animal Behaviour*, **11**(1), 1–20.

Hoyle, G., 1955. Neuromuscular mechanisms of a locust skeletal muscle, *Proc. Roy. Soc. (London)*, **B-143**, 343–67.

Hoyle, G., 1957a. The nervous control of insect muscle, in *Recent Advances in Invertebrate Physiology*, B. T. Scheer, ed., Eugene, Ore.: Univ. Oregon Pubs., 73–98.

Hoyle, G., 1957b. *Comparative Physiology of the Nervous Control of Muscular Contraction*, Cambridge, Eng.: Cambridge Univ. Press.

Huber, F., 1960. Untersuchungen über die Funktion des Zentralnervensystems und Insbesondere des Gehirnes bei der Fortbewegung und der Lauterzeugung der Grillen, *Z. vergl. Physiol*, **44**, 60–132.

Huber, F., 1962. Central nervous control of sound production in crickets and some speculations on its evolution, *Evolution*, **16**, 429–42.

Kennedy, J. S., 1954. Is modern ethology objective? *Brit. J. Animal Behaviour*, **2**, 12–19.

Lehrman, D. S., 1953. A critique of Konrad Lorenz's theory of instinctive behavior, *Quart. Rev. Biol.*, **28**, 337–63.

Lorenz, K., 1950. The comparative method in studying innate behavior patterns, in *Physiological Mechanisms in Animal Behavior*, (Symposia of the Society for Experimental Biology, No. 4), New York: Academic, pp. 220–69.

Mittelstaedt, H., 1957. Prey capture in mantids, in *Recent Advances in Invertebrate Physiology*, B. T. Scheer, ed., Eugene, Ore.: Univ. Oregon Pubs., pp. 51–71.

Mittelstaedt, H., 1962. Control systems of orientation in insects, in *Annual Review of Entomology*, vol. 7, Palo Alto, Calif.: Annual Reviews, pp. 177–98.

Roeder, K. D., L. Tozian, and E. A. Weiant, 1960. Endogenous nerve activity and behavior in the mantis and cockroach, *J. Ins. Physiol.*, **4**, 45–62.

Seyffarth, H., 1940. The behavior of motor-units in voluntary contraction, *Norske vid. akad.*, **4**, 1–62.

SOKOLOV, E. N., 1960. Neuronal models and the orienting influence, in *CNS and Behavior*, M. A. B. Brazier, ed., New York: Josiah Macy, Jr. Foundation.

TINBERGEN, N., 1950. The hierarchical organization of nervous mechanisms underlying instinctive behavior, in *Physiological Mechanisms in Animal Behavior*, (Symposia of the Society for Experimental Biology, No. 4), New York: Academic, 304–12.

TINBERGEN, N., 1951. *The Study of Instinct*, Oxford: Oxford Univ. Press.

THORPE, W., 1956. *Learning and Instinct in Animals*, London: Methuen.

VOWLES, D. M., 1961. Neural mechanisms in insect behavior, in *Current Problems in Animal Behavior*, W. H. Thorpe and O. L. Zangwill, eds., Cambridge, Eng.: Cambridge Univ. Press.

Vowles, D. M., 1961. The physiology of the insect nervous system, *Int. Rev. Neurobiol.*, **3,** 349–72.

WEISS, P., 1950. Experimental analysis of co-ordination by the disarrangement of central peripheral relations, in *Physiological Mechanisms in Animal Behavior*, (Symposia of the Society for Experimental Biology, No. 4), New York: Academic, pp. 92–110.

WIERSMA, C. A. G., 1957. On the number of nerve cells in a crustacean central nervous system, *Acta Physiol. Neerl.*, **6,** 135–42.

WILSON, D. M., 1961. The central nervous control of flight in a locust, *J. Exptl. Biol.*, **38,** 471–90.

WILSON, D. M., 1962. Bifunctional muscles in the thorax of grasshoppers, *J. Exptl. Biol.*, **39,** 669–77.

19. Models and the Insect Brain

D. M. VOWLES, *Oxford University*

One of the more remarkable features of Oxford is that it possesses 67 philosophers who are mainly exercised in linguistic analysis. A point they often make is that when one uses a word one should be clear both about its possible meanings and about which meaning one is adopting at the time. Although we all use the word *model*, it is often far from clear, in present day discussions, precisely what we mean by it. I am not sure when the term first became fashionable in science; although many early workers undoubtedly used models, the word itself seems a fairly recent coin in intellectual currency. In psychological literature, at least, it seems to be a post-war phenomenon.

One obvious question is whether models are synonymous with hypotheses, and if not, what the distinction is. When I was at school it was very clear to me what the various concepts and procedures of science were. First, one collected some data. Having examined this, one might be lucky enough to notice some regular pattern in it; a formal description of this pattern was then regarded as a hypothesis. For example, when Boyle observed various regularities in the behavior of gases he was able to describe these regularities by the equation $PV = K$. These formal descriptions constitute a hypothesis based on a limited number of specific observations and suggest the collection of more facts under a wider variety of conditions in order to test the generality of the description. If, as a result of further investigation, the hypothesis remains uncontradicted, it is regarded as a theory. Should this fortunate state of affairs persist for many observers over a long period of time, the equation receives a respectable title: it is called a law.

Like many of the things we learn about science at school, this analysis seems to be wrong. Scientists seldom work in this way, and if they did they would not progress very far. First, nothing is said about one's motives in collecting particular observations in the first place. One may perhaps assume that Boyle has already noticed the qualitative effect of squeezing or warming an inflated pig's bladder before he started to measure pressures, volumes, and temperatures. Second, it should be noticed that, on the above basis, hypotheses are

The experimental work reported in this document has been sponsored in part by the Research and Technology Division of the Air Research and Development Command, United States Air Force, through its European Office, under Contracts AF 61(052)-420 and AF 61(052)-114. I also wish to express my gratitude to the Royal Society and the Nuffield Foundation for their generous support.

only descriptions relating to previously observed events and that they do not suggest any new *way* of making observations, merely a repetition of the same *sort* of observation as had been previously made. For further advances one relies upon the intuitive and irrational suggestion of some underlying activities which could be correlated with the observed phenomena. Thus if one makes the surprising assumption that gases are made up of a large number of small spheres bounding about freely in space, one can derive the basis of the Kinetic Theory of Gases, and from this deduce the approximate generality of Boyle's Law. One then has a model representing the structure of a gas. This model was not deducible from previously known observations on gases, and it leads to a different sort of description of their behavior. If this analysis is correct, then devising a "model" refers to a different sort of activity than that previously described as making a "hypothesis" (see Deutsch, 1960, for a further discussion). I am aware that other sorts of models are used in other ways, but I wish to restrict the usage in this paper as indicated.

The same sort of distinction could perhaps have its parallel in the study of animal behavior. For example, Skinner (1938) has suggested that the main task of psychology is to discover lawful relationships between externally observable variables, such as the operations one performs on an animal and the overt responses the animal makes. If a rat learns more rapidly to press a bar to get food when it has been deprived of food, then the concept of reinforcement refers merely to the sort of finding which is stated in those terms. In this sense no model is employed, and the Skinnerian results could be compared with Boyle's attempt to describe the behavior of gases. On the other hand, ethologists, such as Lorenz (1950), developed a psycho-hydraulic model of drive which, though later proved inadequate (see, e.g., Hinde, 1959), did explain some observable changes in the responsiveness of an animal's behavior, and from which predictions about new types of experiments could be made. In this sense, therefore, the ethologists had a model, whereas Skinner and his behaviorist school had a hypothesis.

One of the problems which arises when one uses models in this way is to decide whether they are merely analogies or whether the model is intended to represent a real state of affairs. In the case of the Lorenzian model it is doubtful if anyone really believed that an animal contains any fluid reservoir except its bladder; but in the Kinetic Theory, for example, one may be intended to believe that a gas is really composed of molecules moving about freely in space. Pure operationalists would say that this intention is misplaced, that it is merely useful for some purposes to treat the gas as if it were made in this way. But many of us, I feel, have a sneaking hope that our models really exist somewhere in the world, and that it is not unreasonable to study how the models may be justified in themselves as well as on account of the predictions for which they can be used.

Hence a model can be used in two different ways: for example, by making assumptions about such things as molecular mass and velocity, predictions can be made about the behavior of the gas as a whole, or alternatively, one can assume the existence of moving molecules and try to study them directly. In such a direct study, however, a different sort of method is needed. For instance, the spatial distribution of atoms within a molecule was postulated to explain a chemical's gross reactions; to test this distribution one can observe the molecule by means of X-ray diffraction or electron microscopy. Kékulé dreamed up the ring structure of benzene many years before its hexagonal pattern could be directly demonstrated.

It is clear that if one postulates a model which would explain some observations, the study of the model itself requires a type of method different from the one used to obtain the original observations. This analysis can be extended to animal behavior and nervous mechanisms. On the basis of animal behavior one may suggest a type of mechanism which might underlie behavior, and may use the model to make testable predictions about that behavior. This is the sort of fruitful approach well exemplified by Mittelstaedt and Kunze in this volume. However, direct examination of the mechanism may involve studying the properties of the nervous system itself by electrophysiological, or surgical, or other procedures. In attempting this, two main steps have to be taken. The first is to decide the sort of model which may be most fruitful, and the second is to translate the active properties of the model into appropriate neurophysiological terms.

It is perhaps relevant for a psychologist to point out that since many individuals have different interests and preferences for the way they work, the usefulness of any model will often depend upon the individual who uses it rather than upon its absolute merit. These differences may extend to more than mere technical aptitudes. However, for various reasons there has been a tendency over the last few years to apply the methods of servo-analysis to the study of behavioral systems. This study was given impetus by the publication of Wiener's book on cybernetics in the forties, suggesting parallels between brain function and control systems. Since servo theory deals largely with control systems which coordinate the activities of simple or complex machines, its relevance to behavior and nervous activity appears obvious.

One of the advantages of black-box theory, which is often stressed, is that the theory applies to any control system irrespective of the contents of the boxes. I sometimes think that to physiologists whose major interest may be in looking at the contents of the black boxes, this advantage appears rather dubious; the engineer's valid retort to this criticism could be that a black box contains merely infinite sequences of smaller black boxes, so that the whole system is rather like one of those Chinese puzzles which, for a physiologist, may become a Chinese torture. The point is that each worker must decide which black box he is working with, even to the deliberate refusal to open any more. In studying the neural mechanisms of behavior, for example, one has to decide whether to deal with populations of neurons, or single cells, or nuclear enzymes, or some other phenomenon.

Servo Models and the Nervous System

One of the problems for the physiologist is to decide how to translate black-box models into physiological terms. Various difficulties immediately arise, the first of which is to locate a black box. It is often stated by theorists that when they refer to a "center" they do not mean to imply that this center is anatomically restricted in location, but it may consist of a network of nerve cells whose properties are such that it functions as a centrally controlling mechanism. Thus the physiologist may be faced with recording from a net of no known anatomical position, which raises some interesting technical difficulties.

Another type of problem is to translate the input-output functions of black boxes into physiologically identifiable forms. The remainder of this paper will be concerned with this problem, but first it is necessary to consider a further point about servo-type models. There seem to be two especially relevant features of such models. The first of these is the logic of the model, and the second might be called *analogic*. In the approach advocated in this volume by Mittelstaedt, it is insisted that some features of a model are logical necessities enforced by the data that have to be explained: alternatives are just not possible. When this is true, it is obviously a useful guide for the neurophysiologist in starting his search for the mechanism of the model. However, most models include features proposed because they fit the data better than any alternatives which have been suggested. In such cases an alternative may later be proposed which fits the data equally well, or more often, equally badly. For example, Bliss (in this volume) has suggested an alternative model to that proposed by Hassenstein and Reichardt (see Reichardt, 1961) to explain movement perception in beetles, and these models differ considerably in terms of the type of data processing involved. Where two or more equally good or bad models are available, a criterion is needed to decide on the most acceptable. Hassenstein (1960) proposes the criterion of simplicity; but simplicity, like innocence, is a quality which tends to evaporate under critical tests. One measure which Hassenstein has proposed is that of the smallest number of channels involved. This may, however, be unsatisfactory, since it does not allow for the simplicity of the functional centers between which the information channels pass. For example, in the model for insect movement perception, one part performs the essentially multiplicative function of correlation, but we might envisage multiplication as being performed in sequential additions, which could involve several more information channels.

When alternative models are possible, neurophysiology could play a crucial role in deciding between them. In neurophysiological techniques, with reference to such models, two problems are involved: first, locating the functional centers and their connecting channels, and second, deciding which parameters of neuron activity to measure. For the rest of this paper I shall discuss specific methods for studying the insect nervous system, together with some specific results obtained with the honeybee. In the light of this, some requirements for future research will be suggested.

Experimental Techniques and their Limitations. For localization and analysis of function the classic methods of experimental neurology can be used. These are as follows.

1. *Neuroanatomical* (the study of the histological structure and connections of the nervous system). This needs little comment, except that it is difficult, laborious, and slow. Good anatomists are hard to come by.

2. *Surgical.* The interpretation of ablation studies is difficult for many reasons, well known in vertebrate studies. Briefly these are as follows.

a. The effect varies with the age of the animal.
b. The effect changes with time after a lesion.
c. The effect may be due to irritation in tissue adjacent to the lesion.
d. The lesion may interfere with vascular or respiratory supply to other regions.
e. The lesion may produce an effect at a distance not by direct influence but indirectly via its influence on another mechanism.

In insects, ablation studies have suggested that the motor pattern of many activities such as courtship, respiration, feeding, and cleaning can be coordinated in segmental ganglia, while the brain (supra- and sub-esophageal ganglia) exerts excitatory and inhibitory influences upon these. Also in the brain are the major sensory centers for the eyes, the antennae and mouth parts, and the motor centers for the head appendages. In arthropods nature seems to have evolved a nervous mechanism suitable for the experimenter. The central nervous system is not only largely composed of segmental ganglia, but the sensory centers seem decently discrete and separate, with a comparable separation of function. This functional layout appears to allow individual coordination mechanisms to be studied in reasonable isolation. One may wonder, however, if such isolation is real, unless surgically caused, and, if surgical, how far the artificiality influences the result.

3. *Localized Electrical Stimulation and Recording.* Although it is very easy to pick up nerve impulses in insects, even with coarse electrodes, it is difficult to isolate single neurons except in peripheral nerves and large neurons, such as giant fibers. When recording from the brain of the honeybee, even with microelectrodes having a tip diameter of approximately 0.5 microns, there are many regions such as the outer optic ganglion (retina intermediaria) where clear single units are very difficult to obtain. Electron-microscope studies on arthropods* show that very fine fibers (less than 0.1 microns in diameter) may be in the majority. If this is so, then even fine microelectrodes may record from a highly selective minority of the larger cells, and this may seriously bias one's results.

Electrical-stimulation experiments using typical coarse electrodes suffer from various difficulties of interpretation. The most important may be summarized as follows.

1. The shape of the stimulating pulse, and its frequency, may influence the efficiency with which different types of neuron are affected. Adequate ranges of pulse shape have not yet been tried; nor has chemical stimulation.

* Camougis, personal communication.

2. Without direct observation it is not possible to tell whether the effect of electrical stimulation is physiologically excitatory or inhibitory, or merely has an interfering effect on some on-going activity.

3. The effective part of the stimulus field may not lie near the electrode tip; the geometry of the field will depend on many unknown tissue properties such as orientation or volume-conductor effects of tracts.

4. The stimulus may be acting physiologically at a distance by virtue of exciting conducting pathways, such as tracts, rather than centers.

Although these difficulties must obviously make one extremely cautious about the interpretation of stimulation experiments, they are useful for detecting which parts of the nervous system may influence particular behavioral activities.

The technique for applying localized electrical stimuli to the insect brain was developed by Huber (1960a and b) for use with crickets. It has been used largely in studying the properties of the mushroom bodies and the central complex, and (by Rowell, 1963) for the tritrocerebrum in locusts. I have also used similar methods for studying the mushroom bodies in the honeybee. Since the methods have produced some interesting results, I shall consider them in a little detail, in an attempt to correlate the results of anatomical, electrophysiological, and stimulation experiments.

The Structure and Function of the Corpora Pedunculata

Electrical Stimulation. Anatomically, the mushroom bodies are composed of three lobes (Vowles, 1955). On each side of the brain there is a pair of cup-shaped structures (the lateral and the medial calyces), which are connected by a short peduncle to the alpha lobe and the beta lobe. The cell bodies lie within the calyx and overflow it, and each cell has a branch to all three lobes. The results of stimulation experiments, which are fully reported elsewhere (Vowles, 1964), may be summarized as follows.

1. In regions distant from the mushroom bodies, stimulation results in simple phasic movements such as twitching of the antennae.

2. Stimulation in or close to the mushroom bodies produces complex coordinated patterns of movement such as locomotion, often with particular orientation, or activities like cleaning, feeding, regurgitation, aggression, comb building, and so on (see Fig. 1).

3. The activities are elicited with a long latency (from 2–3 up to 30–90 sec) and a prolonged after effect (many seconds long).

4. Locomotor activity can be produced from a large number of loci scattered throughout and around the mushroom bodies. Sometimes the activity appears in a pure form, while in others it may be associated with activities such as feeding behavior or aggression.

5 Complex activities other than locomotion are elicited by stimulation within or near the alpha and beta lobes (see Fig. 1), but not the calyces.

6. Sometimes stimulation suppresses, rather than elicits, an activity, and very often an activity recurs vigorously after the end of such an inhibitory

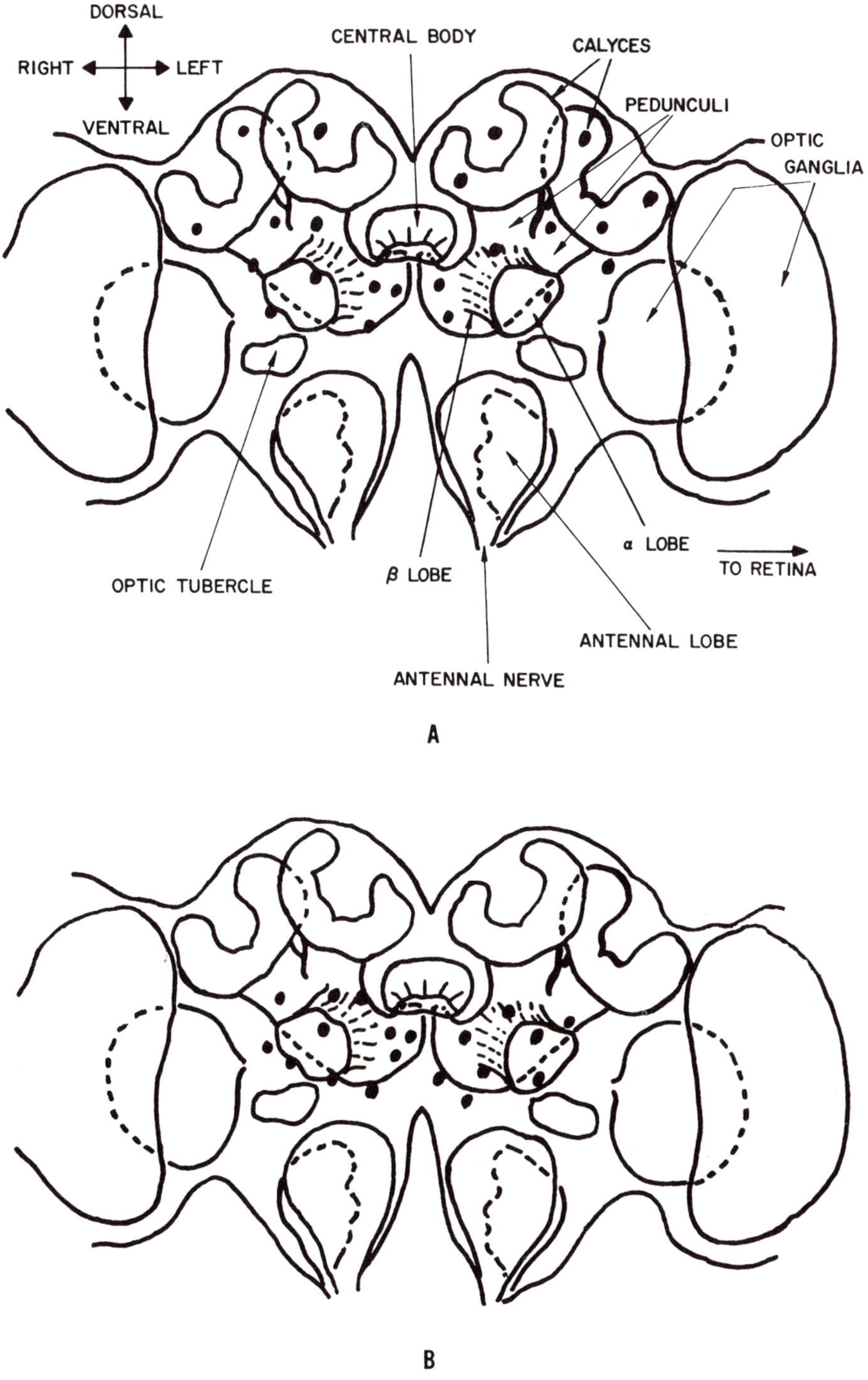

FIG. 1. A frontal view of the brain of the honeybee, showing points (identified by solid circles) from which coordinated walking (A) and cleaning (B) could be elicited by electrical stimulation.

stimulus. If two activities are simultaneously elicited by stimulation, these usually alternate for a short time, after which one becomes dominant.

These results are, of course, very similar to those already obtained by Huber (1960a and b). It is apparent from such studies that the mushroom bodies are involved in the control of orientated movement and instinctive activities, but are not necessary for their coordination. Thus the mushroom bodies may be envisaged as selecting which behavior pattern will occur, controlling its intensity and duration, and suppressing other activities. In attempting to discover how they may do this, one may look at their electrical activity under conditions of both natural and electrical stimulation and see whether these can be correlated with the bodies' anatomical structure and with the bee's behavior.

Electrophysiological Recording. (See Fig. 2.) The study of the connections of the mushroom bodies (Vowles, 1955) reveals that the calyces and the alpha lobes are connected to sensory centers, while the beta lobe gives connections to motor centers. Thus, when recording in the calyces and alpha lobe, one might anticipate detecting both nerve impulses associated with sensory stimulation and a degree of convergence of afferent fibers upon mushroom-body fibers; from stimulation experiments one would predict a correlation of discharge with motor activity. Microelectrode recordings were obtained from all parts of the mushroom bodies, in response to various types of sensory stimulation. The general characteristics of these responses are as follows.

1. They have a very long latency. The shortest latent period between sensory stimulation and the arrival of impulses is about 40 msec, but it is commonly 2 or 3 sec or longer. The latency is very variable.

2. The discharges are very variable in both duration and frequency, sometimes consisting of only a few impulses and sometimes involving a prolonged burst of impulses outlasting the stimulus by many seconds. When such prolonged bursts occur, it is often obvious that the bee is performing motor movements such as waving the antennae, struggling with legs, "flying" with its wing stumps, and so on. When this happens, it is difficult to decide whether one is recording impulses from sense organs excited by this activity, or whether the activity is caused by the impulses, or both. If it were possible to sever the motor nerves involved without damaging sensory structures and nerves, this problem could be resolved. Unfortunately the experiment is technically impossible. However, two related experiments can be done.

1. Where bending the antenna produces movements correlated with discharges in the mushroom bodies, this movement can be prevented by cutting the ring of chitin and its attached muscles around the base of the antenna. The antenna then hangs limply. This operation increased the frequency of impulses recorded in the mushroom body.

2. In some preparations, discharges were recorded which were correlated with movements of the abdomen, wings, and legs. Such activity can sometimes be elicited by light flashes. When it is, sectioning the prothoracic connectives, or the ventral nerve cord at the front of the abdomen, abolishes struggling in response to visual stimulation, and the evoked discharge to light is then reduced, though not abolished.

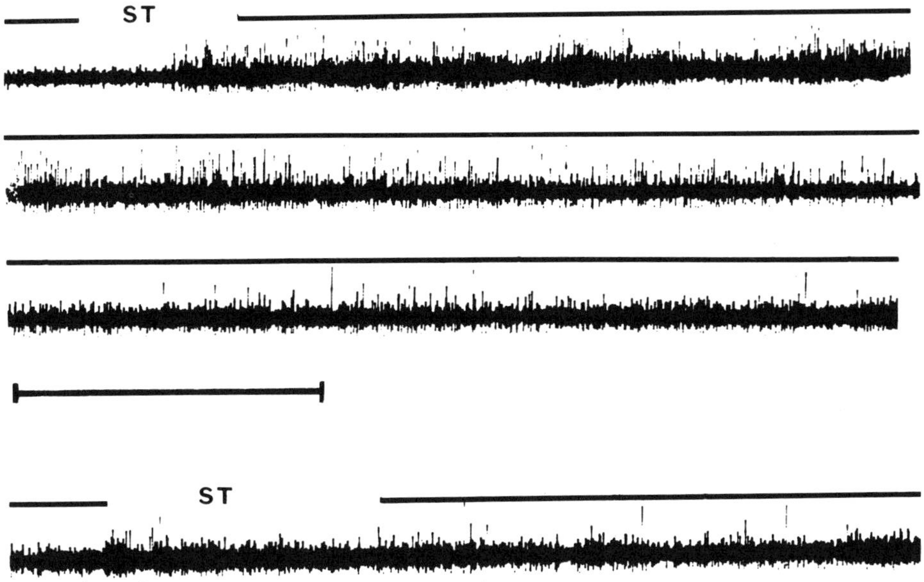

FIG. 2. Recording from the left medial calyx in the honeybee in response to abdominal stretch. Two records are shown from the same preparation in response to similar stimuli. The top three lines are continuous. The stimulus is indicated by the interruption, labeled ST, of the continuous black line. Note the difference in duration and vigor of response under two similar conditions. Time calibration 5 sec.

It is clear, therefore, that the prolonged discharges in the mushroom bodies are partly due to "reafference" (the sensory changes resulting from the movements with which they are correlated) and are partly self-sustaining, possibly by reverberation between mushroom bodies and sensory centers. In some cases the reafference may reduce the activity.

3. Inhibitory relationships can be observed in which the exposure to one stimulus may suppress the response elicited by another.

One problem which arises is to decide the direction in which information passes through the mushroom-body system; on histological grounds alone one cannot decide whether impulses pass into both the calyx and the alpha lobe, or whether they go in at one place and out at the other. There is, however, reasonably strong evidence that the direction of flow of impulses is from the sensory centers to the calyx, from the calyx to the alpha and beta lobes, and from the alpha lobes back to the sensory centers. The various lines of evidence which suggest this are as follows.

1. As Maynard (1956) showed (with cockroaches), the application of an electric shock to the antennal nerve produces large electrical waves in the mushroom bodies, presumably resulting from the synchronous discharge of a number of cells. The latency of these waves in the calyx and the alpha and beta lobes suggests that they originate in the calyx and then spread to the other lobes.

2. It is possible to interrupt the tracks between the antennal lobes and the

alpha lobes before damaging the deeper tracks to the calyces. Stimuli caused by touching and bending the ipsilateral antenna can still evoke impulses in both the alpha lobe and the calyx following superficial sectioning, but recordings from both regions are abolished following deep sectioning.

3. By stimulating in the calyx one can record easily, from both the alpha lobe and the beta lobe, bursts of impulses which have a short latency (about 2 msec); however, when stimulating in the alpha and beta lobes or just outside them, it is very difficult, although not impossible, to record from the calyx any impulses of short latency, although bursts of impulses may arise some time later. This is rather surprising, since one would expect considerable antidromic stimulation to occur. This is perhaps due to the arrangement and configuration of the fibers in the alpha lobe.

4. Impulses can be recorded from the optic lobes after stimulating in the calyx or the alpha lobe. Superficial section of tracks between the alpha lobes and the optic ganglia abolishes this discharge.

5. A similar experiment can be done with the antennal lobes.

6. Impulses can be picked up in the optic ganglion upon stimulation of other sense organs. These are abolished following superficial section as in 4.

It will be seen, therefore, that there is good evidence in favor of a feedback loop from the mushroom bodies to sensory centers.

Histology. If one looks at the microscopic structure of the mushroom bodies, one finds an apparent difference in the dendritic arrangements in the calyces and the alpha lobes and beta lobes (Kenyon, 1896; Sanchez y Sanchez, 1941). Although the cells of the mushroom bodies have diffuse dendritic fields in the calyx, their axons, after passing down the peduncle, divide into two

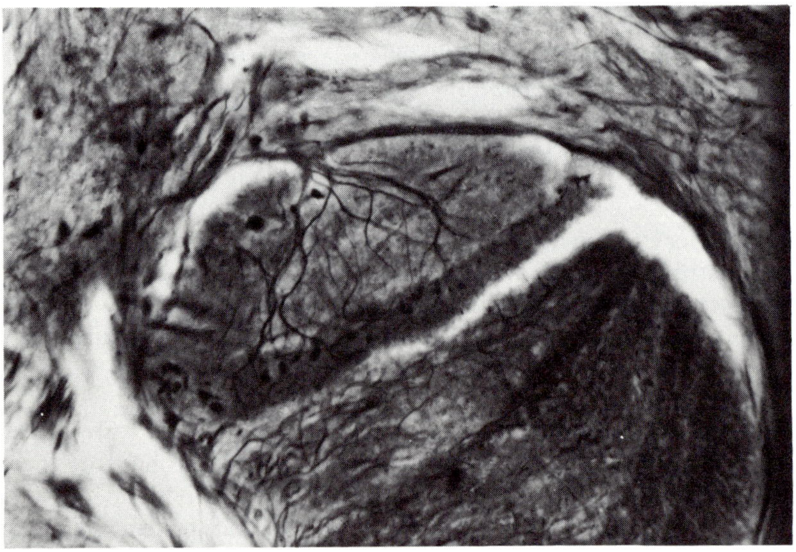

FIG. 3. A transverse section of the alpha lobe of the honeybee, showing transverse striations. The section is 15 microns thick and stained with silver pyridine.

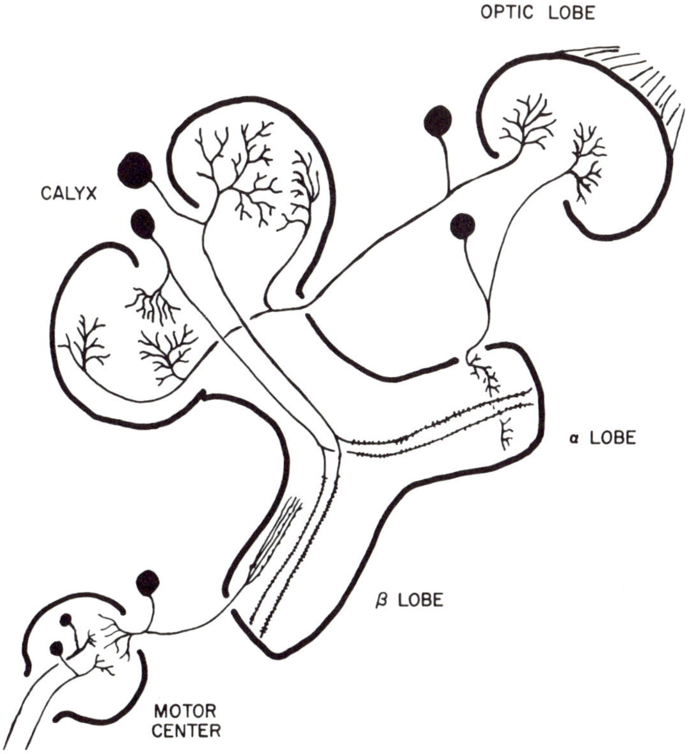

FIG. 4. Diagram to illustrate some suggested connections in the mushroom bodies.

equal branches, each branch passing throughout the alpha and beta lobes. In Golgi preparations these two branches do not seem to form spreading dendritic arrangements but have very fine tooth-like projections on each side. The dendritic arrangements of neurons connecting the mushroom bodies with other regions of the nervous system also differ in the different lobes. In the calyces, the afferent fibers branch profusely, and a single axon may pass around the calyx, giving off a large number of dendritic branches. There seems to be ample anatomical basis here for convergence and sensory interaction in the calyces.

In the alpha and beta lobes, however, the synaptic arrangements are not so clear. It has repeatedly been observed, both by classical authors and in my own stains, that curious dark striations appear within the alpha lobes and the beta lobes. In the alpha lobes these run across from side to side, while in the beta lobes they run the length of this lobe, like the letters in a stick of rock (see Fig. 3). It has been suggested that these striated regions represent the areas of synaptic contact between internuncial neurons and the mushroom-body neurons. This can partly be seen in the illustrations. If this is so, then one may regard the internuncials in the alpha lobe as sampling the activity of neurons across the lobe, while in the beta lobe they are sampling the activity in small, closely related bundles of fibers. This is summarized in Fig. 4.

Hypothesis of Function of Mushroom Bodies. If this interpretation is correct (and it is undoubtedly an over-simple account of the connections and functions of the mushroom bodies), then one might make some predictions about the electrical responses of the mushroom body fibres.

Firstly, if afferent fibers are coming from many sensory centers to the calyces, where they converge upon mushroom body cells, one would anticipate picking up impulses specific to the modality from these afferent fibers as well as recording from the mushroom-body cells themselves. On the other hand, one would anticipate fewer, if any, discharges in the alpha lobe specific to a given modality and relatively more fibers which can be excited by combinations of stimuli. These predictions seem to be fulfilled. In a series of experiments, recordings were made from the calyces and alpha lobes in response to a battery of different stimuli in different modalities. The only recordings used were those showing a clear response of only one or two units.

The number of units which discharged to a single mode of stimulation and the number which responded to a variety of stimuli are shown in Table 1, together with the loci of the recordings.

It was observed that in the electrophysiological records the characteristic discharges of a specifically sensory nature are more often brief bursts of a regular sort than are the discharges from units responding to mixed stimuli. The sensory specific discharges are also more commonly monophasic or biphasic with a large negative-going spike and a small positive after-potential. If the numbers for the calyces may be taken as an indication of the relative number of afferent neurons to mushroom-body neurons, then it is clear that there is a high degree of convergence here.

Another suggestion which might be made on the basis of histological studies is that the bundles of fibers giving rise to striations in the beta lobe are correlated with specific motor activities. When mushroom-body fibers discharge in response to different stimuli, one might then anticipate that the stimuli to which they discharge are functionally correlated in some way with some normal pattern of behavior; e.g., if one detects the antenna moving in response to a moving visual stimulus, and one finds correlated discharges in the mushroom bodies, one might anticipate that the units firing will do so in response both to visual stimuli and to mechanical deflection of the antennae. Similarly, one might anticipate units concerned with flying which will discharge both to antenna deflection and to wing displacement. We tested this prediction, using the battery of stimuli, and there seems to be a large number of possible combinations of stimuli to which mushroom-body units respond. The most

TABLE 1

	Calyces	Alpha Lobe	Beta Lobe
Sensory specific responses	25	5	3
Responses to a variety of stimuli	12	17	8
Totals	37	22	11

common are the following. Out of a total of 36 units, 7 responded to stimulation of the mouth parts by touch and sugar; this might be correlated with feeding behavior in the bee. In bees which moved their antennae in response to visual stimulation that consisted of moving a small object near the eye, five units discharged both to the visual stimulus and to deflection of the antennae. Five other units responded to bending the wings and legs and stretching the abdomen; this could be correlated with flying.

On the basis of the above results one may propose a tentative suggestion about the way in which the mushroom bodies work. It seems possible that the inputs from various sensory centers to the calyces interact in these lobes, exciting and inhibiting mushroom-body fibers. As a result of this interaction, impulses are initiated which pass away down the peduncle and thence to the alpha and beta lobes. The information passing along the alpha lobe is a copy of that passing to the beta lobe. The activity of specific fiber bundles (or groups of fibers) in the beta lobe is supposed to determine which motor activities shall occur, either by facilitation or inhibition of lower centers. Which fibers are excited in the calyx, and hence which are excited in the beta lobe, will obviously depend upon the pattern of excitation coming into the calyx. If the output from the beta lobe is not the correct one for making the right response to the sensory stimuli impinging on the insect, then information about this will be fed back from the alpha lobe to the sensory centers. If the connections within this feedback loop between mushroom body and the sensory centers are appropriately arranged, corrective signals could then arise in the sensory centers and pass back to the calyx, which would remedy the error in output. A similar correction factor could also be fed directly from sensory centers to motor centers.

In addition to the closed loops existing between the mushroom bodies and the sensory centers, there is also a loop which involves the external environment. Thus when movements are produced under the influence of the mushroom bodies, this produces a change in the sensory environment (called reafference by von Holst and Mittelstaedt, 1950). Information about these self-induced sensory changes is also fed back to the mushroom bodies and thence to other sensory centers. Thus a correction factor could again be applied to the mushroom bodies, which can continue to exert their control in the face of sensory changes induced both by the bee's own responses and by external independently produced changes in the environment.

One may also ask why there should be many points from which stimulation elicits the same behavior; this result seems a commonplace of brain-stimulation experiments. There might be three possible and compatible explanations.

1. The different groups of fibers eliciting the same behavior might be correlated with different combinations of stimuli which can produce that behavior.

2. The different pathways might be correlated with different orientations of a particular behavior. Thus one point might elicit aggressive movements towards the right and another aggressive movements towards the left. This certainly applies to stimuli applied on different sides of the brain, but whether it is also true of different loci on the same side of the brain is not clear.

3. The number of different groups of fibers might have some influence upon the intensity of the behavior elicited. The greater the number of active groups, the stronger could be the behavior evoked. When a variety of stimuli fall on the animal, competing responses have to be eliminated, and the bee must make the most adaptive response in the face of conflicting choices. When different stimuli compete for the final common motor pathways used in different behavior patterns, the behavior which finally emerges would depend upon the relative strengths of the influences coming from the mushroom bodies to lower centers. These might depend upon the total number of mushroom-body fibers active which elicit one kind of behavior rather than another. The mushroom body could therefore be regarded as a decision-making mechanism which bases its choice upon the strength of various inputs coming to it. It would obviously be necessary in such a system to have different weightings given to different inputs. Thus stimuli which elicit fleeing or attacking might be supposed to override stimuli giving other behavior.

The Analysis of Visual Stimuli

If the above hypothesis is accepted, one implication is that the analysis of the stimuli falling upon the animal and the weighting of the significant characteristics of those stimuli (i.e., the identification of the nature of stimuli and their adaptive significance) should be determined by the sensory centers. Information coming from the sensory centers to the mushroom bodies would then be fairly specifically related to the perceptual activity of the bee rather than to mere mosaics of sensory stimuli. One would expect information arriving at the mushroom bodies to be about highly specified stimulus characteristics.

Some light is thrown upon this by some work Mrs. Crossley and I did on the optic ganglia of the bee (Crossley and Vowles, 1964). We would like to stress, however, that our investigation was of a preliminary nature because, although we spent a full year on the study, the results were so complex that a considerably greater time is needed to extend the initial results. The experiments were started partly in response to a suggestion by Bullock that one might be foolish, when studying electrophysiological responses in higher centers, to use simple stimuli such as short flashes of light. He stressed that, as Maturana *et al.* (1960) had shown, the frog's retina already performs quite a complex analysis of the visual world, and the optic nerve carries specific information about various high-order stimulus characteristics. Similarly, work by Hubel and Wiesel (1959, 1962) on the cat's visual cortex gives some indication of high-level processing of information. We therefore thought it might be interesting to record from the bee's optic ganglion in response to various stimuli which were known to be of importance in the life of the bee. The experiments had three aims:

1. to describe the types of response given by the optic ganglia of the bee;
2. to identify the types of stimuli to which these responses were given;
3. to pinpoint the locations in the optic ganglion of particular types of responses to particular stimuli.

If localization, as in (3), could be made, then one might get some clue to the nature of the nervous mechanisms involved by looking at the structure and connections of the nerve cells in those regions.

Types of Response. The types of response were classified as follows.

Transient bursts lasting less than a second in response to changing stimuli. These are typical on-off responses.

Rhythms. Some units produced, in an apparently spontaneous fashion, fairly regular rhythms throughout the course of recording. There were some fast rhythms of about 70 per second which seemed almost undisturbed by any stimuli and some slow rhythms of about 1–10 per second. The frequency seemed to depend upon conditions of illumination. They discharged more frequently in the dark than in the light.

Bursts. The main feature of many recordings was the occurrence of bursts of impulses. These were surprisingly regular and occurred at a frequency of about 0.3 per second, with a duration of about 1.2 sec. These bursts were either synchronized (i.e., developed from a previously irregular background discharge) or desynchronized (i.e., changed from regular bursts to irregular discharge by stimulation). One interesting result was that stimulation of the mushroom bodies could cause desynchronization of discharges in the optic ganglia.

Irregular Discharges. These were discharges of no apparent regularity and often of varying duration from 2 or 3 sec up to 30 sec or even longer. Often the discharge was maintained for as long as the stimulus was presented.

It is characteristic of transient responses, irregular discharges, and bursts that they showed ready habituation to repeated stimuli. The transient responses and irregular discharges were most notable in this respect. Quite often only two or three repetitions of a stimulus were sufficient to abolish a response, and it was necessary to wait several minutes before recovery occurred. That this was true habituation rather than fatigue was indicated by the fact that a novel stimulus could result in the immediate reappearance of the response.

Types of Stimuli. The method of stimulating the bee was to project either a white or a colored background (Kodachrome floral scene) onto a screen and to place, or move, against either of these backgrounds various types of stimulus objects. Two types of stimulus object were used: large contours which occupied the whole screen, and figures which occupied only a small part of the screen. The figures were those used in earlier workers' training experiments (see von Frisch, 1956), when it had been found that bees could distinguish easily between objects with a large amount of contour and those with relatively small amounts of contour, but not between objects which had different shapes but similar amounts of contour. The shapes used are shown in Fig. 5. The objects were made of black cardboard and subtended an angle of about 7 deg at the bee's eye. A small black object simulating a wasp in outline was also used. This object subtended about 2 deg at the bee's eye and bobbed up and down in a manner typical of a robber wasp, a stimulus to which bees respond by aggressive movements. (Free, 1954). All the stimuli were presented both

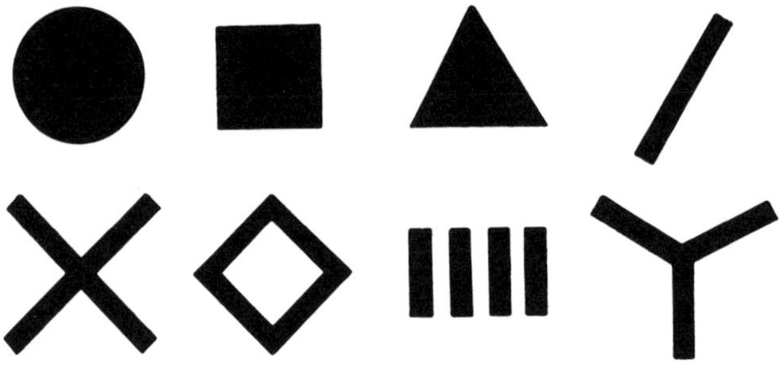

Fig. 5. Shapes used in behavioral and physiological tests. Bees easily distinguish objects in the top row from those in the bottom, but do not distinguish between objects within the rows. (from von Frisch, 1956). The top row of objects are referred to in the text as solid figures, and the bottom row as contoured figures.

stationary and moving against the two types of background, and at different places in the field of view.

When responses were evoked in the bee's optic ganglia, it was noted which stimuli produced them, whether they were specific for stationary or moving stimuli, for the location of the stimulus, and for the direction of movements of the stimulus. When fairly clear unit discharges were obtained (sometimes two units could be simultaneously detected, showing different responses), it was usually found that clear indications either of complete or of totally incomplete specificity were obtained; i.e., there was either very marked discharge or very small response. In cases where preparations gave biased responses toward some stimuli rather than others, these were usually obtained in preparations where clearly separate units were not discernible. In the subsequent analysis, figures are quoted only for cases where clear unit recording was obtained. Some results are shown in Table 2, and it is clear that units can be identified which respond to specific stimulus characteristics.

In our study the precise depth of the electrode tip was not ascertained, but an accurate record of the point on the surface of the ganglia at which the electrode entered was obtained, thus some degree of localization of the electrode tip was possible. Owing to the shape and arrangement of the ganglia within the optic lobes, the precise point at which the electrode lay was not

TABLE 2

Responses of Bee's Optic Ganglia to Test Stimuli

Stimulus Eliciting a Response	Number of Single Units Detected
Wasp only	4
Solid figures only	9
Contoured figures only	16
All figures	58
Large contours only	11
Both figures and large contours	66

identifiable. However, it was clear that the further toward the midline of the brain one moves, the greater the specificity of the stimulus required to excite the particular unit. This is shown in Fig. 6.

The horizontal axis represents the axis of the optic ganglia passing from the outer ganglion (retina intermediaria) on the right to the inner ganglion on the left of the diagram. The histogram was constructed as follows. In each region of the ganglia corresponding to a single column on the histogram the number of single units which responded to figures was counted (whether or not they also responded to other stimuli). These numbers are shown under each

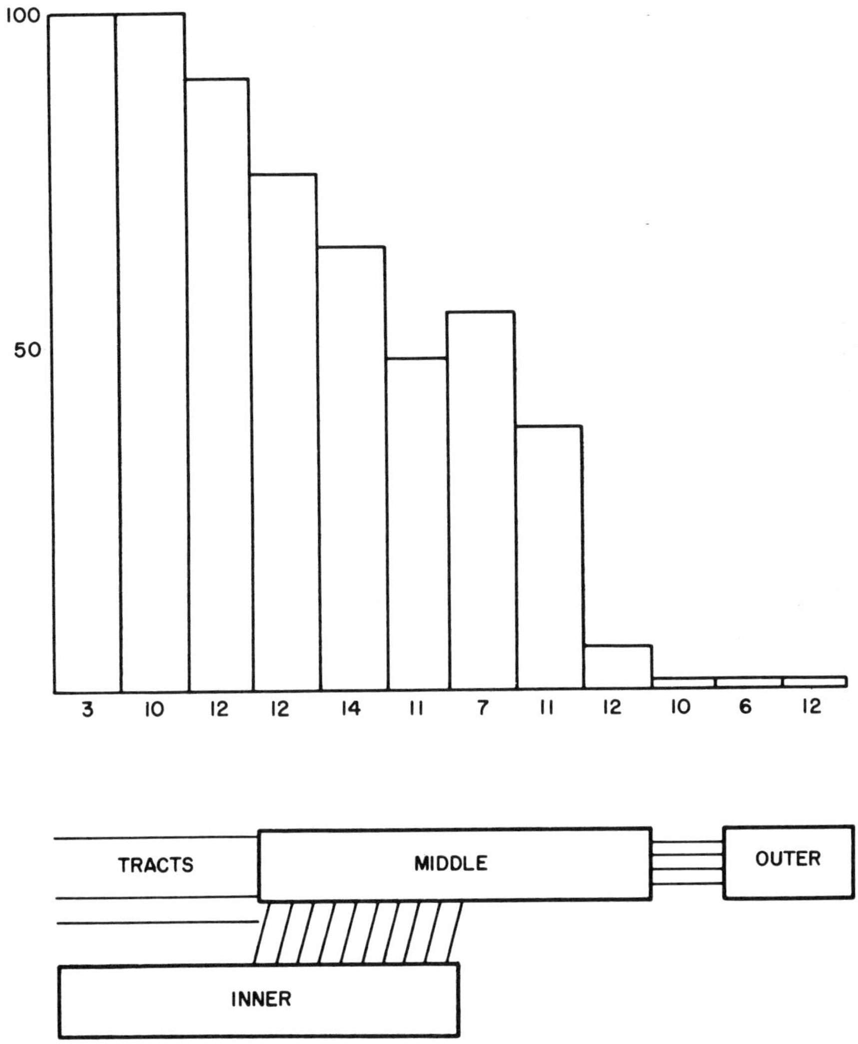

FIG. 6. A histogram to show the specificity of stimulus characteristics required to excite single neurons in different regions of the optic ganglia of the honeybee.

column. For each column the number of units were counted which responded

(a) to figures only;
(b) *either* to solid *or* to contoured figures, but *not* to both;
(c) *either* to stationary *or* to moving figures, but *not* to both (in fact no units responded only to stationary figures);
(d) to figures moving in any preferred direction.

The numbers of units in these four categories were added and then expressed as a percentage of four times the total number of units in each column. Thus the more specific the stimulus needed to excite a unit, the greater the number of categories in which it appears, and the higher its contribution to the percentage scores. Hence the higher the score, the more specific must be the stimuli and the greater the number of units responding only to specific stimuli. For example, a score of 100% indicates that all the units respond *only* to figures, moving in a preferred direction, and these figures must be either solid or contoured, but not both.

General Comments

It will be clear from the above account that all the results obtained were the outcome of fairly classical methods of functional anatomy and physiology. The only novelty was perhaps the size of the animal to which they were being applied. No particular type of model was employed in this research, but only the qualitative ideas which arose from a study of behavior and of the anatomy of the bee's brain. I wish to discuss now the possibility that the methods of servoanalysis might have been more fruitful than the classical approach, and also whether they are likely to prove more fruitful in the future.

Many technical difficulties have been mentioned in the text, but, as stated earlier, a neurophysiologist trying to use a servo-theory type of model postulated to explain behavior is faced with two basic problems. First, he has to localize the black boxes and their connections, and second, he has to translate the input-output relationships of the boxes, and the information flow in their connecting links, into neurophysiological terms. In the light of our present know edge of the insect brain, one may comment upon these possibilities with particular reference to methods and techniques.

The problems involved in locating black boxes are of various sorts. The box has been defined as a functional, rather than as a discrete anatomical entity, and this implies that it is not restricted to a specific position in the brain but may be scattered throughout the brain tissue. In insects it superficially appears that this problem may not be as acute as in other animals, since, in gross terms, functions seem to be localized in different parts of the nervous system. It is necessary, however, when using the concept of "function," to be sure that it has the same terms of reference in a servo model as in an anatomical study. Unless one can define what one means by "localization" and "function," this is not an easy correlation to establish. When one says a function is localized in a part of the nervous system, one usually means merely that interference with that part of the nervous system by, for example, stimulation or ablation, disturbs some aspect of normal behavior, or that recording in that region can be correlated with that behavior or both. This is obviously a different sort of definition from that implied in the servo model, where function refers to the processing the box performs on the information it handles: a specific statement of which is the transfer function.

If the servo model is to be of use to the neurophysiologist, it must make two sorts of fairly detailed prediction about

1. the different types of change in behavior which result from ablation or stimulation of particular parts of the system;
2. the type of changes which occur in the pattern of nervous signals as they pass from one part of the system to another, particularly under different stimulus conditions.

The first sort of prediction will be useful only if the different parts of the system are accessible for direct experimental manipulation. Surgery will obviously be most valuable if used to interrupt tracts, since lesions in integration centers seem to produce either slight or fairly unspecific behavioral changes (see Lashley's Law of Mass Action). The method is based on the assumption that related information is carried in restricted tracts rather than spread over many pathways. To apply it, a refined technique of precision surgery would need to be developed, linked with stereotaxic studies of the insect brain's structure and connections. This is not impossible, and insects are cheap, numerous, and hardy.

Stimulation could be useful when applied both in tracts and in centers. However, it is necessary to know precisely what sort of physiological effect is produced by the stimulus, and this can only be checked by direct examination, such as recording, both in the neighborhood of the stimulating electrode and at a distance.

In this context one might suggest the use of chemical stimuli, applied locally through cannulae, either in fluid or in crystalline form (see, e.g., Grossman, 1960; Epstein, 1960). Multiple stimulating electrodes placed in different parts of the brain might make possible the controlled study of the effect of both temporal and spatial patterns of stimulation. This could be a valuable technique when dealing with a diffuse mechanism of widely separate components. With fine electrodes (electrodes 12 microns in diameter are effective in the bee) quite a number could be implanted, even in an insect.

The classical types of experiment serve to indicate where one might search for mechanisms with specific transfer functions. That search, however, raises the second problem of the neurophysiologist: how to translate the input-output relationships of the model into neurophysiological terms. As Bullock (1958) has pointed out, we have too many rather than too few possibilities even in a single neuron. One must distinguish somatic from axonal changes and changes in responsiveness from changes in sensitivity, and identify the effects in space and time of one event upon another. All these may involve potentials of short or long duration, or of negative or positive sign, and may be either local or propagated. Hence an experimenter should try to record as many types of changes as possible. This may require the use of different types of electrodes and amplifiers for different signals, since these produce different degrees of distortion of the signal. Ideally, one should record different types of potential at the same time, and if no suitable single electrode and placement can be found, multiple electrodes will be required.

A further complication arises if the significant information depends upon spatial as well as temporal patterns of activity. In terms of patterns in space and time, one would wish to use simultaneously a number of electrodes placed in different parts of the system, and to correlate their activity. The experiment's limits should be imposed by the preparation and the experimenter rather than by electronic equipment. If it proves possible, within these limits, to record simultaneously from a large number of multiple electrodes, a further problem must be faced. This problem is the limitation of the human observer who must examine the resulting data and attempt to discern complex patterns arising in a multiple display, both in space and time. One solution is, of course, to make a permanent record of the whole experiment, e.g., on a pen oscillograph or by using an electronic memory from which experimental data can be played back for display on an oscilloscope. However, with multiple channels even a permanent display may prove too complex for a human observer. As any psychologist will tell you, a human is only good at identifying unfamiliar patterns when they are simple, and only too good at seeing order in random noise. The obvious remedy would be to use a computer to process the data coming in over different channels and to display the result of the processing in a way which a human observer can easily assimilate. If this can be done rapidly in real time, the experimenter could be guided by the result throughout the course of a single experiment. The crux of this technique would lie in programming the computer, and in this one could use the original model as a guide. However, it might also be possible, by using a variety of programs, to extend the basic descriptions of neural transformations almost indefinitely. Whether such a project is feasible cannot be answered by a neurophysiologist.

The above comments have been concerned with general methods and techniques which could, in theory, be used to study the insect brain. In practice many difficulties arise, and the next several paragraphs are concerned with some technical problems discovered empirically from our work on the bee. These problems, their consequences, and their possible solutions are discussed below.

1. It was mentioned earlier that it is more difficult to obtain clear single-unit recordings in insect brains than in vertebrates. Even if one does so, one may be sampling an atypical population of larger cells. Some of the smaller potentials appearing in oscilloscope recordings possibly come from smaller cells. Since these cannot be isolated, however fine the electrode, it might be fruitful to expand the use of pulse-height analysis as a means of separation between units. Perhaps we have gone too far in our attempts to isolate single units, when electronic techniques could be as effective as making electrodes smaller and smaller, besides being more economical.

2. It is clear from the work on the bee's mushroom bodies that the mechanisms underlying different behavior patterns are far from independent; they interact in an inhibitory manner in most cases, and may possibly be facilitatory in others. This has clear repercussions when we are using restrained preparations. A bee pinned down with its brain exposed is obviously trying to escape or attack, and any experimental results are obtained against that background. Two methods have been used in vertebrates to overcome these difficulties. The first involves the use of anesthetics, and is based on the assumption that the system one is studying is not seriously affected by the treatment: this assumption seems reasonably safe when we are dealing with the primary afferent systems, but in higher centers (even primary projection areas) responses are considerably modified by anesthesia. Hence, even if it proved practicable to use this technique in insects, it would be valuable only for a limited range of phenomena, such as the responses of peripheral systems.

The second method is to use chronically implanted electrodes in free-moving animals. Although Rowell (1963) has achieved this in locusts, the flexibility of the head chitin, and the fact that the brain moves in the head when, for example, the jaw muscles contract, make the further refinement of this technique unlikely.

Surgery and dissection have commonly been used to separate components of the nervous system for isolated study. The technical difficulties of sufficiently precise surgery in the insect brain have already been mentioned. Even when successful, the results would require very cautious interpretation, since it is not known how the responses of isolated parts of the neurons system are affected by isolation. Clearly, interrupting feedback loops will distort normal functioning, and although the nature of the distortion may itself give important information, it is still necessary to study the intact system. In any case, surgical methods of isolation are more applicable to isolating parts of one system, rather than to separating different systems.

One possible way of reducing the apparent interference between systems is to apply statistical processes, such as averaging or correlating, to the behavioral or physiological data. These techniques are commonly used to improve signal-to-noise ratios. As McCann (*this volume*) has shown, mere averaging can clearly reveal significant turning tendencies in a fly which is also showing other movements. Such techniques obviously need repeated experiments and will be most useful if the processing can be done in real time and the results displayed to the experimenter immediately in some easily assimilable form.

3. The experiments on the optic ganglia of the bee showed that specific neurons may react to very specific stimulus characteristics. Thus, for optimum responses, particular stimulus features are required; sub-optimum stimuli may produce less reaction. The optimum stimulus can be determined only by empirical tests, possibly tests based on knowledge of those stimulus characteristics which are important in the animal's normal behavior. The use of stimuli whose features fall into several different categories, dealt with by separate neural mechanisms, might lead to confused results which are difficult to

interpret. A further difficulty arises when the animal shows similar behavior in response to different stimuli; e.g., a bee responds by turning movements to several different stimuli, such as a light beam, the moving background, a flower, or a robber wasp. The results suggest that the only common factor involved is turning, whereas identifying the stimulus requires that the data be processed in different ways to categorize the stimuli. Thus, when stimulating for one experimental purpose (e.g., to study turning), one may record responses from the wrong mechanism, and the result could then be confusing. It seems that the only way of overcoming these difficulties is to make a detailed behavioral study to identify the stimulus categories which the insect uses, followed by a detailed descriptive neurophysiological study to detect whether these or other characteristics have optimal effects in the nervous system.

Summary

It has often been suggested that the methods and concepts of servo theory and information theory should be applied to the nervous system. It seems likely that, if the suggestion is to be other than a woolly, analogical pipe-dream, it must be done by the use of quantitative procedures. But there has been no coherent statement of the methodological difficulties involved or the data one should study. In this paper an attempt has been made to indicate some of the difficulties which might arise, and how they might be solved. It is clear that a great deal of simple formal description of behavioral acts and unitary neurophysiological responses is a necessary preliminary. After this, many problems can be realistically tackled only by developing refined surgical, stimulating, and recording techniques, probably using multiple electrodes and sophisticated computer programs for the analysis and display of data. In 40 years neurophysiology has come a long way from Adrian working in a back room with a capillary electrometer. But it is not surprising that, if the study of a simple entity like the atom should require so much time, money, and effort, a really complex mechanism like the brain of an ant will need considerably more.

References

Bullock, T. H., 1958. Parameters of integrative action of the nervous system at the neuronal level, *Exptl. Cell. Res.*, Suppl. 5, 323–37.

Crossley, S., and D. M. Vowles, 1964. An electrophysiological analysis of visual responses in the honeybee, in manuscript.

Deutsch, J. A., 1960. *The Structural Basis of Behavior*, Chicago, Ill.: Univ. Chicago Press.

Epstein, A. N., 1960. Reciprocal changes in feeding behavior produced by intrahypothalamic chemical injections, *Am. J. Psychol.*, **199**, 969–71.

Free, J., 1954. The behavior of robber honey bees. *Behaviour*, **7** (2–3), 233–40.

Frisch, K. von, 1956. *Bees, Their Vision, Chemical Senses, and Language*, Ithaca, N.Y.: Cornell Univ. Press.

Grossman, S. P., 1960. Eating and drinking elicited by direct chemical stimulation of the hypothalamus, *Science*, **132**, 301–2.

Hassenstein, B., 1960. Die bisherige Rolle der Kybernetik in der biologischen Forschung, *Naturwiss. Rundschau* **13**(9–11) 3–24.

Hinde, R. A., 1959. Unitary drives. *Animal Behaviour*, **7**, 130–40.

Holst, E. von, and H. Mittelstaedt, 1950. Das Reafferenzprinzip (Wechselwirkungen zwischen Zentralnervensystem und Peripherie), *Naturwiss.*, **37**, 464–76.

Hubel, D. H., and T. N. Wiesel, 1959. Receptive fields of single neurons in the cat's striate cortex, *J. Physiol.*, **148**, 574–91.

Hubel, D. H., and T. N. Wiesel, 1962. Receptive fields, binocular interaction and functional architecture in the cat's visual cortex, *J. Physiol.*, **160**, 106–54.

Huber, F. 1960a, Experimentelle Untersuchungen zur Nervösen Atmungsregulation der Orthopteren, *Z. vergl. Physiol.*, **43**, 359–91.

Huber, F. 1960b, Untersuchungen über die Funktion des Zentralnervensystems und Insbesondere des Gehirnes bei der Fortbewegung und der Lauterzeugen der Grillen, *Z. vergl. Physiol.*, **44**, 60–132.

Kenyon, C. F., 1896. The anatomy of the corpora pedunculata in the honey bee. *J. Comp. Neurol.*, **6**, 133.

Lorenz, K., 1950. The comparative method in studying innate behavior patterns, in *Physiological Mechanisms in Animal Behavior*, (Symposia of the Society for Experimental Biology, No. 4), New York: Academic, pp. 220–69.

Maturana, R., J. Lettvin, W. Pitts, and W. McCulloch, 1960. Anatomy and physiology of vision in the frog (*Rana pipiens*), *J. Gen. Physiol.*, **43** (6, part 2), 129–75.

Maynard, D. M., 1956. Electrical activity in the cockroach cerebrum, *Nature*, **177**, 529–30.

Reichardt, W., 1961. Autocorrelation, a principle for the evaluation of sensory information by the central nervous system, in *Sensory Communication*, W. A. Rosenblith, ed., Cambridge, Mass: M.I.T. Press, pp. 303–317.

Rowell, C. H. F., 1963. A technique for chronically implanting electrodes in the brains of locusts, *J. Exptl. Biol.*, **40**, 271–84.

Sanchez y Sanchez, D., 1941. Contribution à la connaissance des centres nerveux des insectes, *Trab. Inst. Cajal. Invest. Biol.*, **33**, 165–213.

Skinner, B. F., 1938. *The Behavior of Organisms: An Experimental Analysis.* New York: Appleton-Century-Crofts.

Vowles, D. M., 1955. The structure and connections of the corpora pedunculata in bees and ants, *Quart. J. Microscop. Sci.*, **96**(2), 239–55.

Vowles, D. M., 1964, (in press).

20. Receptive-Field Organization and Response Patterns in Neurons with Spatially Distributed Input

DONALD KENNEDY and DEFOREST MELLON, JR.,
Stanford University

It has long been known that the property of impulse propagation is distributed along the length of the nerve fiber; indeed, the major part of peripheral-nerve electrophysiology of the twentieth century has been devoted to the remarkable property of all-or-none conduction at finite velocity and to the interpretation of its ionic basis. More recently, attention has shifted from the axon—the all-or-none region of the neuron—to other regions in which graded electrical phenomena take place. In addition to the biophysical questions raised because of their electrical inexcitability (Grundfest, 1957), these graded-response regions interest neurophysiologists whose primary concern is with the flow of information through systems of neurons, for these are the special regions where all the variables of synaptic transfer (graded excitatory and inhibitory synaptic potentials, temporal effects such as facilitation and anti-facilitation, after-potentials, and so forth) exert their integrative actions upon incoming sets of nerve impulses. Bullock (1959) has recently surveyed this "quiet revolution" in neurophysiology, and there is no need to do so here.

In differentiating between graded-response and all-or-none membrane, neurophysiologists have widely assumed that the former type occurs in single specific areas of neuron topography, in contrast to the *distributed* nature of the spike-conduction property. This viewpoint at least partially stems from the impressive amount of information we have about certain kinds of central vertebrate neurons—especially spinal motoneurons (Eccles, 1957), but also others (see, e.g., Spencer and Kandel, 1961). Models provided by single receptor cells have also contributed their share to a picture of the neuron as comprising a graded, input-transducing region, a specific site of impulse initiation, and a long spike-propagating process. Dendritic processes on the proximal side receive synaptic input, as does the soma itself in many cases; depolarizing current from activity at these sites then initiates impulses where it first encounters low-threshold, electrically excitable membrane, often at the axon hillock region. The spikes so produced propagate distally along the axon and may also move back toward the input region for a distance which depends

Present address of the second author: Department of Biology, University of Virginia, Charlottesville, Va.

upon the local safety factor and upon conditions of membrane excitability. In a way, this new view of impulse initiation, which owes so much to the use of intracellular recording techniques in the past decade, is the ultimate expression of the view of neuron polarity first developed by Cajal on purely anatomical grounds. The spatial restriction of graded activity to input-terminal regions is pleasingly consistent with the long-known fact that for all reflex actions in the most familiar kinds of neurons, all-or-none activity begins at one fixed point and propagates to one or more others.

However, a large number of the real neurons in this world are quite different. In invertebrates—among which the arthropods will here be used as the major example—the majority of central neurons are monopolar ganglion cells whose somata are far removed from the synaptic field. The single process which leaves the soma may branch in a complex fashion within the neuropile of one or more ganglia; these branches are engaged in synaptic connections with other fibers, but in all but a few exceptional cases the somata are entirely innocent of input connections. It is profitless to try to label "dendrites" and "axons" in the branching system of such a cell; all one can do is to specify that some parts of the cell have afferent endings on them and that other parts conduct all-or-none spikes to more distant regions. In some animals, notably certain molluscs, the spikes initiated in such a branching system invade the soma; this has made possible some elegant studies on the topography and mechanism of impulse excitation in such neurons (Tauc, 1962; Tauc and Hughes, 1963). But in crustacean ganglia, for example, some somata are normally not invaded by impulses (Kennedy and Preston, 1960). The entire interplay between presynaptic excitation and postsynaptic response must therefore be studied by blindly penetrating elements within the neuropile. Despite this handicap, it has at least been possible to show that these junctions exhibit the by-now-expected relationships between graded excitatory postsynaptic potentials (epsp's) and spike initiation, and that the branching character of postsynaptic cells allows for separate all-or-none activity in different branches at the same time—sometimes with and sometimes without invasion of the "main" axon passing out of the ganglion (Preston and Kennedy, 1960).

It is apparent that the complex branching character of these invertebrate neurons raises interesting problems. For example, linear arrays of presynaptic endings may require special integrative solutions not demanded by the global morphology of the vertebrate motoneuron. It has already been shown (Preston and Kennedy, 1960) that when interneurons respond to stimulation of sensory fibers entering the two sides of a single ganglion the relationships between epsp's and spikes for the two sides may be markedly different. Thus within a *single* ganglionic neuropile, the location of synapses along a postsynaptic fiber has important consequences for input/output relations. It is only a step from such complexities to the situation which will occupy our attention in the rest of this discussion, namely that of interneurons which respond to stimuli originating in several abdominal segments. It is quite easy to show that in many such cells, afferent fibers synapse with the postsynaptic neuron in the neuropiles of several ganglia: each such "integrative" input region is separated from the neighboring one, in a medium-sized crayfish, by about 5 mm of all-or-none axon. The consequences of such organization—which is ubiquitous in the crayfish central nervous system—are a little disquieting to those used to thinking in terms of vertebrate central neurons. It is meaningless to talk of "antidromic" and "orthodromic" directions of propagation in such a neuron; both are normal. Neither can one conveniently describe neurons like these as belonging to "ascending" or "descending" systems; they are part of both at once and distribute information to a variety of local and distant sites in both directions. Finally, one must for the first time worry seriously about what happens when a synaptic region is invaded by spikes originating far away—a circumstance which neurophysiologists had previously regarded as an artificially produced experimental convenience (antidromic stimulation of motoneurons) or at most as a rather exotic physiological reality (the dorsal-root reflex).

Our present work grew out of an analysis of the synaptic and integrative properties of single interneurons in the abdominal nerve cord of the crayfish. In a beautiful series of investigations, Wiersma and his colleagues (Wiersma, 1958; Hughes and Wiersma, 1960; Wiersma and Hughes, 1961) have clearly demonstrated the feasibility of identifying single functional entities in this nervous system and, in addition, have given us a great deal of information about how particular sensory regions are represented centrally. The major principle may be illustrated by the following example. A single patch of hairs on the dorsum of abdominal segment 4 may activate a dozen or so interneurons; among these one may serve that segment alone, one collects from 4 in combination with 5, another from 4 and 3, another collects from 2, 3 and 4, another from 3, 4, and 5, another from 4, 5, and 6, another from 2, 3, 4, 5, and 6, and finally, another from all tactile hairs on the entire abdomen. The picture is thus one of parallel representation of particular sensory areas in varying spatial combinations with each other. In the example chosen, by far the greater part of the representation (especially of that reaching the brain) is in interneurons that are called *multisegmental* because they collect from a receptive field involving more than one segment.

Two possible basic modes of connection may account for such receptive fields. The first is one in which sensory fibers synapse only with the interneuron in their ganglion of entry, thus producing a ladder-like pattern of input connections at spatially separated locations. The second is one in which sensory fibers from different segments turn and pass up or down the nerve cord, all connecting with the interneuron in one ganglion. Hughes and Wiersma (1960) showed that in some cases, by causing impulse collision along the interneuron, one can demonstrate that the synaptic inputs are in separate ganglia. In at least one case, in which the interneuron collects from all homolateral stretch-receptor fibers, the synapses are almost certainly concentrated in one ganglion by the routing of afferent fibers to a single central location.

One of our interests was to evaluate the relative abundance of these two modes of connection. It was clear at the outset that each had possible adaptive advantages. The first mode, involving synaptic connections in separate ganglia, should make for shorter reflex times, since it eliminates long, central, conduction paths for small-diameter sensory fibers. However, such neurons might be expected to lose information because of impulse collision. The second mode, although it may involve longer reflex times, does allow for subthreshold summation (by addition of epsp's which cannot interact over the long interganglionic distances involved). For interneurons connected in this way, the receptive field can thus show summation over its entire extent. It is interesting that the stretch-receptor interneuron—the only one known to connect exclusively in this way—receives the largest-caliber sensory fibers found in the abdominal region, so that the sacrifice in conduction time is virtually nil.

Methods

The critical technical issue in this work was to establish *where* in a single unit (i.e., in which ganglion) a given impulse had been initiated. In these experiments we recorded from single interneurons that respond to stimulation of dorsal tactile hairs. Single nerve fibers were dissected from a connective (usually between the second and third abdominal ganglia) and the cut, rostral end was raised into an oil layer, located just above the ganglion, for extracellular recording. From well-isolated fibers graded epsp's as well as spikes could be recorded, provided that the locus of initiation was in the nearby ganglion; the presence or absence of epsp's allowed discrimination between the nearby ganglion and any other as a source of impulses. In later experiments it became possible to identify (among a series of three or four ganglia) the specific source by another means: An appropriate fiber was isolated from the anterior end of the abdominal cord, and then bundles were dissected from the homologous region of a distant, posterior connective until the same fiber was isolated in that region. Verification that both electrodes were indeed recording from the same interneuron was obtained by making the anterior recording lead a stimulating electrode, recording posteriorly, and then reversing the procedure. Invariably a single direct impulse resulted which had identical conduction velocities in the two directions. When presynaptic roots were electrically stimulated, spikes were always propagated to *both* recording sites. Dual recording from both ends of the same interneuron allowed the arrival times of all spikes to be compared; this measurement was used directly to identify the locus of impulse initiation in terms of its nearness to either of the two recording sites. This technique proved useful for other kinds of analysis as well.

Results and Discussion

Modes of Afferent Connection. The results of our investigations into modes of connection between primary tactile afferent fibers and multisegmental interneurons may be summarized as follows: In rare cases, an interneuron can be excited in only a single ganglion by sensory fibers which enter the cord in other ganglia, travel through the connectives, and converge upon it. By far the most common arrangement, however, is a complex one in which some sensory fibers innervate a multisegmental interneuron directly in the ganglion

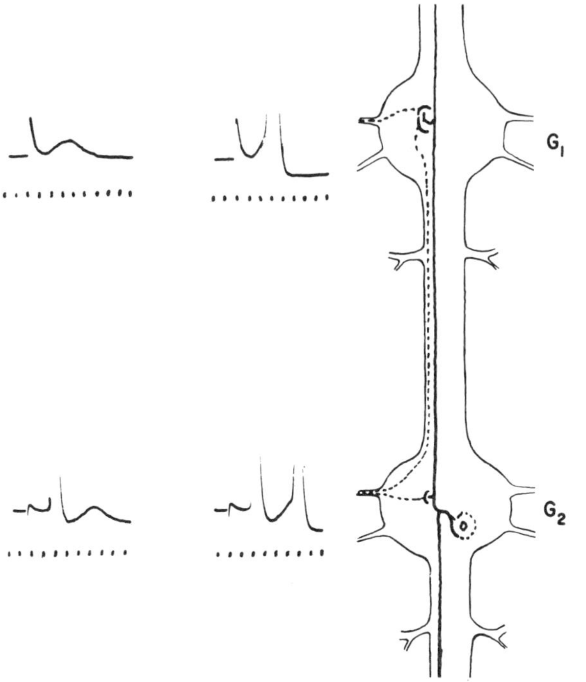

FIG. 1. Stimuli of increasing intensity, left to right, delivered to afferent roots of anterior ganglion (G_1), top row, and posterior ganglion (G_2), bottom row. Response to anterior-root (G_1) stimulation consists, after the stimulus artifact, of a graded epsp (upper left trace) which, if suprathreshold, generates a spike (upper right trace). Threshold stimulation of posterior root (G_2) evokes a spike unaccompanied by an epsp (lower left trace, top of stimulus artifact not visible) followed by a delayed epsp which will produce a spike at higher stimulus intensities (lower right trace). At right is a diagram of the anatomical arrangement. Time calibrations, 2 msec. The rising phases of the spikes are not reproduced in these high-gain recordings.

they enter, while others may turn and synapse with the same interneuron in an adjacent or subadjacent ganglion. The presence of the accessory innervation was first shown by experiments like that illustrated in Fig. 1. Electrical stimulation of the afferent root fibers of the proximate ganglion (G_1) produces an epsp which, if of sufficient amplitude, evokes a spike. Stimulating the roots of the next ganglion below (G_2), as one would expect, yields a spike which is clearly generated in G_2, since it is without an underlying epsp. Following this impulse, however, a subthreshold potential is recorded which looks very much like an epsp; at slightly higher stimulus intensities it evokes a spike. This late epsp also exhibits summation with an epsp generated by stimulating roots to G_1. It is thus clear that there is a presynaptic pathway which leads from roots entering the posterior ganglion to the interneuron in the anterior one. Some evidence exists that this pathway does consist of sensory fibers which turn up the cord and pass directly to the anterior ganglion. First, stimulation of the nerve cord between the two ganglia produces epsp's with approximately the same degree of gradation as in those generated by stimulating the roots of the posterior ganglion. Second, for both modes of stimulation, measurements indicate that the synaptic recovery cycles are about the same. These results make it unlikely that a special relay set of interneurons is transmitting the excitation between ganglia; in fact, conduction time in the connective is long, as one would expect for relatively thin, primary afferent fibers.

It was possible to confirm these results with greater certainty by recording from both ends of the same fiber simultaneously (see "Methods"), so that the ganglion of origin for any spike could be identified immediately. Figure 2 shows the results of one such experiment.

The discharges shown in the figure arise by summation (at appropriate intervals) of volleys generated in two different ganglionic roots. The stimulus to the posterior ganglion precedes in each case; it is set to produce a volley suprathreshold for spike initiation in the ganglion of entry, and is followed by a subthreshold volley to the roots of the adjacent anterior ganglion. In both summation records (S_2–S_1 and S_3–S_2) the second spike arises in the anterior ganglion (S_1 or S_2). Stimulating in the reverse order did not produce summation; therefore a posteriorly directed accessory pathway must not exist.

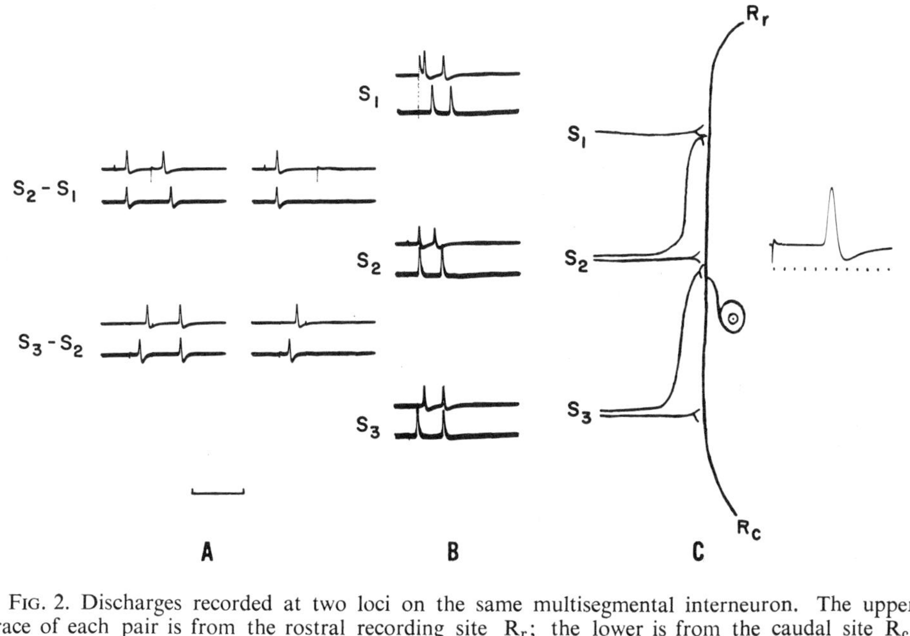

FIG. 2. Discharges recorded at two loci on the same multisegmental interneuron. The upper trace of each pair is from the rostral recording site R_r; the lower is from the caudal site R_c. Stimuli in the summation experiments (A) are in the indicated sequence, in the first column at effective summation intervals and in the second column at longer intervals but with identical stimulus intensities. In each of the summation experiments, the first stimulus is suprathreshold for initiating a spike in the posterior (caudal) ganglion; the second is subthreshold for the anterior (rostral) ganglion. The summed second spikes in each case are shown by the timing relationship to have arisen in the anterior of a pair of ganglia. In B, higher-intensity stimuli are delivered to each root; the second impulses on S_2 and S_3 stimulation arise from the next anterior ganglion. At the right of the diagram, C, a trace of the impulse evoked by stimulating at R_r and recording at R_c confirms that the two recording sites are indeed located on the same neuron. Time calibration, line for A and B, 25 msec; marks for C, 1 msec.

Some of the different arrangements of connection actually found for such interneurons in this study are shown in Fig. 3. It will be noted that the accessory innervation always is directed anteriorly or is bidirectional; it never goes posteriorly alone. Furthermore, when it is bidirectional, the posterior innervation is never so effective as that directed anteriorly; in most of the cases, the pathway going posteriorly is not sufficiently dense to evoke discharge by itself, though it may sum with other inputs to do so. Frequently, however, the anteriorly directed pathway has nearly as low a threshold as that of the ganglion of entry; the single-segment stimuli (S_2 and S_3) in Fig. 2 show that for an input volley restricted to a single segmental root, the locus of impulse initiation can shift.

What general implications does this mode of connection have for receptive-field organization in multisegmental interneurons? First, it means that the receptive field may exhibit summation over much or even all of its extent, even though it is subdivided into spatially discrete input blocks. In this way, reasonably strong stimuli initiated in a particular abdominal segment have the shortest possible route to the fast-conducting interneuronal pathway, but at the same time subthreshold stimuli spaced over two or more segments may sum to produce discharge. The interneuron may be viewed as a linear array of excitable sites. Subthreshold stimulation of the sensory field in any given segment produces over this array a distribution of excitation which is strongest in the ganglion of entry and weakens at successively more distant ganglia; it is, furthermore, generally strongly skewed in the anterior direction. The

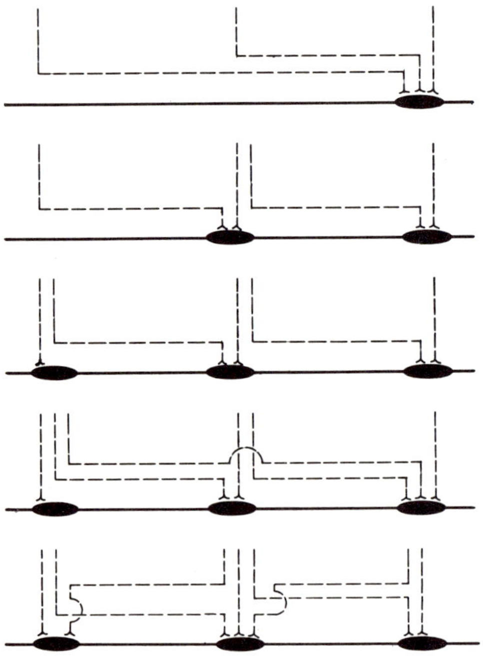

FIG. 3. Configurations of afferent connection found from experiments such as those illustrated in Figs. 1 and 2. The dashed lines indicate afferent fibers; each one actually represents the population of afferents entering in a particular segment.

most strongly stimulated segment will always serve as the focus for excitation, and its ganglion will be facilitated considerably by virtue of being in the "subliminal fringe" for the ganglia of adjacent segments, especially the posterior ones. If all segments are precisely synchronously stimulated at suprathreshold intensity, impulse cancellation will occur along the interneuron between ganglia marking the periphery of the receptive field, and the impulses that escape to other regions of the central nervous system will have their origin at these outpost ganglia. The anterior polarization of the accessory innervation ensures that the most strongly stimulated ganglia from multisegmental stimulation will be at the anterior end of the chain; this suggests the relatively greater importance of anterior versus posterior propagation. Finally, it is clear that only with rather precise synchrony will a large amount of information be lost by collision. The average distance between abdominal ganglia in these crayfish is 5 mm; typical conduction velocities for these multisegmental interneurons range from 3 to 5 m/sec. Thus an impulse is typically "on the line" between any two adjacent ganglia for a period of slightly more than 1 msec. One really should add to this "cancellation time" another 1 to 2 msec for refractoriness at the next set of ganglionic synapses (see below). Even this total figure, however, still gives the impulses from center-field ganglia a substantial chance to escape—especially since there is significant variation in latency and afferent conduction time of the sensory endings and since the natural tactile stimuli are apt to be imprecise in timing. The existence of collision, however, must introduce considerable flattening of the curve relating discharge frequency to stimulus *area*.

Discharge Pattern. In an interneuron the sequence of impulses resulting from synchronous activation of a number of presynaptic fibers is most often a regular one. When it is not, irregularities of patterning may usually be attributed to irregular conformation of the epsp due to asynchrony of arrival of presynaptic impulses. In the present case, another explanation may be invoked: the interposition of different conduction delays due to shifts in the locus of initiation between different ganglia. Figure 4 shows impulse patterns recorded from two ends of the same interneuron in response to high-intensity stimulation of a *single* ganglionic root. Comparison of the traces shows that impulses 1 and 2 are initiated in the ganglion of entry; subsequently, arrival of presynaptic impulses along the slower accessory afferent pathway initiates impulse 3 in the anterior ganglion, after which the locus shifts back to the posterior ganglion. Such patterns are the rule in multisegmental interneurons connected in this

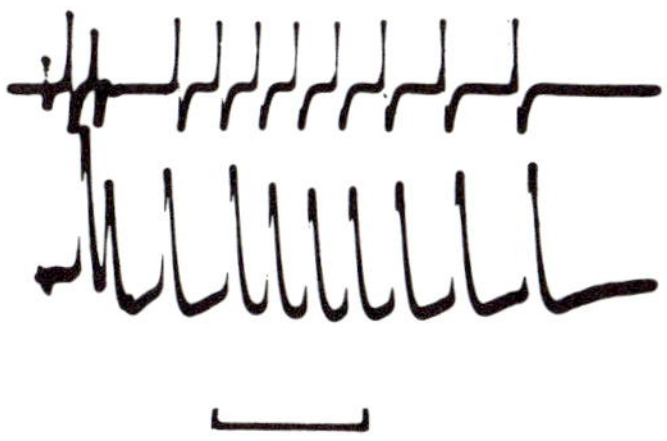

FIG. 4. Simultaneously recorded patterns of impulses from a strongly repeating interneuron produced at sites two segments apart by single, brief-pulse electrical stimulation of roots to the *posterior* ganglion. Upper trace, posterior recording site; lower trace, anterior recording site. Time calibration, 25 msec.

fashion. They could be important in that they form a specific pulse-train microstructure which may be critical for generating activity in postsynaptic units (Segundo *et al.*, 1963; personal communication from Professor Bullock.) It is also possible, though unlikely, that certain patterns of this kind may be specific for certain segments, thus providing a spatial-discrimination capacity. Finally, it is clear from the illustrated record that the patterns formed are different for the two ends of the neuron. Although the output connections of these interneurons are difficult to study, it is known that other interneurons in the crayfish central nervous system (e.g., the giant fibers) make connections with motor neurons in several ganglia, and it seems possible that the multisegmental interneurons do so as well. If this is true, then the same peripheral stimulus is capable of delivering different pulse-coded messages to different places at the same time along one neural channel.

Response to Sequence and Velocity of Moving Stimuli. In a variety of central systems for processing sensory data, higher-order neurons perform complex feats of analyzing the rate and direction of moving stimuli. Outstanding among these, for obvious reasons, are neurons of the visual system, both in vertebrates (Hubel and Wiesel, 1962; Maturana *et al.*, 1960), and arthropods (Wiersma, Waterman, and Bush, 1961). How are such interneurons connected with lower-order elements to allow for such sequence- or rate-sensitive responses? Although such complex responses are not as obviously significant for a segmental "somatosensory" system as they are for a retina, there are good reasons for suspecting that they may be important; at any rate, the

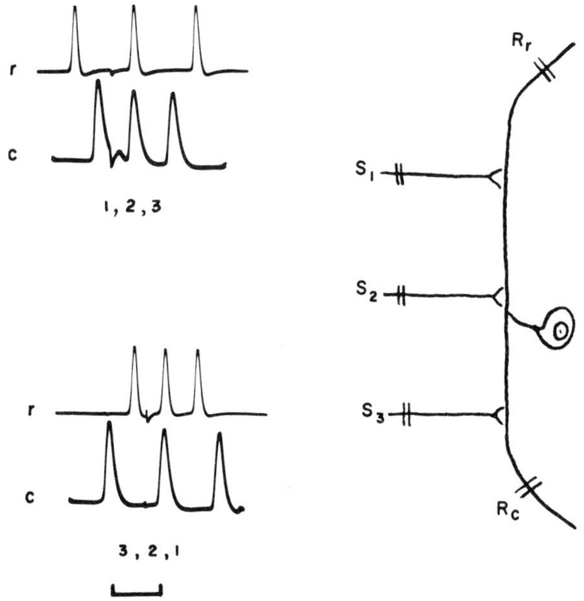

FIG. 5. Sequential stimulation of the afferent roots to three segments. The stimuli are not at a precisely constant interval; instead, the sequences were made *as short as possible* so as to produce the maximum frequency in each case. The intervals are approximately 5 msec (c.f. Fig. 6): the stimulus artifacts are not all visible. Time calibration, 5 msec.

multi-segmental interneurons of crayfish may be used as simplified models to show how such complex sensitivities *may* be achieved elsewhere.

If a multisegmental interneuron is stimulated by electrically evoked volleys delivered through a descending series of segmental roots, each incoming volley evokes a single interneuron impulse at each ganglion (provided the stimulus is not sufficiently strong to evoke second responses mediated by an accessory pathway to some adjacent ganglion). The mode of stimulation mimics that produced by a tactile stimulus of roughly constant strength moving through the receptive field. In all cases tested, each impulse propagates in both directions from the ganglion of initiation and can thus be recorded by electrodes located at either end of the interneuron. Figure 5 illustrates the results of recording from two ends of a multisegmental interneuron with three ganglia separating the recording sites. A sequence of stimuli is delivered, in either the anterior-posterior sequence or the reverse, with an average interval of about 5 msec. The highest-frequency discharge always results when the sequence progresses *toward* a given recording point, obviously because for that sequence the distance from stimulus to recording point gets shorter. The general statement can be made that there is always a direction of movement which has greater stimulating effect, if one considers a particular propagation direction for the interneuron as the relevant one.

If the interval between segmental stimuli is now decreased, an interval value is reached at which the second impulse fails. Figure 6 illustrates this phenomenon

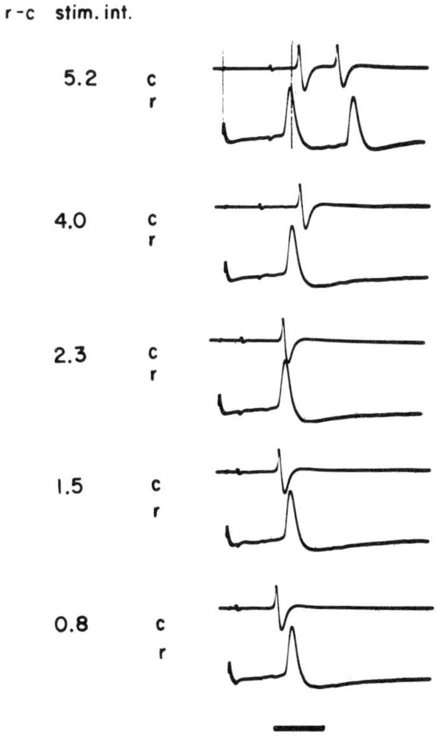

FIG. 6. Effect of decreasing the interval between two stimuli delivered to the afferent roots of adjacent segments in a rostral–caudal sequence. The upper trace of each pair is from the caudal recording site, the lower from the rostral site; there are two ganglia between these recording stations. Time calibration, 5 msec.

for a case in which the two recording sites are two segments apart. Failure of the second impulse is initially due to refractoriness at the synaptic area in the second ganglion stimulated; at this point the temporal relationship of the first spike at the two recording sites is preserved, indicating that its source is the anterior ganglion. At shorter intervals, the failure has a different cause, namely, collision of spikes generated at the two ganglionic sites. Within this interval range, a decrease in interval is accompanied by a shift in the arrival time of spikes at the two recording sites consistent with their separate origin at anterior and posterior ganglia. Such neurons, then, are velocity-sensitive: For increases in rate of movement up to a point, they respond with increased output frequency, but at a particular velocity value there is an abrupt cutoff, the output dropping from n spikes (where n is the number of stimuli and segments in the sequence) to one spike.

This phenomenon also makes possible *absolute* coding of direction within certain frequency values. It may be seen that at a particular point the maximum frequency (F_{max}) due to approaching sequential stimulation is given by

$$F_{max} = \frac{1}{d/v + r}, \tag{1}$$

where d is the distance between ganglia, v is the conduction velocity, and r is the duration of refractoriness.

For the reverse sequence (*away* from the point at which frequency is measured), the frequency is given by

$$F_{max} = \frac{1}{2d/v + r}. \tag{2}$$

The average conduction velocity of 37 multisegmental interneurons sampled in a previous investigation was 3.8 m/sec. Average refractory periods for synaptic activation following direct impulses are around 2 msec, somewhat longer than for paired stimuli delivered to the same single axons. The interganglionic distances in the abdominal cord in crayfish of 7 cm. body length are about 5.0 mm. By using these values, the maximum frequency for impulses generated by approaching sequences is calculated to be 301 per second, and that for receding sequences to be 219 per second. Thus all frequencies between these values reaching the brain, for example, would unambiguously indicate a posterior-to-anterior direction of movement.

One assumption underlying these statements requires comment: that stimuli restricted to a single segment would be unlikely to produce frequencies which could be confused with ones from an approaching sequence. This assumption is based upon the fact that for most (but not all) of the multisegmental interneurons we have studied, even massive electrical volleys delivered to afferent roots of a single segment produce only a single postsynaptic spike over most of the intensity range. Natural stimuli, of course, would be expected to yield more iterative and temporally dispersed presynaptic excitation, but even with this type of stimulation one rarely is able to produce high-frequency repetitive responses from a single segment *in this type of neuron.* This failure is probably due to two factors; first, the tendency toward rather long synaptic recovery cycles for specific sites and second, the abolition of maintained epsp's by reinvasion of generated spikes. Although we cannot at this time regard it as proved, it seems quite likely that in such neurons the region of highest possible frequency response may unambiguously identify the direction of moving stimuli. It may be noted here that even if single segments did normally respond with two or three high-frequency spikes to natural stimuli, the over-all frequency and the burst-and-gap pattern would still show direction-dependent differences.

Effect of Impulses on Adjacent Synaptic Regions. As pointed out in the introductory statement, the spatial separation of synaptic areas by regions of all-or-none conduction compels one to be concerned with the effects of through-propagating impulses on synaptic efficacy at adjacent synaptic sites. Depending upon the nature of such effects, they might exert powerful integrative influences on the regional distribution of excitation and hence on the pattern and frequency of response.

One such influence has already been touched on: It is the obvious fact that, until it is over, refractoriness will block initiation of an impulse from a particular site. More interesting because of their longer duration are two other categories of effect, one of them facilitatory and the other inhibitory. Figure 7A shows an intracellularly recorded spike from a multisegmental interneuron; it exhibits a pronounced depolarizing after-potential. This behavior is quite typical of many crayfish interneurons. As long as such after-potentials last at a spike-initiation zone, they may be expected to sum with locally generated epsp's and boost them toward the firing level. Such interactions have also been demonstrated on several occasions with dual extracellular recording (Fig. 7B). In the latter type of experiment, one end of the isolated single fiber is stimulated and the other end is employed as a recording station. Impulses are thus initiated by single-fiber stimulation, so that no accessory pathways in the system influence the results. The ganglion nearest the recording electrode may be stimulated by near-threshold presynaptic volleys at varying intervals after the passage of the direct spike, in order to analyze effects on the synaptic threshold. The results show that after producing a brief refractory wake, direct spikes may lower synaptic thresholds for more than 25 msec (in some cases) after the impulse has passed the site. The reliance of this effect on the presence of depolarizing after-potentials like that shown in Fig. 7A may be demonstrated by the following experiment. If an impulse is evoked elsewhere so that it passes over the direct-stimulating cathode, the electrical threshold is lowered for about as long as the synaptic threshold was lowered.

At some junctions, however, nothing of this sort occurs. Instead, the synaptic threshold is raised for up to 10 msec following the direct impulse. We have no explanation for this effect; it is unlikely to be due to positive after-potentials, since on occasion an interneuron has shown the enhancement described above in one segment but inhibition in another. A good guess is that the synaptic morphology is the critical factor, inhibition possibly resulting whenever the

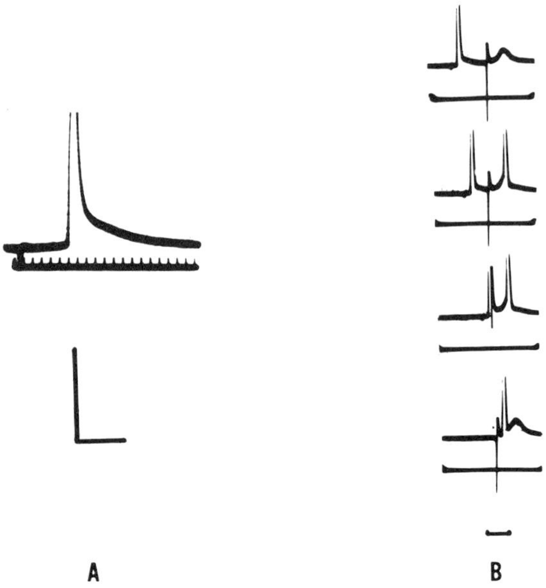

FIG. 7. Effect of impulses on adjacent synaptic regions. A. Intracellular recording from a multisegmental interneuron showing pronounced depolarizing after-potential. Calibration lines, 10 msec, 50 mV; top of spike does not appear in record. B. Lowering of discharge threshold by after-potentials. The neuron was originally isolated at both ends. First response is a direct spike produced by single-fiber stimulation at one end; following it is the epsp produced by subthreshold stimulation of a ganglionic root. Powerful enhancement occurs when the stimulus interval is 10 msec or less; at very short intervals the trailing epsp becomes affected by refractoriness. Time calibration, 10 msec.

entire branching system within the neuropile is invaded by the by-passing spike; this would be expected to reduce the epsp's whenever the synaptic zone is distant from the locus of spike initiation.

Summary

The majority of central interneurons which respond to tactile stimulation of the dorsal abdomen in crayfish serve more than one abdominal segment; a given sensory field is thus represented many times centrally, but in varying combinations with other fields. The usual pattern of peripheral connection with these multisegmental interneurons is one in which incoming afferent fibers in each ganglion synapse in the segment of entry with the interneuron, but also pass anteriorly for one or more segments and make accessory synaptic contacts with the same interneuron in other ganglia. Such accessory connections are also sometimes made in the posterior direction, but they are always weaker. The receptive field of the interneurons is therefore capable of broad summation, but it is at the same time fractionated into segmental subareas which have local entry to the interneuron and hence short reflex delays.

Since the strongest synaptic connections made by afferent fibers are in their own ganglion of entry, the single interneuron is a network of spatially segregated input zones with spike-conducting regions between. Impulses therefore may interact by colliding with one another, or by producing refractoriness or facilitation, or both, by after-potentials in ganglia through which they propagate. It is possible to show that in response to sequential activation (which mimics the effect of a moving natural stimulus) the frequency of discharge may be sensitive to both the *direction* of the sequence and to its *velocity*. Because of the anteriorly directed polarity of the accessory innervation and the effect of depolarizing after-potentials, input into the ganglion of a particular segment strongly facilitates anterior ganglia. Ascending trains of impulses always show higher impulse frequencies for input sequences in the posterior-to-anterior direction, and the number of spikes composing such trains decreases sharply when the sequence interval is reduced to a critical value at which refractoriness and then collision occur. These neurons may thus be used as models for one possible mechanism of achieving "complex" responses to patterns of peripheral stimuli of the kind known to exist in more complex systems. Whether in fact this method is employed in nervous systems other than those of arthropods is, at the moment, a matter for speculation.

References

BULLOCK, T. H., 1959. Neuron doctrine and electrophysiology. *Science*, **129,** 997–1002.

ECCLES, J. C., 1957. *The Physiology of Nerve Cells*, Baltimore: Johns Hopkins Press.

GRUNDFEST, H., 1957. Electrical inexcitability of synapses and some consequences for the central nervous system, *Physiol. Revs.*, **37,** 337–61.

HUBEL, D. H., and T. N. WIESEL, 1962. Receptive fields, binocular interaction and functional architecture in the cat's visual cortex, *J. Physiol.*, **160,** 106–54.

HUGHES, G. M., and C. A. G. WIERSMA, 1960. Neuronal pathways and synaptic connexions in the abdominal cord of the crayfish, *J. Exptl. Biol.*, **37,** 291–307.

KENNEDY, D., and J. B. PRESTON, 1960. Activity patterns of interneurons in the caudal ganglion of the crayfish, *J. Gen. Physiol.* **43,** 655–70.

MATURANA, H. R., J. Y. LETTVIN, W. S. MCCULLOCH, and W. H. PITTS, 1960. Anatomy and physiology of vision in the frog (*Rana pipiens*), *J. Gen. Physiol.*, **43** (6, part 2), 129–75.

PRESTON, J. B., and D. KENNEDY, 1960. Integrative synaptic mechanisms in the caudal ganglion of the crayfish, *J. Gen. Physiol.*, **43,** 671–81.

SEGUNDO, J. P., G. P. MOORE, L. J. STENSAAS, and T. H. BULLOCK, 1963. Sensitivity of neurones in *Aplysia* to temporal pattern of arriving impulses, *J. Exptl. Biol.*, **40,** 643–67.

SPENCER, W. A., and E. R. KANDEL, 1961. Electrophysiology of hippocampal neurons: IV. Fast prepotentials, *J. Neurophysiol.*, **24,** 272–85.

TAUC, L., 1962. Site of origin and propagation of spike in the gaint neuron of *Aplysia*, *J. Gen. Physiol.*, **45,** 1077–98.

TAUC, L., and G. M. HUGHES, 1963. Modes of initiation and propagation of spikes in the branching axons of molluscan central neurons, *J. Gen. Physiol.*, **46,** 533–49.

WIERSMA, C. A. G., 1958. On the functional connections of single units in the central nervous system of the crayfish *Procambarus clarkii* Girard, *J. Comp. Neurol.*, **110,** 421–71.

WIERSMA, C. A. G., and G. M. HUGHES, 1961. On the functional anatomy of neuronal units in the abdominal cord of the crayfish *Procambarus clarkii* Girard, *J. Comp. Neurol.*, **116,** 209–28.

WIERSMA, C. A. G., T. H. WATERMAN, and B. M. H. BUSH, 1961. Impulse traffic in the optic nerve of decapod crustacea, *Science*, **134,** 1435.

21. Coelenterate Conducting Systems

ROBERT K. JOSEPHSON, *University of Minnesota*

I am reasonably confident that a coelenterate—a jellyfish, a sea anemone, or one of their relatives—will be the first animal whose complete behavior will be explicable in terms of known properties and organization of its receptors, coordinating systems, and effectors. I say this despite the fact that coelenterates are not especially easy animals to work with. A typical coelenterate contains a reasonably large number of nerve and muscle cells, which, because of their small size, present serious problems for standard electrophysiological techniques. Coelenterate neurons have been, and continue to be, difficult subjects for the neurohistologist (see, for example, Pantin, 1952; Batham, Pantin, and Robson, 1961). Nor do the neurons of coelenterates operate with fundamentally different and simpler principles than those of higher animals. According to present evidence, the neurons of coelenterates, like those of higher animals, conduct all-or-nothing impulses. They usually interact synaptically, show both excitatory and inhibitory interaction, and are sometimes spontaneously active. The reason for my confidence lies in the relative simplicity of the neural organization and behavior of coelenterates. Coelenterates offer the physiologist the most simply organized nervous systems in the animal kingdom. Although, as more is known about coelenterates, it becomes increasingly difficult to term their behavior simple; their behavior is certainly less complex than that of higher animals.

The bulk of the nervous system of a coelenterate occurs as diffuse, essentially two-dimensional assemblages of neurons termed nerve nets. The neurons of a net are bipolar or multipolar and presumably interact where they cross one another. These intersections are usually synaptic, the neurons being contiguous and not continuous (Pantin, 1952; Horridge and MacKay, 1962), although in at least one net there is evidence for protoplasmic fusion where neurons meet (Mackie, 1960).

Coelenterate nerve nets primarily serve two functions, the conduction of excitation between various regions of the system and the activation of muscles. In some cases (for example, in muscle-free areas of medusae) their function must be solely that of conduction (e.g., Horridge, 1956). There are many cases where coelenterate conducting systems have been identified physiologically but not yet histologically. It seems prudent to refer to these as conducting systems and reserve the term nerve net for those instances in which there is also histological evidence. Although conducting systems are but one of the components involved in coelenterate behavior, they are important components, and an understanding of their functioning is necessary before the behavior of any coelenterate can be completely explained.

Through-Conducting Systems

Certainly the best known conducting systems in coelenterates are the through-conducting systems. These conduct excitation in an all-or-nothing manner; the responses mediated are not graded in intensity with distance from the point of stimulation, and the distance of excitation spread is independent of the total number and frequency of stimuli. Through-conducting systems are known from all three classes of coelenterates, and in some cases there is strong evidence that they are nerve nets. They are involved in such responses as the coordination of the swimming beat of jellyfish and the protective retraction in corals, sea anemones, and hydroids.

A nerve net in which a large proportion of the neurons are linked by junctions which transmit each arriving impulse functions as a through-conducting system. But it is clear from model experiments that through-conduction does not mean ubiquitous conduction; when some of the junctions between conducting elements do not transmit each impulse, excitation can spread to all borders of a net and yet leave many of the conducting elements unfired (Horridge, 1957; Josephson, Reiss, and Worthy, 1961). Passano (1958) has shown that individual neurons in a through-conducting net in jellyfish may be active during only some of the passing excitation waves. This leads to the possibility that a conducting system may be all-or-nothing in its role of spreading excitation to distant parts but not in its role of activating the musculature over which it lies. It is quite common in coelenterates to find increasing increments of muscular tension following each stimulus of a series, a phenomenon attributed to neuromuscular facilitation. But it is possible that a portion of this increase in tension could be due to an increase in the density of activated neurons at any point in a nerve net during repetitive stimulation and a consequent recruitment of the muscle fibers innervated by the newly activated neurons.

Local Conducting Systems

Not all coelenterate conducting systems respond in an all-or-nothing manner. In some, the wave of excitation initiated by a single stimulus stops before it reaches the boundaries of the system. This is evidenced by the further spread obtained if another stimulus follows shortly after the first. Such conducting systems can be termed local systems. The distance to which excitation spreads in a local system is usually a function of the total number and frequency of stimuli. This property is in agreement with the concept of a nerve net in which many of the junctions between neurons do not transmit impulses unless "primed" by previous impulses, a process Pantin (1935) termed "interneural facilitation."

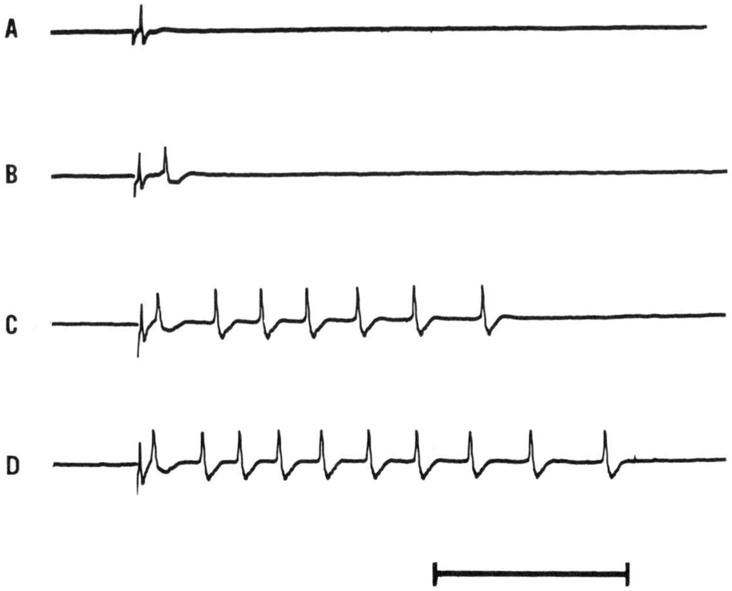

FIG. 1. Potentials recorded with a metal microelectrode from a *Cordylophora* stolon following single 0.5-msec shocks. The initial downward deflection in each case is the shock artifact. Stimulus is, at just subthreshold intensity in A; in B, C, and D it is at 1.14, 1.43, and 2.14 times threshold. Calibration, 2 sec.

Some local systems show repetitive firing like that in Fig. 1 when stimulated mechanically or electrically (Josephson, 1961). A single stimulus applied to these systems may be equivalent to a number of stimuli applied to local systems which do not fire repetitively. Since the number of pulses in a burst following a stimulus increases with increasing stimulus strength, the distance to which excitation spreads is also a function of stimulus strength (because of the dependency of excitation spread on number and frequency of stimuli, as discussed above). And since more waves of excitation pass effectors near the stimulated point than at some distance, local systems which fire repetitively can be expected to show effector responses which are graded with distance from the stimulated point, quite unlike responses mediated by through-conducting systems.

Much of our information on coelenterate conducting systems, especially local systems, has come from colonial animals. Colonial animals are useful for two reasons. First, they offer large expanses of tissue through which conducting systems can be demonstrated and studied; a single coral colony, for example, can be several meters in diameter. Second, the polyps scattered over the surface of a colony can serve as small measuring devices, responding only to activity in limited areas of the conducting system. Merely by watching the polyps of a colony, one can learn much about happenings in the underlying conducting system.

Horridge (1957) studied conduction in a number of colonial corals. In some species all of the polyps are joined by a through-conducting system; in other

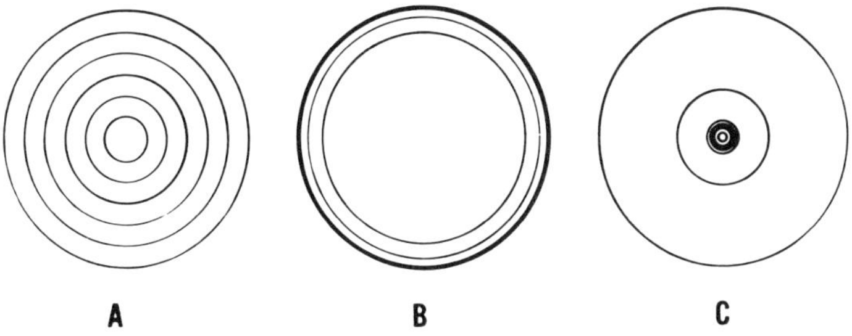

FIG. 2. The three patterns of increasing response with repetitive stimulation found by Horridge in local conducting systems of colonial corals. Scaled so that total spread is the same in each case following six stimuli (after Horridge, 1957). A. Excitation spreads in equal increments. B. Excitation spreads in progressively smaller increments. C. Excitation spreads in equal increments at first, then in progressively larger increments.

species conduction is strictly local, and the response area cannot be appreciably increased by increasing the stimulus strength or by giving repetitive stimuli. In the remaining species, and these are the most interesting for an understanding of the functioning of coelenterate conducting systems, conduction following a single shock is local but the response area can be increased by repetitive stimulation. Horridge distinguished three types of response-area increase during constant-frequency repetitive stimulation (Fig. 2). In some species (Fig. 2A) the excitation spread in concentric circles, all stimuli causing approximately equal increases in the radius of the responding area. In other species (Fig. 2B) the radial increments due to successive shocks were progressively smaller, and still other species (Fig. 2C) gave approximately equal increments of spread at first, followed by progressively increasing increments until the response was bounded only by the size of the colony. Horridge proposed a statistical model which he felt could account for these classes of excitation spread during repetitive stimulation on the basis of a mesh of conducting elements which are joined by a mixed population of transmissive and nontransmissive junctions. Among the features of this model are a postulated progressive increase in the number of elements initially activated by each shock in a series and, in the case of spread in increasing increments, a physiologically undefined mechanism for spatial summation.

A Simulation Model of Nerve Nets

We have used digital-computer simulation to study the behavior of a model net based on the structure and physiological properties known or presumed for coelenterate nets. Our interest in this study centered upon an investigation of patterns of excitation spread elicited by repetitive stimulation. Details on the simulation technique and the results of this study have been presented elsewhere (Josephson and Worthy, 1960; Josephson, Riess, and Worthy, 1961). It must be admitted that there is some circularity in our approach (making a model to see how a number of coelenterate neurons should function

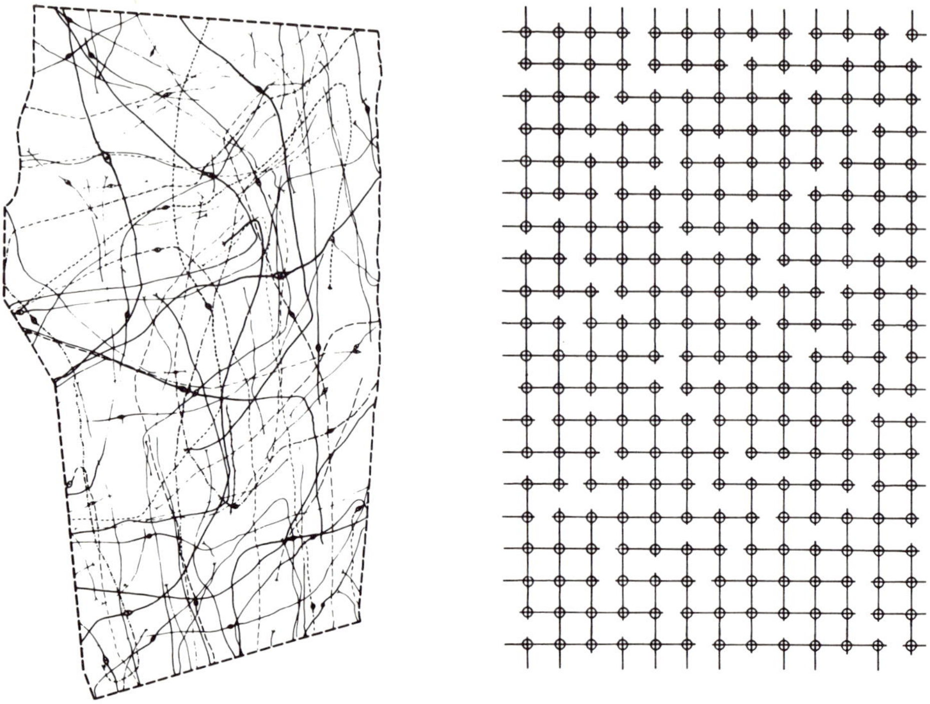

A B

FIG. 3. A. The nerve net in one face of a mesentery from the sea anemone *Metridium senile* (from Batham, Pantin, and Robson, 1960). B. A small portion (less than 3%) of a net used in the simulation studies. This particular net had conducting elements which ranged from 2 to 6 units in length, the average element being 4 units long.

together as a system), since most of the properties of the neurons have in fact been inferred from the system properties of coelenterate nerve nets. For example, that coelenterate neurons conduct all-or-nothing impulses was first inferred from the all-or-nothing nature of conduction in some anemone conducting systems (Pantin, 1935) and only subsequently confirmed by direct recording (Horridge, 1954). Similarly, the presence of interneural facilitation was postulated to account for increasing response area with repetitive stimulation (Pantin, 1935). But the identification of the properties of coelenterate neurons and types of interaction between neurons from specific cases does not indicate the range of behavior which these properties might allow. Our purpose was to see what sorts of behavior could be obtained by quantitative changes in the known parameters and what sorts of behavior could not be obtained. If examples of the latter were found in coelenterates, it would imply that different and unsuspected processes were occurring in the living systems.

The model has linear conducting elements in a two-dimensional array (Fig. 3). These elements conduct all-or-nothing impulses in either direction and interact where they intersect. All elements have the same conduction velocity and have sufficiently long refractory periods to prevent any element from firing more than once in response to a single stimulus, i.e., to preclude the possibility of having reverberating circuits. Unlike coelenterate nets where the neurons meet at a wide range of angles, the conducting elements in the model are in a rectangular array, some vertical, some horizontal, and they always meet at right angles. This approximation greatly simplifies the simulation procedures. The distance between adjacent intersections is constant; it is the unit used in measuring the lengths of the conducting elements. Thus the longer the element, the more intersections it makes with other elements.

A variable proportion of the junctions between conducting elements can transmit each arriving impulse; the others require facilitation by previous impulses before they become transmissive. In the latter junctions, the state of facilitation following an arriving impulse decays with time until they are again not transmissive, the decay rate being different at different junctions. The frequency distribution of conducting-element lengths and of the properties of the junctions between elements is specified by the experimenter, but the length of each conducting element and the properties of its junctions are randomly selected from within the given distributions.

Activity is initiated in the net by starting impulses from a specified group of crossings. The activity is followed until all impulses die out, either by reaching the end of a conducting element or by colliding with an impulse coming in the opposite direction along the same element so that each runs into the other's refractory period. Several types of output are available: a two-dimensional print-out of all crossings reached by impulses, the number of crossings reached by impulses, and the square root of the number of crossings reached. The number of crossings reached gives the area of spread, and the square root of this is an approximate measure of the distance of spread. Since this is in part a stochastic model, a number of trials are needed with each set of parameter values before typical responses for that set can be determined.

With the model it is possible to make nets which are through-conducting and nets in which conduction is local following "stimulation" of a group of crossings. Spread in the local systems after a single stimulus can be quite limited or over considerable distance. Whether a net is through-conducting or local and how far conduction will spread in the local systems depend on the average number of junctions each conducting element makes with other elements and the proportion of junctions in the net which are transmissive without prior facilitation. Increase in either the average number of junctions per element or the proportion of transmissive junctions results in greater spread.

In local systems, during constant-frequency stimulation, excitation spread occurs in essentially equal radial increments or in decreasing increments (Fig. 4). In many cases the excitation spreads in increasing or decreasing increments for the first few stimuli but thereafter proceeds in equal increments. However, spread in increasing increments for more than the first two or three stimuli has not been seen in the variety of nets tested. Whether a particular net responds with equal or decreasing increments depends on the importance of facilitation of normally non-transmissive junctions. Excitation spreads in

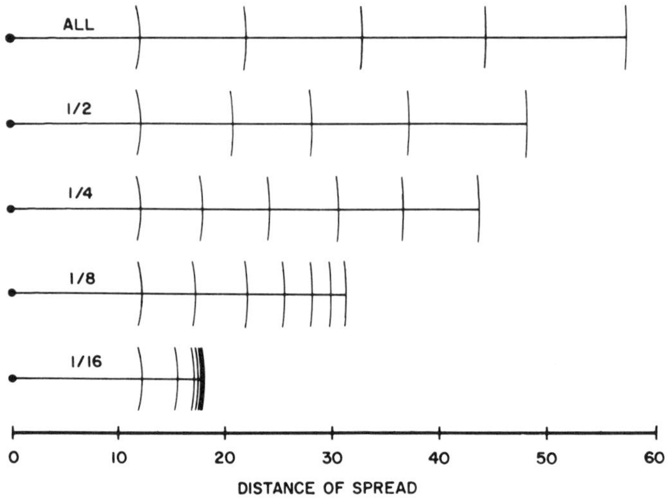

FIG. 4. An example of patterns of spread with repetitive stimulation found in the simulation experiments. Each upright line represents part of the circumference of a circle, the center of which is to the left. Each point is the average of 15 determinations, each on a different net. The average length of the elements in the nets was 4 units, and $\frac{1}{4}$ of the junctions were transmissive without prior facilitation. The numbers above each line indicate the proportion of normally non-transmissive junctions which would remain facilitated and transmissive until the next stimulus if reached by impulses following a preceding stimulus (from Josephson, Reiss, and Worthy, 1961).

decreasing increments when relatively few non-transmissive junctions facilitated by a stimulus remain facilitated until the next stimulus is given. This condition prevails when the intervals between stimuli are long in comparison with the rate of facilitation decay. When the proportion of junctions which remain facilitated is large, spread is in equal increments. Interestingly, the size of these equal increments does not depend solely on the lengths of conducting elements in the net, for it can be varied in the same net by varying the effectiveness of facilitation. This is perhaps unfortunate, for it means that the average lengths of neurons in a coelenterate net cannot be determined on the basis of excitation spread with repetitive stimulation. Information regarding the maximum lengths of neurons could, however, be determined from such physiological tests.

These results on local systems show that it is not necessary to invoke an increase in the number of neurons initially activated by each shock of a series to account for spread in coelenterate nets in either equal or decreasing increments. Such spread is what should be expected following repeated stimulation of the same neurons in a net operating according to the principles now known for coelenterate systems. Spread in increasing increments, however, requires some property or properties not included in the model. This might be spatial summation, as suggested by Horridge, but there are other possibilities (for example, a tendency toward repetitive firing which becomes more pronounced as more stimuli are given). Repetitive firing which is a function of the stimulation history is known for coelenterate conducting systems (Josephson, 1961), and the addition of this feature to the model would greatly increase the types of response which could be obtained. An elucidation of the mechanisms operating in living systems which exhibit increasing increments of spread will have to wait until the species showing this type of behavior are more completely studied.

Conclusions

It is clear that much of the behavior of coelenterate conducting systems can be simulated by a model based on present knowledge of the structure and elemental physiology of coelenterate nerve nets. With appropriate adjustments of the available parameters, it is possible to make a through-conducting system or a local system. In a local system the area of response may increase in equal or decreasing radial increments during a period of repetitive stimulation. The addition of other properties to the model (for example, a tendency toward repetitive firing which itself might be a function of the stimulation history) could account for spread in increasing increments during repetitive stimulation. But the fact that we are able to mimic the response of a living system does not necessarily mean that the mechanisms involved in the model are the same as those in the living system, unless it can also be demonstrated that there is a unique way of achieving this response. This is usually impossible. The question one can ask with simulation is "If the system were to operate with these rules, what would be its behavior?" We have shown that properties known or presumed for coelenterate nets are sufficient to account for the responses of coelenterate conducting systems. Whether or not all coelenterate conducting systems are nerve nets and whether all of the significant parameters of such nets have yet been discovered are questions for the zoologist's laboratory and not the computing room.

The success of the simulation model allows some speculation about the functional organization of coelenterate nerve nets. The neurons in a coelenterate net appear to be haphazardly arranged, rather like a handful of straw scattered loosely over a surface. This apparent lack of detailed organization, however, may be only a reflection of our ignorance about the system. One can well imagine that a cannibal, on first seeing the underside of a television set, would conclude that it consisted of a number of haphazardly arranged and randomly connected cylinders and wires. The fact that it is possible to duplicate the functioning of coelenterate conducting systems with a model in which the details of the connections are not specified, in which randomness is purposely introduced, makes it seem more likely that the randomness of coelenterate nets is real and not a product of our ignorance.

References

Batham, E. J., C. F. A. Pantin, and E. A. Robson, 1960. The nerve-net of the sea-anemone *Metridium senile:* the mesenteries and the column, *Quart. J. Microscop. Sci.*, **101**, 487–510.

Batham, E. J., C. F. A. Pantin, and E. A. Robson, 1961. The nerve-net of *Metridium senile:* artifacts and the nerve-net, *Quart. J. Microscop. Sci.*, **102**, 143–56.

Horridge, G. A., 1954. The nerves and muscles of medusae: I. Conduction in the nervous system of *Aurellia aurita* Lamarck, *J. Exptl. Biol.*, **31**, 594–600.

Horridge, G. A., 1956. The nervous system of the ephyra larva of *Aurellia aurita*, *Quart. J. Microscop. Sci.*, **97,** 59–74.

Horridge, G. A., 1957. The coordination of the protective retraction of coral polyps, *Phil. Trans. Roy. Soc.* (*London*), **B-240,** 495–529.

Horridge, G. A., and B. MacKay, 1962. Naked axons and symmetrical synapses in coelenterates, *Quart. J. Microscop. Sci.*, **103,** 531–41.

Josephson, R. K., 1961. Repetitive potentials following brief electric stimuli in a hydroid, *J. Exptl. Biol.*, **38,** 579–93.

Josephson, R. K., and R. M. Worthy, 1960. "A Digital Computer Simulation Study of a Diffuse Nerve Net," in Laboratory for Automata Research, Semi-Annual Report No. 1, Glendale, Calif. General Precision, Inc., Librascope Division, pp. 16–43.

Josephson, R. K., R. F. Reiss, and R. M. Worthy, 1961. A simulation study of a diffuse conducting system based on coelenterate nerve nets, *J. Theoret. Biol.*, **1,** 460–87.

Mackie, G. O., 1960. The structure of the nervous system in *Velella*, *Quart. J. Microscop. Sci.*, **101,** 119–31.

Pantin, C. F. A., 1935. The nerve net of the Actinozoa: I. Facilitation, *J. Exptl. Biol*, **12,** 119–38.

Pantin, C. F. A., 1952. The elementary nervous system, *Proc. Roy. Soc.* (*London*), **B-140,** 147–68.

Passano, L. M., 1958. Intermittent conduction in scyphozoan nerve nets, *Anat. Rec.*, **132,** 486.

Index